Elementary Parametric Techniques

Standard score forms
$$z_{\text{obs}} = \frac{X - \mu}{\sigma_X}, \qquad z_{\text{obs}} = \frac{\overline{X} - \mu}{\sigma_{\overline{X}}} = \frac{\overline{X} - \mu}{\sigma_X / \sqrt{N}}$$

$$t_{\text{obs}} = \frac{\overline{X} - \mu}{s_{\overline{X}}} = \frac{\overline{X} - \mu}{s_x / \sqrt{N}}$$

Confidence intervals for the mean
$$\overline{X} \pm t_\alpha(s_x)$$

Confidence intervals for the correlation coefficient
$$z_r \pm z_\alpha \sqrt{\frac{1}{N - 3}}$$

Difference between two independent groups
$$t_{\text{obs}} = \frac{\overline{X}_1 - \overline{X}_2}{\sqrt{\left[\dfrac{(N_1 - 1)s_1^2 + (N_2 - 1)s_2^2}{N_1 + N_2 - 2}\right] \cdot \left[\dfrac{1}{N_1} + \dfrac{1}{N_2}\right]}}$$

$$df = N_1 + N_2 - 2$$

$$\text{est. } \omega^2 = \frac{t_{\text{obs}}^2 - 1}{t_{\text{obs}}^2 + N_1 + N_2 - 1}$$

Difference between two correlated groups
$$t_{\text{obs}} = \frac{\Sigma D_i}{\sqrt{\dfrac{N\Sigma D_i^2 - (\Sigma D_i)^2}{N - 1}}} \qquad df = N - 1$$

Significance of the correlation coefficient

See Table C, Appendix 2

Difference between two independent correlation coefficients
$$z_{\text{obs}} = \frac{z_{r1} - z_{r2}}{\sqrt{\dfrac{1}{N_1 - 3} + \dfrac{1}{N_2 - 3}}}$$

For z_{r1} and z_{r2}, see Table D, Appendix 2

Simple analysis of variance

$$\textbf{(I)} = \frac{T^2}{N} \qquad \textbf{(II)} = \sum_{j=1}^{p}\left(\sum_{i=1}^{n_j} X_{ij}^2\right) \qquad \textbf{(III)} = \sum_{j=1}^{p}\left(\frac{T_j^2}{n_j}\right)$$

Source	df	SS	MS	F
Between Groups	$p - 1$	**(III)** − **(I)**	SS_b / df_b	MS_b / MS_w
Within Groups	$N - p$	**(II)** − **(III)**	SS_w / df_w	
Total	$N - 1$	**(II)** − **(I)**		

$$\text{est. } \omega^2 = \frac{SS_{\text{between}} - (p - 1)MS_{\text{within}}}{SS_{\text{total}} + MS_{\text{within}}}$$

Confidence intervals for $\overline{X}$
$$\overline{X}_j \pm (t_{\alpha;df\text{within}}) \sqrt{\frac{MS_{\text{within}}}{n_j}}$$

Individual comparisons
$$F_{\text{comp}} = \frac{(\overline{X}_1 - \overline{X}_2)^2}{\left(\dfrac{1}{n_1} + \dfrac{1}{n_2}\right)MS_{\text{within}}}$$

(continued on back endpaper)

FUNDAMENTAL STATISTICS FOR BEHAVIORAL SCIENCES

Seventh Edition

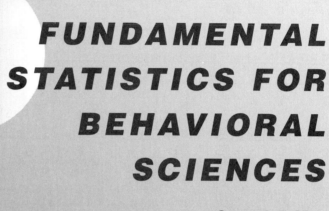

FUNDAMENTAL STATISTICS FOR BEHAVIORAL SCIENCES

Seventh Edition

Robert B. McCall
University of Pittsburgh

Brooks/Cole Publishing Company

I(**T**)**P**® *An International Thomson Publishing Company*

Pacific Grove • Albany • Belmont • Bonn • Boston • Cincinnati • Detroit • Johannesburg • London
Madrid • Melbourne • Mexico City • New York • Paris • Singapore • Tokyo • Toronto • Washington

Sponsoring Editors: Curt Hinrichs and Denis Ralling

Marketing Team: Christine Davis, Deborah Petit, and Michael Campbell

Assistant Editor: Cynthia Mazow

Editorial Assistants: Robert Sean MacBeth, Martha O'Connor, and Stephanie Andersen

Production Editor: Karen Ralling

Manuscript Editor: Susan Reiland

Interior Design: Donna Davis

Interior Illustration: Lori Heckelman

Cover Design: Roy R. Neuhaus

Cover Photo: J. P. Fruchet/FPG International

Art Editor: Jennifer Mackres

Typesetting: Impressions Book and Journal Services, Inc.

Cover Printing: Phoenix Color Corporation

Printing and Binding: The Courier Company, Inc.

For more information, contact:

BROOKS/COLE PUBLISHING COMPANY
511 Forest Lodge Road
Pacific Grove, CA 93950
USA

International Thomson Publishing Europe
Berkshire House 168–173
High Holborn
London WC1V 7AA
England

Thomas Nelson Australia
102 Dodds Street
South Melbourne, 3205
Victoria, Australia

Nelson Canada
1120 Birchmount Road
Scarborough, Ontario
Canada M1K 5G4

International Thomson Editores
Seneca 53
Col. Polanco
11560 México, D. F., México

International Thomson Publishing GmbH
Königswinterer Strasse 418
53227 Bonn
Germany

International Thomson Publishing Asia
221 Henderson Road
#05–10 Henderson Building
Singapore 0315

International Thomson Publishing Japan
Hirakawacho Kyowa Building, 3F
2-2-1 Hirakawacho
Chiyoda-ku, Tokyo 102
Japan

Printed in the United States of America.

10   9   8   7   6   5   4   3   2   1

Library of Congress Cataloging-in-Publication Data
McCall, Robert B.
  Fundamental statistics for behavioral sciences / Robert B. McCall.
  —7th ed.
    p.   cm.
  Includes index.
  ISBN 0-534-52371-4 (hardcover)
  1. Statistics. 2. Social sciences—Statistical methods.
I. Title.
QA276.12.M399   1998   97-20694
519.5—dc21   CIP

ABOUT THE AUTHOR

Robert B. McCall is Professor of Psychology and Co-Director of the University of Pittsburgh Office of Child Development. He is the author of 22 editions of textbooks and guides and several hundred books, chapters, and articles for scholarly audiences and the general public. As a methodologist, he has published numerous articles and handbook chapters on statistics and longitudinal design, and he has been on the permanent editorial boards of a dozen journals and book series, including *Psychological Methods*. He is currently an Action Editor of *Child Development*. As a developmental psychologist, Dr. McCall is a specialist in infant attention and cognition as predictors of later IQ and in developmental changes in general mental performance. He was honored with the 1994 American Psychological Association Distinguished Contribution to Public Service for communicating psychology to the general public through the media and for creating the Office of Child Development, which promotes and conducts interdisciplinary university-community educational, research, service, and policy projects.

Preface

A veteran scholar of statistics once remarked that there are two ways to teach statistics—accurately or understandably. The challenge of disproving that "either-or" statement has guided my writing of this book.

Purpose and Approach

As a textbook for the first course in applied statistics, *Fundamental Statistics for Behavioral Sciences* is used primarily by students majoring in psychology, education, and other social and behavioral sciences. In writing for this audience, my earliest and most basic decision was to emphasize the purpose, rationale, and application of important statistical concepts over rote memorization and the mechanical application of formulas. I believe that students at the introductory level, whether or not they plan to take advanced courses in statistics, are served better by a book that fosters an understanding of statistical logic than by one that stresses mechanics.

When understanding is emphasized, elementary statistics is neither dull nor mathematically difficult. *Fundamental Statistics for Behavioral Sciences* does not require much background in mathematics. The student need be familiar only with the thinking patterns learned in high school algebra and geometry; all relevant terms and operations are reviewed in Appendix 1. To be sure, the book contains many computations and problems to solve, but most statistical formulas rely heavily on simple arithmetic—addition, subtraction, multiplication, division, and the taking of square roots—and can be worked out quickly with the aid of a hand calculator. In addition, I have kept the data for computational problems simple so that the emphasis remains on the rationale and outcome of techniques instead of on calculation for its own sake.

The goal of understanding concepts is not hard to reach if students also understand as much of the mathematical reasoning behind the concepts as is within their grasp. Whenever possible, I have explained in mathematically simple terms the logic that undergirds the basic concepts and techniques, although a few items require advanced mathematics and must therefore be taken on faith at this level. Beginning in Chapter 3, optional tables that show full algebraic derivations and proofs supplement the text explanations. These tables can be omitted without loss of continuity.

For the sake of students, and contrary to the traditional practice of mathematical writers, I have included and explained every step in proofs, however "obvious." I have also avoided excessive use of symbols, since symbols require an extra mental translation and thus often confuse students. Derivation scores $(X_i - \overline{X})$, for instance, are *not* abbreviated by x. Further, each new symbol is carefully introduced and is frequently accompanied by its verbal equivalent.

Anyone who has analyzed his or her own data knows the anticipation that accompanies the final calculation of the *r, t,* or *F* hidden in a mass of numbers that took months to gather. Students, too—even though it is not their own data they are analyzing—can experience the excitement of seeing meaning emerge as they manipulate an apparently patternless collection of numbers. Yet they sometimes fail to see the fascination of statistical analysis because it is presented more as a numbers game played in a vacuum than as a crucial part of the scientific investigation of real phenomena.

For this reason, many end-of-chapter exercises and examples in the text are drawn from actual studies (modified for numerical simplicity). For example, the distinction between a correlation and a difference between means is demonstrated through findings that related the IQs of adopted children to those of their biological and adoptive mothers. In another case, Freedman's work on the feeding behavior of dogs reared under indulged and disciplined conditions is used to illustrate interaction effects in the two-factor analysis of variance. Although most of these studies were performed by psychologists, many of them concern developmental and educational issues of interest to future teachers and school administrators.

As another means of showing that numbers can have real applications, I have tried to give students a feel for the behavior of a statistic by providing several data sets that display obvious contrasts. In Chapter 10, for example, I present a set of random numbers and calculate an analysis of variance on them, then I add a main effect to the data and recalculate, and finally I add an interaction effect and recalculate. This shows students how various effects in the data are reflected in the statistical quantities being studied. In addition, I ask in some exercises for the student to alter a set of scores in some way and observe the effect on the value of a statistic.

Study Tips

Despite these attempts to help students, some find this course more difficult than many other courses. Apart from differences in quantitative abilities, I suspect that much of the reason is that students try to study it in the same manner as a "reading course." A course in history or introduction to psychology, for example, typically uses a textbook and, except for new vocabulary, everything one reads is quite understandable the first time through. Also, the material in tables and figures is largely illustrative and not crucial to understanding the main text. The task is largely one of memorizing vocabulary and concepts and, in more advanced courses, integrating material across chapters.

Statistics also has its own vocabulary, some of which is in symbols; once introduced, a symbol is often used repeatedly thereafter. While I have tried to repeat the definition or verbal label with the symbol an extra time or two to help the learning process, eventually the symbol will be used by itself. Therefore, if the student does not commit the symbol's meaning to memory when it is first encountered, everything that subsequently uses that symbol—and it is sometimes a good deal of material—will not be truly understood. So, students should pause to learn each vocabulary term or symbol when it is first introduced.

Also, statistics often requires the student to follow a sequence of logical steps or elementary mathematical operations. Some students should pause in the middle of the sequence to review it after the first reading to be sure the entire sequence is understood. Also, the basic material of statistics is mathematical, symbolic, and conceptual, and it is therefore quite abstract without the concrete examples given. But students need to read and especially *do* the examples, step by step, along with the text. Skipping them can be disastrous for some students, and simply reading but not studying them is insufficient for other students. They must actually go through the calculations or the steps in the proof to understand it. And some examples, proofs, and illustrations are presented in separate tables and figures. Students should stop reading the text when such material is cited, turn to that table or figure, and study it until it is understood. If it is a calculation, they should do it on a piece of scratch paper along with the text. These are not pictures in a newspaper, which typically can be skipped; they are integral to understanding what is being presented.

Another tip is to compute by hand at least one — but two or three would be better for some students — computational example or exercise of a given type before rushing to a fancy one-button calculator or computer program. Students often need to get a "computational feel" for a formula before they can understand and interpret it thoroughly. And, of course, it helps to do as many exercises as possible in the text and *Study Guide.* Some students assume that once they understand the presentation in the text, they will be able to work a problem without practice. Of course, a few can, but a very few, in my experience. Most of us need to work one or two or even more problems in the text and *Study Guide* to become proficient at them, even if we understand the text material quite well.

Lastly, the answers and even some of the intermediate steps for most of the exercises are given in the backs of the text and *Study Guide.* This is done so that students can check their work immediately and study more independently. But if students "peek" before finishing, they often deprive themselves of the opportunity to figure the problem out on their own. The exams will not have the answers on them, so it is best to practice without them, too.

What's New

This Seventh Edition incorporates a number of changes and additions, mostly suggested by users, reviewers, and students. First, the textual material has been thoroughly reviewed and revised in many places to improve the clarity of presentation. I hope this will help bring understanding to more students.

Additional exercises have been added to each chapter under the heading "For Homework." Answers for these exercises are not provided in the text as they are for all other exercises, but they are included in the *Instructor's Manual.* With the conceptual questions and exercises in the text, the self-test and exercises in the *Study Guide,* and these new homework exercises, there should be more questions and exercises available than most students need and have the time to do. Also, I have checked the answers to every exercise in the text and guide. I, and other authors, strive for accuracy and are chagrined and perplexed at what seem to be inevitable errors. I have sincerely tried to eliminate them, but if not, please write or e-mail your correction so that it may be implemented in the next printing.

While I have steadfastly resisted adding much new material over the years, trying to avoid having the book become encyclopedic rather than a good introduction to the basic concepts of applied statistics, the field is changing and the text must evolve with it. Consequently, the exploratory data analysis concepts of stem-and-leaf displays, five-number summaries, and boxplots have been added to the traditional descriptive indicators and graphing techniques in Chapters 2 and 3. Also, the calculation of the median has been simplified to be the value of the middle score or the average of the middle two scores. The previous formula that interpolated in cases in which duplication of scores existed near the median has been eliminated. This simpler definition and calculation not only make determining the median easier for students but conform to many other advanced texts and computer programs and are consistent with the exploratory data analysis definition. However, the determination of percentiles with a formula that does interpolate has been retained, because this is the approach given in many advanced texts and it permits the determination of any percentile rank or point regardless of whether it actually appears in the distribution. I believe these advantages outweigh the disadvantages that most computer programs do not interpolate and that the use of the formulas to determine the 50th percentile point may not produce the same value as the simplified definition of the median if duplication of scores exists near the median.

I have used $\hat{Y}$ instead of $\tilde{Y}$ because it is more common, easier to distinguish from $\tilde{Y}$, and can be written by many word processors by instructors who provide students with their own exercises and additional materials. Also, I have used intermediate quantities in the computation of the slope, the standard error of estimate, and other statistics to simplify calculations. However, the intermediate quantities, symbolized by **(I), (II),** etc., sometimes with subscripts, are defined differently for regression and correlation than they are for the analysis of variance. Since students tend to follow the computational procedures of the guided examples rather than memorize formulas, this should not be a problem.

In Chapter 7 on sampling, I have increased the discussion of randomness as a process rather than as an outcome, and in Chapter 9 I have included a discussion of the difference between using independent vs. correlated groups in research design.

Consistent with the movement away from sheer statistical significance and toward describing the nature and size of differences and effects, I have added confidence intervals for r (Chapter 9) and the means in a simple analysis of variance using the pooled MS_{within} to estimate the standard errors for each group (Chapter 10). Further, I include estimates for the proportion of variance associated with the differences between two means (Chapter 9) and effects in the simple (Chapter 10) and two-factor (Chapter 13) analyses of variance. I chose to present omega squared as the estimate, because it is commonly used and is most consistent with the proportion of variance concept reflected in r^2, but I recognize and describe the limitations of this approach.

In Chapter 11 on research design, I have added a brief discussion of ethics and scientific integrity, and I have added more commonplace illustrative examples in Chapter 12 on probability. Finally, I have added brief descriptions of the application of chi square to goodness of fit problems and the testing of the significance of a proportion.

The Study Guide

The *Study Guide* in general is intended to be a workbook. All the basics are presented in stripped-down fashion and without much conceptual explanation. But the *Guide* is intended to be a supplement to the text and it assumes that the text has been read before the student begins the review and practice that the *Guide* offers. Some students find that reading the text, then doing the *Guide,* and then rereading the text is an effective strategy, because the *Guide* reviews the text, and then the concepts in the text are more understandable after the vocabulary and the concrete computational mechanics are mastered.

More specifically, the main section of each chapter in the *Guide* is a semi-programmed unit that reviews the basic terms, concepts, and computational routines described in the corresponding chapter of the text. Often students are required to actually perform the computations in a step-by-step manner in an example, making it more difficult for them to skip the illustration. Also, they may be asked to compute a statistic for a small data set, then recompute it after the data are changed in some way to illustrate how the statistic reflects those changes in the data. Then a guided computational exercise is presented that shows students how to lay out the information and perform the calculation of a particular statistic. Of course, this is most useful when calculating a statistic by hand or with a hand calculator, but I feel it is helpful to most students to do at least one problem by hand before using a computer. Then a set of exercises with answers is presented, followed by the boxed instructions for using MINITAB, SPSS, and StataQuest computer programs.

The instructions for these statistical computer programs are separated from the main presentation so that they can be skipped or another statistical program can be used by the instructor. Obviously, each instructor must provide instructions on how to access the programs on their systems. I urge instructors to consider carefully whether they want to teach computer usage at this stage. Many do, feeling it is almost as important in modern times for students to learn how to use the computerized statistical packages as it is to learn statistics. Other teachers feel that it is best for their students to understand the basic concepts, that most students in their course will not become researchers and will not need to actually analyze data, and that it takes more time for students to learn the computer packages than they save in computation time. At least now one has a more convenient choice in this matter.

Selecting Topics

Individual instructors emphasize different aspects of elementary statistics, and many must select a subset of chapters that can be covered in the available time. The organization of the text is straightforward. Part 1 (Chapters 1–6) presents descriptive statistics, including central tendency, variability, relative position, regression, and correlation. Part 2 (Chapters 7–10) deals with elementary inferential statistics, including sampling distributions, the logic of hypothesis testing, elementary parametric tests, and simple analysis of variance. For most courses, these are the core chapters. Part 3 contains special topics that instructors can select to supplement the core material. Most commonly, the chapter on nonparametrics is assigned, but those instructors teaching a more intense course may also add the chapters on research design, probability, and two-factor analysis of variance.

While the material is sequenced and cumulative, I have tried to write some chapters so that they could be inserted nearly anywhere in a course. Chapter 11 on research design and Chapter 12 on probability could be taught anywhere in a sequence after Chapter 1, and some instructors do teach them at the beginning of their courses. Nonparametrics (Chapter 14) could be taught any time after Chapters 7 and 8, and regression and correlation could be taught later in the course (but significance tests for *r* appear in Chapter 8). Also, material on confidence intervals and effect sizes usually appears near the ends of chapters and could be omitted.

Thanks

In addition to the colleagues and friends who contributed their advice to earlier editions, I want to thank Mark Appelbaum of the Department of Psychology at the University of California at San Diego. He was my chief technical consultant, helping to blend my personal style with the demands of formal theory. I am indebted to the many whose commitments to the course and to the book led them to communicate to the publishers over the years; their comments, criticisms, and suggestions stimulated many of the improvements in this new edition.

Thanks also to the following statistics scholars, who reviewed the manuscript for this edition and provided helpful comments and suggestions: Mark Appelbaum, Vanderbilt University; Kathleen Bloom, University of Waterloo; Drake Bradley, Bates College; James Chumbley, University of Massachusetts–Amherst; Anna Napoli, University of Redlands; Robert Patterson, Washington State University; Thomas J. Thieman, College of St. Catherine; Linda Woolf, Webster University; Martha Shalitta, Eastern College; Leonard Kroeker, United States International University; Glenn Sauders, SUNY at Albany; Jack Fyock, University of Maryland; David A. Frieske, Western Kentucky University; and William Addison, Eastern Illinois University (IL).

I appreciate the assistance of Richard Sass with the *Study Guide.* Thanks also to the following contributors to the *Instructor's Manual:* William Ghiselli, University of Missouri, Kansas City; Linda Juang, Michigan State University; Alexander von Eye, Michigan State University; and Richard Sass. And thanks to Catherine Kelley for preparing the manuscript, which can only be described as a clerical nightmare.

I am grateful to the Literary Executor of the late Sir Ronald A. Fisher, F.R.S., to Dr. Frank Yates, F.R.S., and to Longman Group Ltd., London, for permission to reprint Tables III, IV, VIII, and part of XXXIII from their book *Statistical Tables for Biological, Agricultural and Medical Research* (Sixth Edition, 1974).

Special thanks go to my wife, Rozanne, for her encouragement and for the absence of complaints while she was temporarily widowed for this cause.

Robert B. McCall

Contents

DESCRIPTIVE STATISTICS

One use of statistical methods is to organize,

summarize, and describe quantitative information.

Such techniques are called descriptive statistics.

Baseball batting averages, the rate of inflation, and

the degree of relationship between heredity and IQ

test performance are three examples. This part of

the text presents some elementary descriptive

statistics.

The Study of Statistics

Т he news is full of statistics: "One-third of American seniors in high school read five or fewer pages per day for school or homework. . . . Only 36% of Americans are satisfied with the quality of the nation's health care and only one in nine is satisfied with the costs, which are expected to rise to $14,000 by the year 2000. . . . Last month inflation rose a seasonally adjusted .3% to an annual rate of 4.5%. . . . Pitcher Randy Johnson's ERA is now 2.95. . . . The probability of rain today is 60%."

Such quantities are "statistics," and an elementary knowledge of them is useful to everyone. But the academic study of statistics is much more than these common percentages, rates, averages, or probabilities, and a detailed knowledge of this field is essential to anyone who wants to read or, especially, to conduct research in psychology, education, sociology, economics, or any of a number of other subjects.

As a field of study, statistics consists of a set of procedures for organizing, describing, and interpreting measurements and for drawing conclusions and making inferences about what is generally true for an entire group when only a few members of the group are actually measured. This chapter explains why the study of statistics is important, discusses the purpose of statistics, and presents the kinds of measurements typically made in the social and behavioral sciences.

WHY STUDY STATISTICS?

A knowledge of statistics is useful for two main purposes: to help understand the common use of statistics in the news, in our jobs, and elsewhere in our daily lives, and to be able to read and understand scientific articles and conduct research in the social and behavioral sciences.

Everyday Use

All of us have a basic understanding of simple percentages, rates, and averages. For example, if we read that the average household income in the United States in 1995 was $44,938, we assume that someone added up the incomes for all the households in the country and divided that total by the number of such households to give the average household income of $44,938.

But many more statistical quantities exist than averages. For example, suppose the same newspaper article reports that the *median* household income was only $34,076. What is the *median?* Why is it different from the average? Which value, the average or the median, is the better index of typical family income? (See Chapter 3.)

Sometimes two simple statistics seem to lead to opposite conclusions. For example, you might read in a magazine that the IQs of adopted children are more closely correlated with those of their biological parents than with those of the parents who reared them. This seems to suggest that heredity plays a more important role in the development of IQ than environment. But the magazine may also report that the same study using the same parents and children found that the IQs of the adoptive children were closer in average value to those of their rearing parents than to those of their biological parents, a fact that seems to suggest that environment is more important than heredity. How can both these facts be true, and what do they say about the contributions of heredity and environment to IQ? (See Chapter 6.)

Or you may wake up one morning and hear on the radio that the probability of rain is 40%, but you look outside and it is already raining. Or you might hear that the probability of rain that afternoon is 70%, so you decide not to use the free tickets you were given to the football game. But, in fact, it doesn't rain at all. In both cases you decide the weather forecaster is incompetent, but actually the forecaster has been properly predicting the weather and using the concept of probability. (See Chapters 7 and 12.)

So, even the statistics commonly used in everyday life are often more complicated than they first appear. An introduction to statistics will help you understand and interpret these statistics.

Scientific Use

While statistics are useful to everyone, they are crucial if you want to read a scientific paper, study a science, or conduct scientific research, and this is especially true if the science is social or behavioral. Undergraduate majors in psychology, for example, are required to take a course in statistics—which may be the main reason you are reading this—and many graduate programs in education, sociology, economics, and other sciences also require statistical training. Why?

First, as we will see below and throughout this text, the social and behavioral sciences are indeed sciences—they collect data about the topics of interest in systematic and objective ways. Those data must be summarized, described, and analyzed by statistical methods, and the conclusions reached form the knowledge base of these disciplines. So it is necessary for anyone wanting to study these disciplines to know some statistics to understand what is presented in textbooks and articles in these fields, even if one never conducts an experiment or makes a scientific observation. Quite simply, you will not be able to read with complete understanding in these areas much past an introductory level without some background in statistics, and the more advanced your studies, the more knowledge of statistics you will need. This is why even those graduate programs in psychology that are designed to train professionals to diagnose and treat behavioral problems, not to conduct research, still typically require students to take one year of advanced statistics.

Second, if you are going to make any scientific observations in the social and behavioral sciences, even for an undergraduate thesis or research project, you will need to know and perform some statistical operations. This is so, because almost all numerical information, or data, in the social or behavioral sciences contains variability. **Variability** refers to the fact that the scores or measurement values obtained in a study differ from one another, even when all the subjects in the study are assessed under the same circumstances. If your teacher were to give a test to all the students in your statistics class, you would not all obtain the same score. Similarly, some children diagnosed as hyperactive improve their classroom behavior while others do not after being placed on medication or after being put through an intensive behavioral regimen at home and at school. This dissimilarity in scores and outcomes is variability.

Variability exists in social and behavioral data for at least three reasons. First, the units (usually people) that such scientists study are rarely identical to one another. A chemist or a physicist assumes that each of the units under study (molecules, atoms, electrons, and so forth) is identical in its composition and behavior to every other unit of its kind. But the social and behavioral scientist cannot assume that all people will respond identically in a given situation. In fact, the behavioral scientist can count on the fact that they will *not* respond in the same way. The first major source of variability, then, is **individual differences** in the behavior of different subjects.

Second, social and behavioral scientists cannot always measure the attribute or behavior they wish to study as accurately as they would like. Scores on a classroom examination are supposed to be a measure of how much students have learned, but even a good test has flaws that make it an imperfect measure of learning. Perhaps some questions are ambiguous, or maybe one part of the course material is emphasized more on the test than other parts, thereby favoring those students who happened to study more thoroughly the relevant parts. These and other factors that influence scores or performance and that are associated with how the measurements are made constitute **measurement error.** Almost all behavioral measurements contain some measurement error; this error favors some people more than others, thus contributing to variability.

Third, even a single unit (person, animal) usually will not respond exactly the same way on two different occasions. If you give a person several opportunities to rate the attractiveness of advertising displays or the degree of aggressiveness in filmed episodes of each of 15 preschool children, he or she will most likely not assign the same scores on each occasion. A basketball team may perform very well in one game and poorly in another, even though the two opposing teams are equally skilled. This source of variability is called **unreliability.**

Individual differences, measurement error, and unreliability are the major reasons that it is often difficult to draw accurate conclusions from social and behavioral science data. It is the task of statistics to quantify the variability in a set of measurements; to describe the data for a group of subjects, despite the variability inherent in it; and to derive precisely stated and consistent decisions about the results by quantifying the uncertainty produced by variability. Below we consider the details of some of these functions of statistics.

DESCRIPTIVE AND INFERENTIAL STATISTICS

Statistics is the study of methods for describing and interpreting quantitative information, including techniques for organizing and summarizing data and techniques for making generalizations and inferences from data. The first of these two broad classes of methods is called descriptive statistics, and the second is called inferential statistics.

Descriptive statistics refers to procedures for organizing, summarizing, and describing quantitative information, which is called data. Most people are familiar with some statistics of this sort. The baseball fan is accustomed to checking over a favorite player's batting average; the sales manager relies on charts showing the sales distribution and cost efficiency of an enterprise; the head of a household may consult tables showing the average domestic expenditures of families of comparable size and income; and the actuary uses charts outlining the life expectancies of people in various professions. These are relatively simple statistical tools for characterizing data, but additional techniques are available to describe such things as the extent to which measured values deviate from one another and the relationship between individual differences in performance of two different kinds. For example, one can describe statistically the degree of relationship between the scores of a group of students on a college entrance examination and their later grades in college.

The second major class of statistics, **inferential statistics,** includes methods for making inferences about a larger group of individuals on the basis of data actually collected on a much smaller group. For example, suppose a researcher wanted to know whether administering the medication Ritalin helps children who have been diagnosed as having attention-deficit hyperactivity disorder (ADHD) to accomplish more of their schoolwork in class. Without medication, such children typically have difficulty paying attention to tasks over even moderate periods of time, and therefore they often do not finish assignments. They also have special difficulty with problems that require them to perform a sequence of steps to obtain the answer. Ritalin presumably slows them down so they can sustain their attention over longer periods, complete more work, and work more accurately.

Suppose, with their parents' permission, 40 third-grade children with the ADHD diagnosis are randomly divided into two groups. One group is given Ritalin treatment under the care of a physician while the other group is administered a **placebo,** a look-alike drug that actually has no effect on the children. Then, after a period of one month to adjust to the treatment, the number of correctly worked problems assigned in class is counted over a four-week period. The Ritalin group worked an average of 63 problems correctly while the placebo group worked an average of 56. The research question is, Does Ritalin improve school performance, at least under these circumstances?

Notice that the question is, "*Does* Ritalin improve performance?" not "*Did* Ritalin improve performance?" The word *does* is much more general than the word *did*. The results already show that the Ritalin group *did* better than the placebo group, 63 vs. 56. The scientific question is whether this difference is likely to happen again if two new groups of children are studied, and then again with two more groups, and again and again until all children with ADHD are studied. In short, *does* Ritalin have this effect for children in general? "All children with ADHD" constitutes the **population** of children with ADHD, while the specific group of 40 children studied constitutes a **sample** from this population. So the scientific question translates into whether Ritalin has an effect in the population, a conclusion that the scientist must *infer* from the results of the sample.

Of course, inferring what would be true for the population of all ADHD children on the basis of results from a small sample of 40 represents something of a guess, and we feel some uncertainty about drawing such a conclusion. For example, 63 problems is definitely more than 56, but we know from the discussion of variability above that if we repeated the experiment, the results for the two groups would not be precisely the same. Perhaps the averages for the treated and untreated groups would be 60 vs. 58, 67 vs. 54, 60 vs. 60—or even 59 vs. 64, the opposite result. Obviously, if the observed difference between the two sample groups was very large, say 80 vs. 20, we would feel more confident that the Ritalin group would do better sample after sample—that is, in the population. Conversely, if the actual observed difference between the two sample groups was quite small, say 64 vs. 62, then we would not feel confident that this result would occur consistently in the population. But most of the time the differences between the groups are not this large or small.

How do we make a decision in the face of this uncertainty? The strategy in science is similar to what you might do to decide whether a coin is fair. First, you tentatively suppose that the coin is indeed fair. If it is, then over countless flips one should expect that half, or 50%, would turn up heads. Second, you cannot flip the coin countless number of times, so you take a sample—say, 100 flips—to see how close the results are to 50–50. If the result is actually 52 vs. 48, you might conclude, "If the coin is fair, this result is quite likely to occur, so I have no compelling reason to believe the coin is biased." But, if the result is 65 vs. 35, then you might conclude, "If the coin is fair, this result is very unlikely to occur, so I believe the coin is biased."

The scientist essentially follows the same strategy. First, it is presumed that Ritalin *has no effect,* just as we presumed above that the coin was fair. Second, we take a sample, conduct the experiment, and observe the difference between the Ritalin and the placebo groups, just as we flipped the coin 100 times. If the observed group difference is small and likely to occur, then there is no compelling evidence to think Ritalin has an effect. On the other hand, if the observed difference is very large and unlikely to occur if the drug had no effect, then perhaps Ritalin does improve performance in the population.

But how "unlikely" does the difference have to be before the scientist concludes that Ritalin has an effect in the population? This is where statistical procedures, some of which are taught in Chapters 8–14, are used. Such procedures quantify that likelihood—they assign it a number, called **probability,** that ranges from 1.00 (it is certain) to .00 (it is impossible)—that the observed result could occur even if no difference actually exists in the population. That is, what is the likelihood—probability—that such a result could occur *by chance?* If the probability is high that such a result could occur by chance, then one has no evidence that Ritalin improves performance. But if the probability is small that the difference could occur by chance, then perhaps Ritalin does improve performance.

The probability determined by statistical procedures is almost never 1.00 or .00; it is usually a value in between. So behavioral scientists have agreed on some rules to decide when the probability is sufficiently small to conclude that a difference exists in the population. When this approach is taken, this probability value is .05. (See Chapter 8.) Therefore, the conclusion will be that Ritalin improves performance only if the probability is less than .05 that the observed result could occur even if no difference actually exists in the population.

This ability to draw inferences about the population on the basis of observing only a relatively small sample is crucial to the progress of science. Certainly, it would be impossible to give *all* ADHD children either Ritalin or the placebo. Further, if we first studied only 40 children, we would probably want to repeat the study on another sample and then another and another if we could not use statistics to obtain a numerical estimate of the likelihood that a single observed result would occur if no differences existed in the population. Therefore, inferential statistics is a powerful tool for behavioral science, because it permits scholars to draw conclusions about what exists generally—that is, in the population—on the basis of studying only a relatively few cases, sometimes in a single experiment.

To summarize: Statistics is the study of methods for describing and interpreting numerical information or data. Descriptive statistics includes procedures for organizing and summarizing data, whereas inferential statistics includes procedures for making inferences about what exists in a population on the basis of a smaller sample of cases.

MEASUREMENT

The data of research are measurements. Which statistical procedures are used to describe and draw inferences from those measurements depends in part on how the measurements are made. Therefore, before discussing statistical procedures in detail, it is first necessary to discuss some basic measurement terminology as well as some of the characteristics of the measurement process itself.

Scales of Measurement

> **Measurement** is the orderly assignment of a numerical value to a characteristic.

If we measure height, for example, we assign a number—perhaps the number of inches—to the height of an individual by some systematic process. For example, the person may be asked to stand barefoot against a vertical wall, a perpendicular level is lowered to the top of the head, and the height is read in inches off a tape measure affixed to the wall.

> A **scale of measurement** is the ordered set of possible numbers that may be obtained by the measurement process.

The tape measure in the above example has a scale of possible values written on it.

Properties of Scales   All scales have certain characteristics or properties. Suppose you were assigned the task of measuring the heights of children in a nursery-school class. If you have no tape measure, you could simply line the children up from the smallest to the tallest and assign them numbers 1, 2, 3, and so on for all children. This is called a **rank order** of height. Alternatively, you could make a judgment that each child fits into one of five categories: very short, short, average, tall, or very tall. This is called a **categorization** or **rating** of height. If, on the other hand, you *did* have a tape measure, you could measure each child's height in inches or centimeters.

While the first two methods seem rather crude, they appear that way only because we all know how to measure height rather precisely. But these approaches might not be so crude if you were given the task of measuring aggressiveness in preschool children. For example, you could observe the children in class for a week and then rank order them, categorize them into five levels of aggressiveness, or count the number of aggressive acts committed by each child during the week.

Each of these approaches to measuring height or aggressiveness produces a scale that possesses certain properties. We will consider three such properties: magnitude, equal intervals, and an absolute zero.

> When a scale has **magnitude,** one instance of the attribute being measured can be judged greater than, less than, or equal to another instance of the attribute.

If the observer in the preceding example on preschool aggressiveness assigns a rank of 8 to Jane and a rank of 5 to Harry, this scale of measurement reflects a difference in the magnitude of aggressiveness of the two children—Jane is more aggressive than Harry (if ranking is from least to most aggressive).

> When a scale has **equal intervals,** the magnitude of the attribute represented by a unit of measurement on the scale is the same regardless of where on the scale the unit falls.

Take the example of measuring height with a tape measure. One is confident that the difference in height between someone measuring 61 inches and someone measuring 60 inches is the same magnitude as the difference in height between 75 and 74 inches. That is, an inch of height reflects the same amount of height regardless of where that inch falls on the scale. This scale, then, is said to possess the property of equal intervals.

But, suppose one ranked the children from 1 (the shortest) on up to the tallest. The child given rank 1 might be just a little shorter than the child ranked 2, but the child ranked 2 might be very much shorter than the child ranked 3. Here the difference in height between ranks 1 and 2 on the scale is not the same as the difference between ranks 2 and 3 on the scale. One unit is not the same amount of height at different points on the scale. This scale, then, does not possess the property of equal intervals.

> An **absolute zero point** is a value which indicates that nothing at all of the attribute being measured exists.

Thus, if height is measured with a tape measure, 0 inches of height is a real scale value that indicates no height whatsoever—it is absolute zero. So height measured in inches has an absolute zero.

However, in the case of ranking children, the lowest score is the rank 1 and that child certainly has some height. The rank of 0 is not used, and even if it were, it would not mean, "zero height at all." There is no rank or scale value that reflects zero height, so a rank-order scale of height does not possess an absolute zero.

Types of Scales Many of the measurement scales one uses in everyday life possess all three of these properties. As explained above, a scale used to measure height in inches has magnitude, equal intervals, and an absolute zero point. So does a scale used to measure weight in ounces or grams, or a scale used to measure time in minutes and seconds.

> Any scale of measurement possessing magnitude, equal intervals, and an absolute zero point is called a **ratio scale.**

The scale is termed *ratio* because the collection of properties that it possesses allows ratio statements to be made about the attribute being measured. If an adult is 70 inches tall and a child is 35 inches, it is correct to infer that the adult is *twice* as tall as the child. Such ratio statements can be made only if the scale possesses all three properties.

Not all scales used in research in the social and behavioral sciences are ratio scales. Many attributes cannot be measured with scales that reflect all three of the properties under discussion.

> An **interval scale** possesses the properties of magnitude and equal intervals but not an absolute zero point.

The most common example of an interval scale is the Fahrenheit (or Celsius) scale for measuring temperature. For all practical purposes, the temperature scales lack an absolute zero, because neither 0° Fahrenheit nor 0° Celsius denotes a point at which there is no temperature at all (that is, absolute zero). Further, if the temperature today is 30° and yesterday it was 15°, it is not correct to say that it is twice as hot today as it was yesterday. Ratio statements cannot be made without an absolute zero point. The temperature scale does possess the properties of magnitude and equal intervals, however. For example, 25° is a warmer temperature than 19°, and the difference between 40° and 50° represents the same difference in temperature as the difference between 90° and 100° (although people may not *perceive* or feel these differences as equal). Since the temperature scale has the properties of magnitude and equal intervals, but not of absolute zero, it is an interval scale.[1]

---

[1] The Kelvin scale of temperature *does* have an absolute zero and is therefore a type of *ratio* scale.

Some scales have only one of the three properties discussed.

> An **ordinal scale** reflects only magnitude and does not possess the properties of equal intervals or an absolute zero point.

For example, ranking children according to their heights, with the shortest child receiving a rank of 1, the next tallest a rank of 2, the next tallest a rank of 3, and so forth, produces an ordinal scale of height. This scale clearly has the property of magnitude, because the ranks indicate whether a child is taller or shorter than another child. But it does not possess equal intervals as described above, because the difference in heights between children ranked 1 and 2 might be substantially larger or smaller than that between children ranked 2 and 3. Further, the ranking scale does not possess an absolute zero point: there is no ranking that expresses no height at all. Hence, the ranking of height (in contrast to a scale that measures height in inches or centimeters) produces an ordinal scale, because it possesses magnitude but not equal intervals or an absolute zero.

It is possible to have a scale that possesses none of the three properties just discussed, although the term *scale* is usually reserved for measurements that imply differences, at least in magnitude.

> A **nominal scale** results from the classification of items into mutually exclusive groups that do not bear any magnitude relationships to one another.

For example, if a person were to stand on a busy street corner and name cars, the classifications might be Ford, Chevrolet, Plymouth, and so forth. The dimension of classification is *make of car.* However, one would not say that Ford is more or less of a make of car than Chevrolet. It may be more or less expensive or appealing to the eye, but not more or less of a make of car. Hence, grouping cars according to brand name represents a nominal scale, which does not possess the properties of magnitude or equal intervals, or an absolute zero point. Examples of nominal scales used in scientific research include the classification of plants and animals by order, genus, and species in biology and the classification of human beings by personality type in psychology.

Variables

Variables vs. Constants   The measurements made with the scales described above produce values of a variable.

> A **variable** is the general characteristic being measured on a set of people, objects, or events, the members of which may take on different values.

If the characteristic being measured is the height in inches of members of a class, then height is a variable because the numerical values of the heights of members of the class may differ from one to another. Gender is also a variable in a sample of male and female individuals, even though it has only two nominal "values" and it is not "measured" in the same way as height or aggressiveness might be. It is a variable because it is a characteristic of people who have one of two nominal categories or "values."

> A **constant** is a quantity that does not change its value within a given context.

The mathematical symbol π is a constant because it always equals 3.1416 . . . ; its value does not change. Also, if one measures the heights of people in a class in inches but wishes to report the measurements in centimeters, height is a variable, but the conversion factor of 2.54 centimeters to the inch is a constant; it is the same value in each case. The average height of people in the class would be another constant, but only with respect to that class. Across several classes within a school, the average height of each of the classes would be a variable.

Discrete vs. Continuous Variables   Variables may be either discrete or continuous.

> A **discrete variable** is one that can assume only a countable number of values between any two points.

For example, a family can have only 0, 1, 2, 3, or more children. One does not think of a family as having $1\frac{1}{2}$ children. Thus between the values 1 and 3, there is only one other possible value, 2. Similarly, the number of points scored in a basketball game is a discrete variable because a team can score 90 or 91 points, but not 90.5 points.

> A **continuous variable** is one that theoretically can assume an infinite number of values between any two points on the measurement scale.

For example, consider measuring weight in pounds. Even between 100 and 101 pounds there is an infinite number of values possible: 100.1, 100.247, and so on. Time in minutes or seconds is another example of a continuous variable.

The discrete–continuous distinction actually applies to the *variable* being assessed and not to the *measurement* of that variable. In practice, continuous as well as discrete variables are almost always measured in discrete scores. For example, we measure weight in ounces, pounds, or tenths of an ounce, even though there are, theoretically, an infinite number of values between any two weights.[2]

---

[2] Technically, the definitions for discrete and continuous variables are for random variables, not variables in general. Such theoretical distinctions are beyond the level of this text.

Continuous variables and some discrete variables may be measured with ratio, interval, or ordinal scales. For example, intelligence is a continuous variable because, given a sufficiently fine instrument, theoretically there are an infinite number of values between any two points. But the current measurement of intelligence is relatively crude, and IQ test scores probably constitute an ordinal scale if the extremely low and high ends of the scale are included.

Real Limits

Since a continuous variable is one for which an infinite number of values potentially exist between any two points on the scale, an actual measurement, which is usually discrete, represents all those theoretical values that would be rounded off to that particular score value. For example, the continuous variable of time may be measured in years, months, days, hours, minutes, seconds, milliseconds, and so on. If one measures to the nearest second, it is clear that more refined measurements could be made with more sophisticated timers. Consequently, if a child is asked to solve a given mathematics problem and does so in "33 seconds," the value 33 technically represents all those theoretically possible times, if more precise measurements were taken, that would be rounded off to 33—for example, 32.8 seconds, 33.3 seconds, and so on. More generally, 33 seconds represents all theoretical values between 32.5 and 33.5 seconds. These two extreme values are called the **real limits** of 33 seconds. The **lower real limit** is 32.5 because any number lower than this (for example, 32.4) would be rounded to a whole number other than 33 (that is, 32), and the **upper real limit** is 33.5 because any number greater than this (for example, 33.7) also would be rounded to a whole number other than 33 (that is, 34).

> The **real limits** of a number are those points falling one-half measurement unit above and one-half measurement unit below that number.

To illustrate, if measurement is being made in whole seconds, the unit of measurement is 1 second. The real limits of 33 seconds are 32.5 seconds and 33.5 seconds because these values are one-half unit (.5 second) below and one-half unit above 33 seconds, respectively.[3]

However, note that the definition states that the limits fall one-half *measurement unit* above and below the number. Therefore, the real limits of 33 seconds are different for measurements made in units other than whole seconds. Suppose that a stopwatch is available and that the length of time to solve a problem is recorded in tenths of a second. In this case, the unit of measurement is .1 second and one-half of this unit is .05 second. Consequently, when measuring in tenths of a second, the real limits of 33.0 are 32.95 and 33.05. Similarly, if measurement is made in hundredths of a second, the real limits of 33.00 are 32.995 and 33.005. Several of these points are presented graphically in Figure 1-1.

---

[3] Technically, the upper real limit of 33 is not 33.5 but 33.49999. . . . That degree of accuracy will not be necessary here.

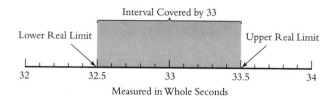

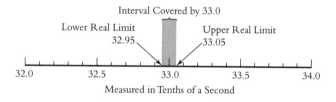

Figure 1-1 Real limits of 33 when the measurement unit is whole seconds (top) and when it is tenths of a second (bottom).

Rounding

If measurements are taken in tenths of a second but reported in whole seconds, they are said to be **rounded** to the nearest whole second.

Numbers are rounded according to the following rules:

1. If the remaining decimal fraction is less than .5 unit, drop the remaining decimal fraction.

2. If the remaining decimal fraction is greater than .5 unit, increase the preceding digit by 1.

3. If the remaining decimal fraction is exactly .5 unit, add 1 to the preceding digit if that digit is an odd number but drop the remaining decimal fraction if the preceding digit is an even number.

For example, if the unit of measurement is a whole second and the value 33.4 is obtained, round this to 33 seconds in accordance with rule 1 above. That is, .4 is less than .5, so it is simply dropped. If the obtained value is 32.6, round up to 33 in accordance with rule 2. That is, .6 is greater than .5, so the preceding digit, 2, is increased by 1 to 3. If the obtained value is 34.5, rule 3 dictates rounding down to 34 because the preceding digit (the 4) is even; but if 37.5 is to be rounded, the result is 38 because the preceding digit (the 7) is odd. Note, however, that 34.51 is rounded up to 35 because the remaining decimal fraction is more than .5 unit (that is, .51 is more than .50). Rule 3 applies only when the remaining fraction is *exactly* .5000 . . . unit.

The reason that some numbers with remaining decimal fractions of exactly .5 are rounded up and some are rounded down is so that, over many instances of rounding such numbers, approximately half will be rounded up and half rounded down. If a simple rule of always rounding .5 up were invoked, more numbers would be rounded up than down. By rounding some of these cases down according to rule 3, this bias is minimized.

Rounding numbers may seem a trivial issue, but it is not. Whether and how one rounds numbers can have a significant effect on the outcome of certain calculations. For example, suppose one were required to determine the amount of money that is needed to be set aside in legislation to subsidize the salaries of early childhood care providers. Suppose the legislator knows that it is necessary to subsidize existing pay by \$1.5349 per hour in her state to raise total wages above the poverty line. But there are 5000 workers needing salary assistance, each working 2000 hours per year or 10,000,000 person-hours requiring subsidies. So the total amount of money that should be set aside is 10,000,000 $\times$ \$1.5349 = \$15,349,000. But if \$1.5349 were properly rounded to \$1.53, then the amount would be 10,000,000 $\times$ \$1.53 = \$15,300,000, a difference of \$49,000, which would mean that there would not be enough money to pay 16 workers. And if the subsidy were rounded to \$1.50, the shortage would be \$349,000, leaving 114 workers without subsidies. The point is that whether and how numbers are rounded can make a big difference, especially if those numbers are later multiplied by large numbers.

Special care is sometimes needed when several numbers must be rounded in a single problem. Some years ago, the custom was to round each number in a series of computations to one more decimal place than the number of places desired in the final answer. This is still good advice. However, when using a small calculator that carries several decimal places and permits the result of one calculation to be automatically entered into the next operation, it is laborious to round all numbers to a certain number of decimal places after each calculation. Because some students or calculators may carry a different number of decimal places than others, not all students will obtain the same numerical answers for the exercises in this book. Different rounding practices may produce slightly different answers. All of the answers provided in this book and in the accompanying *Study Guide* have been performed with as many places as small computing aids would carry, and final answers have been rounded to two decimal places (except when more are required). Occasionally a range of answers is given when rounding practices would make a great difference. Students who are calculating by hand should try to carry three or four decimals in intermediate calculations (the exercises have been kept as numerically simple as possible), but all students will at times obtain answers slightly different from those given here. Intermediate values as well as the final answers are offered to help determine the source of a possible error, rounding or otherwise. Also note that not all calculators and computers follow rule 3 above when rounding.

SUMMATION SIGN

Certain calculations in this book are made not just on one score but on all the scores in a particular group. Therefore, it is convenient to have some symbolic terminology to represent groups of scores and operations on groups rather than single scores and operations on single scores.

Notation for Scores and Summation

Consider the group of five scores below:

Subject	Score Symbol	Score Value
1	X_1	2
2	X_2	3
3	X_3	8
4	X_4	4
5	X_5	7

Suppose we let the capital letter X symbolize the variable reflected by these scores. In actuality, it could be any variable—response time, number of correct answers, eye movements per minute, and so on. But since formulas can apply to any variable, it is convenient to symbolize the variable by a capital letter (such as X, Y, or V). In contrast, when the value of a constant is unknown, it is symbolized by a lowercase letter, often c or k.

There are five scores in the above group, and each score represents a specific measurement of the variable under investigation, X. To distinguish one specific score from another, each X symbol is given a **subscript** corresponding to the number of the subject who made that score on the variable. Customarily, in a group of five scores, these subscripts would be 1, 2, 3, 4, 5. Thus the score on variable X for the first subject is represented by X_1. In general, the total number of scores is represented by N and so the subscripts run from 1 to N. The N scores comprise a **distribution.** Frequently, it is useful to be able to refer to a single score, but not necessarily any particular one—just any single score in the group of scores or distribution. This single score is referred to as the ith score in the distribution, and it is written X_i. In this example, the distribution of all the X_i (read "all the scores on variable X") contains five scores. Any one of them is referred to as X_i, but the particular score of the third subject is represented by X_3 (which has a score value of 8), the score of the fourth subject is represented by X_4 (score value of 4), and so on.

One of the most frequently performed operations in statistics consists of summing all or some of the scores in a group. For example, in computing the average of a group of measurements, one sums all the scores and divides by the number of scores. However, when writing the formula for the average it is cumbersome to use

$$\text{average} = \frac{\text{sum of all scores}}{\text{number of scores}}$$

The operation of summing the scores, which is called for in many formulas, is given a symbolic abbreviation. The Greek capital letter *sigma,* Σ, is employed to indicate the operation of summing, a letter (such as X) is used to indicate the variable to be summed, and N represents the number of scores. Thus, another way to write the formula for the average is

$$\text{average} = \frac{\Sigma X}{N}$$

The operation of summing all the scores on variable X, written above as ΣX, can also be written out more fully, as

$$\sum_{i=1}^{N} X_i$$

The small notations under and over the Σ in this fuller expression are called the **limits of the summation.** The entire symbol is read "sum of the X_i from $i = 1$ to N." It means to add X_1 plus X_2 plus ... plus X_N. In symbols,

$$\sum_{i=1}^{N} X_i = X_1 + X_2 + X_3 + \cdots + X_N$$

Thus $\sum_{i=1}^{5} X_i$ signifies the sum of the first five scores and $\sum_{i=2}^{4} X_i$ means the sum of the second through the fourth scores, inclusive. In terms of the above data,

$$\sum_{i=1}^{5} X_i = 2 + 3 + 8 + 4 + 7 = 24$$

and

$$\sum_{i=2}^{4} X_i = 3 + 8 + 4 = 15$$

Often, when all the scores in a distribution are to be summed (as in the formula for the average shown above), the limits of the summation are not written and/or the subscript i is omitted:

$$\Sigma X_i \text{ and } \Sigma X \text{ both mean } \sum_{i=1}^{N} X_i$$

Operations with the Summation Sign

There are three shortcuts for using the summation sign with variables and constants in algebraic operations.

> **1.** The sum of a constant times a variable equals the constant times the sum of the variable. If c is a constant and X_i a variable,
>
> $$\sum_{i=1}^{N} cX_i = c\sum_{i=1}^{N} X_i$$

Consider first a numerical example. Suppose a father, mother, and child ran a mile in 7, 8, and 10 minutes, respectively. How many *seconds* did it take the family to run the mile? There are two ways to solve the problem. One way is to multiply each person's time in minutes (X_i) by 60 seconds (c) and then add the times together:

$$\sum_{i=1}^{N} cX_i = 60(7) + 60(8) + 60(10) = 1500 \text{ seconds}$$

The other way is to multiply the total number of minutes by 60:

$$c\sum_{i=1}^{N} X_i = 60(7 + 8 + 10) = 1500 \text{ seconds}$$

Both procedures give the same answer, illustrating that the sum of a constant times a variable equals the constant times the sum of the variable, or, in symbols,

$$\sum_{i=1}^{N} cX_i = c\sum_{i=1}^{N} X_i$$

This equivalence is demonstrated to be generally true in Optional Table 1-1.

2. The sum of a constant taken N times is N times the constant. If c is a constant,

$$\sum_{i=1}^{N} c = Nc$$

Suppose the constant is $c = 5$ and the number of scores is $N = 3$. There are two ways to obtain $\sum_{i=1}^{3} 5$. One is to add 5 three times: $5 + 5 + 5 = 15$. The other is to recognize that adding the same number N times is identical to multiplying that number by N: $3(5) = 15$. Again, the two approaches lead to the same result, illustrating that

$$\sum_{i=1}^{N} c = Nc$$

This equivalence also is demonstrated to be generally true in Optional Table 1-1.

3. The summation of a sum of variables is the sum of each of these variable sums. If X and Y are variables,

$$\sum_{i=1}^{N} (X_i + Y_i) = \sum_{i=1}^{N} X_i + \sum_{i=1}^{N} Y_i$$

1-1 Demonstration of Three Principles of the Summation Sign

1. **The sum of a constant times a variable equals the constant times the sum of the variable.** If c is a constant and X_i a variable,

$$\sum_{i=1}^{N} cX_i = c\sum_{i=1}^{N} X_i$$

Consider what the expression $\sum_{i=1}^{N} cX_i$ means algebraically:

$$\sum_{i=1}^{N} cX_i = cX_1 + cX_2 + cX_3 + \cdots + cX_N$$

But the series of terms to the right of the equals sign may be factored in the same manner as

$$ca + cb = c(a + b)$$

with the following results:

$$\sum_{i=1}^{N} cX_i = cX_1 + cX_2 + cX_3 + \cdots + cX_N$$

$$= c(X_1 + X_2 + X_3 + \cdots + X_N)$$

$$= c\left(\sum_{i=1}^{N} X_i\right)$$

Since the expression within the parentheses is what has been defined to be $\sum_{i=1}^{N} X_i$, the sum of a constant times a variable is the constant times the sum of the variable:

$$\sum_{i=1}^{N} cX_i = c\sum_{i=1}^{N} X_i$$

2. **The sum of a constant taken N times is N times the constant.** If c is a constant,

$$\sum_{i=1}^{N} c = Nc$$

Write out the expression under consideration:

$$\sum_{i=1}^{N} c = \underbrace{c + c + c + \cdots + c}_{N \text{ terms}}$$

The symbol $\sum_{i=1}^{N} c$ calls for adding N c's together. However, the operation of multiplication is precisely this repetitive addition, so that adding N c's is identical to multiplying c by N. Therefore, the sum of a constant c taken N times is Nc.

(continued)

1-1 **Continued**

3. **The summation of a sum of variables is the sum of each of these variable sums.** If X and Y are variables,

$$\sum_{i=1}^{N} (X_i + Y_i) = \sum_{i=1}^{N} X_i + \sum_{i=1}^{N} Y_i$$

Again, writing out the expression,

$$\sum_{i=1}^{N} (X_i + Y_i) = (X_1 + Y_1) + (X_2 + Y_2) + \cdots + (X_N + Y_N)$$

removing parentheses,

$$= X_1 + Y_1 + X_2 + Y_2 \qquad + \cdots + X_N + Y_N$$

and regrouping produces

$$= \underbrace{X_i + X_2 + \cdots + X_N}_{} \qquad + \qquad \underbrace{Y_i + Y_2 + \cdots Y_N}_{}$$

$$\sum_{i=1}^{N} (X_i + Y_i) = \qquad \sum_{i=1}^{N} X_i \qquad + \qquad \sum_{i=1}^{N} Y_i$$

This result may be generalized to any number of terms. For example,

$$\Sigma(X_i + Y_i + W_i) = \Sigma X_i + \Sigma Y_i + \Sigma W_i$$

This expression merely says that if the scores of N students on a midterm and a final exam are to be added together, they can be added in any order. For example, first you can add the two exam grades for each student separately ($X_i + Y_i$) and then add all the students' combined totals: $\sum_{i=1}^{N} (X_i + Y_i)$. Alternatively, you can add the sum of the scores on the first exam to the sum of the scores on the second exam: $\sum_{i=1}^{N} X_i + \sum_{i=1}^{N} Y_i$.

Suppose the scores for three students were:

Student	X_i (Test 1)	Y_i (Test 2)	Student Totals
1	12	16	28
2	8	11	19
3	9	7	16
Test Totals	29	34	63

One can add either the Student Totals

$$28 + 19 + 16 = 63$$

or the Test Totals

$$29 + 34 = 63$$

to get the answer. This illustrates that the summation of a sum of variables is the sum of each of the variable sums:

$$\sum_{i=1}^{N} (X_i + Y_i) = \sum_{i=1}^{N} X_i + \sum_{i=1}^{N} Y_i$$

This equivalence also is demonstrated to be generally true in Optional Table 1-1.

An Important Distinction

Throughout this text the student must be attentive to the difference between two summation quantities, ΣX^2 and $(\Sigma X)^2$.

ΣX^2 is the sum of each squared score.

$(\Sigma X)^2$ is the square of the sum of the scores.

Suppose one had a sample of the following three subjects:

Subject	X	X²
1	4	16
2	5	25
3	2	4
	$\Sigma X = 11$	$\Sigma X^2 = 45$
	$(\Sigma X)^2 = 121$	

The symbol ΣX^2 directs that each X score be squared first and then the squared values summed, whereas $(\Sigma X)^2$ tells one to sum the X scores first and then to square that sum. Stated another way, ΣX^2 means the sum of the squared X scores, whereas $(\Sigma X)^2$ means the squared sum of the X scores. Generally, these are not equal quantities. That is,

ΣX^2 is not equal to $(\Sigma X)^2$

Students should notice that some other quantities that look similar are also not usually equal. For example, $\Sigma(X_i Y_i)$ is *not* the same as $\Sigma(X_i + Y_i)$, and $\Sigma(X_i Y_i)$ is *not* equal to $(\Sigma X_i)(\Sigma Y_i)$. The expression $\Sigma(X_i Y_i)$, or simply ΣXY, instructs one to sum a group of **cross products.** To do this, one first obtains a cross product for each subject by multiplying each person's first score (X_i) by that person's second score (Y_i). Then one adds the individual cross products of all people in the group. Notice the difference between ΣXY and $\Sigma(X + Y)$ in the following table.

Subject	X	Y	X + Y	XY
1	4	6	10	24
2	2	5	7	10
3	9	3	12	27
			$\Sigma(X + Y) = 29$	$\Sigma XY = 61$

Understanding and being able to use the summation sign are necessary to follow the mathematics and to calculate many formulas presented in the rest of this text. The student is advised to practice the exercises for mastery in the use of the summation sign. In compensation, no new mathematics other than elementary algebra and geometry will be required to understand the material presented in this text. Moreover, most mathematical material will be presented in optional tables, which may be skipped by some instructors. Appendix 1 of this text and the Appendix of the *Study Guide* both present brief reviews of basic algebra for the interested reader.

SUMMARY

Statistics is the study of methods for describing and interpreting quantitative information, including techniques for organizing and summarizing data and techniques for making generalizations and inferences from data. The first of these two broad classes of methods is called descriptive statistics, and the second is called inferential statistics. A knowledge of statistics is useful in understanding statistical information commonly presented in the media and in other aspects of everyday life, and it is essential in understanding and conducting social and behavioral research. Statistics is necessary because almost all numerical information in these sciences has variability, and statistics helps to interpret and draw conclusions from data that contain variability.

This numerical information consists of measurements made along scales that have various properties, including magnitude, equal intervals, and an absolute zero point. Depending on which properties they possess, the scales may be ratio, interval, ordinal, or nominal. A variable is a general characteristic being measured on a set of objects, people, or events, the members of which may take on different numerical values. Variables may be discrete or continuous. A constant, however, is a quantity that does not change its value within a given context. Various procedures for rounding and summing are also described.

Exercises

For Conceptual Understanding

1. Indicate the scale of measurement used in the studies described below, and justify your choice.

   a. A public opinion poll was conducted of a random sample of registered voters one week before an election. The respondents were asked whether they intended to vote for (1) the Democratic candidate, (2) the Republican candidate, (3) one of the other candidates, or (4) none of the candidates.

   b. A psychologist studied the physiological effects of crowding by housing laboratory rats in conditions that were crowded (10 rats per cage) or uncrowded (2 rats per cage). After two months the adrenal glands of all the animals were removed and weighed to the nearest .01 gram (the sizes of these glands indicate sensitivity to stress).

   c. A military training officer ranked each of 20 cadets on leadership, with the rank of 1 indicating the greatest leadership.

   d. Ms. Smithson, a nutritionist in an elementary school, believes that a good breakfast helps children perform better in school. She gives an examination containing 40 questions on different subjects to 30 low-income children, then provides them with a nutritious breakfast and tests them again.

   e. Mr. Bates rated the 30 children in his preschool class on readiness for kindergarten by rating them on a five-point scale, 1 for least ready and 5 for most ready.

   f. A college teacher gave an essay exam and then graded the papers by using five letter grades A, B, C, D, F that represented equally spaced categories of competence.

   g. Students were asked whether they (1) favored an increase in the student activities fee to be eligible to attend more events, (2) favored no increase in fee to attend the current number of events, or (3) had no opinion.

2. Discuss the differences in the scales of temperature produced by the Celsius versus the Kelvin methods. (The Kelvin method uses $-273°C$ as its zero point. This is the point at which all molecular movement ceases.)

3. The accompanying table shows a set of scores for seven students on an examination and the rank ordering of these students on the basis of their performance. Using students 4, 5, and 6, illustrate the nature of the information lost when one resorts to ordinal scales rather than interval (or ratio) scales of measurement.

Student	Rank	Score
1	7	79
2	6	52
3	5	46
4	4	41
5	3	25
6	2	24
7	1	21

4. Which of the following variables are discrete and which are continuous?

   a. temperature as measured on a Celsius scale

   b. number of students in a class

   c. the Dow-Jones Industrial Average (average dollar value of 30 stocks)

   d. ratings from 1 to 10 of owner satisfaction with their automobiles

   e. the number of babies born to families

   f. the speed in miles per hour of cars passing a certain point on a highway

   g. height in inches

   h. academic rank in class

5. Identify three potential sources of variability in behavioral assessments, and tell how statistics can help to draw scientific conclusions in the presence of such variability.

6. How do inferential statistics help to make the scientific process more precise?

For Solving Problems

7. What are the real limits of the following numbers? (The unit of measurement is the smallest decimal place in the number.)

a. 21.2   f. .41   k. 6.0001

b. −2.22   g. 21.20 l. −93.42

c. 99.99   h. 14.001 m. 27.70

d. 21   i. −1

e. 1.5   j. 15.8

8. Round the following numbers to tenths:

a. 1.56   d. 9.45   g. 11.45001

b. 3.22   e. 10.05   h. 11.45000

c. 5.95   f. 6.251   i. 11.45

9. Given the following data, determine the numerical value of each of the expressions below.

Subject (i)	X_i	Y_i
1	6	9
2	4	5
3	6	3
4	0	7
5	7	8

a. $\sum_{i=1}^{5} X_i$ f. $\sum_{i=1}^{5} X_i Y_i$

b. $\sum_{i=1}^{3} Y_i$ g. $\sum_{i=1}^{5} X_i^2$

c. $\sum_{i=3}^{5} X_i$ h. $\left(\sum_{i=1}^{5} X_i\right)^2$

d. ΣY i. ΣX

e. $\sum_{i=1}^{5} (X + Y)$ j. ΣY^2

k. $(\Sigma Y)^2$

l. $\Sigma X + \Sigma Y$

10. Given the data from Exercise 9 for X and Y and that c, a constant, equals 3, determine the numerical value for the following expressions:

a. $\sum_{i=1}^{5} cX_i$ d. $\sum_{i=1}^{5} c$

b. ΣcXY

c. $\Sigma c(X + Y)$ e. $\sum_{i=1}^{3} (X + c)$

11. Simplify the following expressions (W and Z are variables, k and c are constants):

a. $\dfrac{\Sigma c(kW + W)}{\Sigma W}$

b. $\dfrac{\Sigma(W + k) + \Sigma(Z - k)}{\Sigma(W + Z)}$

c. $\dfrac{\Sigma(k - W) + \Sigma(Z + W) + (\Sigma k)(\Sigma Z)}{N\left(\Sigma kZ + \dfrac{\Sigma k}{N}\right)}$

For Homework

12. Distinguish between *descriptive* and *inferential* statistics.

13. A scale of measurement can be described by the presence or absence of each of three characteristics. Discuss these three properties and describe and illustrate four scales that can be distinguished on the basis of these properties.

14. Determine whether each of the following is an example of a discrete or of a continuous variable.

a. height in inches

b. number of children per family

c. ounces of fluid

d. number of chairs in a room

15. Round the following to one decimal place; to two decimal places.

a. 7.0549   f. 2.9950

b. 7.0550   g. 2.9940

c. 7.0551   h. 4.0010

d. 1.925   i. 4.0080

e. 1.915

16. Name the type of scale created by each of the following classifications.

a. a 1-to-10–point rating of friendliness

b. brand of cereal

c. ranking of first, second, and third in a race

d. number correct on a test

17. Determine the numerical value of each of the following expressions.

i	X	Y
1	3	2
2	5	1
3	0	3
4	2	1
5	4	4

a. $\displaystyle\sum_{i=1}^{5} (X_i + 3)$

b. $\displaystyle\left(\sum_{i=1}^{5} X_i\right) + 3$

c. $\displaystyle\sum_{i=1}^{3} X_i Y_i$

d. $\displaystyle\left(\sum_{i=1}^{3} X_i\right)\left(\sum_{i=1}^{5} Y_i\right)$

e. $\displaystyle\sum_{i=1}^{3} X_i X_{i+1}$

f. $\displaystyle\sum_{i=1}^{5} Y_i^2$

g. $\Sigma cX \ (c = 2)$

h. $\displaystyle\sum_{i=1}^{3} c \ (c = 2)$

i. $\displaystyle\left(\sum_{i=1}^{3} X^2\right) + \left(\sum_{i=1}^{5} Y^2\right)$

18. Differentiate between ΣX_i^2 and $(\Sigma X_i)^2$. Given $X_1 = 2$, $X_2 = 4$, and $X_3 = 6$, determine the values for $\displaystyle\sum_{i=1}^{3} X_i^2$ and $\displaystyle\left(\sum_{i=1}^{3} X_i\right)^2$.

19. Write the following in summation notation:
    a. $X_1 + c + X_2 + c + X_3 + c + X_4 + c$
    b. $X_1 + X_2 + X_3 + X_4 + c$
    c. $X_1 Y_1 + X_2 Y_2^2 + X_3 Y_3^3 + X_4 Y_4^4$
    d. $X_1 + Y_5 + X_2 + Y_6 + X_3 + Y_7 + X_4 + Y_8$

20. What are the real limits of the following numbers?
    a. 20 cm
    b. 20.1 cm
    c. 20.12 cm
    d. 20.124 cm

Frequency Distributions and Graphing

hen behavioral scientists design a simple research study, they select a variable and a scale of measurement and then make observations on a group of subjects. Once a collection of scores has been obtained, how can one describe it efficiently? How can the group of scores as a whole be summarized to obtain its general nature and then this general result be communicated simply to others? Frequency distributions accomplish these goals.

TYPES OF FREQUENCY DISTRIBUTIONS

There are several types of distributions, the most straightforward of which is called simply a frequency distribution.

> A **frequency distribution** indicates the number of cases observed at each score value or within each interval of score values in a group of scores.

Suppose a short history quiz is given to a class of 10 students, and the essay is graded on a 10-point scale. The scores are presented in column A of Table 2-1. To obtain a clearer picture of this small group of scores, list the scores in descending order, as in column B of the table. Next, observe that there is one score with a value of 10,

---

2-1 Development of a Frequency Distribution for Scores on a History Quiz

A. Scores (X)	B. Scores Arranged in Descending Order	C. Frequency Distribution of Scores		
		X	Tally	f
8	10			
10	9	10	/	1
9	9	9	//	2
7	8	8	///	3
9	8	7	///	3
8	8	6	/	1
7	7			N = 10
6	7			
8	7			
7	6			

two of 9, three each of 8 and 7, and one of 6. If a score value is symbolized by X and a tally is made next to the X for each occurrence of that value, the result is a frequency distribution, which is presented under section C of the table. There, f indicates the numerical frequency for any given value of the variable X.

Consider another example of the development of a frequency distribution. An opinion researcher administers a questionnaire that asks a sample of people to what extent they approve of the way the president of the United States is performing his duties. The researcher provides five possible responses and arbitrarily assigns a number to each level of response:

1. Disapprove greatly
2. Disapprove (generally disapprove but approve of some policies)
3. Ambivalent (disapprove of about half the president's actions and approve of half)
4. Approve (generally approve but disapprove of some policies)
5. Approve greatly

Suppose 80 people were questioned and were asked to indicate one of these five opinions. The frequency of their responses is presented in Table 2-2. It is clear from this frequency distribution that people's opinions about the president are somewhat divided. A sizable group generally disapproves while another sizable group approves of the president's actions. Few people are either ambivalent ($X = 3$) or adamantly positive or negative ($X = 1$ or 5). Such descriptive conclusions would be difficult to reach if all 80 scores were written down without the assistance of the frequency distribution. Hence, not only does a frequency distribution save time in displaying data, it also organizes the numbers in a way that efficiently summarizes the data and conveys their meaning.

Sometimes, as in the opinion poll described above, it is more useful to know the proportion or percentage of cases that are of each score value than to know the actual number of such cases. Such a distribution is a relative frequency distribution.

> A **relative frequency distribution** is a distribution that indicates the *proportion* of the total number of cases (that is, scores) observed at each score value or interval of score values.

2-2   **Frequency Distribution of Opinions on Presidential Performance**

Opinion	X	f	*Rel f*
Approve greatly	5	9	.11
Approve	4	30	.38
Ambivalent	3	10	.12
Disapprove	2	25	.31
Disapprove greatly	1	6	.08
		$N = 80$	1.00

An example of a relative frequency distribution (*Rel f*) is given in the rightmost column of Table 2-2. Because the numbers there are proportions, they are expressed as decimal fractions; multiplying these proportions by 100 (that is, moving the decimal point two places to the right) will change the proportions to percentages. The advantage of a relative frequency distribution is that it expresses the pattern of scores in a manner that does not depend on the specific number of cases involved. Thus, an opinion pollster would not say that 9 people greatly approved of the president's actions, but rather that 11% did. This is because the frequency of 9 is specific to a sample of 80 people, but 11% has meaning regardless of the size of sample. Presumably, 11% of American adults, for example, might be expected to greatly approve of the president's performance.

The need for a frequency distribution is most obvious when one has a large number of scores and a wide range of score values. To illustrate, suppose 150 eighth-grade students were given a mathematics ability test prior to taking an algebra course. The scores for these students are presented in Table 2-3. This display emphasizes the fact that listing all the scores does not provide much immediate information. Further, with so many score values, a frequency distribution constructed by counting the number of cases of each possible score value between 0 and 100 would be only slightly more informative. Therefore, to describe the entire set of scores efficiently, the researcher must group the score values into sets, called class intervals.

2-3	**Mathematics Ability Test Scores for 150 Eighth-Grade Students**								
79	51	67	50	78	80	77	75	55	65
62	89	83	73	80	67	74	63	32	88
88	48	60	71	79	79	47	55	70	34
89	63	55	93	71	81	72	68	75	93
41	81	46	50	61	72	86	66	54	58
59	50	90	75	61	82	73	57	87	41
75	98	53	79	80	64	67	51	36	52
70	37	42	72	74	78	91	69	95	76
67	73	79	67	85	74	70	62	76	69
91	73	77	36	77	45	39	59	63	57
53	67	85	74	77	78	73	61	47	43
76	43	42	96	83	83	84	67	81	75
70	92	59	86	53	71	49	68	42	46
32	67	67	71	71	59	80	66	39	49
82	68	30	72	57	92	50	38	73	56

A **class interval** is a segment of the measurement scale that contains more than one possible score value.

Table 2-4 was constructed by ordering the scores listed in Table 2-3 and by grouping score values into class intervals of five values each. From Table 2-4, one can see that there were only three cases of a score between 95 and 99 inclusive (namely,

2-4	**Ordering of 150 Mathematics Ability Test Scores**	
—	74, 74, 74, 74	49, 49
98	73, 73, 73, 73, 73, 73	48
—	72, 72, 72, 72	47, 47
96	71, 71, 71, 71, 71	46, 46
95	70, 70, 70, 70	45
—	69, 69	—
93, 93	68, 68, 68	43, 43
92, 92	67, 67, 67, 67, 67, 67, 67, 67, 67	42, 42, 42
91, 91	66, 66	41, 41
90	65	—
89, 89	64	39, 39
88, 88	63, 63, 63	38
87	62, 62	37
86, 86	61, 61, 61	36, 36
85, 85	60	—
84	59, 59, 59, 59	34
83, 83, 83	58	—
82, 82	57, 57, 57	32, 32
81, 81, 81	56	—
80, 80, 80, 80	55, 55, 55	30
79, 79, 79, 79, 79	54	
78, 78, 78	53, 53, 53	
77, 77, 77, 77	52	
76, 76, 76	51, 51	
75, 75, 75, 75, 75	50, 50, 50, 50	

95, 96, and 98), seven cases within the interval of 90 to 94 inclusive, nine between 85 and 89 inclusive, and so on. A summary of this accounting is presented in the frequency distribution shown in the two leftmost columns of Table 2-5. The only difference between this frequency distribution and previous ones is that, rather than having frequencies stated for each possible score value, the score values are grouped into class intervals (for example, 95–99, 90–94) and frequencies are determined for each interval. When scores are presented this way, they are sometimes called **grouped data.**

As with ungrouped data, grouped data can be examined with a relative frequency distribution as well as a frequency distribution, as shown in the third column ($Rel f$), or with cumulative frequency ($Cum f$) and cumulative relative frequency ($Cum Rel f$) distributions.

A **cumulative frequency distribution** is one in which the entry for any score value or class interval is the sum of the frequencies for that value or that interval plus the frequencies of all lower score values.

A **cumulative relative frequency distribution** is one in which the entry for any score value or class interval expresses that value's or that interval's cumulative frequency as a proportion of the total number of cases.

2-5   **Distributions for 150 Mathematics Ability Test Scores**

Class Interval	f	$Rel f$	$Cum f$	$Cum Rel f$
95–99	3	.02	150	1.00
90–94	7	.05	147	.98
85–89	9	.06	140	.93
80–84	13	.09	131	.87
75–79	20	.13	118	.79
70–74	23	.15	98	.65
65–69	17	.11	75	.50
60–64	10	.07	58	.39
55–59	12	.08	48	.32
50–54	11	.07	36	.24
45–49	8	.05	25	.17
40–44	7	.05	17	.11
35–39	6	.04	10	.07
30–34	4	.03	4	.03
$N = 150$		1.00		

That is, in the cumulative versions of frequency and relative frequency distributions, the entries are progressively accumulated starting from the lowest class interval. In Table 2-5, for example, the lowest interval (30–34) has 4 frequencies and the next lowest interval has 6 frequencies. So the cumulative frequency (*Cum f*) of the lowest interval is simply 4, but the cumulative frequency of the next lowest interval is 4 + 6 = 10. Similarly, the relative frequency of these two intervals is .03 and .04, respectively, so the cumulative relative frequency (*Cum Rel f*) of these intervals is .03 and .03 + .04 = .07, respectively.

Cumulative distributions provide a means for rapidly determining the number or proportion of scores that fall within and below a given class interval. For example, suppose that Johnny had a score of 64 on this mathematics test. The cumulative proportion (*Cum Rel f*) of scores within and below the interval 60–64 in Table 2-5 is .39, which means that 39% of the scores were equal to or below the score of 64. Johnny could be said to be at the 39th percentile. Percentiles will be discussed in more detail in Chapter 4.

A distribution of scores, then, may be displayed as a frequency, relative frequency, cumulative frequency, or cumulative relative frequency distribution. The advantage of such distributions is that they provide an efficient method of organizing and presenting a large group of scores so that certain characteristics of the group as a whole become apparent.

CONSTRUCTING FREQUENCY DISTRIBUTIONS WITH CLASS INTERVALS

Number of Class Intervals

It is somewhat easier to read and understand a frequency distribution than it is to actually organize a set of scores into that form because certain decisions must be made about the nature of the class interval to be used. In the above case, an interval such as 30–34 was selected, but would 30–32 or 30–47 have done just as well? Table 2-6 illustrates these two alternatives.

Consider the examples of Table 2-6 from the standpoint of the general goal of frequency distributions, which is to summarize data in a form that accurately depicts the group as a whole. In the first instance (left-hand distribution in Table 2-6), the summarization advantage of the frequency distribution is lost because there are too many class intervals.

Conversely, the right-hand distribution in Table 2-6 has too few class intervals. It is clear that most of the scores fall between 66 and 83, but considerable information and detail have been lost by grouping the data into only four classes. For example, one cannot tell whether the 22 cases between 30 and 47 fall nearer to the value of 47, nearer to 30, or evenly within the interval.

Therefore, the summarization and descriptive purposes of frequency distributions will be lost if there are too few intervals each with many cases, or too many intervals

	Two Distributions for 150 Mathematics Ability Test Scores		
2-6	**Using Class Intervals of Different Sizes**		

Class Interval	f	Class Interval	f
96–98	2	84–101	20
93–95	3	66–83	71
90–92	5	48–65	37
87–89	5	30–47	22
84–86	5		N = 150
81–83	8		
78–80	12		
75–77	12		
72–74	14		
69–71	11		
66–68	14		
63–65	5		
60–62	6		
57–59	8		
54–56	5		
51–53	6		
48–50	7		
45–47	5		
42–44	5		
39–41	4		
36–38	4		
33–35	1		
30–32	3		
	N = 150		

each with only a few cases. You will need to judge how many intervals are needed to accurately summarize each particular distribution. However, to get you started on selecting the number of class intervals, turn to Table 2-7. This table presents a guide to the maximum number of intervals for distributions of different numbers of cases. For example, if you have a distribution of $N = 40$ cases, Table 2-7 suggests that the maximum number of class intervals is approximately 13. If $N = 100$, start with a maximum of 20 intervals.

2-7	Maximum Number of Class Intervals for Distributions of Size N		
N	Number of Intervals	N	Number of Intervals
10	6	80	18
15	8	90	19
20	9	100	20
25	10	125	21
30	11	150	22
40	13	200	23
50	14	350	25
60	15	500	27
70	17	1000	30

To determine an approximate size of class interval, divide the range of score values in the distribution (i.e., the largest score value minus the smallest score value) by the maximum number of intervals given above and round this result *up* to a number that is convenient to use for an interval size.

The above maximum numbers of intervals use $2\sqrt{N}$ for $N \leq 100$ and $10 \log_{10} N$ for $N \geq 100$ as discussed by Emerson, J. D., and Hoaglin, D. C. (1983). Stem-and-Leaf Displays. In Hoaglin, D. C., Mosteller, F., and Tukey, J. W. (eds.), *Understanding Robust and Exploratory Data Analysis* (New York: Wiley, 1983), pp. 7–32.

Size of the Class Interval

Once a tentative decision on the approximate number of class intervals has been made, the size of the interval must be determined. A good approach is to subtract the smallest score from the largest. This provides a range of values covered by the group of scores that must be broken into intervals. If this result is divided by the maximum number of intervals (from Table 2-7), an estimate of the smallest size of interval is obtained. For example, for the mathematics ability data in Tables 2-3 and 2-4, the largest score value is 98 and the smallest score value is 30. Therefore, the range of values is $98 - 30 = 68$. Since there are 150 cases in this distribution, Table 2-7 suggests that the maximum number of class intervals for distributions of $N = 150$ is 22. Now divide the range (i.e., 68) by the maximum number of intervals (i.e., 22) to obtain an estimate of the smallest interval size, which is $68/22 = 3.1$. However, there is an informal preference for intervals to have sizes that are "round numbers"—for example, 1, 2, 5, or 10. Therefore, since 3.1 is an estimate of the *smallest* interval size, we round it *up*— in this case generously to the "round number" 5. Thus, each interval will include five adjacent score values, such as (30, 31, 32, 33, 34) or (95, 96, 97, 98, 99). With an interval of size 5, the range of 68 actually will be covered with $68/5 = 13.6$ or 14

2-8	**Steps in Constructing Frequency Distributions with Grouped Data**

1. Obtain an estimate of the **maximum number of class intervals** from Table 2-7.

2. Calculate the **range** of score values by subtracting the smallest score value from the largest score value in the distribution.

3. Obtain an estimate of the **smallest size of class interval** by dividing the range (obtained in step 2) by the maximum number of class intervals (obtained in step 1).

4. **Round up** the estimate of the smallest interval size (obtained in step 3) to the next "round number" (i.e., round up to .1, .2, .5, 1, 2, 5, 10, 20, 50, 100, etc.).

5. Determine the **lowest class interval** so that its lowest *stated* limit is evenly divisible by the size of the interval (obtained in step 4).

6. Place the lowest interval at the bottom of the frequency distribution.

7. After constructing the distribution, determine whether the distribution accurately describes the data, and adjust the size and number of intervals if appropriate.

intervals. If, however, after seeing this distribution, we think that there are too many cases in some intervals to describe the distribution accurately, we could try an interval of size 4 instead of 5.

Lowest Class Interval

Once the number and size of the class intervals have been decided, all that remains is to specify the first interval; then the scale will be completely determined. Obviously, the first interval must include the lowest score. There is a custom that the first score value in the lowest class interval should be evenly divisible by the size of the interval. Here the size is 5, the smallest score is 30, and it happens that 30 is evenly divisible by 5; therefore, 30, 31, 32, 33, 34, or 30–34, is the first interval. If the lowest score in a distribution is 49 and an interval of size 4 is selected, the first interval would be 48–51 because 48 (but not 49) is evenly divisible by 4.

The steps in constructing a frequency distribution that have been described above are summarized in Table 2-8.

Limits, Sizes, and Midpoints of Class Intervals

Once a frequency distribution has been constructed, it is sometimes useful to determine the stated and real limits, the size, and the midpoints of the class intervals.

First, determine the stated limits of the intervals:

The **stated limits** of a class interval are the highest and lowest score values contained in the interval.

Therefore, the class interval including the scores 30, 31, 32, 33, and 34 has the stated limits of 30 and 34, because 30 is the lowest and 34 is the highest score value contained in the interval.

Second, determine the real limits of the score intervals:

The **upper real limit** of a class interval is the upper real limit of the highest score value contained in that interval; the **lower real limit** of the interval is the lower real limit of the lowest score value contained in the interval.

For the stated interval 30–34, the real limits are 29.5 to 34.5 because the lower real limit of 30 is 29.5 and the upper real limit of 34 is 34.5.

Third, determine the size of an interval by its real limits:

The **size of a class interval** is obtained by subtracting the lower real limit of the interval from its upper real limit.

The size of the stated interval 30–34 is 5 because

$$34.5 - 29.5 = 5$$

It may seem a bit puzzling at first that the size of the interval 30–34 is 5 and not 4. But if the scores contained within the interval are listed (30, 31, 32, 33, 34), there are clearly 5 of them, not 4. This notion is further illustrated in Figure 2-1, which shows a segment of a linear scale of measurement and the real limits for each score in the interval.

Fourth, determine the midpoint of an interval:

The **midpoint of a class interval** is the precise center of that interval—that is, the point halfway between the interval's real limits. It can be determined by adding one-half the size of the interval to its lower real limit.

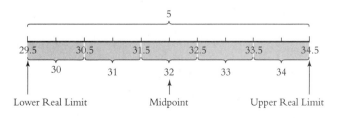

Figure 2-1   Demonstration that the size of the interval 30–34 is 5, and that its midpoint is 32.

For the interval 30–34, whose size is 5 and whose lower real limit is 29.5, the midpoint is

$$\begin{aligned} \text{midpoint} &= \text{(lower real limit)} + .5 \text{ (interval size)} \\ &= 29.5 + .5(5) \\ &= 29.5 + 2.5 = 32 \end{aligned}$$

For an interval with the stated limits 45.6–48.3, whose size is 2.8 and whose lower real limit is 45.55, the midpoint is

$$\begin{aligned} \text{midpoint} &= 45.55 + .5(2.8) \\ &= 45.55 + 1.4 = 46.95 \end{aligned}$$

The important thing to remember is to add one-half the size of the interval to the lower *real* limit (in this case, 45.55), not the lower *stated* limit (45.6).

Table 2-9 shows the real limits, size, and midpoint of each interval for the grouped frequency distribution of mathematics test scores given in Tables 2-4 and 2-5.

2-9　**Frequency Distribution for the Data in Tables 2-4 and 2-5 with Real Limits, Interval Size, and Midpoint**

Class Interval	Real Limits	Interval Size	Midpoint	Frequency
95–99	94.5–99.5	5	97	3
90–94	89.5–94.5	5	92	7
85–89	84.5–89.5	5	87	9
80–84	79.5–84.5	5	82	13
75–79	74.5–79.5	5	77	20
70–74	69.5–74.5	5	72	23
65–69	64.5–69.5	5	67	17
60–64	59.5–64.5	5	62	10
55–59	54.5–59.5	5	57	12
50–54	49.5–54.5	5	52	11
45–49	44.5–49.5	5	47	8
40–44	39.5–44.5	5	42	7
35–39	34.5–39.5	5	37	6
30–34	29.5–34.5	5	32	4
				$N = 150$

GRAPHS OF FREQUENCY DISTRIBUTIONS

Frequency Histogram

A frequency distribution can be described pictorially by drawing a graph. One type of graph is a **frequency histogram.** Figure 2-2 presents the distribution of math ability scores shown in Tables 2-5 and 2-9 as a frequency histogram.

To construct this histogram, a horizontal scale was drawn corresponding to the scale of math ability scores. The horizontal dimension of such a plot is called the **abscissa.** Notice that the numbers along the abscissa are the midpoints of the intervals, as shown in Table 2-9. The vertical dimension is called the **ordinate.** Typically, the ordinate is drawn to be approximately three-fourths as long as the abscissa, and in the case of simple frequency distributions it corresponds to f, frequency. The abscissa and ordinate are called **axes.** Note that the axes are clearly marked with the numbers and the names of their respective scales. Notice also in the lower left corner that the abscissa is broken with two small vertical lines. This break is used because the two axes must intersect at the origin of each scale—that is, at the point $X = 0$, $Y = 0$, often written $(0, 0)$—and the break indicates that part of the math ability scale has been omitted. Such a break could occur on the ordinate as well as the abscissa if part of its scale were omitted. This is more likely to be needed when the ordinate represents score values, not frequencies. Ordinarily, a break is used only between the origin and the segment

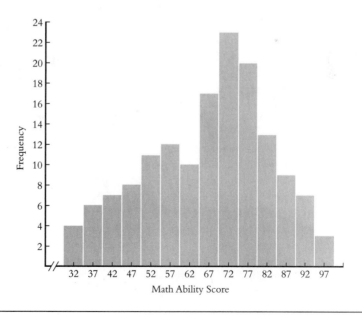

Figure 2-2   Frequency histogram for data in Tables 2-5 and 2-9.

of the scale corresponding to available score values, although occasionally it is used between the main body of scores and a few extreme values.

The width of each bar of the histogram covers the entire range of its class interval, from lower to upper *real* limits. Therefore, each bar covers an entire interval and exactly straddles the midpoint of the interval as designated along the abscissa. The height of a bar corresponds to the frequency of scores for that interval, as indicated on the ordinate. A summary of the steps in constructing a frequency histogram as well as the graphs described below is presented in Table 2-10.

Frequency Polygon

Another type of graph of a frequency distribution is a **frequency polygon,** an example of which is presented in Figure 2-3. The graph is constructed by placing a point above the midpoint of each class interval corresponding to the frequency within that interval. The adjacent points are connected by straight lines. Note that when drawing a frequency polygon, one adds an "empty" interval with zero frequencies to the left of the first and also to the right of the last intervals that contain scores. The line that connects the points on the graph intersects the abscissa at the midpoints of these empty intervals, thus closing the polygon, from which this graph derives its name.

Relative Frequency Histogram and Polygon

Figure 2-4 presents a relative frequency histogram and polygon for the same data. Note that these plots are graphically similar to the previously illustrated plots, except that the ordinate is relative frequency and the values are taken from the column labeled *Rel f* in Table 2-5.

Cumulative Frequency Histogram and Polygon

Just as frequency and relative frequency distributions have been plotted in the form of histograms and polygons, the two cumulative distributions in Table 2-5 can also be graphed in these ways. Steps for constructing both these graphs can be found in Table 2-10, and an example of a cumulative frequency polygon is given in Figure 2-5.

One difference between the graphs of a frequency distribution and a cumulative frequency distribution is that the ordinate changes from frequency to cumulative frequency. A second difference is that the point on a cumulative polygon is placed over the *upper real limit* of each interval, rather than over its midpoint as in a frequency polygon. This is because this point must indicate that up to the *end* of that interval, a certain number or proportion of cases has occurred. Since the scores that fall within a given interval may be located anywhere within that interval, the point representing the accumulation of all frequencies within and below this interval is placed at the upper real limit of the interval.

Histogram vs. Polygon

When should one use a histogram and when should one use a polygon, or does it matter?

2-10   **Steps in Constructing Histograms and Polygons**

Frequency Histogram

1. Draw two axes, with the ordinate approximately three-fourths as long as the abscissa. Mark off the abscissa with values corresponding to the midpoints of the class intervals and mark off the ordinate in frequencies. Label the axes appropriately.

2. Construct each bar of the histogram over its class interval so that the bar's width covers the class interval from its lower to its upper real limits (not from midpoint to midpoint), and its height corresponds to the frequency of scores in the interval. There should be no space between bars (except if the abscissa is a nominal scale).

Frequency Polygon

1. Mark off and label axes as for a frequency histogram, but add one interval below the lowest and one above the highest class interval and assign them 0 frequencies.

2. Place points corresponding to the frequencies of each interval (including the two 0-frequency intervals) directly over the midpoints of each class interval. Connect all adjacent points (including the 0s) with straight lines.

Relative Frequency Histogram or Polygon

These are plotted in the same way as above, but the labels on the ordinate (and the heights of the bars or points) show relative frequency, not frequency.

Cumulative Frequency Histogram or Polygon

Follow the steps for constructing a frequency histogram or polygon, except:

a. Mark off and label the ordinate for cumulative frequency rather than frequency.

b. In drawing a cumulative frequency polygon, the points are placed over the upper real limit of each class interval, including the lowest interval of 0 accumulated frequencies (note that there is no upper 0-frequency interval).

Cumulative Relative Frequency Histogram or Polygon

These are plotted in the same way as the cumulative distributions described above, except that the labels on the ordinate (and the heights of the bars or points) show cumulative relative frequency.

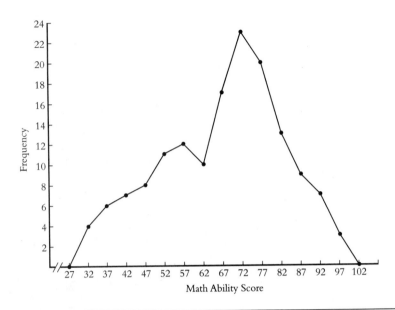

Figure 2-3 Frequency polygon for data in Tables 2-5 and 2-9.

Technically, histograms should be used when the scale of measurement is nominal or ordinal, or the variable is discrete. This is because each bar of a histogram is the same height across the width of a bar, which represents the entire score value or class interval; the histogram is not used to estimate values between the discrete values actually observed (that is, within the bar's width). For example, gender (male, female), frequencies of personality types (that is, a nominal scale), or ratings of aggressiveness of preschool children on a five-point scale are best presented as histograms.

Polygons, however, should be used when the measurement scale is interval, ratio,

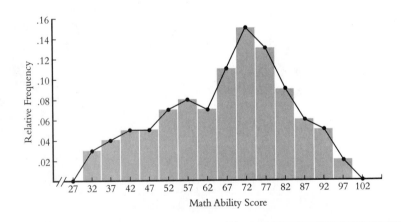

Figure 2-4 Relative frequency polygon and histogram for data in Tables 2-5 and 2-9.

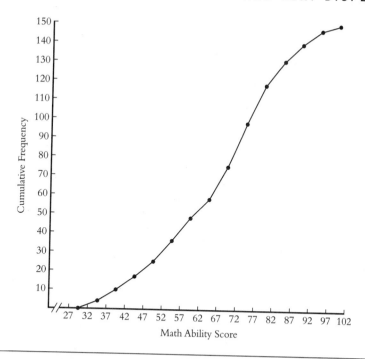

Figure 2-5 Cumulative frequency polygon for data in Tables 2-5 and 2-9.

or continuous. In the extreme, as the number of score values increases, the polygon would approach a smooth curve, reflecting the number of frequencies for any score value on the continuous scale. The frequency distribution for times in a marathon for 15,000 runners is best displayed with a polygon.

When the distinction between discrete and continuous scales becomes blurred, so does the choice of histogram vs. polygon. For example, the scores on the mathematics ability test discussed earlier reflect a discretely measured scale (that is, only whole number scores are given, no fractions), but conceptually it might be interpreted to reflect a continuous scale of mathematical ability. In this case, either a histogram or a polygon might be used.

STEM-AND-LEAF DISPLAYS

The distributions and graphs presented above are the traditional means of describing a set of data, but they are not the only methods. In the 1970s, John Tukey initiated a movement called **exploratory data analysis,** which has developed alternative techniques for describing data. These approaches attempt to display the structure and general nature of data in ways that are straightforward and that produce summary characteristics that are less likely to be influenced by one or two scores.

> **Exploratory data analysis** is an approach and a set of tools that are used, often in an unplanned and exploratory manner, to describe and understand the meaning of a set of data.

Exploratory data analysis is a bit like detective work. Different tools and approaches are used to solve different types of crimes, but they all seek to uncover evidence that lets the detective understand what happened. Often detective work is only partly planned—as the detective uncovers clues and information, he or she may pursue unanticipated leads until a comprehensive picture of the events in question forms. Similarly, the statistician using exploratory data analysis attempts to describe the data in a partly planned fashion, following up on interesting characteristics of the data that are uncovered, all for the purpose of gaining an understanding of what the data reveal about the phenomenon they measure. Later, other statistical techniques (e.g., inferential statistics, presented in this text) may be used to confirm characteristics uncovered during the exploration.

For example, a **stem-and-leaf display** is an alternative to a frequency distribution for grouped data. It performs all of the summarization functions that the traditional methods described above accomplish, and in addition it has the advantage of combining both a frequency distribution and a histogram in one display while preserving more of the information in the original data than does a frequency distribution.

This approach relies on breaking up a score value into a stem and a leaf.

> A score value may be broken into a **stem** and **leaf.** The leaf typically consists of the smallest digit and the stem typically consists of the remaining larger digits.

For example, the number 23 is composed of a tens digit (i.e., the 2) and a ones digit (i.e., the 3):

score value	→	stem	and	leaf
23	→	2		3

Larger numbers may have more than one digit in the stem, but leaves usually consist of only one digit, the smallest digit. For example, the score value 47.5 has a tens digit, a ones digit, and a tenths digit, and it may be broken down as follows:

score value	→	stem	and	leaf
47.5	→	47		.5

Usually decimal points are not written for stems and leaves, but we will do so here for clarity.

A stem-and-leaf display is simply an ordered listing of the stems and leaves of the numbers in a set of data, which is often called a **batch** instead of a distribution. For example, consider the data for two batches at the top of Table 2-11. Suppose these data represent the number of attempts to mentally stimulate children in a class of four-

2-11	Stem-and-Leaf Displays of Attempts to Mentally Stimulate Preschool Children by Untrained and Trained Student Teachers

Untrained	Trained
Raw Data	*Raw Data*
64, 48, 98, 66, 21, 61, 81, 63, 59, 70, 66, 54, 76, 61, 55	80, 72, 94, 99, 66, 85, 73, 91, 91, 52, 96, 82, 75, 99, 85, 69
Ordered Data	*Ordered Data*
21, 48, 54, 55, 59, 61, 61, 63, 64, 66, 66, 70, 76, 81, 98	52, 66, 69, 72, 73, 75, 80, 82, 85, 85, 91, 91, 94, 96, 99, 99

Stem-and-Leaf Displays

Attempts to Stimulate— Untrained			Attempts to Stimulate— Trained		
Depths (N = 15)	(Unit = 1 Attempt)		Depths (N = 16)	(Unit = 1 Attempt)	
	Stem	Leaves		Stem	Leaves
1	9	8	6	9	1 1 4 6 9 9
2	8	1	(4)	8	0 2 5 5
4	7	0 6	6	7	2 3 5
(6)	6	1 1 3 4 6 6	3	6	6 9
5	5	4 5 9	1	5	2
2	4	8			
1	3				
1	2	1			

year-olds performed by two groups of early childhood student teachers during a two-hour observation period. An attempt to mentally stimulate might consist of asking a child how many balls are in a box, insisting that a child verbally request a toy rather than just pointing at it, having a child identify an object that the teacher names, and so forth. The group of 15 presented at the left of the table did their student teaching before receiving special training on encouraging mental development and so are "Untrained," whereas the group of 16 at the right of the table had several sessions of instruction on how to promote mental development before their student teaching experience and are labeled "Trained."

The first task in constructing a stem-and-leaf display is the same as for creating a frequency distribution—order the data by score value, beginning with the smallest number. This is done in Table 2-11 under "Ordered Data." Next, determine generally the definition of the stems and leaves for the entire batch. In these two examples, the stems are the tens digits and the leaves are the ones digits. Then create a table composed of ordered stem-and-leaf **lines** (comparable to class intervals), one line per stem, starting with the smallest stem at the bottom. Include in ascending order a line for each possible stem even if no score with that stem exists in the batch. For example, for the Untrained group, the lowest score is 21, which has a stem of 2. Place the 2 as the stem for the bottom line of the display. Above it, list each possible stem in order (e.g., 3, 4, 5, 6, . . .) until you reach the stem (i.e., 9) of the largest score in the distribution (i.e., 98). Now, to the right of each stem, list in increasing order all the leaves for scores, including duplicates in the batch having that stem. Specifically, the score 21 has a stem of 2 and a leaf of 1; no scores have a stem of 3; the score 48 has a stem of 4 and a leaf of 8; the scores 54, 55, and 59 have leaves 4, 5, and 9, which are listed on the line for stem 5 in increasing order and with a single space, not a comma, between them.

Most stem-and-leaf displays also contain depths.

The **depth** of a case is its rank order from the top or bottom of the distribution, whichever rank is smaller.

Depth corresponds roughly to cumulative frequency, but it is calculated from both the top and the bottom of the distribution until the accumulated total equals half the total number of cases [i.e., $(N + 1)/2$]. Depths can be determined for individual cases and for lines in the stem-and-leaf display. For example, for the Untrained group at the left, the depth of the bottom line is the number of cases (i.e., leaves) with the stem of 2, which is 1 case. For stem 3, no cases exist, so the cumulative frequency working up from the bottom is still 1. An additional case is present for stem 4, making the depth for that line $1 + 1 = 2$; and three cases exist for stem 5, making the depth $2 + 3 = 5$ for that line. The stem of 6 has six cases, but the depth for that line would be $5 + 6 = 11$, which is more than half the total cases of $15/2 = 7.5$. Now starting from the top of the display, the depths moving down the lines are 1, 2, 4, and then $4 + 6 = 10$, which is more than half (i.e., 7.5) the number of cases. The line that contains the middle score (i.e., the $N/2 + 1$ case, which is the 8th case in this example) is given a depth equal to the number of cases that exist only on that line, and this value is put in parentheses [i.e., (6) in this case] to indicate that it is a frequency, not a cumulative frequency. If a line has a depth that exactly equals half the number of scores in the distribution, there will be two such lines with depths $N/2$ adjacent to one another (this can happen only if the total N is an even number). In this event, there is no need for a special depth to indicate the center of the distribution.

It should be noted that when counting depths for individuals in a stem-and-leaf display, one counts up from the bottom and *within a line one counts from left to right* because score values increase from left to right. For example, in Table 2-11 for the Trained group at the right, to locate the score value with a depth of 5 from the bottom, one would count up through the lowest two lines which have three scores and then within the third line count from left to right to identify the score of 73 as the fifth score (i.e., having a depth of 5). However, when counting down from the top, *within a line one counts from right to left* because score values decrease from right to left. In the same display, if one sought the score with depth of 5 from the top, one would count within the first line from right to left, that is, from the scores of 99, 99, 96, and 94 to the score of 91, which is the fifth score from the top.

Finally, the stem-and-leaf display should be labeled. This includes a title giving the measure and perhaps the group (e.g., "Attempts to Stimulate—Untrained"), a designation of the unit of measurement placed in parentheses (i.e., "Unit = 1 Attempt"), column headings (i.e., "Depth," "Stem," "Leaves"), and the total number of cases in parentheses (e.g., "$N = 15$"). For clarity, we have included here many more labels than are often used in practice, in which only the heading of depth, N, and the unit might be presented.

Notice that the two displays in Table 2-11 can be used to summarize and compare the two groups. For the Untrained student teachers, the scores are more spread out with at least one (i.e., 21) extremely low and maybe one (i.e., 98) extremely high score. The center of the batch is at stem 6, and the shape of the distribution is roughly symmetrical. In contrast, after training, the number of attempts to stimulate children increased, with the center of the Trained batch at stem 8. There are fewer extremes, and scores are asymmetrically bunched more at the high end than in the middle of the distribution. Notice also how the stem-and-leaf display represents visually a horizontal histogram while also being a frequency distribution, and since the value of each score is retained in the display, no information is lost by putting different scores on the same line, as happens when they are put into the same class interval.

A line in a stem-and-leaf display is the same as a class interval in a frequency distribution, and the number of possible different leaves per line defines the line width, which is equivalent to the size of the class interval in a frequency distribution.

> A **line** in a stem-and-leaf display is a single stem, and the **line width** is the number of possible leaves for that stem.

In the above examples, the number of possible different leaves per stem is 10 (i.e., 0, 1, 2, 3, 4, 5, 6, 7, 8, 9). What happens if a line width (size of class interval) of 10 is too big? Perhaps all the data fall under two or three stems; such a display would not accurately portray the nature of the data, just as too few class intervals at the right of Table 2-6 did not summarize accurately the set of 150 mathematics ability test scores. How does one have stem-and-leaf displays with line widths less than 10?

Suppose we try to create a stem-and-leaf display for the 150 mathematics ability scores in Table 2-4. Following the steps in Table 2-8 for constructing a frequency distribution, we first turned to Table 2-7 to get an estimate of the largest number of intervals, then divided the range by that estimate to obtain the smallest size of interval, and finally rounded that value up to obtain an interval size that was a "round number," which might be 2, 5, or 10. In the case of the math ability scores, this interval size was 5. Stem-and-leaf displays handle line widths of 5 (rather than 10) by listing the same stem twice, once followed by an asterisk (e.g., 3*) to signify that only leaves 0, 1, 2, 3, 4 are included on that line, and once followed by a dot (e.g., 3•) to signify that only leaves 5, 6, 7, 8, 9 follow on that line. The stem-and-leaf display for these data, analogous to the frequency distribution in Table 2-5, is given in Table 2-12.

	Stem-and-Leaf Display for the Mathematics Ability Test Scores
2-12	**in Tables 2-3 and 2-4**

Math Ability Scores

Depths (N = 150)	(Unit = 1 Point)	
	Stem	**Leaves**
3	9•	5 6 8
10	9★	0 1 1 2 2 3 3
19	8•	5 5 6 6 7 8 8 9 9
32	8★	0 0 0 0 1 1 1 2 2 3 3 3 3 4
52	7•	5 5 5 5 5 6 6 6 7 7 7 7 8 8 8 9 9 9 9 9
75	7★	0 0 0 0 1 1 1 1 1 2 2 2 2 3 3 3 3 3 3 4 4 4 4 4
75	6•	5 6 6 7 7 7 7 7 7 7 7 7 8 8 8 9 9
58	6★	0 1 1 1 2 2 3 3 3 4
48	5•	5 5 5 6 7 7 7 8 9 9 9 9
36	5★	0 0 0 0 1 1 2 3 3 3 4
25	4•	5 6 6 7 7 8 9 9
17	4★	1 1 2 2 2 3 3
10	3•	6 6 7 8 9 9
4	3★	0 2 2 4

A similar strategy of repeating a stem followed by a sign restricting the possible leaves on that line is used if a line width of 2 is desired. In this case, a stem plus an asterisk, t, f, s, or dot is used, in which

an asterisk designates leaves 0 and 1,

a t designates leaves 2 and 3 (**t** for **t**wo and **t**hree),

an f designates leaves 4 and 5 (**f** for **f**our and **f**ive),

an s designates leaves 6 and 7 (**s** for **s**ix and **s**even), and

a dot designates leaves 8 and 9.

In contrast to frequency distributions, which potentially could have class intervals of any size, stem-and-leaf displays may have line widths of only 2, 5, or 10 times a power of 10 (e.g., .1, 1, 10, 100)—that is, for example, widths of .2, .5, 1, 2, 5, 10, 20, 50, 100, etc. This is one reason why step 4 in Table 2-8 suggests rounding up the interval size to these "round numbers."

When actually constructing a stem-and-leaf display, one generally follows the steps for frequency distributions given in Table 2-8. The only difference is that one is restricted in a stem-and-leaf plot to the above line widths. For example, Table 2-13 gives the possible score values for a few particular stems plus the different possible leaves for that stem for line widths of 10, 5, 2 (at the left) and 1.0, .5, and .2 (at the right) when the data are all positive numbers (at the top) and when the data include both positive and negative numbers (at the bottom). Once you have determined the line width (size of class interval) and lowest interval, locate the width in Table 2-13 and select stems and leaves beginning with the lowest interval appropriate for the data. Notice in Table 2-13 that when both positive and negative numbers are involved, a line with a stem of 0 is used for positive leaves (e.g., for scores 0 to 9) and a line with a stem of -0 is used for negative leaves (e.g., for scores 0 to -9).

HOW DISTRIBUTIONS DIFFER

Some of the most important ways distributions differ from one another are with respect to (1) central tendency, (2) variability, (3) skewness, and (4) kurtosis. The general meaning of these concepts is presented here. The next chapter explains how central tendency and variability can be expressed numerically. First, consider central tendency:

> The **central tendency** of a distribution is a point on the scale corresponding to a typical, representative, or central score.

There are several more specific definitions of central tendency, each with its own set of characteristics and implications. Three of these are discussed in Chapter 3: mean, median, and mode.

To illustrate the concept of central tendency, consider the two frequency polygons in Figure 2-6 (which have been smoothed over score values or score intervals). They differ only with respect to central tendency, which is indicated by the vertical lines. They have the same shape but occupy different places on the scale of measurement. For example, the two distributions might be for females (A) and males (B), and the "score value" might be birth weight. On the average, males are a little heavier than females—their distribution has a higher central tendency.

A second characteristic of a distribution is variability:

> **Variability** is the extent to which scores in a distribution deviate from their central tendency.

Figure 2-7 shows two curves with similar central tendencies but different amounts of variability. The scores in distribution A cluster more closely about the central tendency of the distribution (which is indicated by the vertical line) than do the scores in distribution B. There is more variability in B than in A. If score value were age at the eruption of the first tooth, males (B) would be more variable—less uniform—than females.

2-13

Stem-and-Leaf Displays for Data That Are Positive Integers or Tenths and for Lines of Different Widths (Interval Size)

Positive Numbers

Line Width = 10

Possible Scores	Stem	(Unit = 1) Possible Leaves
30–39	3	0 1 2 3 4 5 6 7 8 9
20–29	2	0 1 2 3 4 5 6 7 8 9
10–19	1	0 1 2 3 4 5 6 7 8 9
0–9	0	0 1 2 3 4 5 6 7 8 9

Line Width = 1.0

Possible Scores	Stem	(Unit = .1) Possible Leaves
3.0–3.9	3	.0 .1 .2 .3 .4 .5 .6 .7 .8 .9
2.0–2.9	2	.0 .1 .2 .3 .4 .5 .6 .7 .8 .9
1.0–1.9	1	.0 .1 .2 .3 .4 .5 .6 .7 .8 .9
.0–.9	0	.0 .1 .2 .3 .4 .5 .6 .7 .8 .9

Line Width = 5

Possible Scores	Stem[a]	(Unit = 1) Possible Leaves
25–29	2•	5 6 7 8 9
20–24	2*	0 1 2 3 4
15–19	1•	5 6 7 8 9
10–14	1*	0 1 2 3 4
5–9	0•	5 6 7 8 9
0–4	0*	0 1 2 3 4

Line Width = .5

Possible Scores	Stem[a]	(Unit = .1) Possible Leaves
2.5–2.9	2•	.5 .6 .7 .8 .9
2.0–2.4	2*	.0 .1 .2 .3 .4
1.5–1.9	1•	.5 .6 .7 .8 .9
1.0–1.4	1*	.0 .1 .2 .3 .4
.5–.9	0•	.5 .6 .7 .8 .9
.0–.4	0*	.0 .1 .2 .3 .4

	Line Width = 2 (Unit = 1)			Line Width = .2 (Unit = .1)	
Possible Scores	Stem[b]	Possible Leaves	Possible Scores	Stem[b]	Possible Leaves
20–21	2*	0 1	2.0–2.1	2*	.0 .1
18–19	1•	8 9	1.8–1.9	1•	.8 .9
16–17	s	6 7	1.6–1.7	s	.6 .7
14–15	f	4 5	1.4–1.5	f	.4 .5
12–13	t	2 3	1.2–1.3	t	.2 .3
10–11	1*	0 1	1.0–1.1	1*	.0 .1
8–9	0•	8 9	.8–.9	0•	.8 .9
6–7	s	6 7	.6–.7	s	.6 .7
4–5	f	4 5	.4–.5	f	.4 .5
2–3	t	2 3	.2–.3	t	.2 .3
0–1	0*	0 1	.0–.1	0*	.0 .1

(continued)

2-13 Continued

Both Positive and Negative Numbers

Line Width = 10 (Unit = 1)			Line Width = 1.0 (Unit = .1)		
Possible Scores	Stem	Possible Leaves	Possible Scores	Stem	Possible Leaves
20 to 29	2	0 1 2 3 4 5 6 7 8 9	2.0 to 2.9	2	.0 .1 .2 .3 .4 .5 .6 .7 .8 .9
10 to 19	1	0 1 2 3 4 5 6 7 8 9	1.0 to 1.9	1	.0 .1 .2 .3 .4 .5 .6 .7 .8 .9
0 to 9	0ᶜ	1 2 3 4 5 6 7 8 9	.0 to .9	0ᶜ	.0 .1 .2 .3 .4 .5 .6 .7 .8 .9
0 to −9	−0ᶜ	1 2 3 4 5 6 7 8 9	.0 to −.9	−0ᶜ	.0 .1 .2 .3 .4 .5 .6 .7 .8 .9
−10 to −19	−1	0 1 2 3 4 5 6 7 8 9	−1.0 to −1.9	−1	.0 .1 .2 .3 .4 .5 .6 .7 .8 .9
−20 to −29	−2	0 1 2 3 4 5 6 7 8 9	−2.0 to −2.9	−2	.0 .1 .2 .3 .4 .5 .6 .7 .8 .9

Usually, when the data are in tenths (or any unit other than 1) the decimals are not written in the stem-and-leaf display; only the unit is written above the display to indicate whether the leaves are integers, tenths, etc.

[a] Stems followed by an asterisk (e.g., 0*, 1*, 2*) have leaves in the lower half of this stem (e.g., 0, 1, 2, 3, and 4; or .0, .1, .2, .3, and .4); stems followed by a dot (e.g., 0•, 1•, 2•) have leaves in the upper half of this stem (e.g., 5, 6, 7, 8, and 9; or .5, .6, .7, .8, and .9).

[b] Stems followed by an asterisk have leaves of 0, 1, or .0, .1; stems followed by t have leaves of two and three (e.g., 2, 3, or .2, .3); stems followed by f have leaves of four and five (e.g., 4, 5, or .4, .5); stems followed by s have leaves of six and seven (e.g., 6, 7, or .6, .7); stems followed by a dot have leaves of 8, 9 or .8, .9.

[c] Values that are exactly 0 are both 0 and −0 and could be placed either on line 0 or line −0. Half (or as nearly half as possible) of such values should be placed on line 0 and half on line −0.

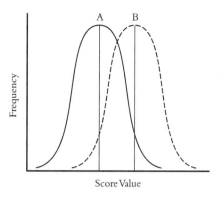

Figure 2-6 Distributions that differ only with respect to central tendency.

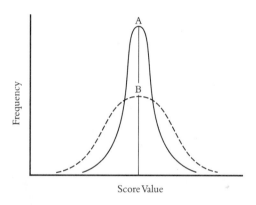

Figure 2-7 Distributions with the same central tendency but different variabilities.

A third characteristic is skewness:

Skewness refers to an asymmetric distribution in which the scores are bunched on one side of the central tendency and trail out on the other.

Figure 2-8 presents two skewed distributions. Both are asymmetrical, but B is more skewed than A because there is a greater tendency for the scores to bunch at one end and trail off at the other. Skewness and variability are usually related: the more the skewness, the greater the variability. But more variability does not necessarily mean more skewness.

Skewness has direction as well as magnitude. In distribution A the scores tend to trail off to the right or positive end of the scale. Distribution A is therefore said to be **skewed to the right** or **positively skewed** (because the scores trail off toward the

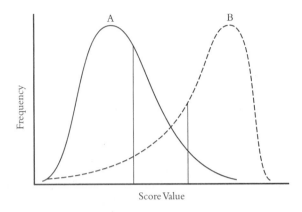

Figure 2-8 Distributions that differ with respect to skewness (and central tendency).

positive end of the scale). Conversely, distribution B trails off to the left or negative end of the scale (recall that the scores to the left of 0 are negative scores). Distribution B is said to be **negatively skewed** or **skewed to the left.** Distribution A might represent family income, with a few families having very high incomes. Distribution B might be scores on an easy statistics quiz, with most students getting very high scores.

A fourth characteristic of frequency distributions is kurtosis:

The **kurtosis** of a distribution is the "curvedness" or "peakedness" of the graph.

Figure 2-9 depicts two distributions with the same central tendencies but different

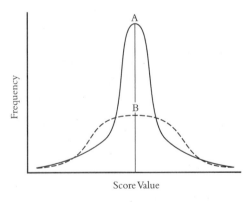

Figure 2-9 Distributions with the same central tendency but different kurtoses.

kurtoses. Distribution A is more peaked than B, and the changes in the height of the curve as the score value increases are more marked for A than for B. Kurtosis is frequently used in a relative sense. Distribution A is said to be more **leptokurtic** than B. The Greek *lepto* means "thin," so leptokurtic implies a thin distribution. On the other hand, B is more **platykurtic** than A. *Platy* means "flat" (for example, *platy*-helminthes—flatworms, or *platy*pus—a flat-billed mammal). Kurtosis is also related to variability—platykurtic distributions tend to have greater variability than do lepto-kurtic distributions.

The terms relating to skewness and kurtosis may be used to describe the general form of a distribution. Thus one might say that a distribution is positively skewed and rather platykurtic. This verbal description gives one some idea of the shape of the distribution in question, but it is not very precise. When precision is called for, as when one is comparing distributions, mathematical indices are needed. Although mea-sures of skewness and kurtosis are available, and sometimes given by statistical computer programs, they are not often used, because their values are not very dependable unless large samples are available. Therefore, only numerical indices of central tendency and variability are taken up in the next chapter.[1]

SUMMARY

A frequency distribution is a tally of the number of scores corresponding to each score value (or each interval of score values) occurring in a group of scores. Its purpose is to neatly summarize these scores as a collection. In addition to the frequency distri-bution, one can determine a relative frequency distribution, a cumulative frequency distribution, and a cumulative relative frequency distribution. Procedures are described for determining the stated limits, real limits, size, and midpoints of the class intervals of these distributions. Such distributions can be graphed, and procedures are given for drawing a frequency histogram, frequency polygon, relative frequency histogram and polygon, cumulative frequency histogram and polygon, and cumulative relative fre-quency histogram and polygon. Stem-and-leaf displays represent an alternative ap-proach that combines a frequency distribution with a histogram in one display and maintains more of the raw data than do frequency distributions. Distributions differ with respect to central tendency, variability, skewness, and kurtosis.

---

[1] The four characteristics of a distribution presented in this chapter—mean, variance, skewness, and kur-tosis—have a specific mathematical basis. They are *moments* of the distribution. Mathematically, a moment is the expected value—roughly the average—of a power of a variable over all cases in a distribution (or over an unlimited number of cases in a theoretical distribution). The mean is the first moment about the origin; it is the average of the variable (raised to the first power) over all cases, $\Sigma X_i^1/N$. The variance is the second moment, but about the mean not the origin, since it is the "average" (divided by $N - 1$ rather than N) of the squares of the differences between each score and the mean $\Sigma(X_i - \overline{X})^2/(N - 1)$. Skewness is the third moment about the mean or the "average" of the cube of those differences, and kurtosis is the fourth moment about the mean or the "average" of those differences, each raised to the fourth power. The entire set of moments (which may include the fifth, sixth, and so forth) will mathematically describe a distribution exactly.

Exercises

For Conceptual Understanding

1. Define and distinguish between a frequency distribution, a relative frequency distribution, a cumulative frequency distribution, and a cumulative relative frequency distribution.

2. Why do the ends of a polygon touch the horizontal or X-axis?

3. When should one use a histogram vs. a polygon?

4. Why should there be no space between the bars of a histogram, except perhaps for a variable that has a nominal scale?

5. What is wrong with plotting a cumulative frequency distribution for a variable that has a nominal scale?

6. What is the relation between variability and skewness? Between variability and kurtosis? Draw some distributions to illustrate your points.

7. What advantages and disadvantages do stem-and-leaf displays have relative to traditional descriptive approaches?

For Solving Problems

8. Following is a set of scores on a final exam in statistics. Construct a frequency distribution appropriate for these data. Present the distribution in the manner of Table 2-9, indicating real limits, interval size, midpoint, and frequency. Then compose a relative frequency distribution, cumulative frequency distribution, and cumulative relative frequency distribution for these data, and present this information in the same table in the manner of Table 2-5.

81	86	92	56	47
81	87	93	61	82
94	82	64	88	82
88	95	67	82	86
95	67	82	85	99
72	82	88	73	82
82	85	73	84	84
77	84	89	77	77
79	89	79	89	89
90	80	84	79	80

9. Construct polygons and histograms for the frequency and relative frequency distributions from Exercise 8. Make a polygon for the cumulative frequency distribution.

10. The numbers below are the number of aggressive acts for each preschool child over a week in a class of 25 students. First, construct a frequency distribution in the manner of Table 2-9, including the real limits, size, and midpoints of the class intervals. For such a small distribution only five or so intervals are necessary. Second, compose a relative frequency, cumulative frequency, and cumulative relative frequency distribution for these data in the manner of Table 2-5. Third, construct polygons and histograms for the frequency and relative frequency distributions and a polygon for the two cumulative distributions.

36	28	47	40	44
44	39	33	33	32
47	34	38	27	40
37	41	42	38	48
43	35	37	37	26

11. The procedures for constructing a frequency distribution were described for data composed of whole numbers. Obviously, many of the measurements made by scientists are in decimal

form. The guidelines for constructing frequency distributions listed in Table 2-8 also apply to decimal data. Perform the tasks required in Exercise 10 with the numbers below. Again, fewer than 10 intervals are appropriate.

2.8	2.5	2.0	2.1
.8	1.9	3.1	2.2
.9	1.3	2.4	1.6
2.3	1.5	1.1	2.0
3.0	3.8	2.2	4.4
1.5	1.2	2.5	1.3
3.1	1.7	2.3	2.7

12. Are the distributions in Exercises 8, 10, and 11 skewed? If so, in what direction(s) are they skewed?

13. Construct a stem-and-leaf display for the data in Exercises 8 and 10.

14. Construct a stem-and-leaf display for the data in Exercise 11, but use a line width of .2 because of the extreme scores.

For Homework

15. Given the data set (4, 2, 7, 3, 3, 5, 8, 7, 7, 4, 6, 4, 9, 5, 6, 4, 6, 5, 5, 8), complete the following table.

X	f	Rel f	Cum f	Cum Rel f

16. What would you use as an interval size (or line width) and lowest stated interval (or stem) in the following circumstances?

a. $N = 20$, range $= 39$, lowest score $= 41$

b. $N = 100$, range $= 182$, lowest score $= 146$

c. $N = 35$, range $= 1.9$, lowest score $= 2.7$

d. $N = 60$, range $= 13$, lowest score $= -6.2$

17. What are the stated limits of an interval with the following sizes and midpoints?

a. size $= 5$, midpoint $= 47$

b. size $= 2$, midpoint $= -2.5$

c. size $= .5$, midpoint $= 11.25$

18. Construct a frequency histogram and frequency polygon for the data in Exercise 15. Construct a cumulative relative polygon for these data.

19. Given the following daily high temperatures in July and August in Pittsburgh, complete the same table as requested in Exercise 15 and construct a stem-and-leaf display.

85	82	74	76	73	81
79	75	64	79	75	80
83	78	86	80	79	79
92	84	83	74	83	72
75	72	89	87	86	74
69	76	76	86	84	77
78	89	79	82	84	69
87	83	80	91	76	73
86	73	81	85	75	75
81	71	87	75	77	81
78	77				

Characteristics of Distributions

All parents want to know at what age their baby will walk, and when the long-awaited event occurs, parents wonder whether their infant's accomplishment is typical or (they hope) advanced. More specifically, they want to know the *typical age* at which infants take their first unassisted step. The answer—roughly—is 12 months.

This answer is "rough" because the situation is actually much more complicated than it appears. Think of what information is necessary to answer this question. One needs a large sample of infants and the age at which each took a first unassisted step. Then one would make a frequency distribution of such ages and graph it. Now, what parents want to know is the "typical age" at which infants take their first step. To statisticians, parents want the **central tendency** of this distribution. But what is "typical" or "central"? It could be the average age, called the **mean.** Or it could be the age at which half the infants have taken their first step and half have not, called the **median.** Or it could be the single age at which more infants take their first step than at any other age, called the **mode.** This is not just academic quibbling about definitions, because these three "typical ages" may not be the same value. The mean is 11.7 months, but the median might be 12.2 months, for example. This may not seem like much, but two weeks is an eternity to some eager-beaver parents. Besides, why are these three "typical values" not all the same? Which is the real typical value?

Once the infant takes his or her first step, parents want to know whether their infant is "advanced" or "precocious." If Sarah walks at 10.0 months is she advanced? Of greater concern is whether Paul, who did not take his first step until 14 months, is somehow developmentally slow? In fact, almost no parent wants his or her baby to be slower than the typical age, despite the fact that half the infants will be slower to walk than the median age.

To answer this question, at least partly, one needs to know how spread out the distribution is—that is, what is the **variability** of the distribution? An easy answer is to give the earliest and the latest ages in the distribution, which might be approximately 8 and 20 months. This is called the **range** of ages. It tells parents between what two ages essentially all infants begin to walk, but it does not say much about whether 10 is truly advanced and 14 is truly slow. Statisticians, however, have developed other indices of variability, such as the **variance** and the **standard deviation,** that reflect the extent to which infants differ from one another and that can be used to characterize ages as advanced or slow.

This chapter describes several numerical indicators of central tendency and variability that can be used to characterize distributions and to compare one distribution to another. The chapter also includes a discussion of statistical estimation, which consists of using such indicators calculated on a **sample** of cases to estimate their corresponding values in a much larger **population** that has not actually been measured. Such estimation is common in statistics and is an important part of inferential statistics, which is presented in Part 2 of this text.

INDICATORS OF CENTRAL TENDENCY

The Mean

The most common indicator of the central tendency of a group of scores is the average or mean:

> The **mean** of scores on variable X is symbolized by $\bar{X}$ (read "X bar") and is computed with the formula
>
> $$\bar{X} = \frac{\Sigma X_i}{N}$$
>
> which instructs one to add all the scores (that is, ΣX_i) and divide by N, which is the number of scores in the distribution.

For example:

$$
\begin{array}{c}
\underline{X_i} \\
8 \\
3 \\
4 \\
10 \\
7 \\
\underline{1} \\
\Sigma X_i = 33 \\
N = 6
\end{array}
\qquad
\bar{X} = \frac{\Sigma X_i}{N} = \frac{33}{6} = 5.5
$$

The mean is also called the **arithmetic average** of the scores. Notice that ΣX_i can be written without the limits $i = 1$ and N (see Chapter 1), which are understood to apply. For example, $\displaystyle\sum_{i=1}^{N} X_i = \Sigma X_i = \Sigma X$. This abbreviated notation is common in the remainder of the text.

Different texts use different symbols to represent the mean. A common alternative to the symbol $\bar{X}$ (used in this book) is the symbol M.

Deviations about the Mean The mean possesses several properties that make it an appropriate measure of central tendency. First,

> The sum of the deviations of scores about their mean is zero. In other words, if the mean is subtracted from each score in the distribution, the sum of such differences is zero. In symbols,
>
> $$\Sigma(X_i - \bar{X}) = 0$$

Consider the following numerical example:

X_i	$\overline{X}$	$(X_i - \overline{X})$
3	5	$3 - 5 = -2$
6	5	$6 - 5 = 1$
5	5	$5 - 5 = 0$
1	5	$1 - 5 = -4$
10	5	$10 - 5 = 5$
$\Sigma X_i = 25$		$\Sigma(X_i - \overline{X}) = 0$
$N = 5$		
$\overline{X} = 5$		

This principle can be proven true of all distributions. The algebraic proof is presented in Optional Table 3-1.

While the sum of the deviations of all the scores about the mean is always zero, the sum of the *squared* deviations about the mean is usually not zero. It is always greater than zero, unless all the scores in the distribution are the same value. This distinction is important because formulas presented later in the chapter (such as those for the variance) use squared deviations.

To summarize:

Whereas

$$\Sigma(X_i - \overline{X}) = 0$$

the expression

$$\Sigma(X_i - \overline{X})^2 \text{ is usually } not \text{ equal to } (\neq) \, 0$$

To illustrate, consider the numerical example given above. If one squares the difference between each score and the mean (the numbers in the extreme right-hand column) and then sums these squared deviations, one obtains

$$(-2)^2 + (1)^2 + (0)^2 + (-4)^2 + (5)^2 = 46$$

which is obviously not zero. The reason squared deviations never add to zero unless all the scores are the same is that squared numbers can never be negative, and thus positive values will not be balanced by negative ones.

Minimum Variability of Scores about the Mean A second property of the mean concerns the squared deviations of scores about their mean:

The sum of the squared deviations of scores about their mean is less than the sum of the squared deviations of the same scores about any other value.

3-1 **Proof that the Sum of the Deviations About the Mean Equals Zero**   Optional Table

Operation	Explanation
To prove $\Sigma(X_i - \overline{X}) = 0$ **1.** $\Sigma(X_i - \overline{X}) = \Sigma X_i - \Sigma \overline{X}$	**1.** The sum of the differences between two quantities equals the difference between their sums.
2. $= \Sigma X_i - N\overline{X}$	**2.** The sum of a constant $\overline{X}$ added to itself N times $\left(\text{that is, } \sum_{i=1}^{N} \overline{X}\right)$ is N times the constant (that is, $N\overline{X}$).
3. $= \Sigma X_i - N\left(\dfrac{\Sigma X_i}{N}\right)$	**3.** Substitution of the definition $\dfrac{\Sigma X_i}{N}$ for $\overline{X}$
4. $\Sigma(X_i - \overline{X}) = \Sigma X_i - \Sigma X_i = 0$	**4.** Cancellation of N's in second term of No. 3 above.

This fundamental principle will be invoked in the explanation of many subsequent concepts. It states that although the sum of the squared deviations of scores about their mean usually does not equal zero, that sum is nevertheless smaller than if the squared deviations of the same scores were taken about any value other than the mean of their distribution. It is in this sense, sometimes called the **least squares sense,** that the mean is an appropriate measure of central tendency: the mean is closer (in terms of squared deviations) to the individual scores over the entire group than is any other single value. For example, in the above illustration the sum of the squared deviations about the mean equals 46. The mean of that distribution is 5.0. The sum of the squared deviations about the number 6.0 equals 51; about the number 4.0 the sum also equals 51; and about the number 7.0 it equals 66. The sum of squared deviations about the mean (46) is less than any of these examples, and in Optional Table 3-2 it is shown that it always will be less than about any other value.

The Median

Another indicator of central tendency is the median:

> The **median,** symbolized by M_d, is the point on the measurement scale that divides the distribution into two parts such that equal numbers of scores (or cases) lie above and below that point.

3-2 **Proof that the Sum of the Squared Deviations About the Mean is Less than About Any Other Value**

Operation	Explanation
1. $(\overline{X} + c)$, $c \neq 0$ and $c \neq \overline{X}$	1. Assumption.
2. (a) The sum of the squared deviations of scores about $\overline{X}$ equals $\Sigma(X_i - \overline{X})^2$	2. Definitions.
(b) The sum of the squared deviations of scores about $(\overline{X} + c)$ equals $\Sigma[X_i - (\overline{X} + c)]^2$	
To prove $$\Sigma(X_i - \overline{X})^2 < \Sigma[X_i - (\overline{X} + c)]^2$$	To prove that the sum of squared deviations about the mean, $\Sigma(X_i - \overline{X})^2$, is less than the sum of squared deviations about any other value, $\Sigma[X_i - (\overline{X} + c)]^2$
3. $< \Sigma[(X_i - \overline{X}) - c]^2$	3. Working with the right side of the inequality, removing parentheses and regrouping
4. $< \Sigma[(X_i - \overline{X})^2 - 2c(X_i - \overline{X}) + c^2]$	4. Binomial expansion of the form: $(a - b)^2 = a^2 - 2ab + b^2$
5. $< \Sigma(X_i - \overline{X})^2 - \underbrace{2c\Sigma(X_i - \overline{X})}_{} + \underbrace{\Sigma c^2}_{}$	5. Distributing the summation sign to all terms within the brackets, the sum (or difference) of several variables is the sum (or difference) of their sums, and the sum of a constant times a variable is the constant times the sum of the variable.
6. $< \Sigma(X_i - \overline{X}^2 - \qquad 0 \qquad + Nc^2$	6. Since $\Sigma(X_i - \overline{X}) = 0$, the second term is 0, and the sum of N c^2's equals N times c^2.
7. $\Sigma(X_i - \overline{X})^2 < \Sigma(X_i - \overline{X})^2 + Nc^2$	7. The expression is true because Nc^2 will always be greater than zero.

It will be helpful in the following discussion to use the term *score value* to refer to the numerical value of a particular score or *case* in a distribution of N cases. Further, since the median divides the number of cases into two groups containing equal numbers of cases, determining the median requires that the cases first be arranged in order of magnitude of score value, from the lowest to the highest score value. Then the way the median is computed depends on whether there is an odd or an even number of total cases (i.e., N) in the distribution.

1. **When there is an odd number of cases in the distribution (i.e., *N* is odd), the median is the score value corresponding to the middle case, which will be the (*N* + 1)/2 case from the bottom of the distribution.** For example, if the distribution is

   (3, 5, 6, 7, 10)

   there is an odd number of cases, namely $N = 5$, so the middle case will be the $(N + 1)/2 = (5 + 1)/2 = 3$ or the third case from the bottom. The score value of the third case is 6, so the median is 6. Note that this score value divides the distribution into two equal parts, since two cases fall below and two cases fall above the score value of 6.

   If the distribution is

   (21, 35, 57, 68, 68, 73, 80)

   there is an odd number of cases (i.e., $N = 7$), the middle case is the $(N + 1)/2 = (7 + 1)/2 = 4$th case, and the value of the fourth case of the distribution is 68. This shows that the median is the value of the middle case even when more than one case has that score value.

2. **If there is an even number of cases in the distribution, the median is halfway between (i.e., is the average of) the score values of the two middle cases, which will be cases *N*/2 and (*N*/2) + 1 from the bottom of the distribution.** Suppose the distribution is

   (3, 5, 6, 7, 10, 14)

   Here there is an even number of cases ($N = 6$), and the two middle cases are the $N/2 = 6/2 = 3$rd case and the $(N/2) + 1 = (6/2) + 1 = 4$th case. The median is the average of the score values of these two middle cases, or $(6 + 7)/2 = 6.5$. Notice that the point 6.5, while not actually in the distribution, would separate the distribution into two equal parts consisting of three cases below and three cases above this point.

   Two other examples are instructive. Suppose the distribution is

   (3, 3, 4, 5, 8, 14, 16, 20)

   Here again, there is an even number of cases, namely $N = 8$, so the middle two scores are cases $N/2 = 8/2 = 4$ and $(N/2) + 1 = (8/2) + 1 = 5$. Their values are 5 and 8, so the median is their average, which is $(5 + 8)/2 = 6.5$. This shows that the median is the average of the two middle scores even if their values are not adjacent on the measurement scale. Now consider the distribution

   (2, 3, 4, 5, 5, 5, 6, 6, 7, 9)

   Here, $N = 10$ is even, and the middle cases are cases $N/2 = 10/2 = 5$ and $(N/2) + 1 = 5 + 1 = 6$. Both these cases have score values of 5, so the median is their average, which is $(5 + 5)/2 = 5$. This shows that the median is still the average of the middle cases even when those cases have the same score value.

The Mode

A third indicator of central tendency is the mode:

> The **mode,** symbolized M_o, is the most frequently occurring score value.

If the distribution is

(3, 4, 4, 5, 5, 5, 6, 8)

the mode is 5. Sometimes a distribution will have two modes, such as the distribution

(3, 4, 4, 4, 5, 6, 6, 7, 7, 7, 8)

In this case, the modes are 4 and 7 and this distribution is called **bimodal.** A distribution that contains more than two modes is called **multimodal.** Notice that the mode is the most frequently occurring *score value, not* the *frequency* of the most commonly occurring score value.

Comparison of the Mean, Median, and Mode

The essential difference between the mean and the median is that the mean reflects the value of each case in the distribution, whereas the median is based largely on where the midpoint of the distribution falls, without regard for the particular value of many of the cases. For example, consider the following illustration:

Distribution	Mean	Median
1, 2, 3, 4, 5	3	3
1, 2, 3, 4, 50	12	3
1, 2, 3, 4, 100	22	3

Only the last number differs from one distribution to the other. The mean reflects these differences, but the median does not. This is because the median is the midpoint of the distribution such that an equal *number of cases* fall above and below it. The particular *values* of the extreme scores do not matter; only the fact that those *cases* are above the midpoint is considered. In contrast, the mean takes into account the value of every case. Thus, changing any score value will likely change the value of the mean. This principle will be important when resistant measures of central tendency are discussed later in this chapter.

The mode reflects only the most frequently occurring score value. It is not used much in the social and behavioral sciences, except to describe a bimodal or highly skewed distribution and to give the central tendency of distributions of categorical (that is, nominal) data.

Because the three different measures of central tendency are sensitive to different aspects of the group of scores, they are usually not the same value in a given distribution. If the distribution is symmetrical and unimodal (having one mode), then the mean, median, and mode are indeed identical. This condition is graphed in part A of Figure 3-1. If the distribution is symmetrical but has two modes as in part B, the mean and median are the same but the modes are different (the distribution is bimodal). In Chapter 2, a skewed distribution was defined as a distribution that is not symmetrical, having scores bunched at one end. Parts C and D of Figure 3-1 show two skewed distributions and the relative positions of the three measures of central tendency. The distribution in part C illustrates a condition in which most cases have moderate values but a few are very high. In this situation, the mean, being sensitive to those extreme values, is somewhat larger than the median, which divides the area under the curve (that is, the total number of cases) into two equal parts. Part D of Figure 3-1 illustrates the relative positions of the measures of central tendency when the skewness of the distribution is in the other direction.

Ordinarily, social and behavioral researchers select the mean as the measure of central tendency. There are several reasons for preferring the mean, but one major consideration is that the mean is required by so many other statistical procedures. However, sometimes the circumstances are such that the median would reflect the central tendency of the distribution more accurately than would the mean. When the

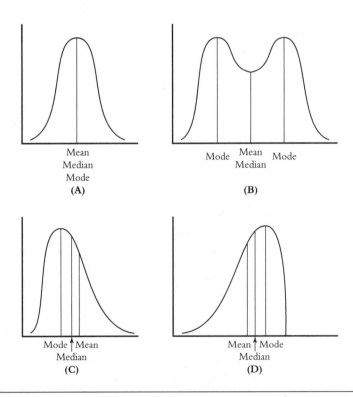

Figure 3-1 Mean, median, and mode in different distributions.

distribution is very skewed, the mean may not be a value that coincides with one's intuitive impression of the typical score. For example, the distribution

(1, 2, 3, 4, 100)

has a mean of 22 and a median of 3. In this situation the median seems to characterize the central tendency more faithfully than does the mean. Thus, for markedly skewed distributions, the median may be preferred.

The most common practical instance in which the median is preferred is in the reporting of annual household incomes. Recall from Chapter 1 that in 1995, the *mean* household income was $44,938, a substantial 32% greater than the *median* income of only $34,076. This difference occurs because incomes are skewed to the right, much like part C of Figure 3-1. That is, a few people, such as the chief executive officers of major corporations, make literally millions of dollars a year, and the values of these few high incomes influence the value of the mean but not the median. Many economists believe that the median is a better measure of central tendency for incomes because it reflects the midpoint of the distribution—half of American households live on less than $34,076/year (and half live on more). For this purpose, the mean of $44,938 is deceiving.

The mode is rarely used by itself to express central tendency, except for categorical data. It is most often reported as a supplement to the mean or median, especially for distributions that are skewed or bimodal. Looking at part C of Figure 3-1, one might imagine that the modal household income would be even less than the median of $34,076.

INDICATORS OF VARIABILITY

In addition to central tendency, an indicator of variability is needed to characterize a distribution more fully.

> **Variability** refers to the extent to which the scores in a distribution differ from their central tendency.

For example, suppose two groups of scores, *A* and *B,* are defined to be

$A = (5, 7, 9)$
$B = (3, 7, 11)$

Although they both have the same mean of 7, set *B* has more variability because the scores differ from that mean more than do the scores in *A*. The purpose of this section is to present numerical indices that reflect the variability of scores in a distribution.

The Range

One indicator of variability is the range.

> The **range** is the largest score minus the smallest score in the distribution.[1]

In the distribution

(3, 5, 6, 6, 8, 9)

the range is $9 - 3 = 6$.

However, the range is limited in its ability to reflect the variability of a distribution. It is not sensitive to the variability of *all* the scores, only to the difference between the two most extreme values. For example,

C = (5, 10, 11, 12, 13, 14, 19)
D = (5, 6, 7, 12, 17, 18, 19)

have the same mean of 12 and range of $19 - 5 = 14$, but D has more variability than C. This can be seen if the distributions are plotted as frequency histograms:

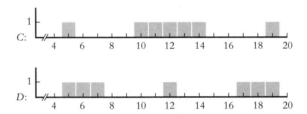

More of the scores in distribution D are a greater distance from the mean of 12 than in distribution C, despite the fact that the ranges of the two distributions are the same. Therefore, although the range is easily computed, it is usually employed only as a crude approximation of variability.

The Variance and the Standard Deviation

The Variance, s^2 An indicator that reflects the degree of variability in a group of scores but that does not have the limitations of the range is the variance:

> The **variance,** symbolized by s^2, is defined to be
>
> $$s^2 = \frac{\Sigma(X_i - \bar{X})^2}{N - 1}$$

---

[1] Technically, the range should be defined as the difference between the upper *real* limit of the largest score and the lower *real* limit of the smallest score, not the difference between these *stated* limits as defined here. However, because the range is an approximate index of variability at best, it does not seem appropriate to insist upon this level of precision for such a crude index.

Notice the numerator of this formula, $\Sigma(X_i - \overline{X})^2$. This quantity is the sum of the squared deviations of the scores about their mean, or, as the phrase is often shortened, the **sum of squares.** This term, sum of squares, is also used in Chapters 10 and 13 on the analysis of variance. Sum of squares can be abbreviated **SS.** So

$$\Sigma(X_i - \overline{X})^2 = \text{Sum of squares} = SS$$

and the formula for the variance is

$$\frac{\Sigma(X_i - \overline{X})^2}{N - 1} = \frac{\text{Sum of squares}}{N - 1} = \frac{SS}{N - 1}$$

Except that the denominator is $N - 1$ instead of just N (which is explained under "Good Estimators" near the end of the chapter), the variance is essentially the "average sum of squared deviations of scores about their mean," or "the average sum of squares."[2]

In the example below, the variance for distribution $A = (5, 7, 9)$ is computed using the definitional formula. Notice that the mean is computed first ($\overline{X} = 7.0$). Then the mean is subtracted from each score ($X_i - \overline{X}$) and this difference is squared [$(X_i - \overline{X})^2$]. The sum of the squared deviations about the mean is divided by $N - 1$ to obtain the variance. Thus, the variance of distribution A is 4.00. Distribution B consists of $(3, 7, 11)$. It has greater variability, and its variance is 16.

X_i	$\overline{X}$	$(X_i - \overline{X})$	$(X_i - \overline{X})^2$
5	7	-2	4
7	7	0	0
9	7	$+2$	4
$\Sigma X_i = 21$		0	$\Sigma(X_i - \overline{X})^2 = 8$
$N = 3$			
$\overline{X} = 7$			

$$s^2 = \frac{\Sigma(X_i - \overline{X})^2}{N - 1}$$

$$= \frac{8}{2}$$

$$s^2 = 4.00$$

The Standard Deviation, s   The variance reflects variability in squared units. If you asked the members of your class how many hours they spent studying yesterday, the mean number would be expressed in hours but the variance would be in "squared

---

[2] Other texts may use a special notation for the quantity $(X_i - \overline{X})$. In this case, the deviation of a single score from its mean is called a *deviation score,* and it is symbolized by a lowercase italicized letter corresponding to the letter used for that variable, such as x for $(X_i - \overline{X})$. The squared deviation score then would be symbolized by x^2, and the sum of squared deviation scores—or the sum of squares—would be represented by Σx^2.

hours." This occurs because the formula for the mean uses the scores as they are (ΣX_i), but the formula for the variance squares the deviations [$\Sigma(X_i - \overline{X})^2$], thereby producing an index in squared units. However, it is also useful to have an index of variability in terms of the original units of measurement, not squared units.

The **standard deviation,** symbolized by **s,** is defined to be the positive square root of the variance:

$$s = \sqrt{s^2} \quad \text{or} \quad s = \sqrt{\frac{\Sigma(X_i - \overline{X})^2}{N - 1}}$$

Since the variance is in squared units, taking its square root accomplishes a return to the original units of measurement.

Just as some other texts use M rather than $\overline{X}$ to symbolize the mean, they sometimes use the symbol SD rather than s for the standard deviation.

Computational Formulas for s^2 and s Formulas like those given above are called **definitional** formulas because they define and reflect the logic behind the concepts they express. However, definitional formulas are frequently inconvenient to use in making calculations, especially when dealing with large amounts of data. An expression for a statistic that is mathematically equivalent to the definitional formula but that is more convenient for calculating is called a **computational** formula.

The computational formula for the variance is

$$s^2 = \frac{N\Sigma X_i^2 - (\Sigma X_i)^2}{N(N - 1)}$$

The computational formula for the standard deviation is

$$s = \sqrt{\frac{N\Sigma X_i^2 - (\Sigma X_i)^2}{N(N - 1)}} \quad \text{or } s = \sqrt{s^2}$$

It is important to realize that these computational formulas yield results that are equivalent (within rounding error) to those calculated by the definitional formulas.

The computational formula has the advantages of requiring only one division, of not requiring that the mean be calculated first and subtracted from each score, and of being easy to compute on a hand calculator. Some hand calculators (and all statistical computer programs) will compute the variance and/or standard deviation "automatically," and the value can be obtained by pressing a single button or special function key. In this case, the computational formula presented here is not needed, but students should note whether the formula used by the calculator or program divides by N or $N - 1$ (see page 84). Some other calculators, however, do not compute s or s^2 but do compute three quantities that are needed in the computational formulas for the variance and standard deviation: ΣX_i, ΣX_i^2, and N.

To illustrate the computation of the variance and standard deviation, consider the following distribution:

X_i	X_i^2
3	9
4	16
7	49
8	64
8	64
9	81
10	100
$\Sigma X_i = 49$	$\Sigma X_2^2 = 383$
$N = 7$	

Variance	Standard Deviation
$s^2 = \dfrac{N\Sigma X_i^2 - (\Sigma X_i)^2}{N(N-1)}$	$s = \sqrt{s^2}$
$= \dfrac{7(383) - (49)^2}{7(7-1)} = \dfrac{2681 - 2401}{42}$	$= \sqrt{6.67}$
$= \dfrac{280}{42}$	$s = 2.58$
$s^2 = 6.67$	

It is very important to distinguish between two quantities used in the computational formulas: ΣX_i^2 and $(\Sigma X_i)^2$. Recall that the first, ΣX_i^2, represents the sum of all the squared scores—*first* square each score, *then* add. The second, $(\Sigma X_i)^2$, represents the square of the sum of the scores—*first* add all the scores, *then* square this sum. Confusion between these two operations is often the source of computational error.

Properties of s^2 and s as Indicators of Variability   The variance is difficult to explain because it cannot be diagrammed or pointed at. Rather, the variance is an abstract numerical index that increases with the amount of variability in the group of scores. But despite its abstractness, the variance does have a number of properties that make it (and its square root, the standard deviation) an appropriate indicator of variability:

1. **Since the mean is the central value of the distribution, it is logical to base a measure of variability upon the extent to which the scores deviate from their mean—that is, on $\Sigma(X_i - \overline{X})^2$.** In addition, recall from the discussion of the mean that the sum of the squared deviations about the mean is less than about any other value. This fact adds to the logic of selecting squared deviations about the mean (as opposed to some other value) as an index of variability.

2. **The values of s^2 and s are always positive.** This is so because squared numbers are always positive (i.e., squaring a negative number results in a positive value) and s is always the positive square root of s^2. If the deviations were not squared, the negative deviations would cancel out the positive and their sum would be zero because $\Sigma(X_i - \overline{X}) = 0$.

3. **The variance is especially sensitive to departures from the mean, because the deviations, when squared, become disproportionately large.** A deviation of 4 units becomes 16 when squared, but a deviation of twice that size—that is, of 8 units—contributes 64 to the total sum of squared deviations. While indicators of variability should reflect the values of all the scores, alternative measures to be discussed later in this chapter are less influenced by very deviant and presumably unusual values.

4. **The variance, which is approximately the average squared deviation of the scores from their mean, is proportional to the average squared deviation of each score from every other score.** The concept of variability refers, in its most general sense, to the extent to which the scores differ from one another. One might express such variability by calculating the difference between each score and every other score, squaring those differences (to make them all positive), adding those squared deviations, and computing their average by dividing by the number of such pairs of scores [there will be $N(N - 1)/2$ such pairs]. This average squared deviation of the scores from *one another* is proportional to the average squared deviation of each score *from its mean*—that is, to the variance. Therefore, the variance reflects the extent to which scores in a distribution deviate from one another as well as from their mean.

5. **As the variability of the scores increases, the statistical variance also increases.** This can be seen in the few examples listed below:

Distribution	s^2
10, 10, 10	0
8, 10, 12	4
6, 10, 14	16
4, 10, 16	36
2, 10, 18	64

   As the scores show more and more variability, the value of s^2 increases, reflecting the extent to which the scores deviate from the mean and from one another. Similar arguments can be made for the standard deviation.

6. **If there is no variability among the scores, that is, if all the scores in the distribution are the same value, the variance and standard deviation are both zero.** This is so, because if all scores are the same value (for example, 5, 5, 5, 5, 5), the mean will also be that value (that is, 5). Then all quantities $(X_i - \overline{X})^2$ and their sum will be zero because X_i and $\overline{X}$ will always be identical. Therefore, when there is no variability among the scores of a distribution (if all scores have the same value), $s^2 = 0$.

7. **Under certain conditions the variance can be partitioned and its portions attributed to different sources.** This capability of being partitioned permits statis-

ticians to ask questions such as, What portion of the variability in a group of scores can be attributed to cause A as opposed to cause B? This aspect of the variance is taken up in greater detail in Chapters 10 and 13.

RESISTANT INDICATORS[3]

The indicators of central tendency and variability presented above are the most traditional and common characteristics used to describe distributions of data. But they have limitations, and other approaches, such as those advocated by proponents of **exploratory data analysis,** are designed to overcome some of those limitations.

One limitation of the mean, variance, and standard deviation, for example, is that their values depend on the value of every score in the distribution. While this feature has certain advantages, it is also troublesome, especially if the distribution includes one or two very high or very low scores that deviate substantially from the remaining body of data. Neither such scores nor the values of the indices of central tendency and variability that are influenced by them are likely to occur again if another sample were selected. In short, the presence of atypical or deviant scores will produce similarly atypical or deviant values for the mean, variance, and standard deviation.

In view of these limitations, it would be useful to have descriptive indicators of distributions that are less influenced by unusual scores and that are likely to accurately characterize the next sample of scores.

> **Resistant indicators** that describe a set of data are those that change relatively little in value if a small portion of the data is replaced with new numbers that may be very different from the original ones.

Such indicators are called "resistant" because they "resist" influence by individual scores, especially atypical and very deviant scores. Instead, they reflect the main body of the data, so they change relatively less when a portion of the data is replaced with new numbers.

A Resistant Indicator of Central Tendency

The median, for example, is a more resistant indicator of central tendency than the mean. As noted on page 65, the value of the mean is influenced by the value of every score in the distribution, whereas the median is determined largely by *how many* cases are higher and lower than it, *not how much* higher or lower they are than it. For example,

---

[3] This material is taken largely from Hoaglin, D. C., Mosteller, F., and Tukey, J. W. (eds.), *Understanding Robust and Exploratory Data Analysis* (New York: Wiley, 1983).

suppose the highest score in a distribution was replaced by another measurement. The value of the mean would change every time this was done (except in the very rare event that the value of the new score was identical to that of the previous one), and the mean likely would change a substantial amount in value. In contrast, the value of the median would change on only approximately half of such occasions, and it would not change very much in value when it did change. So, the median is a more resistant indicator of central tendency than the mean.

Resistant indicators are usually presented in the context of other exploratory data analysis techniques, such as the stem-and-leaf display presented previously on pages 44–49, and the terminology used in such contexts is slightly different. Recall, for example, that a distribution is called a **batch,** and a class interval is called a **line.** The vocabulary used to define the median is also different, and to help make this translation, the two stem-and-leaf displays presented in the previous chapter for Untrained and Trained preschool student teachers are reproduced here at the top of Table 3-3.

Recall that at the left of each display is a column reporting the **depth** of each stem-and-leaf line. As explained previously, the depth of a line is its cumulative frequency, except that frequencies are accumulated in both directions—up from the bottom of the display (i.e., up from the lowest score) and down from the top of the display (i.e., down from the highest score)—until a line is reached at which half or more of the total cases are accumulated. That line, of course, contains the median.

Now recall the method of calculating the median given in this chapter: The median is the value of the central case [i.e., the $(N + 1)/2$ case] if the total N in the distribution is an odd number, or it is the average of the values of the two middle cases (i.e., cases $N/2$ and $(N/2) + 1$) if the total N is an even number. Translating this definition into the vocabulary of stem-and-leaf displays and the concept of depth, we have the following:

The **median** is the value of the score having a **depth** of $(N + 1)/2$. If this depth is fractional (i.e., an integer plus .5), the median is the average of the values of the two cases bordering the fractional depth. More specifically:

1. **If the total N of the batch is an odd number,** this depth will be a whole number, and the median is the value of the case having a depth of $(N + 1)/2$.

2. **If the total N of the batch is an even number,** this depth will be fractional (i.e., an integer plus .5), and the median is the average of the values of the two cases [i.e., cases $N/2$ and $(N/2) + 1$] bordering that fractional depth.

For example, the stem-and-leaf display for the Untrained preschool student teachers presented at the top left of Table 3-3 has a total $N = 15$ (an odd number), and the depth of the median is $(N + 1)/2 = (15 + 1)/2 = 8$, a whole number. Therefore, the median is the value of the 8th score (counting in either direction), which is 63. Turning to the display for the Trained student teachers at the top right of Table 3-3, we find that $N = 16$ (an even number), and the depth of the median is $(N + 1)/2$

3-3   Resistant Indicators of Central Tendency and Variability for Untrained and Trained Batches

Stem-and-Leaf Displays

Attempts to Stimulate—Untrained			Attempts to Stimulate—Trained		
Depths (N = 15)	(Unit = 1 Attempt)		Depths (N = 16)	(Unit = 1 Attempt)	
	Stem	Leaves		Stem	Leaves
1	9	8	6	9	1 1 4 6 9 9
2	8	1	(4)	8	0 2 5 5
4	7	0 6	6	7	2 3 5
(6)	6	1 1 3 4 6 6	3	6	6 9
5	5	4 5 9	1	5	2
2	4	8			
1	3				
1	2	1			

The Median

The median is the value of the case having a depth of $(N + 1)/2$. If this depth is fractional, the median is the average of the values of the two cases bordering the fractional depth.

For Untrained batch, $N = 15$, so the depth of the median is $(N + 1)/2 = (15 + 1)/2 = 8$. The median is the value of the 8th score: $M_d = 63$.

For Trained batch, $N = 16$, so the depth of the median is $(N + 1)/2 = (16 + 1)/2 = 8.5$. The median is the average of the values of the 8th and 9th cases: $M_d = (82 + 85)/2 = 83.5$.

Fourths

The depth of the fourths equals

$$\text{depth of fourths} = \frac{(\text{depth of median}) + 1}{2}$$

in which a fractional depth of the median is first rounded down to the next lower integer.

For Untrained batch, the depth of the median is 8, so the depth of the fourth is $(8 + 1)/2 = 4.5$.

For Trained batch, the depth of the median is 8.5; the fractional part is dropped, so the depth of the fourths is $(8 + 1)/2 = 4.5$.

(continued)

3-3 **Continued**

The lower fourth, F_L, is the average of the values of the 4th and 5th cases from the bottom of the batch:

$$F_L = \frac{55 + 59}{2} = 57$$

The upper fourth, F_U, is the average of the values of the 4th and 5th cases from the top of the batch:

$$F_U = \frac{70 + 66}{2} = 68$$

The lower fourth, F_L, is the average of the values of the 4th and 5th cases from the bottom of the batch:

$$F_L = \frac{72 + 73}{2} = 72.5$$

The upper fourth, F_U, is the average of the values of the 4th and 5th cases from the top of the batch:

$$F_U = \frac{94 + 91}{2} = 92.5$$

The fourth spread is the difference in score values between the upper and lower fourths:

fourth-spread $= F_U - F_L$

For the Untrained batch:
fourth-spread $= 68 - 57 = 11$

For the Trained batch:
fourth-spread $= 92.5 - 72.5 = 20$

Outliers

Outliers are cases whose values fall either:

Below $F_L - 1.5(F_U - F_L)$
Above $F_U + 1.5(F_U - F_L)$

For the Untrained group:
$F_L - 1.5(F_U - F_L) =$
$57 - 1.5(68 - 57) = 40.5$
$F_U + 1.5(F_U - F_L) =$
$68 + 1.5(68 - 57) = 84.5$

Therefore, scores that fall outside the range of 40.5 to 84.5 are outliers. Scores of 21 and 98 are outliers.

For the Trained group:
$F_L - 1.5(F_U - F_L) = 72.5 -$
$1.5(92.5 - 72.5) = 42.5$
$F_U + 1.5(F_U - F_L) = 92.5 +$
$1.5(92.5 - 72.5) = 122.5$

Therefore, scores that fall outside the range of 42.5 to 122.5 are outliers. There are no outliers.

(continued)

3-3   **Continued**

Extreme Scores

The extreme scores are the lowest and highest score values in the batch excluding outliers.

For the Untrained batch, the lower extreme score is 48 and the upper extreme score is 81 (because 21 and 98 are outliers).

For the Trained batch, the lower extreme score is 52 and the upper extreme is 99 (because there are no outliers).

$= (16 + 1)/2 = 8.5$, a fractional number. So the median is the average of the values of the cases bordering this fractional depth, namely the 8th and 9th cases, which is $(82 + 85)/2 = 83.5$.

Resistant Indicators of Variability

The traditional indicators of variability—the range, variance, and standard deviation—are highly influenced by atypical deviant scores. The range, for example, is based entirely on the highest and lowest scores, so it is defined by extreme or deviant score values. Similarly, recall from the discussion on page 72 that the variance and standard deviation are disproportionately influenced by even a single high or low score value. What can be used to indicate variability that is more resistant than these traditional characteristics?

Fourth-Spread   One approach used by exploratory data analysts relies on the concept of fourths, or hinges as they are sometimes called.

The **depth of the fourths** is given by

$$\text{depth of the fourths} = \frac{(\text{depth of median}) + 1}{2}$$

in which a fractional depth of the median is first rounded down to the next lower integer.

This says that one first drops any fractional part of the depth of the median, adds one, and takes half of it. For example, in the batches in Table 3-3, the depth of the median is 8 for Untrained and 8.5 for Trained student teachers. Since we drop the fractional part, the depth of the median is 8 in each case. So the depth of the fourths is $(8 + 1)/2 = 4.5$ for each batch.

There are two fourths. Each has the same depth, but one is above the median and the other is below the median.

> The **lower fourth,** symbolized by F_L, is the score value of the case with a depth of a fourth counted up from the bottom of the batch.
>
> The **upper fourth,** symbolized by F_U, is the score value of the case with a depth of a fourth counted down from the top of the batch.
>
> In each event, if the depth of a fourth is fractional (i.e., integer plus .5), take the average of the values of the two cases bordering the fractional depth.

For example, for the Untrained batch in Table 3-3, the lower fourth, F_L, is obtained by counting up from the bottom and *moving from left to right along a line* in the stem-and-leaf display until one reaches the depth of 4.5, that is, the 4th and 5th cases from the bottom. The lower fourth is their average value: $F_L = (55 + 59)/2 = 57$. The upper fourth, F_U, is obtained by counting down from the top and *moving right to left across a line* (since higher scores are at the right) to locate the 4th and 5th cases from the top. The upper fourth is their average value: $F_U = (70 + 66)/2 = 68$. Similarly, for the Trained batch, $F_L = (72 + 73)/2 = 72.5$ and $F_U = (94 + 91)/2 = 92.5$.

Notice that the lower fourth, F_L, is halfway between the median and the bottom of the batch, and the upper fourth, F_U, is halfway between the median and the top of the batch. So the two fourths border the middle half of the batch, which is called the fourth-spread.

> The **fourth-spread** is the difference in score values between the upper and lower fourths:
>
> fourth-spread = upper fourth − lower fourth
> fourth-spread = $F_U - F_L$

For example, for the Untrained group, the fourths are $F_U = 68$ and $F_L = 57$, so the fourth-spread is $F_U - F_L = 68 - 57 = 11$. For the Trained group, these values are $92.5 - 72.5 = 20$.

Outliers and Extreme Scores The fourths, F_L and F_U, help to define outliers, cases whose values are substantially deviant from the group and would have little likelihood of being obtained again in a new sample. Identifying outliers is important, because it may be useful to know which particular cases are unusually high (e.g., Wayne Gretsky's career scoring in professional hockey) or low, or because we want to eliminate such extreme scores as being bizarre and very unlikely to be observed again.

> **Outliers** are cases whose values fall either:
>
> **1.** Below F_L − 1.5(fourth-spread)
> **2.** Above F_U + 1.5(fourth-spread)
> in which the fourth-spread = $F_U - F_L$.

Basically, outliers are those cases that fall more than plus or minus 1.5 times the fourth-spread (i.e., $F_U - F_L$) beyond the fourths. Specifically, for the Untrained group at the left of Table 3-3, these limits are

$$F_L - 1.5(F_U - F_L) = 57 - 1.5(11) = 40.5$$
$$F_U + 1.5(F_U - F_L) = 68 + 1.5(11) = 84.5$$

So any scores outside these limits are considered outliers, and therefore the scores of 21 and 98 are outliers. For the Trained group, these limits are calculated at the right in Table 3-3 to be 42.5 and 122.5, and no scores fall beyond these values, so there are no outliers.

These limits of the fourths plus and minus 1.5(fourth-spread) separate outliers from another type of designated score, the lower and upper extreme scores.

The **extreme scores** are the lowest and highest scores in the batch excluding outliers. Specifically:

The **lower extreme score,** LE_x, is the lowest score excluding outliers. That is, it is that score most immediately *above* $F_L - 1.5(F_U - F_L)$.

The **upper extreme score,** UE_x, is the highest score excluding outliers. That is, it is that score most immediately *below* $F_U + 1.5(F_U - F_L)$.

For the Untrained student teachers at the left of Table 3-3, the scores of 21 and 98 were determined above to be outliers. Looking at the stem-and-leaf display for this batch, the next lowest score is 48 and the next highest score is 81. So the lower and upper extreme scores for this batch are 48 and 81. For the Trained group, no outliers were present, so the extreme scores are simply the lowest and highest values in the batch, which are 52 and 99 (note again that the *highest scores are at the right of a line* whereas the *lowest scores are at the left of a line* in a stem-and-leaf display). Therefore, the fourths plus and minus 1.5 times the fourth-spread define upper and lower cutoffs that separate the upper and lower extreme scores from the more deviant outliers and represent a spread of values that includes all the scores except the atypical or outliers.

The fourth-spread and the fourths $\pm$ 1.5(fourth-spread) are resistant indicators of variability. The fourth-spread, for example, is a resistant analogue to the standard deviation. It is defined partly by the median, a resistant indicator of central tendency, and it reflects the main body of the data, namely the central half of the batch or distribution. As such, its value will not be influenced by the values of deviant scores, and its value is less likely than the variance or standard deviation to be changed if some cases of the batch are replaced with new data. Further, in certain theoretical distributions (i.e., the theoretical "normal" distribution to be presented in the next chapter), the fourth-spread is directly related to the standard deviation (i.e., $1.349s = $ fourth-spread).

Further, the fourths $\pm$ 1.5(fourth-spread) represent a spread of score values that is a resistant analogue to the traditional range as a measure of variability. Since this spread excludes the outliers, it is more resistant than the range because it is not influenced much by the presence or absence or the particular values of those outliers. At the same time, it includes nearly all of the score values that one might usually expect, excluding very atypical ones.

Table 3-3 presents a summary of the definitions of all the resistant indicators of central tendency and variability discussed above plus illustrations of their calculation.

Displays of Resistant Indicators

These resistant indicators are often displayed in tabular and graphical fashion, which are defined and illustrated in Table 3-4.

Five-Number Summary   A table containing the major resistant indicators of central tendency and variability is called a **five–number summary.** The general form of the five-number summary followed by illustrations for the Untrained and Trained batches is presented at the top of Table 3-4.

The five numbers included in the summary are (1) the value of the median (M_d); (2) the value of the lower fourth (F_L); (3) the value of the upper fourth (F_U); (4) the value of the lower extreme score (LEx); and (5) the value of the upper extreme score (UEx). They are placed under a three-sided rectangular bracket with the name of the measure on the top. To the left of this five-number summary are the total N, the depth of the median (M), the depth of the fourths (F), and the depth of the extreme scores (E), which are reassigned the depths of 1 in all such summaries. The form of the five-number summary used in this text is extensively labeled for clarity. In practice, as with the stem-and-leaf displays, only a few labels are routinely presented in a table, namely N, M, F, and the measurement scale.

Boxplot   A five-number summary can also be presented graphically in a **boxplot,** which is defined in general terms and illustrated with the information for Untrained and Trained groups presented on a single graph at the bottom of Table 3-4.

To construct a boxplot:

1. Mark off the abscissa in units of the measurement scale and label. Then draw an ordinate without a scale high enough to accommodate all the boxplots to be drawn, and label the batches to the left of the ordinate.

2. Draw a rectangular box for each batch that stretches between the lower fourth (F_L) and the upper fourth (F_U).

3. Within the box, draw a vertical line at the median (M_d).

4. Draw horizontal lines in each direction from the box, ending at the lower extreme score (LEx) and at the upper extreme score (UEx).

5. Locate each outlier, if any, with an $\times$ and label that case, if appropriate.

Notice the tidy picture of these batches that the boxplot portrays. In general, the median reflects the central tendency of the batch, and the breadth of the box and lines convey the variability. For example, the box, which spans the fourths, represents the central half of each batch. The spread of score values between the ends of the lines, which are defined by the extreme scores, is an approximation of the

3-4

Displays of Resistant Indicators of Central Tendency and Variability

Five–Number Summary

General form:

N = Number of cases

M = Depth of median

F = Depth of fourths

E = Depth of extreme
     scores = 1

Measure

M_d = Value of median

F_L = Lower fourth F_U = Upper fourth

LEx = Value of lower UEx = Value of upper
    extreme score   extreme score

Untrained Group:

N = 15

M = 8

F = 4.5

E = 1

Stimulation Attempts

M_d = 63

F_L = 57 F_U = 68

LEx = 48 UEx = 81

Trained Group:

N = 16

M = 8.5

F = 4.5

E = 1

Stimulation Attempts

M_d = 83.5

F_L = 72.5 F_U = 92.5

LEx = 52 UEx = 99

Boxplot

General Form:

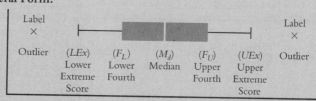

Measurement Scale

For Untrained and Trained Batches:

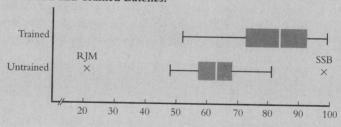

fourths $\pm$ 1.5(fourth-spread) range that represents nearly all the scores, excluding the outliers located on the plots. Asymmetry in a boxplot, in which the median is not located in the center of the box or in which the lines extend from the box farther in one direction than another, implies skewness. Finally, if the box is short relative to the line, the distribution is more peaked or leptokurtic; if the box is broad relative to the lines, then perhaps the distribution is flatter or more platykurtic.

With respect to the two specific groups, one can readily see that the Trained student teachers do much more mental stimulation than the Untrained. For example, the central half of the Trained group (i.e., the box) does not overlap with the central half of the Untrained group, and the upper extreme score of the Untrained group reaches only as high as the median of the Trained group. Also, the Untrained group has more variability if the two outliers are included but not if they are excluded. The distributions are fairly symmetrical, with the Untrained group only slightly skewed to the right (ignoring outliers) and the Trained group slightly skewed to the left.

POPULATIONS AND SAMPLES

Frequently, a scientist performs an experiment on a relatively small group of subjects. At the conclusion of the research, however, the results are generalized to a much larger group of subjects. For example, in Chapter 1 a hypothetical experiment was described in which 40 children with attention-deficit hyperactivity disorder (ADHD) were randomly assigned to one of two groups, one that received medication and one that was given a placebo that had no real effect on the disorder. While only 40 children were actually studied, the results were intended to be generalized to all children with ADHD. The 40 children are said to constitute a sample from the population of all ADHD children.

> A **population** is a collection of subjects, events, or scores that have some common characteristic of interest.[4]
>
> A **sample** is a subgroup of a population.

Typically, the population is the group of scientific interest, but scientists actually study a sample and then generalize the results obtained on the sample to the population. In short, the population is the group to which results from the sample will be generalized. This means that sample and population are relative terms. All students enrolled at State

---

[4] A population is sometimes considered to be composed of an infinite number of cases. As such, the population is a theoretical concept because it can never actually be observed or assessed. The sample, in contrast, is an empirical concept because it can be observed and assessed. The concept of theoretically infinite populations is crucial to mathematical statistics, but it is of less use to students at the level of study reflected in the text.

University might be the population from which a sample of 100 students is drawn for a given study, because the researcher is interested in generalizing the results only to the population of State University students. On the other hand, all students at State University might function instead as a sample of the larger population of all college students in the country.

One obvious reason for using samples rather than populations in research is that populations are usually too large to be studied efficiently. In addition, research results that must be limited to the specific subjects studied are less interesting and less useful than those applicable to a much larger group. Therefore, the scientist designs the study so that generalizations from the sample to the population may be made.

Parameters and Statistics

Frequently, the results that are generalized from sample to population are the statistical quantities, such as the mean and variance, that are calculated on the sample of scores. That is, the mean and variance computed on the sample are used as **estimators** of the mean and variance in the population. Therefore, it is necessary to be able to distinguish between quantities associated with a sample and those associated with the population.

> A quantitative characteristic of a sample is called a **statistic** and is symbolized with a Roman letter.
>
> A quantitative characteristic of a population is called a **parameter** and is symbolized with a Greek letter.

Earlier in this chapter, indices of central tendency and variability were presented. These were calculated on samples of subjects, so the mean and standard deviation, for example, were statistics symbolized by the Roman letters, $\overline{X}$ and s, respectively. Most of the data that researchers analyze are from samples rather than populations. That is why this course is called statistics instead of parameters, and why this book presents all quantitative characteristics of distributions as statistics and not as parameters.

Almost all data sets are considered samples because scientists almost always use statistics to estimate population parameters. For example, if the English Department at State University wanted to estimate the ability of all freshmen at State to guide them in creating an appropriate English curriculum, they might sample 100 freshmen and test them on basic English skills. If they found the average score of the sample to be 82 and the standard deviation to be 7, they could regard these sample statistics as estimates of their respective population parameters—that is, of the mean and standard deviation for all freshmen at State. Of course, sometimes it is possible to assess all the members of a population. One could, for example, test all incoming freshmen at State, rather than just a sample of 100. But even then, the results might be used to estimate the abilities of the *next* freshman class, which would make this year's freshman class a sample of a larger population.

In research practice, then, it is rare to measure an entire population. Instead, sample statistics are typically used to estimate population parameters, so it is necessary to know what the population parameters corresponding to each sample statistic are called and how they are symbolized. This information is summarized in Table 3-5 for the mean, variance, and standard deviation. Of course, all other statistics—including the median, mode, and range—also have corresponding parameters symbolized with Greek letters, but we will not have occasion to use them in this text.

"Good" Estimators

Since a main purpose of calculating statistics is to use them to estimate population parameters, statisticians are concerned that a statistic be a "good" estimator of its corresponding population parameter. The criteria that make an estimator "good" are discussed in Chapter 7, but we have already seen the consequence of this concern in the formula for the sample variance. Recall that the formula for the variance of a sample is

$$s^2 = \frac{\Sigma(X_i - \overline{X})^2}{N - 1}$$

But notice in Table 3-5 that the formula for the variance of a population is

$$\sigma^2 = \frac{\Sigma(X_i - \mu)^2}{N}$$

In particular, the sum of the squared deviations is divided by N in the formula for the *population parameter* σ^2 but it is divided by $N - 1$ in the formula for the *sample statistic* s^2. It turns out that dividing by $N - 1$ makes s^2 a "better" estimator of σ^2 than it would be if the sum of squared deviations were divided by N. In contrast, notice in Table 3-5 that the formula for the sample mean is the same as the formula for the population mean. Therefore, for some statistics the formulas are the same as for their corresponding parameters, whereas for others the formulas for the sample statistics are slightly different than for their corresponding population parameters. This difference often consists of dividing by $N - 1$ (or even by $N - 2$) instead of just N, and this is done to make the statistic a "better" estimator of its corresponding parameter (the criteria for making a statistic a "better" estimator are presented in Chapter 7). Because we almost never have an entire population available, only the formulas for statistics are presented in the remainder of this text.

A NOTE ON CALCULATORS AND COMPUTERS

A great variety of hand calculators and computer programs are now available to perform the calculations needed in this text. Some of them simply require the student to enter the numbers in a distribution, for example, and push one or two buttons or enter a simple instruction and the machine will automatically create one or more different frequency or cumulative frequency distributions; calculate the mean, median,

3-5 Summary of the Names, Symbols, and Formulas for Common Statistics and Parameters

Quantity	Sample Statistic			Population Parameter		
	Symbol	Read As	Formula	Symbol	Read As	Formula[a]
Mean	$\overline{X}$	"X bar"	$\dfrac{\sum X}{N}$	μ	"mew"	$\dfrac{\sum X}{N}$
Variance	s^2	"s squared"	$\dfrac{\sum(X - \overline{X})^2}{N - 1}$	σ^2	"sigma squared"	$\dfrac{\sum(X - \mu)^2}{N}$
Standard Deviation	s	"s"	$\sqrt{\dfrac{\sum(X - \overline{X})^2}{N - 1}}$	σ	"sigma"	$\sqrt{\dfrac{\sum(X - \mu)^2}{N}}$

[a] When populations are of uncountable size or are theoretical, other mathematical procedures are used to define these quantities.

mode, range, variance, and standard deviation; and perhaps even provide additional statistics, such as numerical indices for skewness and kurtosis.

Unfortunately, each machine and computer program is different, so it is impossible to write a textbook keyed to all of these machines or programs. Therefore, you must read the directions for the particular machine or program you have available and learn how to use it with this text. You will, however, likely run into some problems—your machine or program may not do things the same way as described in the text.

For one thing, the symbols might be different. Whereas $\overline{X}$ is used to symbolize the mean in this text, M might be used in the manual or on the keypad of your machine. Also, whereas s^2 and s are used here for the variance and standard deviation, *Var* and *SD* might be used by your machine or program manual.

Second, you may not always get the same answer as in the text or even as a classmate who uses another machine or program. There may be several reasons for this. Perhaps the rules for rounding numbers are not the same. For example, numbers ending in 5 may always be rounded up (or down), regardless of the odd or even nature of the previous digit. Or perhaps the same number of digits is not used during the calculations, causing one answer to be slightly (or sometimes substantially) different from another.

Third, and more serious, perhaps the formulas being used are different. Does the calculator or program use the formula for the statistic or for the parameter? Often it is not obvious which is being used, so consult the manual for the calculator or computer program to determine whether the formula being used to calculate the variance, for example, is that for the sample variance (with a denominator of $N - 1$) or that for the population variance (with a denominator of N). This issue also pertains to the standard deviation, of course, and a few other statistics discussed later, but not to the mean, which is calculated by the same formula in either case. If you cannot locate the manual, test the button or program on the computational examples in the text to determine which formula is being used. Do not rely on the manual or the label on the calculator button to follow the tradition of using Roman letters for statistics and Greek letters for parameters. Just because the symbol s^2 is used, it is no guarantee that the formula for the sample variance is used.

The *Study Guide* that accompanies this text provides optional guided examples of how to input data and identify statistical results for three common statistical computer programs, MINITAB, StataQuest, and SPSS. These computer packages will perform on a personal computer nearly all of the calculations presented in this book and many more. Other programs will perform the same functions—these three were selected because they are most commonly used. If you have a choice, StataQuest and MINITAB are easier, so they might be favored by those students who are less experienced with computers or who do not expect to take another course in statistics. SPSS, on the other hand, will perform a much greater variety of statistical analyses, including many advanced techniques commonly used in research. So, if you expect to take additional statistics courses, go to graduate school in a social or behavioral science, conduct your own research, or analyze research data for a faculty member, SPSS would be the better choice.

However, while specialized statistical calculators and computer programs simplify computations to the press of a button, *they do not teach you statistics.* You will learn more if you work at least some of the problems "by hand." This is why the numbers in the examples and in the problems are kept simple and do not generally require elaborate calculations. Use the calculator or computer primarily to check your answers to the first few problems and to do more complicated calculations.

SUMMARY

Two important characteristics of distributions are central tendency and variability. The mean, or average, is the most common indicator of central tendency, partly because the sum of the deviations about the mean equals zero and the sum of the squared deviations of each score about the mean is less than about any other value. The median (the point that divides the distribution into two equal parts) and the mode (the most frequently occurring score) are also used as indices of central tendency. The median is often preferred when the distribution is skewed, whereas the mode is useful when a distribution is skewed or bimodal. The range, variance, and standard deviation are indices of variability. The variance and standard deviation are frequently used in other statistical formulas.

The study of exploratory data analysis has provided alternative indicators of central tendency and variability that are more resistant to the effects of the values of a few scores, especially atypical or deviant scores. The median is a more resistant index of central tendency than the mean. The fourth-spread, which includes the central half of a group of scores, is a resistant analogue to the standard deviation; and the fourths ± 1.5(fourth-spread) includes nearly all the scores excluding outliers, and is a resistant analogue of the range.

A population is a collection of subjects, events, or scores that have some common characteristic of interest, and a sample is a subgroup of a population. Quantitative characteristics of a population are called parameters, whereas such characteristics of a sample are called statistics. Statistics are often used to estimate parameters.

FORMULAS

1. Mean

$$\overline{X} = \frac{\Sigma X_i}{N}$$

The population mean is symbolized by μ.

2. Median

a. **When there is an odd number of cases in the distribution:** M_d is the score value of the middle case, which is the $(N + 1)/2$ case from the bottom (or top) of the distribution.

b. **When there is an even number of cases**

in the distribution: M_d is the average of the score values of the two middle cases, which are the $N/2$ and the $(N/2) + 1$ cases from the bottom (or top) of the distribution.

3. Mode

The mode, M_o, is the value of the most frequently occurring score if such a value exists.

4. Range

The range is estimated by taking the largest minus the smallest score.

5. **Variance**

$$s^2 = \frac{\Sigma(X_i - \bar{X})^2}{N-1} \qquad \text{(definitional)}$$

$$s^2 = \frac{N\Sigma X_i^2 - (\Sigma X_i)^2}{N(N-1)} \qquad \text{(computational)}$$

The population variance is symbolized by σ^2.

6. **Standard deviation**

$$s = \sqrt{s^2} = \sqrt{\frac{\Sigma(X_i - \bar{X})^2}{N-1}} \qquad \text{(definitional)}$$

$$s = \sqrt{\frac{N\Sigma X_i^2 - (\Sigma X_i)^2}{N(N-1)}} \qquad \text{(computational)}$$

The population standard deviation is symbolized by σ.

7. **Resistant indicators**

   a. **Median.** The median, M_d, is the value of the case having a depth of $(N+1)/2$. If this depth is fractional (i.e., an integer plus .5), the median is the average of the values of the two cases bordering the fractional depth.

   b. **Fourth-spread.**
   The **depth of the fourths** equals

$$\text{depth of fourths} = \frac{(\text{depth of median}) + 1}{2}$$

   in which a fractional depth of the median is first rounded down to the next lower integer. The **lower** (F_L) and **upper** (F_U) **fourths** are the values of the cases having depths of fourths. If the depth of fourths is fractional (i.e., an integer plus .5), the fourth is the average of the values of the cases bordering the fractional fourth.

   The **fourth-spread** is the difference between the upper and lower fourths:

$$\text{fourth-spread} = F_U - F_L$$

   All the scores, excluding outliers, will fall between the fourths $\pm$ 1.5(fourth-spread).

Exercises

For Conceptual Understanding

1. In what sense is the mean an appropriate index of central tendency?

2. In what sense is the variance an appropriate index of variability?

3. Indicate which index of central tendency would be preferred for each of the following distributions, and explain why.

   a. household incomes in the United States

   b. heights of seniors in a public high school

   c. IQ scores of third-graders in a public school

4. Draw the curves for distributions in which

   a. the mean, median, and mode are identical.

   b. the mean and median are identical but the mode is different.

   c. the mean is greater than the median.

   d. the median is greater than the mean.

5. Compose the score values for three distributions that have the same mean and median but differ in their amount of variability.

6. Discuss the limitations of the range as a measure of variability and present some numerical examples to illustrate your point.

7. Discuss the properties, characteristics, and advantages of s as a measure of variability. Does s have any potential advantages over s^2?

8. Discuss resistant measures of central tendency and variability and their advantages.

9. Why does the formula for the variance have $N-1$ in the denominator and not N?

For Solving Problems

10. Compute the mean, median, and mode for each of the distributions below.

A	B	C	D
1	1	2	2
2	1	2	2
2	1	2	4
4	3	3	4
7	4	4	4
8	5	5	5
11	6	8	7
	7	9	7
		10	

11. Show that the sum of the deviations about the mean of distribution C in Exercise 10 is zero. Compute the sum of the squared deviations about the mean and median of distribution C. Which is less, the sum of the squared deviations about the mean or the sum of the squared deviations about the median?

12. Find the median for the following distributions:
    a. 2, 5, 6, 8, 9
    b. 0, 2, 3, 6, 8, 10
    c. 1, 2, 3, 3, 4, 5
    d. 5, 6, 6, 6, 6, 20, 21
    e. 0, 2, 3, 3, 7
    f. 3.1, 3.2, 3.3, 3.4, 3.6
    g. 3.1, 3.2, 3.3, 3.3, 3.3, 3.8
    h. 3.1, 3.2, 3.3, 3.3, 3.3, 3.8, 3.9

13. Calculate the variance and the standard deviation for each distribution in Exercise 10.

14. Compute with both the definitional and computational formulas the variance and standard deviation for each of the following distributions. Also compare the means of the distributions.
    a. 2, 4, 4, 6
    b. 0, 2, 4, 10
    c. −4, −3, −2, 0, 0, 5, 8, 8, 12, 16

15. A lab instructor asked her class of 30 college seniors how many beers or mixed drinks they had last Saturday night. The data are presented below. Construct a frequency distribution and calculate measures of central tendency and variability. Then, using these statistics, how would you characterize the drinking behavior of seniors at this school?

1	0	3	4	0	5
0	0	6	3	3	0
2	1	4	3	4	3
3	2	0	5	0	4
0	2	0	0	5	4

16. At a golf tournament consisting of 72 holes, 40 competitors had the following scores. Create a stem-and-leaf display, a five-number summary, and a boxplot for these data.

293, 300, 280, 302, 294, 288, 284, 290, 299, 306, 305, 292, 296, 285, 289, 301, 287, 294, 298, 293, 303, 299, 286, 287, 302, 305, 291, 295, 288, 290, 296, 300, 306, 285, 282, 315, 291, 293, 319, 297

For Homework

17. What are the different messages conveyed by the mean and median, and when should each be used? Illustrate with numerical examples.

18. Which measure or measures of central tendency are appropriate for each of the following types of data?
    a. nominal data
    b. interval data
    c. ordinal data
    d. ratio data

19. What are the mean, median, and mode of the following distributions? Which measure of central tendency best reflects each distribution and why?
    a. 1, 2, 3, 4, 4, 4, 5, 6, 7
    b. 1, 1, 1, 100, 100, 100
    c. 2, 3, 4, 6, 100
    d. 1, 1, 1, 1, 1, 1, 5, 5, 6, 8, 9

On the following curves, label X, Y, and Z as mean, median, or mode.

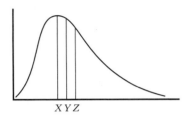

X Y Z

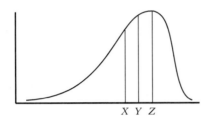

X Y Z

i	X	Y	Z
1	6	0	1
2	2	6	3
3	8	6	0
4	3	0	4
5	1	0	6
6	5	6	8
7	5	0	2
8	4	6	5
9	3	6	7
10	7	0	1

20. Distinguish *briefly* between

   a. sample and population

   b. statistic and parameter

21. Define variability. Explain the rationale behind the variance, s^2, as an index of variability. List some properties of s^2 and s as measures of variability.

22. Is it possible to obtain a negative value for the standard deviation or variance of a distribution? Why?

23. Determine the mean, median, mode, range, variance, and standard deviation for each of the following three distributions.

24. In the distributions for Exercise 23, explain how

   a. the mean of Z can be less than the mean of X although the variance of Z is greater than the variance of X

   b. the range of Y is the smallest of all three distributions, whereas the variance of Y is the greatest of all three distributions.

25. If $\Sigma X^2 = 151$, $N = 5$, and $s^2 = 6.5$, what is the mean of the sample?

26. What characteristics of stem-and-leaf displays and resistant indicators are most similar to a frequency distribution, number of class intervals, size of class intervals, rank order or cumulative frequency, central tendency, standard deviation, and range?

27. Pilots in a flight simulator are presented with a changing set of circumstances on the instruments that eventually indicate a mid-air collision. Below are the time in seconds from the start of this event until each of a group of 30 pilots takes corrective action. Create a stem-and-leaf display, five-number summary, and boxplot for these data.

19.2, 15.7, 20.3, 21.6, 18.9, 16.7, 19.0, 18.5, 28.3, 17.9, 18.3, 19.1, 21.7, 19.9, 17.2, 21.5, 20.5, 20.1, 22.6, 23.0, 17.9, 18.4, 20.4, 19.3, 22.5, 21.9, 19.2, 18.0, 19.0, 20.7

Indicators of Relative Standing

Although the statistics discussed in the previous chapter help describe an entire distribution, they do not provide direct assistance in interpreting individual scores. Suppose a teacher gives a final exam and wants to assign grades on a curve—the top 10% get A's, the next 20% get B's, and so on. The teacher would need to determine what scores are in the top 10%, the next 20%, and so on. Similarly, if you were a student in the class, knowing that you had a score of 88 on the exam would not provide much information on how well you did relative to other students. Knowing the mean, range, and standard deviation for the class would help somewhat, but it also would be informative to know what proportion of the students scored below you.

PERCENTILES

Percentile points and percentile ranks provide such information about the relative standing of scores in a distribution.

> The *P*th **percentile point** is the score value below which the proportion *P* of the cases in the distribution falls. That percentile point is said to have a **percentile rank** of *P*.

If a score of 88 is at the 90th percentile rank, .90 (or 90%) of the people in the group scored less than the percentile point of 88. In symbols, this information is written

$$P_{.90} = 88$$

Students sometimes have difficulty remembering which piece of information is the percentile rank and which is the percentile point. The *percentile rank* is the percentage, which is essentially the *rank* of the score value in a distribution of 100 cases. The percentile *point* is the *point* on the measurement scale—that is, the score value—corresponding to a particular percentile rank. Again in symbols,

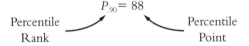

$$P_{.90} = 88$$

Percentile   Percentile
Rank   Point

Some specific percentile ranks have special names. For example, $P_{.50}$ represents the point in the distribution below which .50 or 50% of the cases lie, which is the point or score value that was called the **median** in Chapter 3. Therefore, $P_{.50} = $ the median. Similarly, $P_{.25}$ represents the score value below which .25 or 25% (that is, a "quarter") of the cases lie, which is called the **lower quartile point.** Also, $P_{.75}$ represents the point below which .75 or 75% of the cases lie, which is called the **upper quartile point** (because 25% or a "quarter" of the cases lie *above* it). Finally, these two score values, corresponding to $P_{.25}$ and $P_{.75}$, define the **interquartile range,** the range of score values that includes the middle 50% of the cases. The interquartile range is very similar to the fourth-spread introduced in the previous chapter. Both are intended to designate the central half of the distribution. They differ, however, in the calculation of quartile points vs. fourths, so the two approaches may produce slightly different ranges.

Percentile ranks and points are very similar to the concept of cumulative relative frequency introduced in Chapter 2. Essentially, a percentile rank is approximately the cumulative relative frequency of its corresponding percentile point. This close relationship can be seen in Table 4-1, which presents the scores of 40 students on a statistics exam listed in order of score value plus a cumulative relative frequency polygon of those scores. By drawing horizontal and vertical dotted lines parallel to the axes, one can simply read approximations of percentile ranks and their corresponding percentile points. For example, the percentile rank of $P_{.90}$ represents a cumulative relative frequency of .90, which corresponds approximately to a percentile point or score value of 88. Similarly, the percentile point or score value of 70 corresponds approximately to a cumulative relative frequency of .40 or the percentile rank $P_{.40}$.

Computation of Percentile Points

Determining the percentile points is similar to, but somewhat more complicated than, determining the median, which is the point corresponding to the rank of .50.[1] Consider again the exam scores for 40 students presented in Table 4-1. Suppose the teacher wants to determine the percentile point (score value) corresponding to the rank of .30. To determine this percentile point, recall the definition of percentile point: it is that score value below which proportion P of the cases in the distribution fall. In this problem, P is .30 and the number of cases in the distribution is $N = 40$. Therefore, 30% of 40 or

$$P(N) = .30(40) = 12$$

cases must fall below the required score value. So the desired value must be such that the 12th student from the bottom falls below it while the 13th student falls above it. In Table 4-1, the 12th student scored 59 and the 13th student scored 60, so, in the same manner as was done for the median, the required value is halfway between 59 and 60 or 59.5, which is the upper real limit of 59, thereby placing the 12th student below it, and which is also the lower real limit of 60, thereby placing the 13th student above it.

Similarly, suppose a teacher wanted to put these scores on a grading curve in which the top 10% are given A's, the next 20% are given B's, the next 40% C's, then 20% D's, and the bottom 10% F's. What is the lowest passing score? This is the per-

---

[1] The methods presented in this chapter permit one to estimate the percentile ranks and points for any value, whether actually represented in the distribution or not, and they give an interpolated answer if duplication of scores exists near the desired value. The method of determining the median given in the previous chapter is the most common, but it does not interpolate if duplication of scores exists near the median. Therefore, in such circumstances, the calculation of the median as $P_{.50}$ using the procedures in this chapter may give a slightly different answer than the procedures given in the previous chapter. Also, many computer programs do not interpolate in either situation and may also give slightly different answers.

4-1 Distribution of Scores on a Statistics Examination for a Class of 40 Students

Student No.	Score	Student No.	Score	Student No.	Score	Student No.	Score
40	97	30	80	20	72	10	56
39	93	29	79	19	72	9	55
38	92	28	78	18	72	8	55
37	91	27	78	17	71	7	55
36	88	26	78	16	70	6	55
35	85	25	78	15	65	5	51
34	85	24	76	14	61	4	49
33	84	23	75	13	60	3	49
32	83	22	75	12	59	2	48
31	82	21	74	11	58	1	46

Statistics Exam Score (Table 4–1)

centile point (or score value) corresponding to the percentile rank of $P = .10$. Since $N = 40$, 10% of 40 or $P(N) = 4$ cases must fall below the minimum passing score. As can be seen in Table 4–1, the lowest four scores are 49 or lower, and the next highest score is 51. So, again in the same manner as was done for the median, the required value is taken to be halfway between 49 and 51, which is 50.

The same general logic can be applied to more complicated situations in which there is duplication of scores near the required percentile point. Duplication of scores simply means that more than one case has the same score value and that value is involved in the determination of the desired percentile point. In such cases, it is easier to use a formula to calculate the desired value.

The Formula

The **percentile point** corresponding to the Pth rank is given by

$$\text{percentile point } X = L + \left[\frac{P(N) - n_b}{n_w}\right] i$$

in which

L = the lower real limit of the score value containing the required percentile point

P = the percentile rank of the required point (that is, the proportion of cases falling below the desired point; P ranges between 0 and 1.00)

N = the total number of cases in the entire distribution

n_b = the number of cases falling below L

n_w = the number of cases falling within the score value containing the required percentile point

i = the size of the score value measurement unit ($i = 1$ if the data are in whole numbers; $i = .1$ if the data are in tenths; and so on)

To Use the Formula

1. **Order all the cases by score value.** Start at the bottom with the lowest score value, label it case 1, and list and number all the cases in the distribution, including duplicate score values, ending at the top with the largest score value.

2. **Determine N.** N is the total number of cases in the entire distribution.

3. **Determine the location of the desired percentile point.**
   a. Calculate $P(N)$, the desired percentile rank P times N (round to the nearest whole number).
   b. Locate the $P(N)$th case in the distribution.

4. **Determine other required values.**
   a. n_w is the number of cases having the score value of the $P(N)$th case.
   b. n_b is the number of cases having a score value smaller than the score value of the $P(N)$th case.
   c. L is the lower real limit of the score value of the $P(N)$th case.
   d. i is the size of the score value measurement unit ($i = 1$ if the data are in whole numbers; $i = .1$ if the data are in tenths; and so on).

5. **Substitute these values in the formula for percentile points and calculate.**

Note: This formula does not always provide an accurate answer when the desired $P(N)$th case is fractional [e.g., $P_{.23}$ for the data in Table 4-1 gives $P(N) = .23(40) = 9.2$]. Such situations will not be covered in this text. Also, this formula produces slightly different answers than does a cumulative relative frequency distribution, because the formula takes a score value at its midpoint (e.g., a score value of 88 is exactly 88.0) whereas a cumulative relative frequency assumes a score value occupies the range of its real limits (i.e., 88 is 87.5 to 88.5). Finally, the formula will give only an approximate answer when used with data in class intervals and i is the size of the interval.

To illustrate, here is how the formula is used to determine the percentile point corresponding to the 20th percentile rank (P_{20}) for the distribution in Table 4-1. In this case, $P = .20$ and $N = 40$, so 20% of 40 or $P(N) = .20(40) = 8$ cases must fall below the desired score value. The eighth lowest subject scored 55, but there were four cases having score values of 55, so $n_w = 4$; there were five cases who scored less than 55, so $n_b = 5$; the lower real limit of the score value of 55 is $L = 54.5$; and the data are in whole numbers, so $i = 1$. Thus, the 20th percentile point is given by

$$\text{percentile point} = L + \left[\frac{P(N) - n_b}{n_w}\right] i$$

$$= 54.5 + \left[\frac{.20(40) - 5}{4}\right] 1$$

$$\text{percentile point} = 55.25$$

Computation of Percentile Ranks

The illustrations just presented show how to compute the score value (that is, the percentile point) corresponding to a given percentile rank. However, the question can be reversed. What is the percentile rank corresponding to a given score value?

A student who scored 83 might want to know what proportion of students scored lower. This information is given by the percentile rank of a score value of 83. Looking at Table 4-1, you can see that the score value of 83 was 32nd in the distribution. This means that 32 people scored lower than 83.5, which is the upper real limit of 83. Similarly, 31 people scored below 82.5, which is the lower real limit of 83. Therefore, 31.5 of 40 people or $31.5/40 = .7875$ of the cases recorded a score below exactly 83.0. Consequently, the score of 83 has a percentile rank of .7875. More conventionally, the proportion is rounded to two decimals, so that the score value of 83 is said to be at the 79th percentile. Thus, 83 corresponds to P_{79} and 79% of the scores were below 83.

Again there is a formula that will produce this result as well as percentile ranks when duplication of score values exists near the desired rank. Note that this formula includes the same elements as the one given above for finding percentile points when percentile rank is given.

The Formula

The **percentile rank** corresponding to a given score value X is determined by

$$P = \frac{n_w(X - L) + in_b}{Ni}$$

in which

P = the desired percentile rank (that is, the proportion of cases falling below X; P ranges between 0 and 1.00)

X = the score value for which the percentile rank is desired

L = the lower real limit of X

n_w = the number of cases having the score value X

n_b = the number of cases having a score value lower than X

N = the total number of cases in the entire distribution

i = the size of the score value measurement unit ($i = 1$ if the data are in whole numbers; $i = .1$ if the data are in tenths; and so on)

To Use the Formula

1. **Order all the cases by score value.** Start at the bottom with the lowest score value, label it case 1, and list and number all the cases in the distribution, including duplicate score values, ending at the top with the largest score value.

2. **Determine the location of X in the distribution,** the score value for which the percentile rank is desired.

3. **Determine other required values.**

   a. n_w is the number of cases having the score value of X.

   b. n_b is the number of cases having a score value lower than X.

   c. L is the lower real limit of X.

   d. N is the total number of cases in the entire distribution.

   e. i is the size of the score value measurement unit ($i = 1$ if the data are in whole numbers; $i = .1$ if the data are in tenths; and so on.

4. **Substitute these values in the formula and calculate.**

Note: This formula does not always provide an accurate answer when the score value for which the percentile rank is desired (i.e., X) is not represented in the distribution. Such situations will not be covered in this text. Also, this formula produces slightly different answers than does a cumulative relative frequency distribution, because the formula takes a score value at its midpoint (e.g., a score value of 88 is exactly 88.0) whereas a cumulative relative frequency assumes a score value occupies the range of its real limits (i.e., 88 is 87.5 to 88.5). Finally, the formula will give only an approximate answer when used with data in class intervals and i is the size of the interval.

To determine the percentile rank for a score value of 83, let $X = 83$, which has a lower real limit $L = 82.5$. There is only one case of 83 in the distribution ($n_w = 1$); 31 cases are below X ($n_b = 31$); there are $N = 40$ cases in the entire distribution; and the data are in whole numbers so $i = 1$. Therefore,

$$P = \frac{n_w(X - L) + in_b}{Ni}$$

$$= \frac{1(83 - 82.5) + 1(31)}{40(1)} = \frac{31.5}{40}$$

$$P = .7875$$

Therefore, $X = 83$ is at the 78.75th or 79th percentile rank ($P_{.79}$).

Suppose one had to determine the percentile rank corresponding to a score of 72. In this case, $X = 72$ with a lower real limit of $L = 71.5$; three cases scored exactly 72 ($n_w = 3$); and 17 cases scored lower than 72 ($n_b = 17$). Consequently,

$$P = \frac{n_w(X - L) + in_b}{Ni}$$

$$= \frac{3(72 - 71.5) + 1(17)}{40(1)} = \frac{18.5}{40}$$

$$P = .4625$$

Thus, $X = 72$ is at the 46.25th or 46th percentile rank ($P_{.46}$).

As with the median, readers should note that not all textbooks and computer programs use the interpolation method described here when duplication of scores exists near the percentile point or rank. This level of accuracy may not be necessary when large numbers of cases are available.

CHANGING THE PROPERTIES OF SCALES

A very powerful and useful method of characterizing the relative position of scores in some distributions is to use **standard scores.** This approach requires that the original scale of measurement be transformed into a single "standard" scale by changing its unit and origin (zero point). This section discusses how to do this and what effect such changes in unit and origin have on the mean and variance of a distribution. Then, in the next section, this information will be applied to determining and interpreting standard scores.

To begin, Celsius and Fahrenheit are two different scales of temperature. They differ in that they have neither the same origin (that is, 0° Celsius and 0° Fahrenheit represent different temperatures) nor the same size units (that is, degrees).

The formula for converting degrees Celsius to degrees Fahrenheit is

$$°F = °C(1.8) + 32°$$

Notice that the formula has two constants, the 1.8 and the 32°. The 1.8 is the factor used to convert the Celsius degree to the Fahrenheit degree. Consider the freezing and boiling points of water on the two scales. For the Fahrenheit scale, these points are 32° and 212°, respectively, whereas for the Celsius scale they are 0° and 100°, respectively. Thus, between the freezing and boiling points of water on the Fahrenheit scale there are 180 Fahrenheit degrees. Between the freezing and boiling points of water on the Celsius scale there are only 100 degrees. Therefore, it takes 1.8 Fahrenheit degrees to equal 1 Celsius degree. This is the reason the 1.8 appears in the conversion formula; it changes the Celsius units into Fahrenheit units.

However, the two scales are not yet equivalent. Even if the Celsius degrees are converted to Fahrenheit degrees, the freezing point is still 0° and the boiling point is $(1.8)(100) = 180°$; both values are 32° short of their corresponding Fahrenheit values. To change the Celsius origin to match the Fahrenheit origin, 32° must be added.

This process can be seen graphically in Figure 4-1. At the top is the Celsius scale with a freezing point at 0° and a boiling point at 100°. To change the size of the Celsius unit (degree) to a Fahrenheit unit (degree), we multiply by 1.8. This is done on the second scale. Note that this has the effect of putting 180° (rather than 100°) between the freezing and boiling points. This is displayed by expanding the scale used to cover the distance between the freezing and boiling points. We have now changed the size of the Celsius degree to a Fahrenheit degree, but the freezing point (0°) and boiling point (180°) are in the wrong place—they are 32° too small. We must change the origin of the scale, so we add 32° to each point, which produces the scale at the bottom of Figure 4-1 with a freezing point at 32° and a boiling point at 212°. This is the Fahrenheit scale. Notice that changing the origin of the scale by adding 32° has the effect of moving the segment covering the freezing and boiling points to the right 32°, but it does not alter the length of the segment (the unit size) between the freezing and boiling points.

Generally:

> To change the *size* of the unit of measurement, multiply or divide the old values by the proper constant (conversion factor). To change the *origin* of a scale, add or subtract the appropriate number of units.

In most transformations common to everyday life, only the size of the unit is converted. For example:

inches = feet × 12 ← conversion factor

$$\text{kilometers} = \frac{\text{miles}}{.62} \leftarrow \text{conversion factor}$$

However, in research in the social and behavioral sciences, transformations from one scale to another frequently involve changing both the unit size and the origin of the scale.

Effects of Scale Changes on the Mean

What happens to the value of the mean of a distribution when the unit size and origin of the measurement scale are changed?

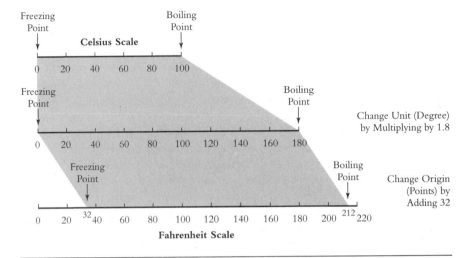

Figure 4-1 Changing the Celsius scale to the Fahrenheit scale.

The origin of a scale is changed by adding a constant to (or subtracting a constant from) every score. If a constant is added to (or subtracted from) every score in the distribution, the mean of the new distribution is the mean of the old distribution plus (or minus) that constant. If the new scores are $X' = X + c$, then the new mean is

$$\overline{X}' = \overline{X} + c$$

Therefore, if the mean of a distribution is 15 and 3 is added to every score, the new mean is $15 + 3 = 18$. If 3 is subtracted from every score, the new mean is $15 - 3 = 12$. This principle is displayed graphically at the top of Figure 4-2, and a formal algebraic proof is presented in Optional Table 4-2.

The unit of a scale is changed by multiplying or dividing every score by a constant. If every score in a distribution is multiplied (or divided) by a constant, the mean of the new distribution is the mean of the old distribution multiplied (or divided) by that constant. If the new scores are $X' = cX$, then the new mean is

$$\overline{X}' = c\overline{X}$$

Therefore, if the mean of a distribution is 15 and every score is multiplied by 3, the new mean is three times the old mean, or $15 \times 3 = 45$. On the other hand, if every score were divided by 3, the mean would also be divided by 3: $^{15}\!/_{3} = 5$. This principle is displayed graphically at the bottom of Figure 4-2, and a general algebraic proof is presented in Optional Table 4-3.

In summary, if the scale of measurement is altered, the mean will change in the same manner and to the same extent as any other value on the scale.

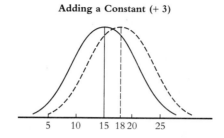

Adding a Constant (+ 3)

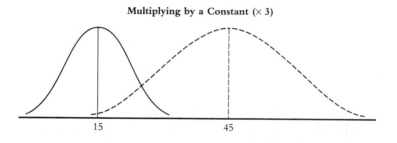

Multiplying by a Constant (× 3)

Figure 4-2 Adding or multiplying by a constant. Notice that when a constant $c = 3$ is added to each score, the mean increases by that constant ($15 + 3 = 18$), but the variance is not changed. The entire distribution simply moves over three units. When each score is multiplied by a constant $c = 3$, however, the mean is multiplied by that constant ($15 \times 3 = 45$) and the variance is multiplied by c^2.

4-2 Change in the Mean with a Change in Origin

Optional Table

Operation	Explanation
To prove $\bar{X}' = \bar{X} + c$	
1. $\bar{X}' = \dfrac{\Sigma(X + c)}{N}$	1. Definition of the mean for a distribution in which c has been added to each X.
2. $\quad = \dfrac{\Sigma X + \Sigma c}{N}$	2. The sum of several terms is the sum of the separate terms.
3. $\quad = \dfrac{\Sigma X + Nc}{N}$	3. The sum (from 1 to N) of a constant is N times that constant.
4. $\quad = \dfrac{\Sigma X}{N} + \dfrac{Nc}{N}$	4. Simplification.
5. $X' = \bar{X} + c$	5. Substitution: $\dfrac{\Sigma X}{N} = \bar{X}$ and cancellation of N's.

4-3 **Change in the Mean with a Change in Unit**   `Optional Table`

Operation	Explanation
To prove $\overline{X}' = c\overline{X}$	
1. $\overline{X}' = \dfrac{\Sigma cX}{N}$	1. Definition.
2. $\overline{X}' = \dfrac{c\Sigma X}{N}$	2. The sum of a constant times a variable is the constant times the sum of that variable.
3. $\overline{X}' = c\overline{X}$	3. Substitution: $\dfrac{\Sigma X}{N} = \overline{X}$

Effects of Scale Changes on the Variance and Standard Deviation

Transformations of scale do not have the same effect on the variance and standard deviation of a distribution as they do on the mean.

> If the origin of a scale is changed by adding a constant to (or subtracting it from) every score in the distribution, the variance and standard deviation are not changed. In symbols, if the new scores are $X' = X + c$ and s_x^2 is the variance of the original distribution, then the variance of the new distribution is
>
> $$s_{x'}^2 = s_x^2$$
>
> and the standard deviation of the new distribution is
>
> $$s_{x'} = s_x$$

If the variance of a distribution is 16 and the standard deviation is 4 and then 3 is added to every score, these statistics will remain unchanged in the new distribution and $s_x^2 = 16$ and $s_{x'} = 4$. This makes intuitive sense if one considers that s^2 and s are functions of the distances between the various score values and the mean, not of the distances between those score values and the origin. For example, consider the distance between scores in the following distribution before and after a change of origin (3 is added to each score and the original mean is 7):

	X					$X' = X + 3$	
	3	7	11			6	10	14
$(X - \overline{X})$:	-4		4		$(X' - \overline{X}')$:	-4		4

Since s^2 and s are based upon the distances between the scores and the mean

4-4 **Proof that No Change Occurs in the Variance (and Standard Deviation) with a Change in Origin**   Optional Table

Operation	Explanation
To prove $s_{x'}^2 = s_x^2$	
1. $s_{x'}^2 = \dfrac{\Sigma[(X + c) - (\overline{X} + c)]^2}{N - 1}$	**1.** Definition: if a constant is added to every score, the mean will also be incremented by that constant.
2. $= \dfrac{\Sigma[X + c - \overline{X} - c]^2}{N - 1}$	**2.** Removing parentheses.
3. $s_{x'}^2 = \dfrac{\Sigma(X - \overline{X})^2}{N - 1} = s_x^2$	**3.** Subtraction.

$(X - \overline{X})$ and since these distances do not change when the origin is changed, adding a constant to each score does not alter s^2 and s. This principle is illustrated at the top of Figure 4-2, where adding a constant $c = 3$ simply moved the distribution up 3 units without changing its form (i.e., variance); a formal proof is given in Optional Table 4-4.

Although adding (or subtracting) a constant does not affect s^2 and s, multiplying (or dividing) by a constant does change these statistics.

> If the unit of measurement is changed by multiplying (or dividing) every score in the distribution by a positive constant,[2] the new variance will equal the old variance multiplied (or divided) by the square of that constant, and the new standard deviation will equal the old standard deviation multiplied (or divided) by that constant. In symbols, if the new scores are $X' = cX$, and s_x^2 is the variance of the original distribution, then the variance of the new distribution is
>
> $$s_{x'}^2 = c^2 s_x^2$$
>
> and the standard deviation of the new distribution is
>
> $$s_{x'} = c s_x$$

Thus, if the variance of a distribution is 16 and the standard deviation is 4 and each score in the distribution is multiplied by 3, the variance of the new distribution will be $(3^2)(16) = 144$ and the new standard deviation will be $(3)(4) = 12$.

The reason that multiplying or dividing by a constant changes the variance and

---

[2] If c is negative, then $s_{x'} = |c| s_x$.

standard deviation can be seen more clearly by examining what happens to the deviations and squared deviations used in the formula for the variance when the scale undergoes a change in unit (as when every score is multiplied by 3):

	X			$X' = 3X$	
	3 7 11			9 21 33	
$(X - \bar{X})$:	-4 4		$(X' - \bar{X}')$:	-12 12	
$(X - \bar{X})^2$:	16 16		$(X' - \bar{X}')^2$:	144 144	

Two important points are made by this example. When the unit of measurement is changed by multiplying every score by $c = 3$, the distances between points are also multiplied by $c = 3$ and the size of the squared deviations increases by a factor of $c^2 = 3^2 = 9$. In other words, because the variance uses squared deviations in its formula, the size of the variance is multiplied by c^2 when each score is multiplied by c. On the other hand, the formula for the standard deviation takes the positive square root of these squared deviations, and thus it is changed by a factor of c (assuming c is a positive constant).[3] This principle is displayed graphically at the bottom of Figure 4–2, in which the variance of the new distribution is clearly increased when each score is multiplied by $c = 3$. A formal algebraic proof is given in Optional Table 4–5. A numerical example illustrating all of these principles is given in Table 4–6.

STANDARD SCORES AND THE NORMAL DISTRIBUTION

The principles of percentile points and ranks and changes in the origin and unit of a scale become important when we consider standard scores and the normal distribution and how they may be used to characterize the relative positions of scores in some distributions.

Standard Scores

Suppose a teacher gives two exams to a class. A score of 88 on the first exam might mean something quite different than the same score on the second exam. For example, 88 could be an extremely high score on the first test but not on the second. One way to describe this difference is to assign the scores percentile ranks as presented at the beginning of this chapter. While percentile ranks provide some idea of relative standing, they indicate only ordinal position within a distribution. That is, percentiles show the proportion of a distribution that falls below a given score, but they do not indicate how *far* below that score the remainder of the distribution is located. To illustrate, assume the results of the two tests are

$A = (78, 81, 87, 88)$
$B = (59, 61, 63, 88)$

---

[3] If c is negative, then $s_{x'} = |c| s_x$.

4-5 **Proof that the Variance Changes by a Factor of c^2 and the Standard Deviation by c with a Change in Unit**   Optional Table

Operation	Explanation
Variance	
To prove	
$s_{x'}^2 = c^2 s_x^2$	
1. $s_{x'}^2 = \dfrac{\Sigma(cX - c\overline{X})^2}{N - 1}$	1. Definition; if each score is multiplied by a constant, the mean is multiplied by that same constant.
2. $\phantom{s_{x'}^2} = \dfrac{\Sigma[c(X - \overline{X})]^2}{N - 1}$	2. Factoring in the manner of $ca - cb = c(a - b)$
3. $\phantom{s_{x'}^2} = \dfrac{\Sigma[c^2(X - \overline{X})^2]}{N - 1}$	3. Simplification in the manner of $(ab)^2 = a^2 b^2$
4. $\phantom{s_{x'}^2} = \dfrac{c^2\Sigma(X - \overline{X})^2}{N - 1}$	4. The sum of a constant times a variable is that constant times the sum of the variable.
5. $s_x^2 = c^2 s_x^2$	5. Substitution: $\dfrac{\Sigma(X - \overline{X})^2}{N - 1} = s_x^2$
Standard Deviation	
To prove	
$s_{x'} = cs_x$	
1. $s_{x'} = \sqrt{s_{x'}^2}$	1. Definition.
2. $\phantom{s_{x'}} = \sqrt{c^2 s_x^2}$	2. Substitution: $s_{x'}^2 = c^2 s_x^2$
3. $s_{x'} = cs_x$	3. Simplification in the manner of $\sqrt{a^2 b^2} = ab$

The score of 88 has the same percentile rank within each distribution—namely, it is the highest score in each distribution. But the 88 on test B represents a considerably greater achievement than the 88 on test A, because in B it is so much higher than the remaining scores. Therefore, it would be desirable to be able to characterize the position of a score within a distribution more precisely than by percentile rank.

The interpretation of a score within a distribution may be based upon its relative standing with respect to the mean as well as the variability of the scores within the distribution. Consider the following example:

$$A = (10, 36, 38, 40, 42, 44, 70)$$
$$B = (10, 12, 14, 40, 66, 68, 70)$$

In both distributions, the range is 60, the mean is 40, and the score of 70 deviates $+30$ points from the mean. But the variability is much greater in the B distribution. Thus, the score of 70 is somewhat less of an achievement in group B, because in the

4-6

Effects of Changes in Origin and Unit on the Mean, Variance, and Standard Deviation

X		Change the Origin $Y = X + 5$		Change the Unit $W = 3X$	
X_i	X_i^2	Y_i	Y_i^2	W_i	W_i^2
3	9	8	64	9	81
5	25	10	100	15	225
6	36	11	121	18	324
10	100	15	225	30	900
$\Sigma X_i = 24$	$\Sigma X_i^2 = 170$	$\Sigma Y_i = 44$	$\Sigma Y_i^2 = 510$	$\Sigma W_i = 72$	$\Sigma W_i^2 = 1530$

X column:

$$N = 4$$
$$\bar{X} = \frac{\Sigma X_i}{N} = \frac{24}{4} = 6$$
$$s_x^2 = \frac{N\Sigma X_i^2 - (\Sigma X_i)^2}{N(N-1)}$$
$$= \frac{4(170) - (24)^2}{4(4-1)} = 8.67$$
$$s_x = \sqrt{8.67} = 2.94$$

Change the Origin $Y = X + 5$:

$$N = 4$$
$$\bar{Y} = \frac{\Sigma Y_i}{N} = \frac{44}{4} = 11$$
$$s_y^2 = \frac{N\Sigma Y_i^2 - (\Sigma Y_i)^2}{N(N-1)}$$
$$= \frac{4(510) - (44)^2}{4(4-1)} = 8.67$$
$$s_y = \sqrt{8.67} = 2.94$$

Change the Unit $W = 3X$:

$$N = 4$$
$$\bar{W} = \frac{\Sigma W_i}{N} = \frac{72}{4} = 18$$
$$s_w^2 = \frac{N\Sigma W_i^2 - (\Sigma W_i)^2}{N(N-1)}$$
$$= \frac{4(1530) - (72)^2}{4(4-1)} = 78.00$$
$$s_w = \sqrt{78.0} = 8.83$$

Summary	X	$Y = X + 5$	$W = 3X$
Mean	6.00	11.00	18.00
Variance	8.67	8.67	78.00
Standard Deviation	2.94	2.94	8.83

B distribution other scores were also likely to deviate greatly from the mean. The scores of 66 and 68 were close behind 70 in distribution B, but the next highest score in A was only 44. Therefore, the interpretation of an individual score is affected by the variability as well as by the mean of the distribution.

One approach to incorporating the variability of a distribution in characterizing the relative positions of scores is to adopt a single "standard" scale of measurement such that all distributions under consideration, despite their different origins and unit sizes and different means and variances, could be converted to that single scale and single distribution for comparison. This is essentially what is done when measures are converted into standard scores (often called z scores). By this standardization process, the relative standing of persons in one distribution may be compared with their relative standing in another distribution, even though the two original distributions had different measurement scales and different means and variances. In a sense, conversion to standard scores permits one to "compare apples and oranges."

To transform one scale of measurement to another, such as from feet to inches or from Celsius to Fahrenheit, one needs to change the origin and unit of the first scale to that of the second by the procedures discussed in the previous section. Specifically, to change the origin, a constant is added to or subtracted from every score, and to change the size of the unit, every score is multiplied or divided by a constant. To transform any scale of measurement into a single standard scale, we select the mean of the original distribution $(\overline{X})$ as the constant to be subtracted from each score (X_i) to change the origin of the scale, and we select the standard deviation of the original distribution (s_x) as the constant with which to divide each score to change the size of the unit of the scale. These two operations are represented simultaneously in the definition of a standard score (z_i):

A **standard score** is defined to be

$$z_i = \frac{X_i - \overline{X}}{s_x}$$

Any single raw score may be transformed into a standard score. For example, if you had a score of 88 on an examination, the class mean was 79, and the standard deviation was 6, the standard score, z, corresponding to $X = 88$ is given by

$$z = \frac{X - \overline{X}}{s_x} = \frac{88 - 79}{6} = \frac{9}{6} = 1.50$$

If the formula is used to standardize each raw score in a distribution, what effect does that process have on the distribution's mean and standard deviation? When a constant (in this case, the mean) is subtracted from each score in a distribution, the constant will also be subtracted from the old mean. Therefore, the new mean will be the old mean minus that constant, that is, minus the old mean, which equals 0. In symbols,

New mean: $\overline{z} = \overline{X} - c$

$c = \overline{X}$

$\overline{z} = \overline{X} - \overline{X}$

$\overline{z} = 0$

Therefore, when scores are standardized, the mean of the standardized scores (z_i) is zero ($\bar{z} = 0$). Note that this implies that the origin of the new scale is 0, and that the z scores in the new distribution will include negative as well as positive numbers.

Recall that subtracting a constant does not alter the standard deviation of a distribution, but dividing by a constant does. So when each score in a distribution is divided by a constant, the standard deviation of the new distribution is divided by that constant. In the formula for a standard score, the constant is s_x, the standard deviation of the original distribution. Thus, the new standard deviation equals the old standard deviation divided by itself, or 1:

New standard deviation: $s_z = s_x/c$

$$c = s_x$$
$$s_z = s_x/s_x$$
$$s_z = 1$$

If the new standard deviation is $s_z = 1$, then the new variance s_z^2 also equals 1.00, because $s_z^2 = (s_z)^2 = (1)^2 = 1.00$. Notice also that dividing by s_x does not change the mean, because the new mean is 0, and $0/s_x = 0$.

But if the mean and origin of the new distribution are zero, what is its unit? Standard scores are said to be in **standard deviation units.** This can be seen most easily by examining a numerical example. Suppose, again, you had a score of 88 on an exam, the class average was 79, and the class standard deviation was 6. Notice that your score of 88 is 9 points above the mean of 79: $X_i - \bar{X} = 88 - 79 = 9$. Since the standard deviation of the distribution is 6, your score is $9/6 = 1.50$ or 1.5 standard deviations above the mean. This logic is exactly the same as is used in the calculation of z scores:

$$z_i = \frac{X_i - \bar{X}}{s_x} = \frac{88 - 79}{6} = \frac{9}{6} = 1.50$$

So z scores are in standard deviation units—the z score indicates how many standard deviations the original score is above (if it is positive) or below (if it is negative) the mean.

To summarize, if every score in a distribution of X_i has the mean of the X distribution subtracted from it, and this result is divided by the standard deviation of the X distribution, *the new distribution of standard scores* (i.e., the z_i) *will have a mean of 0 and a standard deviation and variance of* 1.00.

Recall that a major advantage of standard scores is that they convert each scale of measurement to a single common scale, thereby essentially equating the means and variances of the different original distributions when they are transformed to the standard score distribution. This characteristic of standard scores is especially useful— and fair—when one wants to combine scores from, for example, three tests, each of which has a different mean and variance, to determine a final grade in a course. Merely adding the scores on the three tests could give a biased picture of a student's performance, if the teacher wanted each test to contribute equally to the final grade.

Table 4-7 compares adding raw scores across tests with adding standard scores for the same data. Notice that the final rank order of the students is not the same using standard scores as when raw scores are employed. This is particularly striking when

4-7 Effect of Computing Standard Scores and Adding Across Distributions

	Raw Scores			
Student	Test 1	Test 2	Test 3	Total
A	93	80	85	258
B	81	80	84	245
C	70	85	90	245
D	76	81	85	242
E	65	89	82	236
F	65	90	81	236
G	69	86	79	234
	$\overline{X}_1 = 74.14$	$\overline{X}_2 = 84.43$	$\overline{X}_3 = 83.71$	
	$s_1 = 10.14$	$s_2 = 4.20$	$s_3 = 3.55$	

	Standard Scores			
Student	Test 1	Test 2	Test 3	Total
A	1.86	−1.06	.36	1.17
B	.68	−1.06	.08	−.30
C	−.41	.14	1.77	1.50
D	.18	−.82	.36	−.27
E	−.90	1.09	−.48	−.30
F	−.90	1.33	−.77	−.34
G	−.51	.37	−1.33	−1.46

students B and C are compared. They had identical raw score totals (245), but student B had a total z score of −.30, compared with 1.50 for student C. The difference was primarily a result of student C's scoring 90 on the third test. The 90 was an extremely high score in a distribution that otherwise did not have much variability. Therefore, that performance was weighted more in terms of standard deviation units than in terms of raw scores. A comparison of students B and E shows a case in which two students scored quite differently in terms of raw score but had the same standard score total.

Unfortunately, standard scores are not used often in schools or in other situations, despite the fact that performance is commonly determined on the basis of several tests or tasks. For example, standard scores could be used to score performance in triathlons, races in which participants must swim, bike, and run in immediate succession. Typically, the winner is simply the first to reach the finish line of the last of the three events. But this method of scoring gives an advantage to the athlete who is best at the event that takes the most time of the three. Depending upon the distances, the swim may take less than an hour, the cycling approximately six hours, and the run less than three hours. A good cyclist, then, can easily gain 10 or 20 minutes on most of the field, but an expert swimmer can gain only a few minutes on the competition. Competitors who are good cyclists have a distinct advantage in such a contest. But if standard scores were calculated separately for each event and then added, skill at each event would be weighted equally in determining the winner, but such a "winner" might not be the first to cross the finish line of the last event.

In general, adding scores in standard-score form provides a fairer and more accurate index of performance than does adding raw scores, because the means and variances of the individual distributions are equated. Standard scores take into account not only how much a person's performance deviates from the average of the group on each test or race but also how likely it is that other individuals in the group would score as high (or as low) as that person did. The result is that the standardization process weights each test or task equally in determining a combined result.

The Normal Distribution

Standard scores take on greater meaning when they are interpreted in relation to the normal distribution. The normal distribution is different from the distributions described in the previous chapter. For one thing, it is a **theoretical distribution,**[4] a hypothetical or idealized distribution based on a population of an infinite number of cases. Because it is based on a population, not a sample, the normal distribution has parameters, not statistics. Thus, its mean is symbolized by μ (pronounced "mew"), not $\overline{X}$, and its variance is symbolized by σ^2 (pronounced "SIG-mah squared"), not s^2. A second difference is that the normal distribution is defined by a formula:

The **normal distribution** is defined by

$$Y = \frac{1}{\sqrt{2\pi\sigma^2}}\, e^{-(X-\mu)^2/(2\sigma^2)}$$

in which

Y = the height of the curve at point X

X = any point along the X-axis

μ = mean of the distribution

σ^2 = variance of the distribution

π = a constant, approximately 3.1416

e = the base of Napierian logarithms, approximately 2.71828

[4] This is also called a probability distribution.

A third difference is that there are many normal distributions. Usually, the normal distribution is described as bell-shaped, like the one pictured in Figure 4-3. However, because the formula includes the population mean and variance, a different normal distribution can be drawn for each combination of μ and σ^2. Figure 4-4 shows three normal distributions that have the same mean but different variances.

Although the extent to which the normal distribution looks bell-shaped depends upon the values of its parameters, certain attributes are present in every normal distribution. First, the mean, median, and mode are all the same value, and they divide the distribution into two equal parts. Second, the normal distribution is symmetrical about this central point, so that if the distribution were folded over at the mean, the right and left sides would fall exactly on top of one another. Finally, the "tails" of the distribution get closer and closer to the X-axis as they get farther from the mean, but they never touch it. These tails are said to be **asymptotic** to the X-axis, and thus the normal distribution actually stretches from minus infinity $(-\infty)$ to plus infinity $(+\infty)$ along the X-axis. However, the most important characteristics of the normal distribution can best be understood only after one transforms it into a standardized form.

The Standard Normal Distribution   In the sections above, we saw how any distribution of X_i could be transformed or *standardized* into a distribution of z_i with a mean of 0 and a standard deviation of 1.00. Suppose the original distribution of X_i is a population distribution with a mean μ_x and a standard deviation σ_x, and suppose further that it follows the normal distribution. Then, the formula

$$z_i = \frac{X_i - \mu_x}{\sigma_x}$$

converts this normal population distribution of X_i into a normal population distribution with a mean of 0 and a standard deviation of 1. This special normal distribution with parameters $\mu_z = 0$ and $\sigma_z = 1.00$ is called the **standard normal distribution.**

The real importance of the standard normal distribution derives from the fact that it is a theoretical distribution whose special characteristics allow us to determine percentile ranks and points very easily. Moreover, since normal distributions are common in nature (for example, height, weight, intelligence) and since any normal distribution

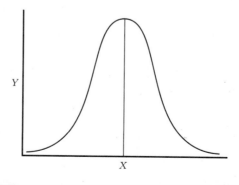

Figure 4-3   A normal distribution.

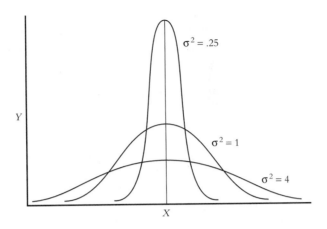

Figure 4-4 Three normal distributions with the same mean but different variances.

can be transformed into the standard normal distribution by employing the conversion formula above, the standard normal distribution can be used as the single reference distribution for comparing a wide variety of otherwise not comparable distributions. Notice that when different normal distributions are converted to the standard normal, each of the transformed distributions has the same mean ($\mu_z = 0$) and standard deviation ($\sigma_z = 1.00$), and thus the percentiles determined for one measure can be directly compared with those for another. For example, if percentile ranks are determined by using the standard normal, a student can be at the 87th percentile rank in math and the 87th percentile rank in English; the student is at precisely the same relative position in the math and English distributions. There is no need to worry about the fact that the original untransformed math and English distributions have different means and standard deviations because these contrasting distributions have been converted into a single distribution, the standard normal distribution.

It is important to realize that one must be able to assume that the original distribution of untransformed scores is normal in form. Converting scores of a nonnormal distribution to standard-score form does not make the standardized distribution normal. Standardization changes only the numerical values of the mean and variance of a distribution—it does not change the distribution's general form. So standardization does not "normalize" a nonnormal distribution. This can be seen in Figure 4-5. Distribution *A* has a somewhat normal shape, *B* does not. Notice that the forms of the *z* distributions (bottom) are nearly the same as the forms of their raw-score distributions pictured above them. Therefore, *although any distribution can be converted to a standardized distribution, only a normal distribution can be converted to a standard normal distribution.*

Some of the special characteristics of the standard normal distribution can be seen by examining Figure 4-6. One can determine percentiles of the standard normal distribution by observing that the proportion of the total area under the curve between two points along the *z*-axis equals the proportion of scores falling between those two *z* values. For example, in the standard normal distribution, the mean, median, and

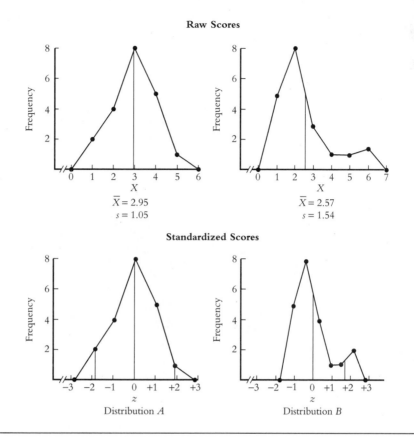

Figure 4-5   Graphic demonstration that standardizing all scores in a distribution does not change the basic form of the distribution. Notice that −1.96 and +1.96 are marked off on the z-axis. They are reasonable estimates of $P_{.025}$ and $P_{.975}$ only for the normal distribution (A). Therefore, the raw-score distribution must be normal for the percentiles of the standard normal distribution to be accurate.

mode are all located at $z = 0$. The area under the curve between $z = 0$ and $+\infty$ is one-half the total area; thus .50 or 50% of the cases fall above $z = 0$. Similarly, .50 or 50% of the cases fall below $z = 0$ because the standard normal is symmetrical about $z = 0$. The important point to remember is that *the area under the standard normal curve implies relative frequency (or proportion of cases).*

With this in mind, certain points can be selected along the z-axis and the proportion of the total area under the curve between these points determined, as has been done in Figure 4-6. Here it can be seen that .3413 of the scores is between $z = 0$ and $z = +1$ and another .3413 is between $z = 0$ and $z = -1$. Further, almost all the scores in a normal distribution (all but .0027) will be included between $z = -3$ and $z = +3$. Since the z scale is in standard deviation units, this implies that in any normal distribution almost all the scores are between three standard deviations above

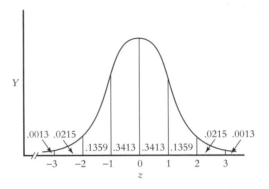

Figure 4-6 Proportions of the total area under a standard normal distribution (i.e., relative frequency) between various values of *z* (i.e., standard deviations).

the mean and three standard deviations below the mean. If the mean of a normal distribution is 55 and the standard deviation is 10, then almost all of the scores are likely to fall between

$$\mu \pm 3(\sigma)$$
$$55 \pm 3(10)$$
$$55 \pm 30$$
$$25 \text{ and } 85$$

Indeed, because the *z* scale is in standard deviation units, *z* scores from normal distributions are sometimes called **standard normal deviates:** they indicate how many standard deviations separate a score from the mean.

The few points along the *z*-axis that are labeled in Figure 4-6 were chosen for convenience. A list of many points along the *z*-axis appears in Table A of Appendix 2 in the back of this book. Turn to this table now. The first column, labeled *z*, corresponds to values along the *z*-axis. The second column gives the proportion of area under the curve (proportion of scores or relative frequency) falling between the mean ($\bar{z} = 0$) and the *z* value for that row. The third column provides the area beyond *z* (that is, between the *z* value and $+\infty$). The areas given by the figures in each column are defined by the shaded areas in the distributions drawn at the tops of each column. Specifically, for example, for *z* = .35, the proportion of area falling between the mean ($\bar{z} = 0$) and *z* = .35 is .1368, and the proportion of area falling between *z* = .35 and *z* = $+\infty$ is .3632. Notice that although the standard normal distribution contains negative *z* values to the left of the mean and positive *z* values to the right of it, the table presents only positive *z* values. Since the standard normal distribution is perfectly symmetrical, the area relating to each positive *z* value is equally appropriate for each negative *z* value. Thus, for example, the third column of the table implies either area to the right of a positive *z* value or area to the left of a negative *z* value.

Application of the Standard Normal Distribution   The steps below show how certain common types of questions regarding relative standing in a distribution can be answered by using the standard normal distribution.

1. In a normal distribution with mean $\mu = 45$ and standard deviation $\sigma = 10$, at what percentile rank does a score of 58 fall?

   a. Determine the z score corresponding to $X = 58$:

   $$z_i = \frac{X_i - \mu}{\sigma} = \frac{58 - 45}{10} = 1.30$$

   b. Draw a picture of a normal distribution similar to that displayed in Figure 4-7. Locate the mean of the distribution, label it $\bar{z} = 0$, and then locate $z = 1.30$. Recall that percentile rank represents the proportion of cases falling below (that is, to the left of) the particular score value ($z = 1.30$, in this case). From your drawing (Figure 4-7), this proportion can be seen as the sum of two proportions, that is, the proportion of the area between the mean $\bar{z} = 0$ and the score $z = 1.30$, plus the proportion of the area between the mean $\bar{z} = 0$ and the left tail, $z = -\infty$. Now look at Table A, Appendix 2. The first of these two proportions (between $\bar{z} = 0$ and $z = 1.30$) is given in the second column in the row for $z = 1.30$. That proportion is .4032. The second proportion is .5000, because the mean, $\bar{z} = 0$, divides the standard normal distribution in half, leaving .5000 of the area to its left. Therefore, the proportion of the total area to the left of $z = 1.30$ is the sum of these two proportions

   $$.4032 + .5000 = .9032$$

   Thus, $X = 58$ corresponds to the 90th percentile rank.

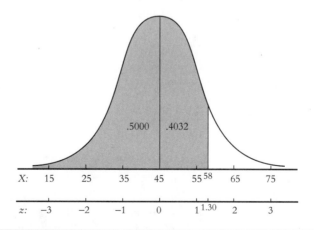

Figure 4-7   Determining the percentile rank of a score of 58 in a normal distribution with $\mu = 45$, $\sigma = 10$.

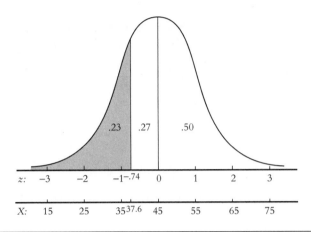

Figure 4-8 Finding the *X* score at the 23rd percentile of a normal distribution with $\mu = 45$, $\sigma = 10$.

2. In the normal distribution described above, what score is at the 23rd percentile? This is the same type of problem as 1, solved in the reverse direction.

a. Draw a picture of the normal distribution, as in Figure 4-8. The question asks for the *X* score corresponding to the 23rd percentile. The *z* score corresponding to P_{23} will have 23% of the cases below it. The third column of Table A, Appendix 2, gives the proportion of cases between *z* and $+\infty$. However, since the distribution is symmetrical, the column also gives the relative frequency between $-z$ and $-\infty$. Therefore, look down the third column until you find .23. The closest figure is .2296, and the *z* corresponding to that area is .74. Since the left side of the distribution is being worked with, the *z* value for the 23rd percentile is $-.74$.

b. Since 23% of the area falls below $z = -.74$, the problem may be solved by converting $z = -.74$ into its corresponding *X* value:

$$z_i = \frac{X_i - \mu}{\sigma}$$

$$-.74 = \frac{X_i - 45}{10}$$

$$X_i = 45 - 7.4$$

$$X_i = 37.6$$

Therefore, $X = 37.6$ represents the 23rd percentile point.

3. What percentage of the cases in a normal distribution falls between $z = -1.00$ and $z = 1.00$?

a. The problem is solved by determining the areas between the mean and $z = 1.00$ and between the mean and $z = -1.00$, and then adding these together. Table A, Appendix 2, shows that the area between the mean and $z = 1.00$ is .3413 of the total. Because the distribution is symmetrical, this is also the area between the mean and $z = -1.00$. Therefore, the total required area is

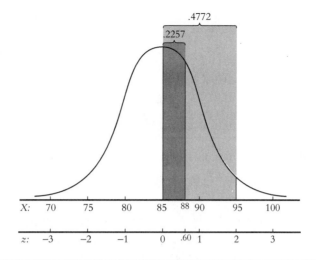

Figure 4-9   Determining the proportion of cases falling between $X = 88$ and $X = 95$ in a normal distribution with $\mu = 85, \sigma = 5$.

$$.3413 + .3413 = .6826$$

and 68.26% of the cases fall between $z = -1.00$ and $z = 1.00$.

4. What proportion of the scores falls between $X = 88$ and $X = 95$ if the normal distribution of X_i has a mean of 85 and a standard deviation of 5?

a. Draw a picture, similar to that presented in Figure 4-9, describing the problem. Locate the mean $\overline{X} = 85$, and shade the desired area between 88 and 95.

b. Convert 88 and 95 to z values:

$$z_i = \frac{X_i - \mu}{\sigma} = \frac{88 - 85}{5} = .60$$

$$z_i = \frac{X_i - \mu}{\sigma} = \frac{95 - 85}{5} = 2.00$$

c. Look in the table under $z = 2.00$ (corresponding to $X = 95$). The area between the mean and this point is .4772. However, the area between the mean and $z = .60$ (corresponding to $X = 88$) is not to be included in the desired answer. Therefore, since this area amounts to .2257, subtract it from the proportion of area between the mean and $z = 2.00$,

$$.4772 - .2257 = .2515$$

to obtain the required proportion.[5]

---

[5] The solution presented for this type of problem assumes that the values of 88 and 95 are exactly 88.000 . . . and 95.000. . . . If not, then their real limits would be 87.5 to 88.5 and 94.5 to 95.5, respectively, in which case the interval of concern would be (87.5–95.5) rather than (88–95).

Problem-Solving Hints Solving the problems described above can be made easier if you will follow a few simple steps that were used in each illustration:

1. Draw a picture of a normal distribution.

2. Put the X scale under the abscissa if X values are given in the problem; put the z scale under the abscissa if z values are given in the problem.

3. Put the other scale (z or X) as an extra abscissa under the previous scale in the manner of Figures 4-7 through 4-9.

4. Locate on the appropriate scale any X or z values given in the problem.

5. Shade in the area under the curve requested by or given in the problem.

6. Convert relevant X scores to z scores or the reverse.

7. Determine relevant areas or z values from Table A, Appendix 2.

8. Use the above information to determine the specific answer to the question (for example, add or subtract appropriate areas).

Samples and the Standard Normal Distribution Because the standard normal distribution is a theoretical distribution, conversions between other normal distributions and the standard normal require μ and σ rather than their corresponding sample statistics, $\overline{X}$ and s. However, the standard normal distribution may be used even if only sample estimates of these population parameters are at hand as long as *each* of two conditions is met:

1. **The population distribution from which the sample is drawn must be normal in form.** One way to examine the tenability of this assumption is to plot the sample distribution and casually observe its general form. If the sample distribution does not depart severely from a normal pattern and there is no reason that it should not be normal, then this condition is satisfied. As Figure 4-5 illustrates, the percentiles of the standard normal distribution are accurate only in the case of a normal distribution.

2. **There must be enough cases in the sample.** A reasonable number of cases is needed for $\overline{X}$ and s to be accurate estimators of μ and σ (and to examine the likelihood that the population distribution is normal—the first condition above). For samples of approximately 30 cases or more, $\overline{X}$ and s are sufficiently good estimators of their respective parameters, but it must be remembered that they are still only estimates or approximations of μ and σ.

If these two conditions are met, one can relate the sample distribution to the theoretical standard normal distribution in the same manner as has been illustrated above for normal population distributions: by using the usual formula but with μ and σ_x replaced by $\overline{X}$ and s_x, respectively. Thus,

$$z_i = \frac{X_i - \overline{X}}{s_x}$$

Although the accuracy of this procedure depends directly on the extent to which the conditions stated above are met, using the standard normal or other theoretical distributions to estimate percentiles in a sample in this manner is a very common procedure.

In fact, most of the procedures of elementary statistical inference use essentially this strategy, as we will see beginning in Chapter 7.

Other Standardized Distributions

The standard normal distribution discussed in the preceding section had a mean of 0 and a standard deviation of 1 because the mean was subtracted from each score and the result divided by the standard deviation. However, the distribution could just as easily be made to have a mean of 100 and a standard deviation of 20 by using the transformation

$$\text{standard score} = 20 \left[\frac{X_i - \overline{X}}{s_x} \right] + 100$$

or a mean of 500 and a standard deviation of 100 by using the transformation

$$\text{standard score} = 100 \left[\frac{X_i - \overline{X}}{s_x} \right] + 500$$

The Army General Classification Test that was used during the Second World War relied on the former standard score, whereas the Scholastic Aptitude Test (SAT) used the latter scale when first developed in the 1940s. The SAT is standardized differently today. In any case, the concept of standard scores is perfectly general and the selection of a mean of 0 and a standard deviation of 1 is convenient but arbitrary.

SUMMARY

Percentiles help locate an individual score within a distribution. The Pth percentile point is the score value below which the proportion P of the cases in the distribution falls. That percentile point is said to have a percentile rank of P.

Scales of measurement are often transformed. The unit of the measurement scale is changed by multiplying or dividing each score by a constant, whereas the origin of the scale is changed by adding or subtracting a constant. Changes in unit and origin directly affect the value of the mean, but only changes in unit affect the variance and standard deviation. Changing the origin and unit of a scale by subtracting the mean then dividing the result by the standard deviation, for example, permits any scale to be transformed into a common standard score scale with mean of 0 and standard deviation of 1.00.

The normal distribution is a theoretical relative frequency distribution that typically has a bell-shaped characteristic when graphed. When its mean is 0 and its standard deviation is 1, it is called the standard normal distribution. Any normal distribution can be transformed into the standard normal distribution, and the tabled percentiles of the theoretical standard normal distribution can be used to estimate the percentage of cases that falls above, below, or between certain points on the scale.

Exercises

For Conceptual Understanding

1. If the mean of a distribution is 50 and the standard deviation is 5, what do the mean, variance, and standard deviation become after each of the following operations?

   a. 3 is added to each score.

   b. 5 is subtracted from each score.

   c. Each score is multiplied by 10.

   d. Each score is divided by 4.

   e. Each score is divided by 2, and then 20 is subtracted from each.

2. If an original set of measurements is made in inches and has a mean of 12 and a variance of 16, what will be the mean, variance, and standard deviation if

   a. the unit is changed to feet?

   b. 2 inches must be added to each measurement to correct for an error?

   c. the measures are converted to centimeters (2.5 centimeters = 1 inch) and then 5 centimeters is subtracted from each score?

3. What are the limitations of percentiles as measures of relative position, and how do standard scores overcome these limitations?

4. Explain and illustrate why it is necessary to consider the variance of a distribution to reflect accurately the relative position of a score.

5. What does the area under a theoretical distribution signify?

6. If your score on a statistics test was 90, and the professor determined grades separately for each of the four sections shown below, which section would you hope was yours? Why?

   a. $\bar{X} = 65$, $s = 13$

   b. $\bar{X} = 75$, $s = 10$

   c. $\bar{X} = 80$, $s = 8$

   d. $\bar{X} = 85$, $s = 2$

For Solving Problems

7. Using the frequency distribution presented in Table 4-1, determine the score values corresponding to the following percentile ranks:

   a. $P_{.45}$ c. $P_{.225}$ e. $P_{.75}$ g. $P_{.925}$

   b. $P_{.20}$ d. $P_{.60}$ f. $P_{.625}$ h. $P_{.50}$

8. Using Table 4-1, determine the percentile ranks corresponding to the following score values:

   a. 51   c. 72   e. 75   g. 84

   b. 58   d. 74   f. 78   h. 97

9. Compute the standard score for each of the members of the following distribution. Then calculate the mean and the standard deviation of the z_i. Would you have anticipated these last two values? Why?

   $X_i = (0, 2, 4, 6, 13)$

10. What proportion of the cases in a normal distribution will fall to the left of $z = 0$? To the right of $z = 0$? To the right of $z = 1.00$? To the right of $z = -1.00$?

11. Determine the proportion of cases falling under the normal curve in the following circumstances:

    a. between $z = -1$ and $z = +1$

    b. between $z = .5$ and $z = 1.5$

    c. to the right of $z = -.75$

    d. to the left of $z = -1.96$ plus to the right of $z = 1.96$

    e. to the left of $z = -2.575$ plus to the right of $z = 2.575$

12. Within how many standard deviation units of the mean would you expect almost all of the scores to fall? What score values would correspond to these points if the mean of a normal distribution is 43 and the standard deviation is 2? Between what two score values would you expect approximately two-thirds of the scores to fall?

13. In a normal distribution with mean 100 and standard deviation 15, at what percentile rank is a score of 70 likely to fall? $X_i = 95$? $X_i = 124$?

14. In a normal distribution with mean 20 and standard deviation 5, what score value is likely to fall at $P_{.925}$? $P_{.166}$?

15. What percentage of the cases is likely to fall between the values of 17 and 21 in a normal distribution with mean 20 and standard deviation 4? Between 21 and 26? Between 19 and 21?

16. Suppose the mean IQ in the population is 100 with a standard deviation of 16. What percentile rank are you if your IQ is

    a. 100?

    b. 116?

    c. 132?

    d. If you must have an IQ of at least 140 to be eligible to join Mensa, a national group of high-IQ individuals, what percent of the population is able to join?

17. Six individuals competed in a mini-triathlon in which they each had to swim 500 meters, bicycle 10 miles, and run 3 miles. Their times (in seconds) in the three events and their total times are given below. Competitor F won the event. But who would have won if performance in the three events were evaluated relative to the other competitors in those events and the events were weighted equally?

Competitor	Swim	Bicycle	Run	Total
A	735	1869	1420	4024
B	785	1814	1393	3992
C	775	1863	1353	3991
D	797	1786	1367	3950
E	791	1817	1435	4043
F	902	1600	1394	3896
$\overline{X}$	797.50	1791.50	1393.67	
s	55.73	98.97	30.88	

18. Given the following distribution of scores:

10	20	33	47	53	67	79	88
10	23	34	49	54	70	80	88
11	23	36	50	55	72	81	91
13	24	36	51	55	75	85	91
17	30	40	53	63	77	86	93

Determine the percentile points for

    a. $P_{.10}$ **b.** $P_{.35}$ **c.** $P_{.90}$ **d.** $P_{.975}$

Determine the percentile ranks for scores of

    e. 24 **f.** 55 **g.** 51 **h.** 91

19. Answer the questions below on the basis of the following data.

X	f	Cum f
5	6	40
4	9	34
3	12	25
2	8	13
1	5	5

    a. Find the 20th percentile point.

    b. What percentile rank does a score of 4 have?

    c. What percentile rank does a score of 2 have?

    d. What score corresponds to the 75th percentile?

20. Student A is told that she has scored at the 90th percentile; student B is told that he has scored at the 45th percentile on a test. Student A exclaims that she has performed twice as well as student B. In what way is student A correct and in what way is she not correct?

21. A teacher wants to determine the final relative standing of each person in a class on the basis of three quizzes and a final exam but wants the final exam to be weighted twice as heavily as each of the quizzes. How can this be accomplished?

22. If a distribution has a mean of 15 and a standard deviation of 5, what happens to $\overline{X}$, s^2, and s if we

    a. add the constant 3 to each score?

    b. multiply each score by 3?

    c. subtract 5 and then divide each score by 5?

    d. add -1 to each score?

    e. multiply each score by -1?

23. Would a score of 30 likely be from a distribution with

    a. $\overline{X} = 50$, $s = 9$?

    b. $X = 50$, $s^2 = 9$?

24. In each of the following, compute z corresponding to each X_i.

  a. $\overline{X} = 26$, $s = 6$, $X_i = 46$

  b. $\overline{X} = 34$, $s = 4$, $X_i = 32$

  c. $\overline{X} = 60$, $s = 8$, $X_i = 70$

  d. $\overline{X} = 80$, $s = 3$, $X_i = 72$

25. The class mean and standard deviation on test A are 60 and 8, respectively, while the class mean and standard deviation on test B are 80 and 8. If a student scores 80 on test A and 92 on test B, which test score is higher relative to the class performance?

26. What would the v, s_v^2, and s_v equal if each X_i in a distribution were transformed to a standardized v_i by the formula

$$v_i = 10 \left[\frac{X_i - \overline{X}}{s_x} \right] + 50$$

27. What proportion of a normal distribution with mean 60 and standard deviation 5 is

  a. between 50 and 70?

  b. greater than 75?

  c. less than 55?

  d. between 40 and 60?

  e. less than 56 or greater than 64?

  f. between 62 and 70?

Regression

n this chapter we will consider procedures that describe the degree of relationship between two variables and the ability to predict a single individual's score on one variable from knowledge of that person's score on the other variable.

For example, an admissions officer at Private College is faced with the task of selecting a freshman class. It is helpful to know whether Scholastic Assessment Test (SAT) scores, for example, are related to grades during the freshman year at Private and, if so, how strong the relationship is. Are SAT scores, for instance, very highly related or only slightly related to freshman grades? If the relationship is strong enough, SAT scores can be used to help select students for admission. In particular, Sara Greenley, an applicant, might have a verbal SAT score of 625. What freshman grade average is she likely to earn if she were to attend Private College?

The question of predicting scores for one variable from scores on another variable is called a problem of **regression.** This term stems from the work of Sir Francis Galton, who studied the relationship between a variety of physical and behavioral characteristics in the nineteenth century. In trying to predict one trait from another, he observed that sons of very tall fathers were tall, but not usually as tall as their fathers. Conversely, sons of very short fathers, while short, were also closer to the average height than their fathers. Such a tendency toward the average is called **regression toward the mean,** which refers to the tendency for people who are extreme on the predictor variable (in this case, height of father) to be less extreme on the trait being predicted (height of son). Of course, exceptions do occur—fathers of average height might have short or tall sons, and short or tall fathers might have exceptionally short or tall sons, respectively—making the total distribution of heights for sons approximately the same as for fathers. In any event, the term *regression* signifies a variety of techniques that describe relationships between variables and that predict one variable from another.

LINEAR RELATIONSHIPS

One of the simplest relationships between two variables occurs when high values on one measure are associated with high values on another variable and low values on one are associated with low values on the other. Drawn on a graph, such a relationship approximates a straight line. Thus, understanding such relationships will be easier after a brief discussion of the algebra and geometry of simple linear relationships.

The Equation for a Straight Line

Suppose two parents want to start their 12-year-old daughter on a baby-sitting career, and hire her to sit with her 3-year-old brother at the rate of a dollar per hour. Consider the relationship between money earned (labeled Y) and hours worked (labeled X). If the baby-sitter does not work at all ($X = 0$), she earns no money ($Y = 0$). If the baby-sitter works two hours ($X = 2$), she makes $2.00 ($Y = 2$). For four hours, she earns $4.00. A table displays these values:

Hours (X)	Dollars (Y)
0	0
2	2
4	4

The basis of this table is the relationship: a dollar earned for each hour worked. Using the symbols Y for dollars earned and X for hours worked, this statement reduces to

$$Y = X$$

This equation will indicate the earnings for any amount of time worked. For example, if the baby-sitter works $3\frac{1}{2}$ hours ($X = 3.5$), then the amount earned is $3.50.

The same information may be expressed by graphing the equation, letting specific pairs of values for X and Y be the coordinates of a set of points. Figure 5-1 shows the X-axis (hours worked), the Y-axis (dollars earned), and the three points from the preceding table [(0, 0), (2, 2), (4, 4)]. The line passing through these points is described by the equation, $Y = X$. The line really represents an infinite number of points, each of which indicates the amount of money earned for a given amount of time worked. For example, if the baby-sitter worked $3\frac{1}{2}$ hours, a vertical line could be drawn at $X = 3.5$, which would intersect the graphed line of the relationship at the height of $Y = 3.50$, as indicated in the figure. Hence, both the equation ($Y = X$) and the graph of that equation in Figure 5-1 describe the relationship between money earned and hours worked and provide a method of predicting one variable (Y) from the other (X).

A **linear relationship** is an association between two variables that may be accurately represented on a graph by a straight line.

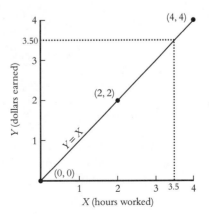

Figure 5-1 Graph of the line $Y = X$.

The relationship in Figure 5-1 is obviously linear—in fact, it is perfectly linear in that all the points fall directly on the straight line. Later in this chapter we will discuss relationships that are essentially linear even though some or all of the points do not fall precisely on the line. Of course, not all relationships are linear; some are **nonlinear.** For example, the relationship between vocabulary and age between one and six years follows an S-shaped pattern, beginning upward slowly, then accelerating, and finally leveling off. Such a nonlinear relationship is said to be **curvilinear.** There are other kinds of nonlinear associations as well. However, we will be concerned only with linear relationships in this text.

The linear relationship in Figure 5-1 is one of the simplest kinds. Other perfect linear relationships can also be described. Suppose after a few successful occasions, the parents mentioned above decide to encourage the daughter and raise her pay to $2 per hour. Once again, for zero hours worked ($X = 0$), pay is 0 ($Y = 0$); for one hour ($X = 1$), earnings are $2 ($Y = 2$); and for three hours, earnings are $6. Putting these values into a table, one obtains

Hours (X)	Dollars (Y)
0	0
1	2
3	6

The table uses the relationship: dollars earned equals $2 times the number of hours worked, which can be expressed by the equation

$$Y = 2X$$

Figure 5-2 illustrates this relationship. Note that the two lines $Y = X$ (Figure 5-1) and $Y = 2X$ (Figure 5-2) are quite similar, differing only in tilt or slope.

> The **slope** of a line is defined to be the vertical distance divided by the horizontal distance between any two points on the line.[1] In symbols, given the two points (x_1, y_1) and (x_2, y_2), the slope b of the line is
>
> $$\text{slope} = b = \frac{\text{vertical distance}}{\text{horizontal distance}} = \frac{y_2 - y_1}{x_2 - x_1}$$

For example, two points located on the line in Figure 5-2 are (1, 2) and (3, 6). Call the first point (x_1, y_1) and the second point (x_2, y_2). Putting these values into the formula for the slope, one obtains

---

[1] Some students define the slope to be rise ÷ run, remembering that in alphabetical order *rise* precedes *run*.

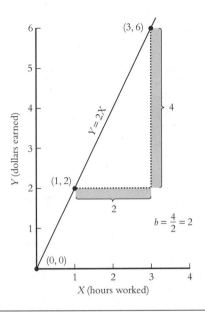

Figure 5-2 Graph of the line $Y = 2X$, indicating the slope of the line, b, to equal 2.

$$b = \frac{y_2 - y_1}{x_2 - x_1} = \frac{6 - 2}{3 - 1} = \frac{4}{2}$$

$$b = 2$$

The derivation of the slope is pictured in Figure 5-2. Recalling that the slope equals the vertical distance between any two points on the line divided by the horizontal distance between the same two points, one can see that this ratio of distances is $\frac{4}{2} =$ 2 for the points (1, 2) and (3, 6). Of course, this ratio will be 2 for *any* two points on the line.

It is not just coincidence that the slope of this line equals 2 and the coefficient of X in the line's equation $Y = 2X$ is also 2. The coefficient of X in an equation of this sort is identical to the slope of the line that the equation describes. In the first illustration, $Y = X$, the coefficient of X is 1 [that is, $Y = (1)X$]. An examination of Figure 5-1 reveals that two points on the graph are $(x_1, y_1) = (2, 2)$ and $(x_2, y_2) = (4, 4)$, yielding a slope of 1:

$$b = \frac{4 - 2}{4 - 2} = \frac{2}{2} = 1$$

The slope of a line may be negative as well as positive. Consider the line in Figure 5-3. Two points on that line, $(x_1, y_1) = (2, 3)$ and $(x_2, y_2) = (4, 2)$, can be substituted into the formula for the slope, b, to yield

$$b = \frac{y_2 - y_1}{x_2 - x_1} = \frac{2 - 3}{4 - 2} = \frac{-1}{2}$$

$$b = -.50$$

To get from the point (2, 3) to (4, 2) in the graph, one must go *down* one unit, equivalent to going -1 vertical units, and to the right two units ($+2$), giving a slope of

$$b = \frac{-1}{2} = -.50$$

The slope of the line signifies whether the relationship is positive or negative.

> A line with a positive slope indicates a **positive** or **direct relationship** and a line with a negative slope represents a **negative** or **inverse relationship.**

Notice that lines with positive slope tend to run upward from left to right on the graph, indicating that low values on one variable are associated with low values on the other, and high values with high values. In the baby-sitting example, few hours worked earn few dollars and many hours worked earn many dollars. Since the slope in such a situation is positive, the relationship is called *positive;* it may also be referred to as *direct* because of the direct correspondence of values on the two dimensions. In contrast, if the slope is negative, the line tends to run downward from left to right on the graph, indicating that low values on one dimension go with high values on the other. Because the slope is negative in these instances, the relationship is called *negative;* it is sometimes also called *inverse* because *high* values go with *low* values.

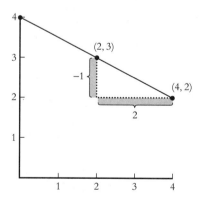

Figure 5-3 Graph of a line with negative slope, $b = -\frac{1}{2} = -.50$. Note that low values of *X* are associated with high values of *Y,* an inverse or negative relationship.

As a final example of a linear relationship, suppose the young girl, after much experience at home, decides to baby-sit for other parents. Her father, a businessman, suggests that she charge only $2 per hour but that she also ask for a $5 "appointment fee" for reserving the date. The use of the appointment fee, which is paid in addition to the hourly rate, means the girl gets something even if a parent cancels an appointment. Also, it gives her a minimum amount in the event that she reserves the whole evening for someone who wants her to baby-sit for only an hour or so. Specifically, if a parent makes an appointment and then cancels it, the girl works no hours ($X = 0$) but earns $5. If the date is completed and the girl sits for 1 hour ($X = 1$), she makes $1 \times \$2$ plus the $5 appointment fee, for a total of $7. A two-hour evening would net $2 \times \$2$ plus $5 = \$9$. In tabular form,

Hours (X)	Dollars (Y)
0	5
1	7
2	9

These values were arrived at by multiplying $2 by the number of hours worked and adding $5, which can be expressed by the general equation

$$Y = 2X + 5$$

The plot of this equation, together with the plots shown in Figures 5-1 and 5-2, are presented in Figure 5-4. Observe first that the coefficients of X indicate that the line $Y = X$ has a slope of 1, while the other two lines both have a slope of 2 ($Y = 2X$ and $Y = 2X + 5$). On the graph, the line for ($Y = X$) clearly has a different tilt or slope, whereas the two lines that have identical slopes of 2 are parallel. But the two parallel lines differ in that one is always five units above the other. If the Y-axis is arbitrarily selected as the place to measure this separation, the line $Y = 2X + 5$ goes through the Y-axis at 5, whereas the line $Y = 2X$ goes through the Y-axis at 0.

> The point at which a line intersects the *Y*-axis is called the **y-intercept,** and its value is symbolized by ***a.***

Note that the value of the *y*-intercept, or *a,* equals the constant that is added to the equation, as follows:

Equation	*y*-intercept
$Y = 2X + 5$	$a = 5$
$Y = 2X + 0$	$a = 0$

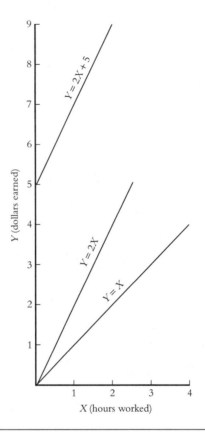

Figure 5-4 Plots of three lines that have different slopes and intercepts.

The y-intercept can be found from the equation of the line by setting $X = 0$, which is, in fact, the point on the X scale where the Y-axis is located. Thus, to compute the y-intercept:

$$Y = 2X + 5$$
$$= 2(0) + 5$$
$$Y = +5 = y\text{-intercept}$$

In summary, one needs to know two values to determine the equation of a straight line: the y-intercept (a) and the slope (b). Further, since a provides one point and b tells one how to locate some other point on the line, a and b completely specify a particular line. Therefore,

Any straight line will have the general equation of

$$Y = bX + a$$

in which a is the y-intercept and b is the slope of the line.

If $a = 5$ and $b = 2$, then the equation

$$Y = 2X + 5$$

describes the line with slope 2 and y-intercept at 5.

The usefulness of the equation of a linear relationship is that it allows one to determine a Y value given any X value. Suppose the appropriate equation is $Y = 2X + 5$, and one wishes to know the Y value corresponding to an X of 4. Substituting $X = 4$,

$$Y = 2X + 5$$
$$= 2(4) + 5$$
$$= 8 + 5$$
$$Y = 13$$

Thus, in the baby-sitting example, if the parents are to be gone for four hours, the girl would earn $13.

The baby-sitting examples illustrate that graphs of linear relationships have a slope and a y-intercept and that they can provide the basis for predicting a Y value given a certain X value. However, all these cases have dealt with perfect linear relationships—that is, all of the points have fallen precisely on the line. None has deviated from it. Most relationships observed in the social and behavioral sciences are not so precise.

A plot of an approximate linear relationship is presented in Figure 5-5. The relationship does have a positive linear trend, as indicated by the straight line drawn through it, but this line obviously is only an approximation. Yet it still might be useful to have its equation to make approximate predictions of Y given X. Suppose this plot represents a relationship between Scholastic Assessment Test (SAT) scores and freshman college grades at a highly selective college in a certain year. If the equation of the line were known, educated guesses concerning the scholastic performance of next year's

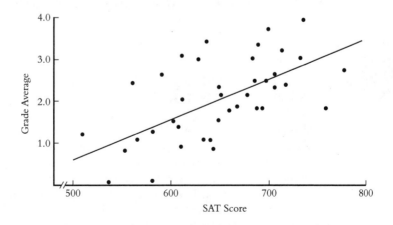

Figure 5-5 A scatterplot of the SAT scores and freshman grades shown in Table 5-1, which shows the underlying relationship to be linear.

applicants could be made on the basis of their SAT scores before they actually get to college and earn grades. Of course, this strategy assumes that the relationship will be much the same from year to year.

Scatterplots

Previously in this chapter, the equation of the line has been rather obvious, but this is not the case when the relationship is not perfect. In fact, it may not be immediately clear whether the underlying relationship is essentially linear or nonlinear. Since it is inaccurate to use a straight line to describe a nonlinear relationship, a necessary first step is to check whether the relationship is indeed linear. This can be done by constructing a scatterplot.

> A **scatterplot** is a graph of a collection of pairs of scores.

Suppose that a college admissions officer has a group of 40 students with the SAT scores and first-year college grades ($A = 4.0$) presented in Table 5-1. This collection of pairs of scores may be treated in the same manner as the small tables of values described in the previous section, and a point may be placed on a graph corresponding to each pair of values. Figure 5-5 is an example of a scatterplot constructed simply by recording a point corresponding to each pair of scores shown in Table 5-1. The variable about which predictions are to be made (the predicted variable, in this case, freshman grades) is always placed on the ordinate. Thus, to plot the first pair of scores (510, 1.3) in Table 5-1, a point is placed directly over 510 on the SAT scale and precisely to the right of a grade average of 1.3.

Although statistical techniques exist that help to make a decision about linearity, for our purposes an observation of the total scatterplot is usually sufficient to determine whether the trend in the data is approximately linear or nonlinear.

REGRESSION CONSTANTS AND THE REGRESSION LINE

Obviously, if a scatterplot tends to be nonlinear, trying to describe it with a straight line would hardly be appropriate. If, however, it is approximately linear, the next task is to determine the equation of the line that best describes the linear relationship between the two observed variables. It is useful to do this even though most or even all the points will not fall precisely on the line that is determined for the scatterplot. Researchers often assume that the line they obtain actually reflects an estimate of the true relationship between the two variables and that the points do not fall on the line because of variability—individual differences, measurement error, and unreliability. For example, in Figure 5-5, the line drawn may be thought of as an estimate of the "real" underlying relationship between grades and SAT scores for students at this

5-1 SAT Scores and Grade Averages for a Sample of 40 College Students

SAT Score	Freshman Grades	SAT Score	Freshman Grades
510	1.3	659	2.1
533	.1	663	1.7
558	.8	670	1.8
565	2.4	678	2.1
569	1.1	679	2.9
580	.1	680	2.3
581	1.3	687	1.8
590	2.6	688	3.2
603	1.5	693	1.8
610	1.3	698	2.4
612	.9	700	3.6
615	2.0	710	2.5
618	3.0	710	2.2
630	2.9	718	3.1
633	1.1	724	2.3
639	3.3	734	2.9
643	1.1	739	3.8
645	.9	750	3.9
651	1.5	767	1.7
654	2.2	778	2.6

college. Students do not all fall exactly on the line because, for example, some are more motivated than others, a few grades may not accurately reflect the performance of some students, and temporary circumstances (e.g., cramming the night before for another course) might have influenced final exam performance for some students in some courses. Therefore, the line describing a linear relationship might actually reflect a true relationship even though the points do not fall precisely on the line because of variability.

Earlier in this chapter, the equation for a straight line, $Y = bX + a$, was given for the case in which a perfect linear relationship existed between two available sets of measures. How can we determine the equation for the line—that is, how can we determine the values of a and b—if the relationship is linear but not perfect? And suppose we want to use the observed relationship between X and Y to predict a person's grade average (Y score) when all we know is his or her SAT (X) score. This prediction is the fundamental task of regression, and the vocabulary of straight lines changes a bit in this new context.

> The line describing the relationship between two variables is known as the **regression line** and is expressed in the form, $\hat{Y} = bX + a$. The variable $\hat{Y}$ is called **Y predicted.** The values **a** (the y-intercept) and **b** (slope) are known as **regression constants.**

The task, then, is to specify the regression constants for the regression line that best describes the relationship. The criterion used to determine which line of the many possible lines is best is the **least squares criterion.**

To examine the meaning of that concept it is necessary to observe a scatterplot in more detail. For purposes of illustration, the following few subjects have been selected from the original sample of 40 students shown in Table 5-1. Their SAT scores (X_i) and grade averages (Y_i) are as shown.

SAT Score (X_i)	Grades (Y_i)
510	1.3
533	.1
603	1.5
670	1.8
750	3.9

A scatterplot of these five points and two possible regression lines are presented in Figure 5-6.

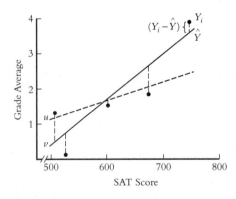

Figure 5-6 Plot of a selected group of points with possible regression lines, *u* and *v*.

Now suppose the task is to select the straight line that best describes the linear relationship in Figure 5-6. If all the points fell precisely on a line, we could specify the line for this relationship by writing its equation using the techniques described at the beginning of this chapter. But the relationship in Figure 5-6 is not perfect; the points do not all fall on a line. The line that best describes an imperfect linear relationship among the points is the one for which the deviations of the points from the line are at a minimum.

In algebraic symbols, the "best" line is represented by $\hat{Y}$ (read "Y hat" or "Yp-redicted") and the actual data points are represented by Y_i. So the deviation of a point from the line is symbolized by $(Y_i - \hat{Y})$. The strategy, then, is to find the line $\hat{Y}$ that minimizes these deviations for the entire set of points. In a sense, the "best" line is the one that is closest to the entire set of points.

The approach to determining the regression line can be understood in the context of using the line to predict grade average from SAT score in our example. Consider the last score in the table of five students, $X = 750$. If we know the line that describes the relationship between SAT scores and grades, we could predict this person's grade average. Graphically, we would predict the point on the line directly over $X = 750$. You can see it pictured for line v (the solid line) in Figure 5-6. That value is $\hat{Y}$. But we know from the table what grade average this student actually received: $Y_i = 3.9$. Now the difference between the person's actual and predicted score from the line is given by $(Y_i - \hat{Y})$, which represents the error or residual in predicting $\hat{Y}$ as this student's grade average. So the task is to select a line that minimizes this error over all cases in the scatterplot.

To illustrate, consider the two possible lines, u and v, in Figure 5-6. It is obvious that, overall, the points cluster more closely about line v than about line u, so line v better characterizes the linear relationship—it results in less error in predictions—than does line u.

The simple distances between points and the line, $(Y_i - \hat{Y})$, could be used as measures of the deviations or error between points and the line. However, if these simple distances are used, some will be positive and some will be negative numbers; if such distances were added for all points, the positive and negative values would cancel each other out, leaving the choice of the best line unresolved. But if these error distances are squared, such cancellation will not happen. In fact, the situation is very similar to that of the mean described in Chapter 3 (pages 60–62). Recall that the sum of squared deviations of scores about their mean was a minimum—that is, less than about any other value. This made the mean closest to all the scores in a least squares sense. Now we seek a regression line that describes a linear relationship such that the sum of squared deviations of scores from the line is a minimum, making the line closest to all the scores in a least squares sense. Thus, one picks the regression line, then, so that $\Sigma(Y_i - \hat{Y})^2$ is a minimum.

The **least squares criterion** for determining the regression line requires that the sum of the squared deviations between points and the regression line, $\Sigma(Y_i - \hat{Y})^2$, is a minimum.

Since the regression line is completely determined if the y-intercept (a) and slope (b) are known, the task of selecting the straight line that best describes an imperfect linear relationship reduces to the selection of these regression constants in accordance with the least squares criterion.

Obtaining an Equation for the Regression Line

Earlier in this chapter the process of determining the equation of a straight line when two points on the line were known was described. The current task is different. Now we are required to determine a and b when we do not know what the line should be and when we do not have two points that we know lie exactly on the line. We need formulas that use the pairs of scores that are available in the data (X_i, Y_i) to provide values for a and b that determine a line that satisfies the least squares criterion. These formulas for a and b, which can be found by calculus, are

$$b = \frac{N(\Sigma XY) - (\Sigma X)(\Sigma Y)}{N\Sigma X^2 - (\overline{X})^2}$$
$$a = \overline{Y} - b\overline{X}$$

The expressions provide a method for computing a and b from the original scores such that when these constants are placed into the regression equation, $\acute{Y} = bX + a$, it will describe the line that best fits the data in the sense of minimizing the squared deviations between the data points and the line [that is, $\Sigma(Y_i - \acute{Y})^2$ will be a minimum].

The use of these formulas will be illustrated with the data provided by the five subjects selected previously. Of course, the admissions director mentioned in the original illustration would use the data from as many college students as possible. However, using just these few subjects will provide the reader with a simple example of the computational routine required. The procedures are presented in Table 5-2. An examination of the formulas reveals that five quantities are needed:

$$N, \Sigma X, \Sigma X^2, \Sigma Y, \text{ and } \Sigma XY$$

These quantities are determined directly from the raw data in the top portion of Table 5-2. Note first that N is the number of subjects, which is the same as the number of X scores, the number of Y scores, and the number of *pairs* of scores. Next, be sure to distinguish between ΣX, which is the sum of the X_i scores; $(\Sigma X)^2$, which is the square of this sum; and ΣX^2, which is the sum of the squared X_i scores. In the last column one determines the XY, which are called **cross products,** because they are the products of each X_i with its corresponding Y_i. This gives the cross products (X_1, Y_1), (X_2, Y_2), (X_3, Y_3), etc., which are then summed to give ΣXY.

In the second section of Table 5-2, two intermediate quantities, $(\mathbf{I_{XY}})$ and $(\mathbf{II_X})$, are calculated. It is necessary in $(\mathbf{I_{XY}})$ to distinguish between the sum of cross products, ΣXY, and the product of the sums of the two variables, $(\Sigma X)(\Sigma Y)$. In $(\mathbf{II_X})$, distinguish between the sum of squared scores, ΣX^2, and the square of the sum of scores, $(\Sigma X)^2$.

In the third section of Table 5-2, the regression constants are determined by using the two intermediate quantities, $(\mathbf{I_{XY}})$ and $(\mathbf{II_X})$, and the two means, $\overline{X}$ and $\overline{Y}$. The

5-2 Computation of the Regression Equation

Subject	X	Y	X^2	XY
1	510	1.3	260,100	663.0
2	533	.1	284,089	53.3
3	603	1.5	363,609	904.5
4	670	1.8	448,900	1206.0
5	750	3.9	562,500	2925.0
$N = 5$	$\Sigma X = 3066$	$\Sigma Y = 8.6$	$\Sigma X^2 = 1{,}919{,}198$	$\Sigma XY = 5751.8$
	$(\Sigma X)^2 = 9{,}400{,}356$			

Intermediate Quantities

$$(\mathbf{I_{XY}}) = N(\Sigma XY) - (\Sigma X)(\Sigma Y)$$
$$= 5(5751.8) - (3066)(8.6)$$
$$(\mathbf{I_{XY}}) = 2391.4$$

$$(\mathbf{II_X}) = N\Sigma X^2 - (\Sigma X)^2$$
$$= 5(1{,}919{,}198) - 9{,}400{,}356$$
$$(\mathbf{II_X}) = 195{,}634$$

Regression Constants

$$\overline{X} = \Sigma X/N = 3066/5 = 613.20 \qquad \overline{Y} = \Sigma Y/N = 8.6/5 = 1.72$$

$$b = \frac{N(\Sigma XY) - (\Sigma X)(\Sigma Y)}{N\Sigma X^2 - (\Sigma X)^2} = \frac{(\mathbf{I_{XY}})}{(\mathbf{II_X})} = \frac{2391.4}{195{,}634} = .012$$

$$a = \overline{Y} - b\overline{X} = 1.72 - (.012)(613.20) = -5.638$$

$$\dot{Y} = bX + a = .012X - 5.638$$

last step is to substitute the resulting constants into the regression equation, which yields

$$\hat{Y} = .012X - 5.638$$

With the regression equation obtained, one may answer the question: If applicant Sara Greenley's SAT score is 625, what would one predict her freshman grade average to be? The question is answered by substituting $X = 625$ into the regression equation and computing $\hat{Y}$:

$$\hat{Y} = .012(625) - 5.638$$
$$\hat{Y} = 1.862$$

On the graph, this is equivalent to drawing a vertical line through $X = 625$ parallel to the Y-axis and determining the Y value at the point of intersection of that vertical line and the regression line. This point of intersection is $\hat{Y}$ for $X = 625$.

The several computations described above may be performed by many small calculators designed for scientific purposes and by statistical programs for personal or larger computers. Typically, you are required to enter X_1 followed by Y_1, then X_2 followed by Y_2, and so on for all pairs of scores. When all data have been entered, a and b can be obtained with the push of one or two buttons or some other simple command. Some programs will even calculate the predicted value for a specific X score. Read the instruction manual for your particular calculator or program, and note that sometimes m rather than b is used to symbolize the slope and the word "constant" may be used to indicate a, the y-intercept. Procedures to use the StataQuest, MINI-TAB, and SPSS programs on a personal computer are given in the *Study Guide* that accompanies this text.

Plotting the Regression Line

It is sometimes necessary to plot a regression line once you have obtained its algebraic equation. Two points determine a straight line, and the regression constants, a and b, will provide those two points. First, mark off a set of axes. Remember, the variable doing the predicting (variable X) goes on the horizontal axis (X-axis), and the variable being predicted (variable Y) goes on the vertical axis (Y-axis). Suppose the regression equation is $\hat{Y} = .75X + 2.00$. The regression constants then are $a = 2.00$ and $b = .75$. Since the value of a is the y-intercept, simply place a point on the Y-axis at the point corresponding to the value calculated for a. In this case $a = 2.00$, which would be a point on the ordinate 2.00 units above 0. If $a = -2.00$, the point would be on the ordinate but 2.00 units below 0.

Now the task is to determine any second point on the line. Recall that the slope represents the amount of vertical distance relative to horizontal distance between any two points. Thus, start with the point a that you just placed on the Y-axis and move up vertically b units. From there, move horizontally 1 X unit to the right. That is, the slope is .75, so move up .75 unit on the Y-axis for every 1 unit you move right on the X-axis. If the slope is $-.75$, move down .75 unit on the Y-axis for every 1 unit you move to the right on the X-axis.

The second point on the regression line may be determined more simply if the means of the X and Y distributions are known, because $(\overline{X}, \overline{Y})$ is always a point on the line. Thus, $(\overline{X}, \overline{Y})$ and a are two points on the line. Figure 5-7 illustrates how to plot a regression line from its equation.

Equation $\hat{Y} = .75X + 2.00$

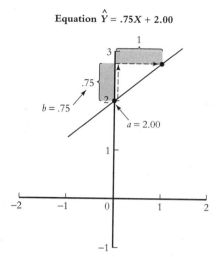

Equation $\hat{Y} = -.75X - 2.00$

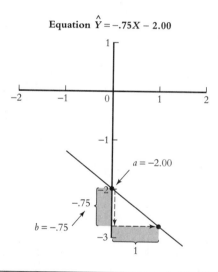

Figure 5-7 Plotting a regression line from the regression equation. At the top, the regression equation is $\hat{Y} = .75X + 2.00$, in which $a = 2.00$ and $b = .75$. Locate one point of the regression line on the vertical axis at $a = 2.00$. Determine the second point by moving up .75 unit ($b = .75$) and then to the right 1 unit. Then draw the regression line by connecting the two points. At the bottom, the regression equation is $\hat{Y} = -.75X - 2.00$, in which $a = -2.00$ and $b = -.75$. Again, locate one point on the vertical axis, but this time at $a = -2.00$. Determine the second point by moving *down* .75 unit (because b is negative; $b = -.75$) and then to the right 1 unit. Connect the two points.

Linearity

The procedures described in the preceding sections are appropriate for data having an underlying linear relationship in the population. A linear relationship probably exists in the population when the form of the scatterplot of the sample data approximates a straight line. Obviously, if it is doubtful that a linear relationship exists, it is foolish to attempt to fit a straight line to data that would be more appropriately described by some curvilinear trend. Techniques exist that can describe curvilinear relations, but they are beyond the scope of this text.

The Range of *X*

When predicting *Y* from a specific *X* score, one must be careful to restrict predictions to those *X* values that fall within the range of *X* values available in the original data. It is this original set of data that one uses to determine whether the relationship is linear. Since one's confidence in the assumption of linearity is based only upon the original range of *X* values, predicting for more extreme X_i exceeds the information on the linearity of the relationship. Consider an example in which rats were placed under food deprivation for 0 to 48 hours prior to performing ten trials in a two-choice problem in which they were rewarded with food for correct choices. The solid line in Figure 5-8 represents the hypothetical relation between hours of deprivation and percentage of correct responses. The equation of this hypothetical line is

$$\hat{Y} = .015X + .02$$

To predict performance for 31 hours of deprivation, simply substitute $X = 31$ into the equation, obtaining a predicted figure of 48.5% correct responses. If $X = 72$,

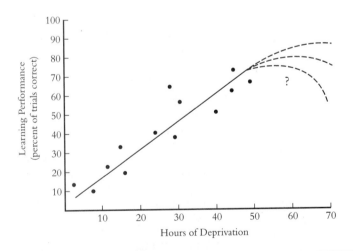

Figure 5-8   A hypothetical relationship between hours of deprivation and learning performance (solid line) and the ambiguity of the form of the relationship for *X* values outside the tested range (dashed lines).

which is outside of the range of original X values, the regression equation predicts that 110% of the trials would be correct! Clearly, the relationship between hours of deprivation and learning performance is not linear over such an extensive range of values. With continued deprivation, eventually the animal will die. Therefore, it is usually appropriate to predict only from X values falling within the range of the original values, because information on the linearity of the relationship at more extreme values may not be available.

The Second Regression Line

All the discussion so far has been directed at predicting Y from X. It would appear that if two variables are measured on each person in a group, the relationship between them should be the same regardless of whether one predicts Y from X or X from Y. It is true that the *degree* of relationship is the same regardless of the direction of prediction, as will be demonstrated in the next chapter on correlation. However, the values of the regression constants will be different depending upon which variable is being predicted. This fact becomes more obvious if one considers the SAT–grades example in terms of the y-intercept. For predicting grades from SAT scores, the y-intercept of the line was measured in the units of the predicted variable, namely, -5.638 grade points. If the direction of prediction is reversed, making SAT scores the predicted dimension and placing those scores along the ordinate (Y-axis), then the y-intercept for the regression line should be in SAT-score points, not grade points. Obviously, the value of a will be different in these two cases because the measurement scale constituting the Y-axis is different.

The fact that the regression constants are different, depending upon which variable is being predicted, may be understood further by examining the manner in which a and b were derived. When predicting grades from SAT scores, $\Sigma(\text{grade} - \text{predicted grade})^2$ is minimized. In contrast, when SAT scores are being predicted, the $\Sigma(\text{SAT} - \text{predicted SAT})^2$ must be minimized. It happens that a line that minimizes one of these sums does not necessarily minimize the other sum. Therefore, there are usually two regression lines, one for predicting Y and one for predicting X. However, if a relationship is perfectly linear, so that all the points fall on the regression line, or if both the X and Y distributions are in standard scores (z), there is only one regression line.

The formulas given above for a and b were for predicting Y from X. This case is sometimes referred to as the **regression of Y on X**. In contrast, when X is predicted from knowledge of Y, it is called the **regression of X on Y**. Although different formulas exist for calculating the regression constants for the second regression line, they are rarely used. All you need to remember is that for any computational situation the variable being predicted is labeled Y and the variable doing the predicting is labeled X. Then, use the formulas presented here for the regression of Y on X.

STANDARD ERROR OF ESTIMATE

The regression procedures described above determine the straight line that predicts Y from X. Except in rare cases, the actual data points will not all fall precisely on the

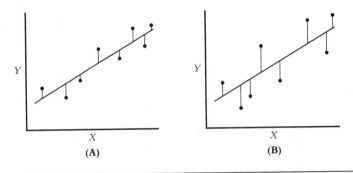

Figure 5-9   Two regression lines with different amounts of error in prediction. Graph B has larger errors than does graph A.

regression line—that is, there will be some error in estimating Y scores. The regression line is selected to minimize such error, but some error almost always exists.

Sometimes the scatterplots for two sets of data show regression lines that are essentially the same, but with a larger amount of variability of points about the regression line, or error, in one plot than in the other. Figure 5-9 portrays such a case. The points in graph A cluster more closely around the line than do those in graph B. Therefore, the sum of the squared deviations between the points and the line—the squared errors of prediction—will be less in graph A than in B. As a result, the typical prediction of Y for an individual X score will be more accurate in A than in B. The standard error of estimate is a measure of the amount of error in prediction.

> The **standard error of estimate,** symbolized by $s_{y \cdot x}$ (read "*s y* dot *x*"), is defined to be the square root of the sum of squared deviations of Y_i about the regression line ($\hat{Y}$) divided by ($N - 2$). In symbols,
>
> $$s_{y \cdot x} = \sqrt{\frac{\Sigma(Y_i - \hat{Y})^2}{N - 2}}$$

The standard error of estimate is abbreviated $s_{y \cdot x}$ to distinguish it from the standard deviation of the Y distribution, which is written s_y (s_x for the X distribution), and to indicate that it is appropriate for the regression of Y on X. Consider the logic of this index.

First, recall the discussion in Chapter 3 of variability of points about the mean. In that case, an index of variability about a constant value, the mean, was developed by taking the sum of the squared deviations of points about that mean and dividing them by $N - 1$ to obtain the variance and its square root, the standard deviation. The quantity $N - 1$ represents a concept called degrees of freedom, which will be explained later. The formula for the variance, then, is

$$s_Y = \sqrt{\frac{\Sigma(\text{score} - \text{mean})^2}{\text{degrees of freedom}}} = \sqrt{\frac{\Sigma(Y_i - \bar{Y})^2}{N - 1}}$$

The present task is similar, except that we want the variability of scores about the regression line, $\hat{Y}$, instead of about the mean, $\overline{Y}$. Further, for the variance, we divided by $N - 1$ (its degrees of freedom) rather than by N to make it a better estimate of the population variance. The standard error of estimate also is made a better estimator of its population value if it is divided by its degrees of freedom, which are $N - 2$. So the definitional formula for the standard error of estimate is

$$s_{y \cdot x} = \sqrt{\frac{\Sigma(\text{score} - \text{regression line})^2}{\text{degrees of freedom}}} = \sqrt{\frac{\Sigma(Y_i - \hat{Y})^2}{N - 2}}$$

The definitional formula above shows algebraically that the standard error of estimate is roughly a standard deviation of the Y_i about the regression line rather than about the mean. This concept is illustrated graphically in Figure 5-10, which presents a theoretical scatterplot containing an uncountable number of cases. The cases represent the range of combinations of X_i and Y_i values, shown simply by an oval circumscribing their location on the plot. If we viewed the scatter three-dimensionally, as if each case were a checker piled up on the paper at its proper location, and if we were to cut that scatterplot at the three X values shown in Figure 5-10 and then looked at the cross-sectional distributions of scores at those three points, we might have the distributions drawn in the figure. These three distributions represent three of the uncountable number of possible distributions that could be conceived—one for each value of X.

Each of these distributions has a mean and a standard deviation. Notice in Figure 5-10 that the means of the three distributions drawn there fall close to, but not exactly on, the regression line. In a sense, the regression line is a single formula that approximates the central tendency of each of the uncountable number of such possible distributions. Similarly, the standard error of estimate is a single value that approximates the variability of the distributions of Y at each value of X. Specifically, the standard error of estimate will be close, but probably not identical, to the standard deviation of the Y for any value of X.[2]

Computing the Standard Error

The expression for the standard error given above is a definitional formula. To use it for computation would be tedious, because a $\hat{Y}$ would need to be calculated for each value of X. Therefore, an alternative formula is used that will give the identical result but that is easier to compute, despite its forbidding appearance:

$$s_{y \cdot x} = \sqrt{\left[\frac{1}{N(N-2)}\right]\left[N\Sigma Y^2 - (\Sigma Y)^2 - \frac{[N\Sigma XY - (\Sigma X)(\Sigma Y)]^2}{N\Sigma X^2 - (\Sigma X)^2}\right]}$$

---

[2] Actually, $s_{y \cdot x}$ will overestimate the variability for X values near the mean but underestimate the variability for X values that deviate substantially from the mean. A special formula exists that can adjust the standard error for this bias. If X_p is the predictor X_i score, then the estimated standard error of $\hat{Y}$ for X_p is

$$\text{Adjusted } s_{y \cdot x} \text{ at } X_p = s_{y \cdot x} \sqrt{1 + \frac{1}{N} + \frac{(X_p - \overline{X})^2}{(N-1)s_x^2}}$$

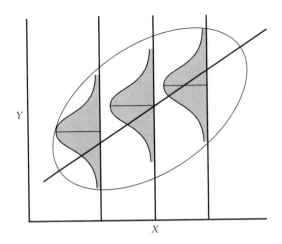

Figure 5-10 Scatterplot with distributions plotted at three values of X.

The computation of $s_{y \cdot x}$ as well as the regression constants are summarized in Table 5-3, which contains the same data and computations introduced previously in Table 5-2 to determine a and b plus those additional quantities needed to compute $s_{y \cdot x}$. The additions to Table 5-2 found in Table 5-3 include an extra column for Y^2, a new intermediate quantity **(III$_Y$)**, and the computation of s_x, s_y, and $s_{y \cdot x}$.

Using the Standard Error

The standard error of estimate is a measure of the amount of error that might be expected when using the regression line to predict Y for a specific value of X. For example, recall that the regression line for the five cases in Tables 5-2 and 5-3 is

$$\hat{Y} = .012X - 5.638$$

and for Sara Greenley, whose SAT score was 625, we predicted a freshman grade average of

$$\hat{Y} = .012(625) - 5.638$$
$$\hat{Y} = 1.862$$

The fact that the standard error for this regression is .77 says that the standard deviation of the freshman grades for individuals who have an SAT score of 625, which includes Sara, is approximately .77. If it can be assumed that the distribution of grade averages at SAT $= 625$ is normal in form, then we could use the percentiles of the standard normal distribution. Recall that approximately 68% of the cases in a normal distribution fall between plus and minus one standard deviation of the mean. Therefore, while one would predict a specific grade average of 1.862 for Sara, 68% of the time the interval between $1.862 - .77 = 1.092$ and $1.862 + .77 = 2.632$ will contain Sara's actual grade average. Similarly, in a normal distribution, 95% of the cases lie

5-3

Computation of the Standard Error of Estimate (Data from Table 5-2)

Subject	X	Y	X^2	Y^2	XY
1	510	1.3	260,100	1.69	663.0
2	533	.1	284,089	.01	53.3
3	603	1.5	363,609	2.25	904.5
4	670	1.8	448,900	3.24	1206.0
5	750	3.9	562,500	15.21	2925.0

$N = 5$ $\quad \Sigma X = 3066 \quad \Sigma Y = 8.6 \quad \Sigma X^2 = 1,919,198 \quad \Sigma Y^2 = 22.4 \quad \Sigma XY = 5751.8$

$(\Sigma X)^2 = 9,400,356 \quad (\Sigma Y)^2 = 73.96$

Intermediate Quantities

$$(\text{I}_{xy}) = N(\Sigma XY) - (\Sigma X)(\Sigma Y)$$
$$= 5(5751.8) - (3066)(8.6)$$
$$= 2391.40$$

$$(\text{II}_x) = N\Sigma X^2 - (\Sigma X)^2$$
$$= 5(1,919,198) - 9,400,356$$
$$= 195,634$$

$$(\text{III}_y) = N\Sigma Y^2 - (\Sigma Y)^2$$
$$= 5(22.4) - 73.96$$
$$= 38.04$$

(continued)

5-3

Continued

Statistical Computations

$$\bar{X} = \Sigma X/N = 3066/5 = 613.20 \qquad \bar{Y} = \Sigma Y/N = 8.6/5 = 1.72$$

$$s_x^2 = (\mathbf{II_x})/N(N-1) = 195,634/5(4) = 9781.7 \qquad s_y^2 = (\mathbf{III_y})/N(N-1) = 38.04/5(4) = 1.90$$

$$s_x = \sqrt{s_x^2} = \sqrt{9781.7} = 98.90 \qquad s_y = \sqrt{s_y^2} = \sqrt{1.90} = 1.38$$

$$b = \frac{N(\Sigma XY) - (\Sigma X)(\Sigma Y)}{N\Sigma X^2 - (\Sigma X)^2} = \frac{(\mathbf{I_{XY}})}{(\mathbf{II_x})} = \frac{2391.40}{195,634} = .012$$

$$a = \bar{Y} - b\bar{X} = 1.72 - (.012)(613.20) = -5.638$$

$$\hat{Y} = bX + a = .012X - 5.638$$

$$s_{y \cdot x} = \sqrt{\left[\frac{1}{N(N-2)}\right]\left[N\Sigma Y^2 - (\Sigma Y)^2 - \frac{[N(\Sigma XY) - (\Sigma X)(\Sigma Y)]^2}{N\Sigma X^2 - (\Sigma X)^2}\right]}$$

$$= \sqrt{\left[\frac{1}{N(N-2)}\right]\left[(\mathbf{III_y}) - \frac{(\mathbf{I_{XY}})^2}{(\mathbf{II_x})}\right]}$$

$$= \sqrt{\left[\frac{1}{5(5-2)}\right]\left[38.04 - \frac{[2391.4]^2}{195,634}\right]} = \sqrt{\left[\frac{1}{15}\right][8.8079]}$$

$$s_{y \cdot x} = .77$$

between -1.96 and $+1.96$ standard deviations (z score units) of the mean. So the interval between her predicted average (1.862) minus $1.96 s_{y \cdot x}$ and her predicted average (1.862) plus $1.96 s_{y \cdot x}$, or between $1.862 - 1.96\,(.77) = .3528$ and $1.862 + 1.96\,(.77) = 3.3712$, will contain Sara's actual score in 95% of such cases.

SUMMARY

Predicting one variable from another is called a problem of regression. Such prediction implies a relationship between the two variables, and typically this relationship is assumed to be linear. A linear relationship is an association between two variables that may be accurately represented on a graph by a straight line. Such a line has a slope, or tilt, and it has a y-intercept, or the point where the line intersects the y-axis. A line may be represented by an equation $Y = bX + a$ in which b is the numerical value of the slope and a is the value of the y-intercept. Typically in social and behavioral research, a graph of a relationship, called a scatterplot, reveals that most points do not fall precisely on a straight line. To determine the best line for such an imperfect relationship, one selects a line such that the sum of the squared distances between all the points and the line is a minimum. This is called the least squares criterion. Since most of the points do not fall precisely on the line, they vary about the line, and the amount of such variability is reflected in the standard error of estimate.

Formulas

1. **Geometry of a straight line**

   a. Equation

   $Y = bX + a$

   b. Slope

   $b = \dfrac{y_2 - y_1}{x_2 - x_1}$ for points (x_1, y_1)

   and (x_2, y_2) on the line

   c. y-intercept

   $a = Y$ value at $X = 0$

2. **Regression of Y on X**

   a. Regression equation

   $\hat{Y} = bX + a$

   b. Slope

   $b = \dfrac{N(\Sigma XY) - (\Sigma X)(\Sigma Y)}{N\Sigma X^2 - (\Sigma X)^2}$

   (raw score formula)

   c. y-intercept

   $a = \overline{Y} - b\overline{X}$

   d. Standard error of estimate

   $s_{y \cdot x} = \sqrt{\dfrac{\Sigma(Y_i - \hat{Y})^2}{N - 2}}$ (definitional formula)

   $s_{y \cdot x} = \sqrt{\left[\dfrac{1}{N(N - 2)}\right]\left[N\Sigma Y^2 - (\Sigma Y)^2 - \dfrac{[N(\Sigma XY) - (\Sigma X)(\Sigma Y)]^2}{N\Sigma X^2 - (\Sigma X)^2}\right]}$ (raw score formula)

3. **Computational approach**

   $(\mathbf{I_{XY}}) = N(\Sigma XY) - (\Sigma X)(\Sigma Y)$ $(\mathbf{II_X}) = N\Sigma X^2 - (\Sigma X)^2$ $(\mathbf{III_Y}) = N\Sigma Y^2 - (\Sigma Y)^2$

   $b = \dfrac{(\mathbf{I_{XY}})}{(\mathbf{II_X})}$

   $s_{y \cdot x} = \sqrt{\left[\dfrac{1}{N(N - 2)}\right]\left[(\mathbf{III_Y}) - \dfrac{(\mathbf{I_{XY}})^2}{(\mathbf{II_X})}\right]}$

Exercises

For Conceptual Understanding

1. If a salesperson receives a base pay of $800 per month and a 5% commission on sales, what is the equation relating monthly sales and income for this person?

2. Recalling that $32°F = 0°C$ and $212°F = 100°C$, determine the linear equation for converting Celsius to Fahrenheit.

3. The squared deviations $\Sigma(Y_i - \hat{Y})^2$ are called squared errors of prediction. Explain why they are "errors" and describe their role in the least squares criterion. Why is the regression line the line of "best fit"?

4. Compare the standard error of estimate with a simple standard deviation. How are the two measures similar and how are they different?

5. Indicate whether each of the following sets of facts could or could not exist simultaneously. If not, why not?

   a. $a = 5$, $b = 1.00$, the regression line crosses the X-axis at 5
   b. $a = 10$, $b = .5$, $s_y = 4$
   c. $b = -.4$, $s_y = 15$, $s_{y \cdot x} = 20$

For Solving Problems

6. For the data below, compute the regression constants for the regression of Y on X, W on X, and W on Y. Also write the regression equation for each of these relationships.

X	Y	W
3	2	7
6	7	4
8	5	4
9	4	0
1	0	7
3	4	6

7. Using the equations from Exercise 6, what would you predict in the following cases?

   a. $\hat{Y}$ for $X = 6$
   b. $\hat{W}$ for $X = 3$
   c. $\hat{W}$ for $Y = 3$
   d. $\hat{Y}$ for $X = 12$

8. Determine the standard errors of estimate, $s_{y \cdot x}$, $s_{w \cdot x}$, and $s_{w \cdot y}$, for the data in Exercise 6.

9. For decades researchers have tried to identify some behavior of human infants in the first year of life that would predict later IQ. For the most part, this has been unsuccessful—even standardized tests of infant development do not predict very accurately. More recently, however, assessments of the infant's ability to study a standard stimulus and then to recognize that stimulus as being familiar when it is presented again have been shown to predict later IQ, at least to a modest degree.[3] Below are the recognition memory scores during infancy and the IQs measured at 6 years of age for a small sample of subjects. Determine the regression of IQ on recognition memory, and calculate the standard error of estimate. What would be the predicted IQ for Louise, who has a recognition score of 7? Between what two values would you expect the IQs of approximately two out of three infants like Louise to fall?

Subject	Recognition Memory Score	IQ
A	4	103
B	10	111
C	12	106
D	8	118
E	5	101
F	5	92
G	7	100
H	9	113
I	6	110

---

[3] McCall, R. B., and Carriger, M. S. (1993). A meta-analysis of infant habituation and recognition memory performance as predictiors of later IQ. *Child Development, 64,* 57–79.

For Homework

10. Given the regression line in the graph determine

  a. b

  b. a

  c. $\hat{Y}$

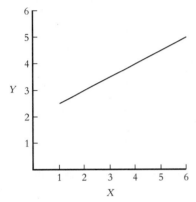

11. Draw the regression line for $a = -3$ and $b = -.33$.

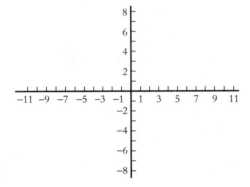

12. Discuss the truth or falsity of the following.

  a. The regression constants will not change systematically with an increase in N.

  b. If $\Sigma(Y_i - \hat{Y})^2$ is minimized, so is $\Sigma(X_i - \hat{X})^2$.

  c. The regression constants will be altered by changing the origin and unit of one of the variables.

13. If $s_y^2 = 0$, what are the values of b and a?

14. What function does the standard error of estimate serve?

15. Below are scores on two variables for a sample of five people. Answer the questions below in reference to these data.

X	Y
1	2
3	3
4	8
2	4
5	9

  a. Find $\overline{X}$, $\overline{Y}$, s_x^2, and s_y^2.

  b. Find b and a.

  c. If $X = 2.5$, what is predicted for Y?

  d. Find $s_{y \cdot x}$.

Correlation

he regression methods described in the previous chapter are useful for predicting one variable from another, such as predicting the likely grade average for a particular applicant to college on the basis of his or her SAT score. As we have seen, however, relationships of this sort are not perfect. That is, we can predict a specific grade average for an individual, but it is only an estimate. We know that students who score high on the SAT test generally do somewhat better in college than those who score less well, but how much better?

One way to answer that question is to express the extent or degree of relationship that exists between grades and SAT scores. The **correlation coefficient** is a numerical index that reflects the degree of linear relationship between two variables. If the correlation coefficient for the relationship between grades and SAT scores is high, a strong relationship exists: students who score well on the SAT are likely to do much better at college than students who score poorly. If the correlation coefficient is low, only a weak relationship or a nonlinear relationship exists: students who score high on the SAT get only slightly better grades than other students.

The correlation coefficient also permits us to compare the degree of linear relationship between different pairs of variables. For example, the correlation coefficient between tested intelligence and grades in school is fairly high. This means that students who have high IQs also tend to get good grades. In contrast, the correlation coefficient between IQ and creativity is quite low, indicating that high-IQ students are not necessarily more creative than other students. This difference between correlation coefficients permits us to say that the relationship between IQ and grades has a stronger linear relationship than the linear relationship between IQ and creativity.

THE PEARSON PRODUCT-MOMENT CORRELATION COEFFICIENT

Suppose a school district is interested in trying to prevent school failure. It knows that early reading difficulties often lead to more serious school problems, so it has designed a special learn-to-read program to prevent early reading problems. But the program is expensive, and the school system can afford to give this program to only a relatively small number of pupils, preferably the ones who are most likely to have early reading problems. It needs to identify one or more risk factors that will predict which pupils will have difficulty learning to read. Such pupils would be enrolled in the special program from the beginning of first grade.

Risk factors are characteristics that predict future problems, although they may or may not actually cause the problem directly. Family income level often relates to school performance, and it is frequently used as a risk factor to decide which children (i.e., those with the lowest family incomes) receive specialized educational services. So a school psychologist is asked to determine whether family income level predicts early reading performance in this particular school district.

The school psychologist has available in the school records two relevant measures for first-grade pupils. One is family income level, which is obtained at the beginning of first grade and which consists of a five-point scale ranging from very low (score of 1) to very high (score of 5) income. The second is a reading test score obtained at the end of first grade. The task is to determine whether income level predicts later reading scores and to what extent.

Derivation of the Correlation Coefficient

Statistically, the differences between scores in any distribution are called variability. Some of the variability in reading scores can be predicted on the basis of the regression of reading scores on income and some of the variability will still remain after that regression. What proportion of the total variability in reading scores is associated with the regression of reading scores on income? That proportion is the square of the correlation coefficient and describes the extent of the relationship between reading and income.

To examine this idea graphically, consider Figure 6-1, which presents an imperfect relationship between reading score and income for five pupils. The reading score for one pupil, Johnny, who had an income level of 5, and its deviations from the mean and from the regression line have been illustrated in detail. Locate three points in Figure 6-1: Y_i, Johnny's actual observed reading score; $\overline{Y}$, the mean of all five reading scores; and $\hat{Y}_i$, the reading score that would be predicted for Johnny on the basis of the relationship between reading and income (that is, the regression line at $X = 5$).

Now recall that the *total variability* in reading scores is reflected in the variance, s_y^2, which is based upon the deviations of each score from the mean of all the scores. In Johnny's case, his contribution to the total variability in reading scores is his deviation from the mean. In Figure 6-1, this is represented by the total dashed vertical line between his score, Y_i, and the mean, $\overline{Y}$. Numerically, this is

$$(Y_i - \overline{Y}) = 100 - 50 = 50$$

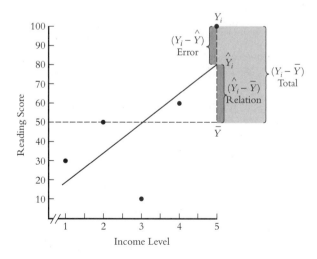

Figure 6-1 Scatterplot showing three deviations: $(Y_i - \overline{Y})$, $(Y_i - \hat{Y}_i)$, and $(\hat{Y}_i - \overline{Y})$.

But notice in Figure 6-1 that this distance is composed of two parts. One part is the distance between Johnny's score, Y_i, and the regression line, $\hat{Y}_i$. Numerically, this is

$$(Y_i - \hat{Y}_i) = 100 - 80 = 20$$

The other part of the total is the distance from the regression line, $\hat{Y}_i$, to the mean, $\overline{Y}$. Numerically, this is

$$(\hat{Y}_i - \overline{Y}) = 80 - 50 = 30$$

So the total distance from Johnny's score to the mean can be partitioned into two parts, the distance from his score to the regression line plus the distance from the regression line to the mean. This relationship can be expressed algebraically as

$$(Y_i - \overline{Y}) = (Y_i - \hat{Y}_i) + (\hat{Y}_i - \overline{Y})$$

and numerically as

$$(100 - 50) = (100 - 80) + (80 - 50)$$
$$50 = 20 + 30$$
$$50 = 50$$

Now consider what these distances mean. To begin, consider what you would predict Johnny's reading score to be *if you had no knowledge of the relationship between reading scores and income.* You would predict that Johnny's reading score would be the mean reading score for all children, $\overline{Y}$. After all, you have no other information, and the squared deviations of reading scores about their mean is less than about any other value. So the mean reading score is the best prediction if no other information is available. But, of course, you would be wrong in making such a prediction. Specifically, you would miss Johnny's score by $(Y_i - \overline{Y})$, or $100 - 50 = 50$ points. Now recall that the variance, s_y^2, is a measure of the variability of scores, and it is composed of the deviations of scores from their mean. So the total distance between Johnny's score and the mean, $(Y_i - \overline{Y})$, not only represents the error in predicting the mean to be Johnny's score but it also represents Johnny's contribution to the total variability in reading scores.

Now consider what you would predict Johnny's reading score to be *if you knew the relationship between reading scores and income.* You would predict Johnny's score to be the point on the regression line corresponding to an income level of 5. That is, you would predict $\hat{Y}_i$, which for Johnny equals 80. The regression line is the best prediction if you know income and its relationship to reading, because the squared deviations of scores about the regression line is less than about any other value, including the mean. Even so, predicting $\hat{Y}$, while more accurate than predicting the mean, would not likely be exactly correct. Specifically, you would be in error by $(Y_i - \hat{Y}_i)$, which equals $100 - 80 = 20$ points. In Figure 6-1, this is the distance between Johnny's reading score, Y_i, and the regression line, $\hat{Y}_i$. Also, it is that portion of Johnny's total contribution to the variability in reading scores that is *associated with the error in predicting Johnny's reading score* to be $\hat{Y}_i$ on the basis of the relationship between reading and income.

Notice that while the error was 50 when predicting the mean, it is now only 20 when predicting with the regression line. In a sense, knowing the relationship between reading scores and income reduced the error in prediction by $50 - 20 = 30$ points.

This is the distance in Figure 6-1 from the regression line, $\hat{Y}_i$, to the mean, $\overline{Y}$. So this distance, $(\hat{Y}_i - \overline{Y})$, is the portion of the total that is *associated with the relationship between reading scores and income levels.* It is that portion of Johnny's contribution to the total variability in reading scores that is associated with the relationship between reading scores and income.

So Johnny's contribution to the total variability in reading scores can be divided into two parts—one that is associated with the relationship between reading and income, and one that is not. This is presented graphically in Figure 6-1, in which the total distance between Johnny's reading score and the mean, $(Y_i - \overline{Y})$, consists of two parts: $(\hat{Y}_i - \overline{Y})$, which is associated with the relationship between reading and income, and $(Y_i - \hat{Y}_i)$, which is not associated with that relationship and which we will call *error.* In symbols:

$$(Y_i - \overline{Y}) = (\hat{Y}_i - \overline{Y}) \quad + (Y_i - \hat{Y}_i)$$
$$\text{Total} = \text{Relationship} + \text{Error}$$

However, measures of variability and proportions of variability usually involve squared deviations summed over all subjects. It happens that when

$$(Y_i - \overline{Y}) = (\hat{Y}_i - \overline{Y}) + (Y_i - \hat{Y}_i)$$

is squared and summed for all subjects, all the cross product terms [that is, $(\hat{Y}_i - \overline{Y})$ $(Y_i - \hat{Y}_i)$] drop from the expression and the result is

$$\Sigma(Y_i - \overline{Y})^2 = \Sigma(\hat{Y}_i - \overline{Y})^2 + \Sigma(Y_i - \hat{Y}_i)^2$$

This states that the total squared deviations of points about their mean can be divided into two parts, one part associated with Y's relationship to X and the other part associated with error. In symbols,

$$\Sigma(Y_i - \overline{Y})^2 = \Sigma(\hat{Y}_i - \overline{Y})^2 + \Sigma(Y_i - \hat{Y}_i)^2$$

Total squared deviations	=	Squared deviations associated with the relationship	+	Squared deviations associated with error

Since $\Sigma(\hat{Y}_i - \overline{Y})^2$ represents the squared deviations in Y that are associated with the relationship to X and $\Sigma(Y_i - \overline{Y})^2$ represents the total squared deviations in Y, their ratio

$$\frac{\text{squared deviations associated with the relationship}}{\text{total squared deviations}} = \frac{\Sigma(\hat{Y}_i - \overline{Y})^2}{\Sigma(Y_i - \overline{Y})^2}$$

constitutes the proportion of squared deviations in Y associated with differences in X. Because variability is defined by squared deviations, this ratio is often said to express the *proportion of variability in Y that is associated with differences in X.*[1] If the value of the

---

[1] Strictly speaking, such an interpretation requires that the conceptual variable X be measured without error. However, in practice this assumption is often disregarded.

ratio was .55 for the reading/income example described above, it would mean that 55% of the total variability in reading scores is associated with differences in income levels.

Although the ratio indeed provides an index of degree of relationship by indicating the proportion of variability in *Y* that is associated with differences in *X*, it does not indicate the direction of the relationship. Is it positive or negative? That is, are high scores on one variable associated with high scores on the other (*positive* or *direct relationship*), or do high scores on one measure relate to low scores on the other (*negative* or *inverse relationship*)? Regrettably, since this proportion is a ratio of two positive quantities, it will always be positive. However, the square roots of a number actually include both a positive and a negative root. For example, the square roots of a^2 are $+a$ and $-a$, because both $(+a)^2$ and $(-a)^2$ equal a^2. Therefore, to indicate the direction of the relationship, one could take the square root of the proportion of variability, accepting the positive root if the relationship between *Y* and *X* is positive or the negative root if it is negative. This square root of the proportion of variability in *Y* associated with *X* is the correlation coefficient.

The **Pearson product-moment correlation coefficient,** symbolized by *r,* is defined as

$$r = \sqrt{\frac{\Sigma(\hat{Y}_i - \overline{Y})^2}{\Sigma(Y_i - \overline{Y})^2}}$$

This index, or coefficient of correlation, was developed by Karl Pearson, one of the fathers of modern statistics. It is a number ranging from -1.00 through .00 to $+1.00$ that reflects the extent of a linear relationship. It is called a *coefficient,* not because it is used to multiply other quantities, but because it is a unitless index. It is unitless because it is not expressed in the units of measurement—it is just a quantity that varies with the direction and degree of linear relationship. It is called a *product-moment correlation* because it is defined mathematically in terms of the product of the moments of the two variables, "moment" being a concept in mathematical statistics.

Other indices of correlation exist. One, the **Spearman rank–order correlation,** is presented in Chapter 14. In addition, other correlation coefficients can be used to express a wide variety of relationships, such as the degree of nonlinear association between two variables (called **eta** and pronounced "ATE-ah"), the degree of relationship between several predictor variables and one criterion (predicted) variable (**multiple correlation**), and the correlation between two variables after each variable's relationship with a third variable has been removed (**partial correlation**). These other correlation coefficients are described in advanced textbooks on statistics.

Computational Formula

Once again the formula that defines a statistic is inconvenient to use in computational work. To use the definitional formula given above requires that $\hat{Y}$ be computed for every value of X_i. However, by substituting the equation ($\hat{Y} = bX + a$) into the definitional formula and then substituting the raw-score formulas for both *b* and *a,* the above quantity becomes

$$r = \frac{N(\Sigma XY) - (\Sigma X)(\Sigma Y)}{\sqrt{[N\Sigma X^2 - (\Sigma X)^2][N\Sigma Y^2 - (\Sigma Y)^2]}} = \frac{(\mathbf{I_{XY}})}{\sqrt{(\mathbf{II_X})\,(\mathbf{III_Y})}}$$

Notice that this formula is entirely composed of the same three intermediate quantities [i.e., $(\mathbf{I_{XY}})$, $(\mathbf{II_X})$, $(\mathbf{III_Y})$] introduced in the previous chapter to compute regression statistics. The distinct advantage of this formula, in addition to its ease of computation when large numbers of subjects are involved (i.e., $\hat{Y}$ does not need to be calculated for each value of X_i) is that the formula for b, the slope of the regression line, was used in its derivation. That means that one does not have to be concerned about selecting the positive or negative square root to indicate the direction of the relationship. The slope is positive for positive relationships and negative for negative relationships; since this formula includes the slope, it provides the appropriate sign for r without further labor.

Table 6-1 presents a numerical example of both formulas; one can see that they yield identical results. The regression line for the data is $\hat{Y}_i = .75X + 1.25$. $\hat{Y}_i$ has been computed for each pair of scores.

The square of the correlation coefficient is $r^2 = .49$. The correlation coefficient, r, is the square root of this value, or .70. In the last three columns of Table 6-1 are the values X^2, Y^2, and XY for use in the computational formula. Again, N is the number of subjects or pairs of scores. These values are substituted into the computational formula, which gives the result $r = .70$. The square of r, .49, is the proportion of variance in Y that is associated with differences in X.

Notice that the square root of a decimal less than 1.00 is *larger*, not smaller, than the original number. The square root of a number greater than 1, for example 4, is the smaller value 2; but the square root of a decimal less than 1.00, for example .49, is the larger value .70. The correlation coefficient, which is the square root of a decimal proportion, will always be a larger value than that proportion (see "The Relation between r and r^2" below).

Hand calculators and computers have made the calculation of the correlation coefficient much easier than the procedure outlined in Table 6-1. As explained in the previous chapter on regression, typically one simply enters the pairs of numbers for each subject and then presses a single button or gives a simple command and the machine will provide the correlation coefficient, r. Computers will often simultaneously calculate the regression coefficients, a and b (sometimes symbolized m), and the standard error of estimate, $s_{y \cdot x}$, and then automatically print out all these values (and perhaps more). Hand calculators require you to push a separate button to obtain each value. See your manual for instructions.

If you must calculate all these regression and correlation values by hand or with a calculator that does not automatically compute these specific quantities, follow the examples outlined in "Computational Procedures" at the end of this chapter. They will save you time over calculating each statistic separately as described in the text, because some of the same quantities are used in several of the formulas.

PROPERTIES OF THE CORRELATION COEFFICIENT

The correlation coefficient can be understood and interpreted better if you know some of its properties.

6-1   Calculation of the Correlation Coefficient

X	Y	$\bar{Y}$	$\hat{Y}$	$(Y_i - \bar{Y})^2$	$(Y_i - \hat{Y}_i)^2$	$(\hat{Y}_i - \bar{Y})^2$	X^2	Y^2	XY
9	10	5	8.0	25	4.00	9.00	81	100	90
7	6	5	6.5	1	.25	2.25	49	36	42
5	1	5	5.0	16	16.00	.00	25	1	5
3	5	5	3.5	0	2.25	2.25	9	25	15
1	3	5	2.0	4	1.00	9.00	1	9	3
25	25			46	23.50	22.50	165	171	155

Definitional Formula

$$r^2 = \frac{\text{variability associated with } X}{\text{total variability}} = \frac{\Sigma(\hat{Y}_i - \bar{Y})^2}{\Sigma(Y_i - \bar{Y})^2} = \frac{22.50}{46} = .49$$

$$r = \sqrt{r^2} = \sqrt{.49} = .70$$

Computational Formula

$$(\mathrm{I_{xy}}) = N\Sigma XY - (\Sigma X)(\Sigma Y)$$
$$= 5(155) - (25)(25)$$
$$(\mathrm{I_{xy}}) = 150$$

$$(\mathrm{II_x}) = N\Sigma X^2 - (\Sigma X)^2$$
$$= 5(165) - (25)^2$$
$$(\mathrm{II_x}) = 200$$

$$(\mathrm{III_y}) = N\Sigma Y^2 - (\Sigma Y)^2$$
$$= 5(171) - (25)^2$$
$$(\mathrm{III_y}) = 230$$

$$r = \frac{N\Sigma XY - (\Sigma X)(\Sigma Y)}{\sqrt{[N\Sigma X^2 - (\Sigma X)^2]\,[N\Sigma Y^2 - (\Sigma Y)^2]}} = \frac{(\mathrm{I_{xy}})}{\sqrt{(\mathrm{II_x})(\mathrm{III_y})}} = \frac{150}{\sqrt{(200)(230)}} = .70$$

$$r^2 = \text{proportion of variance associated with } X = (.70)^2 = .49$$

The Range of *r*

The correlation coefficient may assume values from -1.00 through .00 to $+1.00$. Positive values reflect positive linear relationships; negative values reflect negative linear relationships. Values of -1.00 and $+1.00$ represent perfect linear relationships; a value of .00 represents no linear relationship at all.

Consider first the case of a perfect linear relationship between X and Y in which the points all fall precisely on a line of nonzero slope (that is, the line is not parallel to the X-axis). In this situation, each point Y_i is identical to its corresponding point $\hat{Y}_i$. Because $\hat{Y}_i = Y_i$, then $\Sigma(\hat{Y}_i - \overline{Y})^2$ and $\Sigma(Y_i - \overline{Y})^2$ are identical, and

$$r = \sqrt{\frac{\Sigma(\hat{Y}_i - \overline{Y})^2}{\Sigma(Y_i - \overline{Y})^2}} = \sqrt{1} = \pm 1.00$$

Thus, in the case of a perfect relationship in which all points fall on the regression line ($b \neq 0$), the correlation coefficient will equal $+1.00$ or -1.00, depending upon whether the relationship is positive or negative. This makes sense, because if all points fell on the regression line, 100% of the variability in the scores would be associated with the regression—no error variability would remain.

Can r ever be greater than 1.00 or less than -1.00? No, because it can be shown that $\Sigma(\hat{Y}_i - \overline{Y})^2$ is always less than or equal to $\Sigma(Y_i - \overline{Y})^2$, which means that the fraction

$$\frac{\Sigma(\hat{Y}_i - \overline{Y})^2}{\Sigma(Y_i - \overline{Y})^2}$$

will always be less than or equal to 1.00, and therefore so will r. One can never account for more than 100% of the variability.

Suppose now that there is no relationship between X and Y. The scatterplot may appear to be a circular clustering of points, and the regression line will be parallel to the X-axis and $\overline{Y}$ units high. Since $\hat{Y}$ will equal $\overline{Y}$ at every X value, $\hat{Y}$ is identical to $\overline{Y}$ and $\Sigma(\hat{Y}_i - \overline{Y})^2$ must be equal to zero, leaving

$$r = \sqrt{\frac{\Sigma(\hat{Y}_i - \overline{Y})^2}{\Sigma(Y_i - \overline{Y})^2}} = \sqrt{\frac{0}{\Sigma(Y_i - \overline{Y})^2}} = .00$$

Therefore, when there is no linear relationship, $r = .00$. This also makes sense, because if no relationship exists, the regression will account for none of the variability in the scores—all the variability will remain as error.

Hence, the correlation coefficient ranges in value between -1.00 and $+1.00$. It is $+1.00$ if all the points fall precisely on a line of positive slope, and it is .00 if there is no linear relationship at all. If there is no variability in Y_i (that is, if all Y_i are the same value, as shown in Figure 6-3D), r is not defined.

The Relation between *r* and *r²*

While the square of the correlation coefficient, r^2, may be interpreted as the proportion of variance in Y_i attributable to differences in X_i, the correlation coefficient is the square root of this proportion with the algebraic sign indicating the direction of the relationship. Since r is obviously not the same as r^2, one must be careful to interpret the terms properly. For example, consider the following table:

Correlation: *r*	Proportion of Variance: *r²*
.10	.01
.20	.04
.30	.09
.40	.16
.50	.25
.60	.36
.70	.49
.80	.64
.90	.81
1.00	1.00

Note that a correlation of .10 to .30 suggests that not very much variance in Y_i is associated with differences in X_i (1–9%). In fact, a correlation of .50, which is frequently considered high in social and behavioral research, implies that only 25% of the variance in Y_i is associated with X_i. That means that 75% of the variability in Y_i is associated with factors other than X_i. One needs a correlation of .71 before one can say that half of the variability in Y_i is associated with X_i. The implication is that in terms of proportion of variance in Y_i, the *unsquared* correlation coefficient (r) gives the impression of indicating a higher degree of relationship with X than does the proportion-of-variance interpretation (i.e., r^2).

For example, suppose the correlation between smoking and a certain form of cancer is $r = .60$. This figure gives the impression that smoking is the biggest cause of the disorder. It may be, but actually, smoking represents only about one-third the variability in who gets the cancer, because only $(.60)^2 = .36$ or 36% of the variance is associated with smoking. Other factors, such as, stress, diet, secondhand smoke, other airborne pollutants, etc., collectively account for approximately two-thirds of the variability, and it is possible that one of these factors is more important than smoking.

The Effect of Scale Changes on *r*

In Chapter 4 we saw how adding a constant to every score or multiplying every score by a constant affects the value of the mean and variance of that distribution. It was also demonstrated that adding and multiplying (or subtracting and dividing) really amounts to changing the origin and the unit of the measuring scale. What happens to the correlation coefficient if the unit and/or origin of either scale of measurement (X and/or Y) is altered?

> The size of the correlation coefficient does not change if every score in either or both distributions is increased or multiplied by a constant ($c \neq 0$). Thus, r is not altered by changes in the origin and unit of the measurement scale.[2]

The algebraic proof of this assertion is given in Optional Table 6-2. Adding or subtracting, multiplying or dividing, or both adding (or subtracting) and multiplying (or dividing) X and/or Y by the same or different nonzero constants does not alter the value of r. Therefore, r does not change when one or both scales of measurement are changed.

This result has important implications for the use of the correlation coefficient. Whether measurement is in feet or inches, minutes or seconds, units or dozens, the correlation between the variables will be the same. If r represents the degree of linear relationship between two variables (for example, age and height), it must be the same regardless of whether age is measured in months or years and height is measured in inches or centimeters. Indeed, the fact that r remains the same, even though the origin and/or unit of measurement for one or both of the variables is changed, gives this statistic a large range of applications.

The Relation between Correlation and Regression

Slope and Correlation   By certain algebraic manipulations it can be shown that

$$r = b_{yx} \left(\frac{s_x}{s_y} \right)$$

This expression states that the correlation coefficient is a joint function of the slope of the regression line (b_{yx}) and the standard deviations of the two variables (s_x and s_y). Consequently, the correlation r is usually related but not equivalent to the slope b_{yx} of the regression line. However, if the standard deviations of the X and Y variables are identical, usually by converting both the X and Y distributions to standard scores (so that they both have a standard deviation of 1.00), then

$$r = b_{z_y z_x} \left(\frac{s_{z_x}}{s_{z_y}} \right) = b_{z_y z_x} \left(\frac{1}{1} \right)$$

$$r = b_{z_y z_x}$$

in which the subscripts z_y and z_x indicate values determined on the standardized X and Y distributions.

In short, when both X and Y distributions are in standard–score form (and thus

---

[2] If one distribution is multiplied by a negative number and the other by a positive number (including $c = 1$), the sign of the correlation coefficient will change, but not the magnitude.

6-2 Proof that No Change Occurs in the Correlation Coefficient with a Change in Origin or Unit

Optional Table

Operation	Explanation
Change in Origin (Adding a Constant)	
1. $r = \dfrac{\Sigma(X - \bar{X})(Y - \bar{Y})}{\sqrt{\Sigma(X - \bar{X})^2 \Sigma(Y - \bar{Y})^2}}$	1. Definition of r.
2. $= \dfrac{\Sigma[(X + c) - (\bar{X} + c)][(Y + k) - (\bar{Y} + k)]}{\sqrt{\Sigma[(X + c) - (\bar{X} + c)]^2 \Sigma[(Y + k) - (\bar{Y} + k)]^2}}$	2. Substituting $(X + c)$ for X and $(Y + k)$ for Y; as a result, the means will be $\bar{X} + c$ and $\bar{Y} + k$.
3. $= \dfrac{\Sigma(X + c - \bar{X} - c)(Y + k - \bar{Y} - k)}{\sqrt{\Sigma(X + c - \bar{X} - c)^2 \Sigma(Y + k - \bar{Y} - k)^2}}$	3. Removing parentheses.
4. $r = \dfrac{\Sigma(X - \bar{X})(Y - \bar{Y})}{\sqrt{\Sigma(X - \bar{X})^2 \Sigma(Y - \bar{Y})^2}}$	4. Subtracting the c's and k's within parentheses leaves the formula for r unchanged.
Change in Unit (Multiplying by a Constant)	
1. $r = \dfrac{\Sigma(cX - c\bar{X})(kY - k\bar{Y})}{\sqrt{\Sigma(cX - c\bar{X})^2 \Sigma(kY - k\bar{Y})^2}}$	1. Definition of r with cX and kY substituted for X and Y; as a result, the means will be $c\bar{X}$ and $k\bar{Y}$.
2. $= \dfrac{\Sigma c(X - \bar{X})k(Y - \bar{Y})}{\sqrt{\Sigma[c(X - \bar{X})]^2 \Sigma[k(Y - \bar{Y})]^2}}$	2. Factoring in the manner of $ab - ac = a(b - c)$.
3. $= \dfrac{ck\Sigma(X - \bar{X})(Y - \bar{Y})}{\sqrt{c^2k^2\Sigma(X - \bar{X})^2 \Sigma(Y - \bar{Y})^2}}$	3. In the numerator, $\Sigma cW = c\Sigma W$, and in the denominator, $(ab)^2 = a^2b^2$.
4. $= \dfrac{ck\Sigma(X - \bar{X})(Y - \bar{Y})}{ck\sqrt{\Sigma(X - \bar{X})^2 \Sigma(Y - \bar{Y})^2}}$	4. $\sqrt{a^2b^2WZ} = ab\sqrt{WZ}$
5. $r = \dfrac{\Sigma(X - \bar{X})(Y - \bar{Y})}{\sqrt{\Sigma(X - \bar{X})^2 \Sigma(Y - \bar{Y})^2}}$	5. Cancellation, leaving r unchanged.

have the same standard deviation), the correlation coefficient is precisely equal to the slope of the regression line. Note, however, that the slope of the line computed with raw scores and the slope computed with standardized scores are not likely to be the same value. That is, ordinarily $b_{yx} \neq b_{z_y z_x}$.

Direction of Prediction   Recall from the previous chapter on regression that one must specify which variable is the predictor (that is, X) and which is the variable being predicted (that is, Y). The intercept, slope, and standard error of estimate are all stated in terms of the variable being predicted (that is, Y) and are specific to it. Predicting grades from SAT scores produces different regression results (a, b, $s_{y \cdot x}$) than predicting SAT scores from grades. The correlation coefficient, however, is unitless—it reflects the degree of linear relationship regardless of the direction of prediction. If $r = .62$ for the relationship between SAT scores and grades, then $r = .62$ for the relationship between grades and SAT scores.

$s_{y \cdot x}$ and r The relationship between the square of the correlation coefficient and the square of the standard error of estimate is approximately[3]

$$r^2 = 1 - \frac{s_{y \cdot x}^2}{s_y^2}$$

This equation states that the degree of the relationship as reflected in the square of the correlation coefficient is a function of the ratio of the variability of points about the regression line (as expressed by $s_{y \cdot x}^2$) to the total variability of the Y scores (s_y^2). More specifically, r^2 and r will be higher if the fraction

$$\frac{s_{y \cdot x}^2}{s_y^2}$$

is relatively small. In words, the correlation will be high if the variability of the points about the regression line is small relative to their total variability. That is, the correlation is high if the points cluster tightly about the line and if the line has a nonzero slope (that is, it is not horizontal).

Thus, it is possible to judge the relative sizes of different correlations by observing their scatterplots. Consider the graphs in Figure 6-2. Plots A and B have approximately the same variability in Y_i (that is s_y^2) but differ in the extent to which the points cluster about the line (that is, $s_{y \cdot x}^2$). Since $s_{y \cdot x}^2$ in A is less than in B, the fraction

$$\frac{s_{y \cdot x}^2}{s_y^2}$$

is smaller, and thus the correlation is larger, for A than for B. The converse situation is presented in graphs C and D. The variability of Y_i is greater in D than in C, while $s_{y \cdot x}^2$ is approximately the same in each. Therefore, the ratio

$$\frac{s_{y \cdot x}^2}{s_y^2}$$

---

[3] In the population the relationship is exact, but it is only approximate for samples.

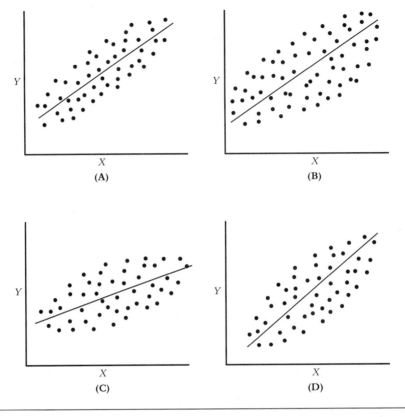

Figure 6-2 Hypothetical scatterplots. In graphs A and B, the s_y are equivalent, but since the $s_{y \cdot x}$ is less in A, its correlation is higher. In graphs C and D, the $s_{y \cdot x}$ are comparable, but since the s_y is greater in D, the correlation is higher.

is smaller in D than in C, and consequently the correlation is higher in D. In brief, then, the closer the points cluster about the regression line and the steeper the slope of the regression line, the higher the correlation.

Figure 6-3 provides several examples of different scatterplots with their respective correlation coefficients. Scatterplots A, B, and C reflect the concepts just discussed. Plots in which the points cluster about a steep regression line reflect high positive or negative r's, whereas a plot in which the points do not cluster tightly about the line and/or the line is nearly horizontal reflects a low r. Scatterplots D, E, and F represent some interesting special cases. Plot D shows a perfect relationship in the sense of clustering, but one that is maximally imperfect with respect to the variability of Y_i. Here the deviations $\Sigma(\hat{Y}_i - \bar{Y})^2$ and $\Sigma(Y_i - \bar{Y})^2$ are both 0, making $r = \frac{0}{0}$, which is best stated r *is undefined*. This result makes intuitive sense also, because prediction of Y_i is not improved by knowing the regression, since regardless of the X value one always predicts Y_i to be $\bar{Y}$. Hence, the correlation is undefined. However, plot E shows

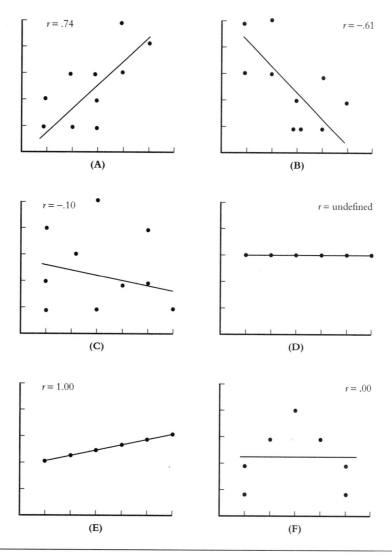

Figure 6-3 Sample scatterplots with their respective regression lines and correlations.

that if all points fall precisely on a regression line, then the correlation is ±1.00, depending on the sign of the slope as long as the slope is not 0 (that is, as long as the line is not parallel to the X-axis). Plot F depicts the case of a perfect but nonlinear relationship. Depending upon the nature of the curvilinearity, one can obtain r's of various sizes. Thus, it is not valid to conclude that if $r = .00$ there is no relationship between X and Y; rather, the conclusion should be that there is no *linear* relationship between X and Y. This is why it is advisable to examine the scatterplot for nonlinear relationships before continuing with the procedures for linear relationships.

SAMPLING FACTORS THAT CHANGE THE CORRELATION COEFFICIENT

Up to this point the discussion has been concerned with developing a measure of the degree of linear relationship between two variables in a sample. However, in practice one usually wishes to use the sample correlation, r, to estimate the correlation that exists in the larger population. This population parameter is symbolized by the Greek letter ρ (read "rho" in English). The topics discussed in this section are concerned with practical problems that arise when one wants to estimate ρ with r. Obviously, the accuracy of any statistic as an estimate of a population value depends upon how representative the sample is of the population. The illustrations below describe in detail how r may be a distorted estimate of the population value when certain biases exist in the sample.

A Restricted Range

As discussed in a previous section, the size of r is a function of the relative values of $s_{y \cdot x}^2$ and s_y^2 such that r becomes large as $s_{y \cdot x}^2$ becomes small relative to s_y^2. Therefore, if the degree of clustering about the regression line is fairly constant over all segments of the line, then as the range and the variance of Y_i are reduced, the correlation is reduced.

Consider the following example. Suppose a Quick Test of Reading Ability is developed. To demonstrate that the Quick Test is valid, a sample of children in grades one through six are given a much more thorough reading assessment to determine their reading competence. The Quick Test is then administered to the children, and its scores are correlated with the children's reading competence. If the correlation is high, the developers of the Quick Test will conclude that it is a valid indicator of reading ability. Suppose the correlation for the entire sample is .76. On the basis of this validity information, someone may propose to use the test on all third graders to single out those students who need special reading instruction. However, when the sample is restricted to third graders, the test may not be nearly as valid as it was for the entire sample. Perhaps the correlation is only .28 for this subgroup, a figure that would certainly discourage using the test for the purpose suggested.

A hypothetical plot of this circumstance is presented in Figure 6-4. For the total group of children in grades one through six, reading competence is rather closely related to their test scores, that is, to the regression line, relative to the total variability of reading competence scores. In short, the plot of points is tightly clustered about a regression line of steep slope. When the sample is restricted to third graders (i.e., the $\times$'s rather than the dots in Figure 6-4), the scatterplot is not so tightly clustered about the regression relative to the total variability. The plot is more like a circle than a thin oval, signifying a lower correlation. This can also be seen by examining the formula for r introduced earlier:

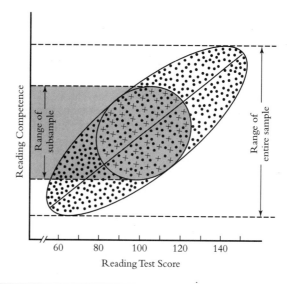

Figure 6-4 Hypothetical scatterplot showing relatively equal clustering of points about the regression line for a group of pupils in grades one through six and for a subsample of third graders, but a marked reduction in the range of the Y_i for the third graders. The correlation is much larger for the total group than for the subsample.

$$r = \sqrt{1 - \frac{s_{y \cdot x}^2}{s_y^2}}$$

Restricting the range of Y scores tends to reduce the size of s_y^2 relatively more than the size of $s_{y \cdot x}^2$, thus the fraction $s_{y \cdot x}^2 / s_y^2$ tends to become larger and r tends to become smaller.

Usually when the range of scores is restricted, the correlation is less than it would be if the complete range were sampled. The safest course to follow is to limit the interpretation of a correlation to the population that is accurately represented by the sample available. Thus, the correlation of .76 between the test and reading competence is appropriate for children in grades one through six, but that figure may not reflect the degree of relationship for children in a portion of that sample, such as third graders.

This situation is not unusual. Many educational tests, for example, are validated on samples that contain children from many grades. Since some of the younger children can barely read, for example, while the older children may be quite accomplished, the correlation for this total sample between the test and some criterion of performance may be very high, giving the impression that the test is quite valid. But most educational tests are used by teachers and schools to select for special services children who are in a specific grade (e.g., the third grade), not children from the wide age range used by the test developer. Without correlations specifically for children from the third grade, the test may appear much more valid for this particular use than is actually the case.

Extreme Groups

The size of *r* is usually different when researchers study only extreme groups of subjects than if they study a sample that is more representative of the entire range of behaviors. For example, suppose an environmental scientist is interested in the degree of relationship between children's lead exposure and their IQs. Lead accumulates in certain bodily tissues, such as teeth, which can be analyzed chemically to determine their lead content and the child's cumulative exposure to lead.[4] Suppose two extreme groups of ten-year-old children were selected, one containing children known to have been exposed to high levels of lead and one known to be relatively free of lead. All children had their teeth analyzed for lead content and were given IQ tests. If lead level and IQ were correlated only for children from these extreme groups, then *r* might equal .84. But if the sample also included children with exposures between these extreme groups, the correlation might be only .66.

This can be seen graphically in Figure 6-5. The dots represent points for the two extreme groups. Notice that the picture of their scatterplot is quite oblong, which we have seen is associated with higher correlations. But if the children between these groups (the ×'s in Figure 6-5) are included, the scatterplot is more rounded in form, indicative of a smaller correlation.

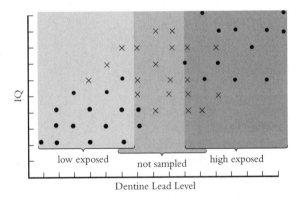

Figure 6-5 Scatterplot for a sample composed of two extreme groups, high- and low-lead-exposed children (that is, the dots). The *r* = .84. The ×'s are points for children with intermediate lead exposure. If they are included, *r* = .66 over all children. Selection of extreme groups usually increases *r*.

---

[4] Inspired by, but not identical to, the original work of Needleman, H. L., et al. (1979). "Deficits in Psychological and Classroom Performance of Children with Elevated Dentine Lead Levels." *The New England Journal of Medicine, 300*, 689–95.

This difference can also be explained by examining the following formula for r, which is different from the definitional and computational formulas given above but which produces the same numerical result:

$$r = \frac{\Sigma(X - \overline{X})(Y - \overline{Y})}{\sqrt{\Sigma(X - \overline{X})^2\Sigma(Y - \overline{Y})^2}}$$

The numerator is composed of the sum of products of the deviation of an X value from its mean multiplied by the deviation of a Y value from its mean. Therefore, r becomes large when there are many subjects whose X and Y scores both deviate markedly from their respective means. Selecting extreme groups eliminates subjects whose scores would be near the means, leaving only those subjects who have large $(X_i - \overline{X})(Y_i - \overline{Y})$ values. This circumstance tends to increase the numerator more than the denominator of the above formula and therefore tends to increase the value of r. As a result, the correlation coefficient is likely to be larger with extreme groups than if random sampling had been employed.

Combined Groups

One must also be cautious when a correlation between two variables is computed for subjects from two groups that differ in their mean values on one or both of the variables. For example, suppose the correlation between mental age and fear of dying is approximately .10 for a group of first graders but $-.40$ for sixth graders. If the two groups of children are combined into one, the correlation reverses to approximately $+.52$. How is that possible?

Figure 6-6 illustrates what may happen when groups that differ in mean values are combined for purposes of correlation. The first graders have lower mental ages

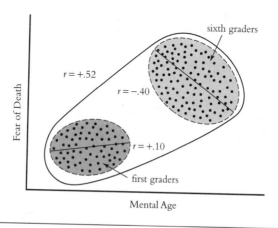

Figure 6-6 Illustration of how the relationship between mental age and fear of death may be approximately $+.10$ for first graders and $-.40$ for sixth graders, but $+.52$ for the two groups combined.

and are also less concerned with death, and therefore points for them cluster in the lower left corner of the scatterplot. The correlation between mental age and fear of death within that group is low positive, $r = +.10$. That is, the smarter first graders are just slightly more fearful. Conversely, the sixth graders have higher mental ages and show considerably more concern about death. Points for them, therefore, cluster in the upper right corner of the plot. Within this group there is a moderately negative association between mental age and fear of death, $r = -.40$. That is, the smarter sixth graders are less fearful. But the scatterplot for these two groups together, as represented by the heavy oval, shows a positive relationship, extending from lower left to upper right with two extreme groups to enhance the correlations to $+.52$, which is a highly unrealistic representation of the true state of affairs.[5]

The potential for having this problem is rather common, because samples frequently contain subgroups (e.g., males and females, racial groups, etc.) that may be different in their means or correlations, thus making the correlation for the combined sample unrepresentative of the subgroups. Also, it is quite possible to have any combination of positive and negative correlations among the subgroups and the combined group. Figure 6-7 presents some of these possibilities. Even if the group means are not very different, as in part D of Figure 6-7, the r for the combined group may still be quite different than for either subgroup. So one should always check the r's within subgroups before interpreting the r for the combined group.

An Extreme Score

Consider the possible influence of a single extreme score on the size of the correlation. In Figure 6-8, most of the scores (dots) cluster in a circular array, but there is one extreme case, $\times$. Without $\times$ the correlation is .05, but with it $r = .48$. Notice that if one of the dots, rather than $\times$, were dropped, the correlation would not change a great deal. Only cases that deviate markedly from the general cluster have a very large effect. Because the numerator of r contains the expression $(X_i - \overline{X})(Y_i - \overline{Y})$, a score that deviates substantially from the means $\overline{X}$ and $\overline{Y}$ will increase or decrease the numerator of r and therefore potentially alter the size of a correlation a great deal.

Ordinarily, if sufficient numbers of cases are sampled in a random manner, such a situation does not occur or a single extreme score has less effect in a large group of otherwise reasonable scores. However, if the sample is small, extreme cases can play a significant role in determining the size of r. In this situation some researchers use other types of correlation coefficients that are not so sensitive to an extreme score (see Chapter 14).

---

[5] Although the data on mental age and fear of death are fictitious, they reflect some general trends found in research reported in the following papers: F. C. Jeffers, C. R. Nichols, and C. Eisdorfer, "Attitudes of Older Persons to Death," *Journal of Gerontology* 16, 53–56; A. Mauer, "Adolescent Attitudes toward Death," *Journal of Genetic Psychology* 105, 75–90; J. M. Natterson and A. G. Knudson, "Children and Their Mothers: Observations Concerning the Fear of Death in Fatally Ill Children," *Psychosomatic Medicine* 22, 456–65.

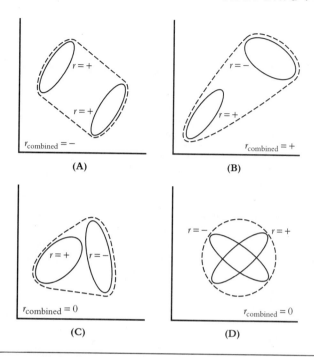

Figure 6-7 Graphs A, B, and C show effects on *r* of combining two groups of subjects that have different means. The solid regions represent the plots of the two groups, while the dashed region represents their combination. Graph D shows that if the two groups have comparable means, the *r* for the combined group may not accurately represent the *r* for either subgroup.

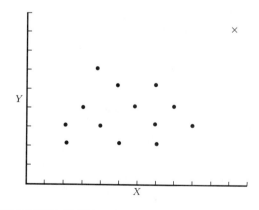

Figure 6-8 Illustration of the effects of a single extreme case, ×. Without × the correlation is .05, but with × it is .48.

CAUSALITY AND CORRELATION

The correlation coefficient represents the degree of observed linear association between two variables, not the extent of their causal relationship. Although having a runny nose correlates with having a cold, one would hardly suggest that the runny nose *causes* the cold. We may say that the cold causes the runny nose, but the correlation between runny nose and cold is the same regardless of which way one states it. Because the sun comes up when we wake in the morning does not prompt the megalomaniacal delusion that our getting up causes the sun to rise. Also, recall in the example given at the beginning of this chapter that risk factors were said to be correlates or predictors of a given problem, not necessarily causes of it. Low income, for example, may correlate with reading and school problems, but it may not cause them directly.

If *A* correlates with *B*, three possible causal relationships exist:

A causes *B*,

B causes *A*, or

C causes both *A* and *B*.

Of course, *C* may be quite remote, with long causal chains interposed before *A* and *B* actually occur, but the point is that there is always the possibility that a third variable (or set of variables) may produce an observed relationship between *A* and *B*. Consequently, one can never infer causality between two variables solely on the basis of their correlation. For example, the physical abuse of children is correlated with the frequency of various problem behaviors in those children during adolescence and adulthood (such as alcohol and drug abuse, suicide, and prostitution). However, many children who are abused are also unloved, rejected, and neglected. It is possible that the physical abuse per se, despite its horror, does not cause these problems but that a third variable or set of variables—lack of love, rejection, and neglect, which are correlated with abuse—actually leads to these unhappy outcomes. Correlation suggests association but not necessarily causality.

COMPUTATIONAL PROCEDURES

This and previous chapters contain several examples for calculating one or another statistic. In practice, it is often the case that many of these statistics, not just one or two, are computed within the context of a single problem. When that is done, certain computational conveniences are available. Therefore, Table 6-3 presents an integrated example that displays the calculation of many of the statistics presented in Chapters 2–6. It adds the calculation of *r* to the calculation of regression statistics given in Table 5-3. The table may be used as a guide to working the problems presented in the exercises.

6-3 Complete Computational Example for Major Statistics in Chapters 2 through 6

Subject	X	Y	X^2	Y^2	XY
1	510	1.3	260,100	1.69	663.0
2	533	.1	284,089	.01	53.3
3	603	1.5	363,609	2.25	904.5
4	670	1.8	448,900	3.24	1206.0
5	750	3.9	562,500	15.21	2925.0

$N = 5$ $\Sigma X = 3066$ $\Sigma Y = 8.6$ $\Sigma X^2 = 1,919,198$ $\Sigma Y^2 = 22.4$ $\Sigma XY = 5751.8$

$(\Sigma X)^2 = 9,400,356$ $(\Sigma Y)^2 = 73.96$

Intermediate Quantities

$$
\begin{aligned}
(\mathrm{I_{xy}}) &= N(\Sigma XY) - (\Sigma X)(\Sigma Y) \\
&= 5(5751.8) - (3066)(8.6) \\
&= 2391.40
\end{aligned}
$$

$$
\begin{aligned}
(\mathrm{II_x}) &= N\Sigma X^2 - (\Sigma X)^2 \\
&= 5(1,919,198) - 9,400,356 \\
&= 195,634
\end{aligned}
$$

$$
\begin{aligned}
(\mathrm{III_y}) &= N\Sigma Y^2 - (\Sigma Y)^2 \\
&= 5(22.4) - 73.96 \\
&= 38.04
\end{aligned}
$$

(continued)

6-3 Continued

Statistical Computations

$$\bar{X} = \Sigma X / N = 3066/5 = 613.20 \qquad \bar{Y} = \Sigma Y / N = 8.6/5 = 1.72$$

$$s_x^2 = (\mathbf{II_x}) / N(N-1) = 195,634/[5(4)] = 9781.7 \qquad s_y^2 = (\mathbf{III_y}) / N(N-1) = 38.04/[5(4)] = 1.90$$

$$s_x = \sqrt{s_x^2} = \sqrt{9781.7} = 98.90 \qquad s_y = \sqrt{s_y^2} = \sqrt{1.90} = 1.38$$

$$b = \frac{N(\Sigma XY) - (\Sigma X)(\Sigma Y)}{N\Sigma X^2 - (\Sigma X)^2} = \frac{(\mathbf{I_{xy}})}{(\mathbf{II_x})} = \frac{2391.40}{195,634} = .012$$

$$a = \bar{Y} - b\bar{X} = 1.72 - (.012)(613.20) = -5.638$$

$$\hat{Y} = bX + a = .012X - 5.638$$

$$s_{y \cdot x} = \sqrt{\frac{1}{N(N-2)} \left[N\Sigma Y^2 - (\Sigma Y)^2 - \frac{[N(\Sigma XY) - (\Sigma X)(\Sigma Y)]^2}{N\Sigma X^2 - (\Sigma X)^2} \right]}$$

$$= \sqrt{\frac{1}{N(N-2)} \left[(\mathbf{III_y}) - \frac{(\mathbf{I_{xy}})^2}{(\mathbf{II_x})} \right]}$$

$$= \sqrt{\frac{1}{5(5-2)} \left[38.04 - \frac{[2391.4]^2}{195,634} \right]} = \sqrt{\frac{1}{15} [8.8079]}$$

$$s_{y \cdot x} = .77$$

$$r = \frac{N(\Sigma XY) - (\Sigma X)(\Sigma Y)}{\sqrt{[N\Sigma X^2 - (\Sigma X)^2][N\Sigma Y^2 - (\Sigma Y)^2]}} = \frac{(\mathbf{I_{xy}})}{\sqrt{(\mathbf{II_x})(\mathbf{III_y})}}$$

$$r = \frac{2391.40}{\sqrt{(195,634)(38.04)}}$$

$$r = .88$$

SUMMARY

The correlation coefficient is an index ranging between -1.00 and $+1.00$ that reflects the direction and degree of linear relationship between two variables. It is the square root of the proportion of variability in Y that is associated with X. The value of this index, known as the Pearson product-moment correlation, is not affected by changes in the scales of either or both of the variables or by the direction of prediction. The size of the correlation coefficient is influenced by a restricted range of values, the presence of extreme groups of subjects, a single extreme score, and a combination of two groups that have different means, variances, or correlations. The correlation reflects the degree of linear association between the two variables; it alone does not indicate that one variable causes or influences the other.

Formulas

1. Pearson product-moment correlation

$$r = \sqrt{\frac{\Sigma(\hat{Y}_i - \bar{Y})^2}{\Sigma(Y_i - \bar{Y})^2}} \quad \text{(definitional formula)}$$

$$r = \frac{N(\Sigma XY) - (\Sigma X)(\Sigma Y)}{\sqrt{[N\Sigma X^2 - (\Sigma X)^2][N\Sigma Y^2 - (\Sigma Y)^2]}} \quad \text{(raw score)}$$

$r^2 =$ proportion of Y variance associated with differences in X

2. Correlation and regression

$$r = b_{yx}\left(\frac{s_x}{s_y}\right) \qquad (r \text{ in terms of slope and standard deviations})$$

$$r = b_{z_y z_x} \qquad \text{(standard-score version of relationship between } r \text{ and slope)}$$

$$r = \sqrt{1 - \frac{s_{y \cdot x}^2}{s_y^2}} \qquad (r \text{ may be estimated approximately in terms of the variability about the line, } s_{y \cdot x}^2, \text{ relative to the total variability in } Y, s_y^2)$$

3. Computational approaches

$$(\mathbf{I_{XY}}) = N(\Sigma XY) - (\Sigma X)(\Sigma Y) \qquad (\mathbf{II_X}) = N\Sigma X^2 - (\Sigma X)^2 \qquad (\mathbf{III_Y}) = N\Sigma Y^2 - (\Sigma Y)^2$$

$$b = \frac{(\mathbf{I_{XY}})}{(\mathbf{II_X})} \qquad s_{y \cdot x} = \sqrt{\left[\frac{1}{N(N-2)}\right]\left[(\mathbf{III_Y}) - \frac{(\mathbf{I_{XY}})^2}{(\mathbf{II_X})}\right]} \qquad r = \frac{(\mathbf{I_{XY}})}{\sqrt{(\mathbf{II_X})(\mathbf{III_Y})}}$$

Exercises

For Conceptual Understanding

1. What squared deviations are said to compose the variability in Y associated with differences in X? Explain the logic of the reasoning.

2. In what way is the ratio of the variability about the regression line to the total variability in the Y_i related to the value of r?

3. If $r = .70$, what proportion of the variance in Y is associated with differences in X?

4. What effect does the direction of prediction (for example, from X to Y or from Y to X) have in regression and correlation?

5. Explain why the correlation coefficient is identical to the slope of the regression line when the scores are in standard-score form but not when raw scores are used.

6. Suppose a researcher interested in the relationship between grade-point average (GPA) and leadership gives a specially designed leadership test to the 40 students with the highest GPA and the 40 students with the lowest GPA in a class of 500 high-school seniors. Why might the correlation between GPA and leadership in these 80 students be unreasonably high relative to the correlation computed on all 500 high-school seniors in the school?

7. If the point $(\overline{X}, \overline{Y})$ is (15, 25), why will the pair of scores (50, 85) probably influence the regression and correlation between X and Y more than the pair of scores (10, 22)?

8. Discuss and explain whether the following combinations of values are possible:
   a. $N = 25$, $b = .43$, $r = .55$
   b. $s_y = 9$, $r = .83$, $b = -.74$
   c. $s_y = 12$, $b = .70$, $s_{y \cdot x} = 15$
   d. $r = .15$, $b = .80$, $s_y = 12$, $s_x = 12$
   e. $N = 20$, $r = 1.00$, $s_{y \cdot x} = 2.5$
   f. $r = b_{z_x z_y}$

For Solving Problems

9. Below are three scores for each of nine subjects. Compute the correlation between A and B, A and C, and B and C. Add 5 to each score in distribution A and then multiply by 2. Recompute the correlation between A and B. Explain the effect of changing scales on the correlation.

Subject	A	B	C
1	2	2	6
2	0	7	3
3	5	3	10
4	6	0	3
5	6	1	6
6	8	3	1
7	0	7	5
8	3	2	7
9	8	1	10

10. Compute the correlation for the following data. Then add the score pair (12, 8) and recompute. Why does adding one pair of scores change the correlation so much? Can you think of other scores that would alter the correlation just as much but in an opposite direction? Illustrate.

X	Y
3	2
5	4
4	5
2	5
0	3
2	0
1	2

11. It is commonly recognized that a "poor environment" (for example, poverty, low education level of the parents) and the presence of "stressors" (for example, divorce of parents, abuse) are associated with lower IQs in children. Recent research indicates that a child's IQ is related to the number of such "risk factors" in the child's

background, regardless of which specific factors are present.[6] Below are the numbers of risk factors and the IQs of 12 children. Determine the regression equation and the correlation for these data. What IQ do you predict for James, who has 5 risk factors, versus Elizabeth, who has only 2?

Child	Risk Factors	IQ
A	3	112
B	6	82
C	0	105
D	5	102
E	1	115
F	1	101
G	5	94
H	4	89
I	3	98
J	3	109
K	4	91
L	2	107

For Homework

12. Explain with an example how it might be possible to obtain a high positive correlation between two variables when in reality there is no causal relationship between them.

13. If $s_{y \cdot x} = 0$ and $b = -.62$, what do you know about r?

14. Explain and diagram how it would be possible to change a correlation of $r = .00$ into a high negative correlation by adding one more subject to the sample.

---

[6] Inspired by Rutter, M. (1979). "Protective Factors in Children's Responses to Stress and Disadvantages." In M. W. Kent and J. E. Rolf (eds.), *Primary Prevention of Psychopathology (vol. 3): Social Competence in Children* (pp. 49–74). Hanover, NH: University Press of New England; Sameroff, A. J., Seifer, R., Barocas, R., Zax, M. and Greenspan, S. (1987). "Intelligence Quotient Scores of 4-year-old Children: Social Environmental Risk Factors. *Pediatrics, 79,* 343–50.

15. Illustrate on a graph how the scatterplots would look to make the following statements true.
    a. Two subgroups have positive correlations, but the combined group has a negative correlation.
    b. There is a very consistent relationship between X and Y, but the correlation coefficient is nearly zero.
    c. There is a high correlation between height and weight for children 4–10 years old but no correlation within any age span of one year.
    d. There is no relationship between the National Merit test and high-school grades for those students who get scholarships to go to college, but there is a relationship for all high-school students.

16. Answer true or false and explain.
    a. If b is negative, r will always be negative.
    b. If a is positive, r will always be positive.
    c. Dividing each X score but none of the Y scores by 4 will change the value of r.
    d. The correlation, r, of variables X and Y will not change if both are multiplied by the variable W.
    e. Increasing the number of cases from 25 to 100 will probably not affect the size of r very much.
    f. Changing all the X's and Y's to standard scores (z scores) will change r somewhat.
    g. An $r = .67$ and an $r = -.67$ indicate the same degree of linear relationship.
    h. An extreme score will always increase the size of the correlation coefficient.

17. Which of the following sets of facts could *not* exist together? Why? (*Hint:* Draw a picture of the facts.)
    a. $b_{yx} = .24$; $r_{xy} = .64$; $r_{yx} = .58$
    b. $s_y = 4.37$; $s_{y \cdot x} = 0$; $r = -.31$
    c. $r^2 = .84$; $s_{y \cdot x} = 4$; $s_y = 10$
    d. $s_{y \cdot x} = 6$; $s_y = 3$; $r^2 = -.80$
    e. $b_{yx} = .5$; $s_x = 4$; $s_y = 5$; $r = .40$
    f. $\overline{X} = 2$; $\overline{Y} = 3$; $a = 1$; $r = -.60$
    g. $b_{z_y z_x} = .20$; $s_{z_y \cdot z_x} = 1.00$; $r = .63$

18. Can $s_{y \cdot x}^2$ ever be greater than s_y^2? Why? What limits does this put on the range of r? Why? Can $s_{y \cdot x}$ ever equal s_y? What would r equal in this case?

19. Discuss the problem in drawing the stated conclusion in each of the following circumstances.

    a. A theory of reading suggests that the performance of certain physical exercises requires the same neurological activity as does reading. A study of 100 children aged 6–14 reveals a correlation.

    b. A certain health club claims that the correlation is .85 between length of time enrolled in its exercise program and pounds lost. This fact is advertised as proof that the program causes weight loss.

20. Answer the questions with the following data.

Subject	A	B	C
a	6	0	3
b	5	9	8
c	3	11	9
d	7	12	7
e	4	3	5

    a. Compute r_{AB}, r_{AC}, and r_{BC}.

    b. If a new subject were added who had scores of $A = 1$, $B = 1$, and $C = 9$, what would the new correlations be?

    c. What percentage of the variance between A and B can be accounted for when the new subject is included? When the new subject is not included?

    d. If the above scores are in terms of minutes and they are converted to seconds, how do the values of r change?

INFERENTIAL STATISTICS

The first six chapters of this book have been

concerned with *descriptive statistics*—procedures

that organize, summarize, and describe

quantitative information. Another purpose of

statistics is to help make inferential decisions

about what is true about a population when only a

small sample from that population is available. This

part of the text presents several elementary

inferential statistics.

Sampling, Sampling Distributions, and Probability

he first six chapters of this book have been concerned with **descriptive** statistics—that is, procedures that describe and summarize groups of measurements. Attention is now turned to **inferential statistics,** which includes techniques for making inferential decisions when only partial information is available.

The task of inferential statistics can be illustrated by the case of Head Start, a preschool program for low-income children. Federal, state, and local governments are faced with the decision of whether to increase funding for Head Start, and this decision is based in part on the ability of Head Start to produce long-term educational gains in the participating children. Suppose an early preschool program involving 100 children has shown that $2 in remedial education costs are saved for every $1 invested in the preschool program. May one infer that such savings will occur across the nation if Head Start is expanded to include more children? Making this inference is the task of inferential statistics, and the answer depends on how similar the sample of children and the early program is to the additional children to be enrolled in Head Start, how much error is associated with the $2-for-$1 estimated savings, and how good an estimate this figure is of what would happen in a much larger population.

METHODS OF SAMPLING

Recall from Chapter 3 that a **population** is a complete set of subjects, events, or scores that have some common characteristic. Characteristics of populations are known as **parameters.** A subset or portion of a population is a **sample,** and quantities computed on a sample are called **statistics.** Often a statistic is used to estimate the value of a parameter. For example, a pollster might select a sample of 1000 cases with which to estimate the percentage of the American adult population that supports the president. A statistic used to estimate a parameter is no better than the sample upon which it is computed.

Simple Random Sampling

There are many different ways of selecting a sample, but one of the most commonly used methods is simple random sampling.

> In **simple random sampling,** all elements of the population have an equal probability of being selected for the sample of N observations.

Suppose you want to have a random sample of students at your university for an opinion poll on the quality and appropriateness of their educational experience. You might obtain a list of all students in the school, go to a **random number table,** and select a sample of 50. A random number table (Table K) is provided in Appendix 2 of this book. It consists of rows and columns of random numbers. The numbers are random in the sense that for any single digit position, each of the 10 numbers from 0 to 9 has an equal chance of occupying that position. This means that every digit was selected independently of every other digit. Further, not only are all the single digits random, but all two-, three-, and N-digit numbers are also random.

To use the table for getting a sample of 50 students from a list of 3000, assign each student a number between 1 and 3000. Then go to the random number table and mentally block the table off into four-digit columns. Read down the columns until you obtain 50 four-digit numbers that fall between 0001 and 3000 inclusive. The students assigned numbers corresponding to these 50 numbers constitute a randomly selected sample.

Another way a random number table is frequently used is to divide subjects randomly into two groups. For example, it may be desired to give individuals in one group a special treatment of interest (e.g., the early childhood preschool program mentioned above) and the individuals in another a comparison treatment (e.g., no special preschool program, health services but no early childhood educational program, etc.). Children might be recruited for the study in general, and then randomly assigned to the two groups. Specifically, children might be placed on a list in the order in which they are recruited into the study. Then columns or rows of the random number table are used for the randomization process. For example, we might choose the last single-digit column in each set of five two-digit columns in Table K. The first such column (i.e., the one before the first double space) begins 4, 6, 7, 1, 8, 9, 4, 0 . . . and the second such column begins 5, 5, 1, 4, 5, 5, 2, ... Then we would adopt some means of defining these numbers to be in two equal-sized groups. For example, these digits might be categorized into odd (1, 3, 5, 7, 9) versus even (0, 2, 4, 6, 8) or into 0–4 versus 5–9. If the odd–even categorization is selected, then the first column of numbers would be categorized into E, E, O, O, E, O, E, E, ... Finally, the ordered list of subjects would then be matched to the sequence of odd–even designations from the random number table to create the two groups. That is, the first subject on the list would be assigned to the E group (e.g., the "experimental group"), the second to the E group, the third to the O group (i.e., the comparison group), the fourth to the O group, etc.

Sometimes you may question whether the numbers in a random number table are really random. For example, the second column selected in the above example was 5, 5, 1, 4, 5, 5. . . . There were four 5s in the first six numbers—that does not look very random. Also, if these numbers were categorized into odd and even numbers, the "random sequence" would be O, O, O, E, O, O . . . , also not very random looking. If you ask people to write down a random sequence of odd and even numbers, they are very reluctant to put four or more of a kind in a row (e.g., EEEE or OOOO). But in truly random sequences, the probability of EEEE is $(\frac{1}{2})(\frac{1}{2})(\frac{1}{2})(\frac{1}{2}) = .0625$ and the probability of OOOO is also .0625, so the probability of a run of four in a row of either kind is .0625 + .0625 = .125 (see Chapter 12 on Topics in Probability). This means that, on the average, one in every eight such sets of four odd–even digits will be all of the same kind, which is much more likely than most people think.

The above example illustrates another important point: *The concept of random sampling refers to the process by which a sample is selected, not the actual characteristics of the sample obtained.* That is, a random selection process, such as illustrated above, does not necessarily produce two groups that are actually identical or even similar on some important characteristic. Children may be assigned randomly to the special treatment and comparison groups, but that process may produce two groups of children who actually differ on family education level, income, motivation to learn, and other factors that might influence the scores of such children on whatever measure of treatment outcome (e.g., a school readiness test) will be used to compare treatment and comparison children. Randomly assigned children each had an *equal probability* of being assigned to one versus the other group, but that process does not necessarily make the groups *equal* on any particular characteristic. This is so, because two atypical cases, for example, have the same chance of being assigned to the same group as being assigned to two different groups. Even so, randomness is an important attribute of the sampling process, because many statistical procedures require it.

Random samples tend to have two other important characteristics in addition to randomness. In the long run, large random samples will be **representative samples** of all aspects of the population. This means that if a population contains 10% Catholics, a random sample will tend toward having 10% Catholics as the sample size increases. But a small, randomly collected sample will not necessarily have 10% Catholics. To ensure that 10% of a sample is Catholic, a random sample may be taken of all Catholics in the population until their total number represents exactly 10% of the intended sample size. Such a sample is called a **proportional stratified random sample.** Political pollsters, for example, attempt to sample the voting public in such a way as to ensure that each area of the country, each ethnic group, each religious group, and so forth, is appropriately represented. The sampling is random and independent within these groups, but not between them. Since all of the procedures outlined in this text are appropriate for simple random samples, stratified sampling and its associated statistical procedures will not be considered further.

In most cases, random sampling also means that each subject is selected independently of other subjects. **Independence** in sampling implies that the selection of any one element, or subject, for inclusion in the sample does not alter the likelihood of drawing any other element of the population into the sample. In almost all cases in which random sampling is required in this text, the implication is that the elements have been independently selected.

Sampling in Practice

In actual practice it is extremely difficult to obtain a truly random sample. Let's suppose, for example, that one wanted to sample the population of a given geographical area. It is possible to purchase a computer program that specifies the area codes and prefixes for all telephones in the defined area and dials telephone numbers randomly within these specifications. But is this a truly random sample of people in the geographic area? It is much closer to that goal than many samples actually used in research, but it is not perfectly random. For one thing, people without telephones are not included. Since such people are usually poor, the random-digit-dialed sample underrepresents the very poor. The representativeness of such a sample also depends on when during the day and week people are called, the number and timing of callbacks to people who do not answer, and many other sampling procedures.

Even when scientists have control over their subjects, such as when rats are used, samples often are not truly random. For example, a group of rats provided by an animal supplier is not a random selection of rats. Rats are raised in cages set on tiers, some of which are closer to the light than others, and the amount of illumination in the rat's rearing experience can influence some types of later behavior. Further, when the rats arrive at the laboratory and are assigned to experimental groups, it is sometimes tempting to place the first 10 in one group, the second 10 in the next, and so on. But it happens that the more curious and active rats frequently come over to the side of the shipping box when it is opened. These rats are more accessible and easier to pick up. They are thus selected first and go into the first group if the above procedure is carried out. It is a good idea to use a random number table to determine group assignments in the manner described earlier or at least alternate the rats by assigning one to the first group, another to the second, and so on.

Another form of bias in sampling occurs when human volunteers are used. College students who volunteer for experiments are probably somewhat different with respect to academic ability, motivation, and so on from students who do not volunteer. Who volunteers can be an interesting issue on its own right, and it determines what kind of research can and cannot be done. For example, some developmental psychologists depend upon parents to volunteer their infants for observation. It is likely that highly educated parents are more receptive to science's "need" for subjects than are less-educated parents. Also, parents of boys are more likely to volunteer than parents of girls. Thus, the sample one obtains is not random because it may contain a preponderance of males and infants from highly educated families.

How can you safeguard against such bias in your sample? The best way is simply to be cautious and aware of what you are doing. In addition, it is advisable to measure the sample you have selected on several dimensions (such as age, education, "normality," and so on) appropriate to the research (as long as the measurement of these traits does not influence the subjects in any way) so that readers can judge whether the sample has the general characteristics of the population being discussed. The goal is to ensure that inferences and generalizations are made to the appropriate population. Further, while the statistical procedures in this book require independent random sampling, they are used even when the samples are *not* perfectly random or independent. The careful researcher does the best job possible and notes when procedures are likely to violate these assumptions and produce bias in the results. Conclusions from the research are generalized only to the population for which the sample was representative (for example, people with telephones).

SAMPLING DISTRIBUTIONS AND SAMPLING ERROR

Even when the sample is random and appropriate, it is not likely to be identical in most respects to the population because it is a sample of considerably fewer cases than are contained in the population. In short, as a method of estimating the population, sampling can be an imprecise process.

An Empirical Sampling Distribution

Suppose there are 20 students in your statistics class and the professor springs a surprise quiz of 10 questions. The scores of the 20 students comprise the population of raw scores and are presented in the left-hand column of Table 7-1. Now suppose the professor forgets to tell the class what the average score was, so you have little idea whether your score is good or bad. So you ask a random sample of four classmates what their scores were for the purpose of estimating the class or population mean (the mean of the 20 students, which is 3.90). The first such sample of four students provides the scores (1, 5, 9, 0), which have a mean of 3.75, as indicated in Table 7-1. But suppose other students also obtain random samples of four students. Looking down the right-hand column in Table 7-1, which shows the means of 10 randomly selected samples from the population, one can see that randomly selected samples from the same population vary in the value of their means. Moreover, not one sample yields a mean that precisely equals the population mean. Why do samples produce means that differ from one another?

In a random sample, each subject in the population has an equal *opportunity* to be drawn into the sample. But, as was discussed above, this is not to say that the sample that is actually drawn will faithfully reflect the population's characteristics. For example, one might randomly select a sample that just happens to contain many exceptionally bright students. One obvious reason why random samples differ from one another is that they are composed of different individuals.

Returning to Table 7-1, we see that the means computed on the 10 samples could themselves be considered scores in a distribution—a distribution of means rather than of raw scores. Such a distribution has a special name and function in statistics.

> The distribution of a statistic determined on separate independent samples of size *N* drawn from a single population is called a **sampling distribution.**

Thus, the distribution of the 10 sample means in the third column of Table 7-1 is a sampling distribution of the mean.

The sampling distribution should be carefully distinguished from the two raw-score distributions introduced earlier. The type of distribution of central concern throughout Part 1 of this text is the **sample distribution,** which is a collection of measured scores obtained from a subgroup of a population. The middle column of Table 7-1 shows 10 sample distributions, each of size $N = 4$. Also introduced in Part 1 (Chapter 3) was the concept of a **population distribution,** which is the full array of raw scores in the population. Now we have a third type of distribution. A **sampling distribution** is a distribution of a statistic, not of raw scores. The third column in Table 7-1 is a distribution of a statistic—the mean—obtained from 10 different samples of size $N = 4$.

Sampling distributions may be of two general types. The right-hand column of Table 7-1 shows an **empirical sampling distribution.** The word *empirical* signifies "experienced" or observed, and these 10 means are observations presumably made by actually collecting 10 samples and computing $\overline{X}$ for each sample. In contrast, a **theoretical sampling distribution** is a theoretical distribution of a statistic, and its characteristics are determined mathematically rather than by repeated observations and pertain to a distribution of a statistic over an unaccountable number of cases.

7-1 Population Distribution of 20 Raw Scores, 10 Observed Sample Means ($N = 4$), and An Empirical Sampling Distribution of the Mean

Population Distribution of Raw Scores	10 Observed Sample Distributions ($N = 4$)	Empirical Sampling Distribution of $\overline{X}$'s
	(1, 5, 9, 0)	3.75
	(0, 3, 1, 5)	2.25
	(5, 8, 3, 0)	4.00
	(1, 5, 0, 7)	3.25
	(7, 6, 1, 3)	4.25
	(3, 2, 1, 7)	3.25
	(2, 0, 3, 5)	2.50
	(1, 2, 1, 1)	1.25
	(2, 7, 1, 7)	4.25
	(9, 7, 6, 2)	6.00

$$\begin{bmatrix} 6, & 9, & 0, & 3, & 1, \\ 5, & 7, & 7, & 1, & 3, \\ 2, & 5, & 1, & 2, & 1, \\ 2, & 7, & 8, & 1, & 7 \end{bmatrix}$$

$\mu = 3.90, \sigma = 2.81$

Mean of $\overline{X}$'s $= \overline{X}_{\bar{x}} = 3.48$
Standard deviation of $\overline{X}$'s $= s_{\bar{x}} = 1.31$

Note that a sampling distribution differs from both a sample distribution and a population distribution in that it is a collection of statistics rather than a collection of raw scores. The statistic whose sampling distribution is shown in the third column of Table 7-1 is the mean, but one can also have sampling distributions of other statistics. For example, one could compute the standard deviations of the 10 sample distributions shown in the table and thus obtain a sampling distribution of the standard deviation.

Just as a distribution of raw scores has certain characteristics, such as a mean and a standard deviation, so too does the sampling distribution of a statistic. The sampling distribution shown in Table 7-1, for example, itself has a mean of 3.48 and a standard deviation of 1.31. Therefore, it will be necessary to have terms and symbols to represent the mean, standard deviation, and variance of the several distributions distinguished above. Table 7-2 summarizes these terms and symbols. First, notice at the left that distributions are composed either of raw scores or of statistics, and in the latter case they are called sampling distributions. In addition, a distribution may be based on a sample or a population. As mentioned above, distributions also are either empirical (that is, based on actual observations) or theoretical (that is, defined in terms of their characteristics—mean, variance, shape—and not in terms of actual observations). In practice, samples are usually empirical, and populations are usually theoretical, but this is not always the case. Similarly, in practice, sampling distributions are almost always theoretical, but Table 7-1 contains an empirical sampling distribution to illustrate the concept.

7-2 **Terms and Symbols for the Mean, Standard Deviation, and Variance in Different Types of Distributions**

Distribution of Raw Scores	Mean	Standard Deviation	Variance
Sample	$\overline{X}$	s_x or s	s_x^2 or s^2
Population	μ_x or μ	σ_x or σ	σ_x^2 or σ^2

Sampling Distribution of the Mean	Mean	Standard Error of the Mean	Square of the Standard Error of the Mean
Sample	$\overline{X}_{\bar{x}}$	$s_{\bar{x}}$	$s_{\bar{x}}^2$
Population	$\mu_{\bar{x}}$ or μ	$\sigma_{\bar{x}}$	$\sigma_{\bar{x}}^2$

Equivalences

$$\mu_x = \mu_{\bar{x}} = \mu \qquad s_{\bar{x}} = \frac{s_x}{\sqrt{N}} \text{ and } \sigma_{\bar{x}} = \frac{\sigma_x}{\sqrt{N}}$$

Each of the four kinds of distributions listed in Table 7-2 has a mean, standard deviation, and variance, but these quantities have different names and symbols depending on the type of distribution to which they pertain. When these quantities refer to samples or empirical distributions, they are symbolized by the letters $\overline{X}$, s, and s^2, when they pertain to populations or theoretical distributions, they are symbolized by the Greek letters μ, σ, and σ^2. A subscript designates the distribution to which the quantity refers.

It is important for students, when reading the remainder of this chapter and the next chapter, to pay special attention to the modifiers of particular statistics and distributions. One must be careful to note whether it is the *sample* mean or the *population* mean that is being discussed, or the *sample* standard deviation or the *population* standard deviation. Also, note whether it is a *sample* distribution, a distribution of *raw scores* (the same as a *sample* distribution), a *sampling* distribution, a distribution of the *means* (the same as the *sampling* distribution of the mean), or a *theoretical sampling* distribution. Some students find it helpful to read the remaining material more slowly to ensure that they are thinking about the right concept.

Sampling Statistics

Statistics may pertain to samples or to sampling distributions.

> **Sample statistics** are quantities that characterize samples of raw scores.
>
> **Sampling statistics** are quantities that characterize sampling distributions of statistics.

While sample statistics are calculated on samples of raw scores, they often are used to estimate characteristics of a population of raw scores. Sampling statistics typically are calculated on a single sample of raw scores but are used to estimate the characteristics of a theoretical sampling distribution of a statistic.

In this chapter we focus on two sampling statistics—the mean and standard deviation of the sampling distribution of means. We consider their special names, how they are related to their corresponding quantities based on a population of raw scores rather than a population of sample means, and how their values may be estimated from a single sample of cases. The material presented below is summarized in Table 7-2.

The first concept is the mean of the population of sample means, symbolized by $\mu_{\bar{x}}$. The Greek μ indicates that this is a parameter, not a statistic, and the subscript $\overline{X}$ signifies that it is based on the population or theoretical distribution of sample means.

The population mean of the sampling distribution of means, $\mu_{\bar{x}}$, can be distinguished from the population mean of the raw scores, μ_x, by the different subscripts. However, it happens that these two population values are identical. Symbolically,

$$\mu_{\bar{x}} = \mu_x = \mu$$

That is, the mean of the population sampling distribution of the means ($\mu_{\bar{x}}$) equals the mean of the population of raw scores (μ_x), and the symbol μ without a subscript is customarily used to indicate this value. It is called simply "the population mean" or "mu" after its Greek symbol.

The value of the population mean may be estimated by the mean of a sample of raw scores. That is, $\overline{X}$ estimates μ. Thus, the population mean of raw scores, μ_x, and the population mean of the sampling distribution of means, $\mu_{\bar{x}}$, can both be estimated by drawing a single sample of raw scores and calculating the mean, $\overline{X}$.

Consequently, the population mean of the sampling distribution of means, $\mu_{\bar{x}}$, can be estimated without drawing several samples and calculating the mean over all such samples as was done above in Table 7-1 when an empirical sampling distribution was created. The mean of a single sample of raw scores, $\overline{X}$, can be used to estimate $\mu_{\bar{x}}$, μ_x, and μ. Of course, it is not a perfect estimate, as discussed below.

Besides a mean, a sampling distribution has a standard deviation.

> The standard deviation of a sampling distribution of a statistic is called the **standard error** of that statistic. Consequently, the standard deviation of the sampling distribution of the mean is known as the **standard error of the mean.** In the population, it is symbolized by $\sigma_{\bar{x}}$.

The standard error of the mean is simply the standard deviation of the sampling distribution of means. Earlier, in Table 7-1, an *empirical* standard error of the mean was actually calculated for the 10 samples and found to be 1.31. In the population or in the theoretical sampling distribution of the mean, however, the standard error of the mean, symbolized by $\sigma_{\bar{x}}$, is not calculated. Again, the Greek letter σ indicates that this is a population or theoretical quantity, not a sample value, and the subscript $\overline{X}$ indicates that it is the standard deviation (or standard error) of the sampling distribution of means. The $\sigma_{\bar{x}}$ can be distinguished from the population standard deviation of raw scores, σ_x, by the different subscripts.

Because of the great importance of the concept of a standard error of the mean in inferential statistics, it is crucial for you to have a firm grasp of its meaning. We have said that the mean of one sample of scores will not likely equal the mean of another sample of scores, even if both samples are randomly selected from the same population. The standard deviation is a numerical index of variability in such a distribution of means. Therefore, the standard deviation of the sampling distribution of means, $\sigma_{\bar{x}}$, is a numerical index of the extent to which means vary from one sample to another, presumably because of error associated with sampling. This is why it is given the special name of the standard error of the mean.

More generally, the standard error of the mean is an index of the amount of error that results when a single sample mean is used to estimate the population mean; that is, it is an index of **sampling error.** The means of samples all drawn from the same population of raw scores vary in value, and their standard deviation—the standard error—reflects the extent of that variation, which is caused by sampling error. For example, if $\sigma_{\bar{x}} = 5$ for samples of 20 males on a reading test but $\sigma_{\bar{x}} = 10$ for samples of 20 females on the same test, there is less sampling error for males than for females. This implies that the random variation between means from one sample to another, for samples of size 20, is less for males than for females.

It was stated above that the mean of the theoretical sampling distribution of the mean is identical to the mean of the population distribution of raw scores, $\mu_{\bar{x}} = \mu_{\bar{x}}$ $= \mu$. In contrast, the standard deviation of the sampling distribution of the mean (the standard error of the mean), $\sigma_{\bar{x}}$, is *not* identical to the standard deviation of the population distribution of raw scores, σ_x, but it is related to it.

> The **standard error of the mean, $\sigma_{\bar{x}}$,** equals the standard deviation of the population of raw scores divided by the square root of the size of the samples on which the means are based:
>
> $$\sigma_{\bar{x}} = \frac{\sigma_x}{\sqrt{N}}$$

If the standard deviation of the population of raw scores is $\sigma_x = 2.81$ and the sample size is $N = 4$, the theoretical standard error of the mean for samples of $N = 4$ is

$$\sigma_{\bar{x}} = \frac{\sigma_x}{\sqrt{N}} = \frac{2.81}{\sqrt{4}} = 1.40$$

How can the value of $\sigma_{\bar{x}}$ be estimated? Just as a sample mean, $\overline{X}$, could be used to estimate μ, the sample standard deviation, s_x, can be used to calculate $s_{\bar{x}}$, which can be used to estimate $\sigma_{\bar{x}}$.

> The population standard error of the mean, $\sigma_{\bar{x}}$, may be estimated by
>
> $$s_{\bar{x}} = \frac{s_x}{\sqrt{N}}$$
>
> in which N is the size of the sample of X's.

Two things should be noticed about the standard error of the mean as expressed by $s_{\bar{x}} = s_x/\sqrt{N}$. First, all that is required to calculate $s_{\bar{x}}$ is the standard deviation and the N from a single sample of cases, yet $s_{\bar{x}}$ represents an estimate of the amount of variability (or sampling error) in means from *all* possible samples of size N from the population of raw scores. Thus, it is not necessary to select several samples to estimate the population sampling error of the mean; $s_{\bar{x}}$ estimates $\sigma_{\bar{x}}$, and all that is required to calculate $s_{\bar{x}}$ is s_x and N from a single sample of raw scores.

Second, observe that the formula for $s_{\bar{x}}$ states that the standard deviation of the sample must be divided by the square root of N; that is, $s_{\bar{x}} = s_x/\sqrt{N}$. Therefore, the variability of means from sample to sample will always be smaller than the variability of raw scores. Also note that as N becomes larger, $s_{\bar{x}}$ becomes smaller. Thus, the variability of sample means decreases as the size of the sample increases. Consequently, for large samples one expects $\overline{X}$, the sample estimator of the population mean μ, to be less variable from sample to sample, and thus a more accurate estimate of μ, than if the sample size were smaller. In short, when parameters must be estimated, it is a good idea to have as large a sample as possible.

Other Standard Errors

The sampling distribution and standard error of the mean have been discussed in detail, but a sampling distribution and a standard error exist for any statistic. In each case the logic is the same. Random samples differ in their characteristics, and any statistic will vary somewhat from sample to sample. The theoretical sampling distribution is the distribution of a particular statistic determined for all possible samples of size N, and the standard deviation of that statistic's sampling distribution is its standard error. Therefore, one can imagine standard errors for the median, the variance, and even for the difference between two sample means. In each case the standard error reflects the relative extent of the error in using that sample statistic to estimate its corresponding population parameter.

Sampling Distributions and Normality

Since statistics calculated on a single sample may be used to estimate parameters of sampling distributions, it is never necessary to actually collect an *empirical* sampling distribution. Empirical sampling distributions are not used in statistics except to help students understand the concept of a distribution of a statistic. From this point forward, *sampling distribution* will refer to a *theoretical sampling distribution*. However, the symbol $s_{\bar{x}}$ will still be used to indicate that $\sigma_{\bar{x}}$ is being estimated by sample observations.

Many of the procedures described in this and later chapters rest on the assumption that the sampling distribution of means is normal in form. This is the case if either one of two conditions is met.

Given random sampling, the sampling distribution of the mean

1. is a normal distribution if the population distribution of the raw scores is normal, or

2. approaches a normal distribution as the size of the sample increases even if the population distribution of raw scores is not normal.

If the population distribution of raw scores is normal, the sampling distribution of the mean will also be normal. However, since the population is rarely available, how can you know whether the population distribution is normal? One way to make an educated guess is to determine whether a random sample of raw scores from the population is normally distributed. Alternatively, some variables are known to be normally distributed in certain groups. Height and IQ among 21-year-old males without specific mental challenges, for example, are likely to be normally distributed. But some variables are usually not normally distributed. For example, while the IQs of all non–mentally challenged 21-year-olds are probably normally distributed, the IQs of all 21-year-old college students are not because low scores are not represented as frequently in college groups as are extremely high scores. Family income, the latency for a rat to move out of a startbox in a maze, and percentage correct on a relatively easy exam are variables that are not usually normally distributed. Notice that these variables are bounded on one end of their scales (for example, $0 income, 0 seconds, 100% correct).

The scores will tend to fall near the bounded end of the scale, and the distribution is likely to be skewed toward the other direction. Fortunately, many of the variables measured in social sciences can be assumed to be normally distributed. When variables are not normal, the statistical techniques described in Chapter 14, rather than those described below, may be used.

A second way to obtain a normal sampling distribution of the mean is to select a large enough sample of raw scores. The sampling distribution of the mean will approach a normal distribution as the size of the samples composing the sampling distribution increases, *even though the population distribution of raw scores on the variable of interest is not normal.* This principle says simply that the sampling distribution of the mean is more likely to be normal in form the larger the size of the samples (i.e., N) on which the means are based. Just how many cases constitute a sufficiently large sample depends upon many factors, one of which is the extent of the departure from normality of the population distribution. If the population distribution does not deviate too much from normality, samples of size $N = 2$ might produce a sampling distribution of the mean that is quite normal, whereas if the nonnormality in the population is severe, N's of 20, 30, or several hundred might be necessary. In short, the sampling distribution of the mean will approach a normal form as the size of the samples upon which the means are based increases. This crucial principle of statistics is called the *Central Limit Theorem.*

The following fact is one of the reasons normality is necessary for the statistical procedures to be described:

> If the population distribution of raw scores is normal and if the sample observations are independent and randomly selected from that population, then the sample mean and sample variance (or standard deviation) are independent of one another across samples.

Any two variables (including the statistics $\overline{X}$ and s^2) are independent if they are unrelated to each other. This means that the value of one tells you nothing about the value of the other, and increasing or decreasing the value of one does not necessarily change the value of the other. So, if the population of raw scores is normal and if the observations in samples from that population are selected randomly and independently, then the means and variances of those samples are independent of each other—that is, knowing that a mean is higher or lower than typical tells you nothing about whether the variance is higher or lower than typical. The independence of the mean and variance of a normal distribution, for example, will be important later.

To review, most of the statistical procedures to be described subsequently depend on several principles:

1. Samples are composed of randomly and independently selected observations.

2. A sampling distribution is the distribution of a statistic; the standard deviation of a sampling distribution (standard error) is an index of the extent to which the statistic varies from one sample to another (that is, of sampling error).

3. The sampling distribution of the mean is normal if the population distribution of raw scores is normal or the size of the samples is large.

4. The mean and standard deviation of a random sample from a normal population are independent.

PROBABILITY AND ITS APPLICATION TO HYPOTHESIS TESTING

The purpose of inferential statistics is to assist in making inferences and judgments about what exists on the basis of only partial evidence. This is accomplished by using probability. In a way, most people use subjective probability every day. You want to go to the football game this afternoon, but someone has warned you that it is going to rain. You look outside an hour before the game and, while there are clouds, the sky is not threatening. You make a judgment about going to the game based on your subjective probability that it will not rain.

Scientists, however, prefer to use numerical probability, rather than their subjective feelings, as an index of the likelihood of events. The numerical probability is public knowledge (all scientists can observe and understand it), the probability of one event can be easily compared with the probability of a different event, and certain rules can be adopted about how high or low the probability must be to justify one decision or another. Therefore, a knowledge of the concept of probability is essential to understanding the process of statistical inference and decision making in science. A more thorough examination of probability is offered in Chapter 12.

Probability and Relative Frequency

The determination of a simple numerical probability implies an **idealized experiment**. In figuring the chances of a head or a tail when flipping a coin, one actually assumes an experiment in which the coin is tossed over and over again. It is assumed that both heads and tails are equally likely. On any single throw, either a head or a tail will occur, but in the idealized experiment of repeated flips of a coin, the ratio of heads to all possible outcomes will approach $\frac{1}{2}$, or .50. In short, the probability of a head, for example, is the relative frequency of heads versus all other possible outcomes in an unlimited number of repetitions of the coin flip, which constitutes an idealized experiment.

> **Probability** is theoretical relative frequency—the relative frequency of score values in a theoretical distribution based upon an unlimited number of cases.

From the standpoint of probability, an idealized experiment consists of an unlimited number of cases, and the probability of a particular outcome is the theoretical relative frequency of that outcome in the distribution of all outcomes of the idealized experiment. For example, suppose Figure 7-1 represents the theoretical relative fre-

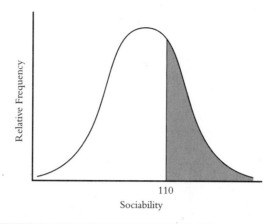

Figure 7-1 Theoretical relative frequency of the sociability scores of salesperson applicants. The probability of the next applicant having a score higher than 110 is indicated by the proportion of the total area under the curve represented by the shaded area.

quency distribution of sociability scores for salesperson applicants at a large corporation over the last 10 years. The higher the score the more social, extroverted, and comfortable the person is with people, a characteristic associated with successful salespersons. In a sense, this distribution represents the results of an "idealized experiment"— the sociability scores of an entire population are available. The many possible numerical scores are the "outcomes" of the idealized experiment. Now, assuming that the applicant pool has not changed in this regard and the next applicant is a random selection, what is the probability that the next applicant would have a sociability score greater than 110? According to the conception of probability as relative frequency discussed above, this probability value should be given by the theoretical relative frequency of scores that exceed 110. The approach to determining the probability here rests on equating the area between the curve and the abscissa with the concept of *theoretical relative frequency,* just as was done in Chapter 4. If the area under the curve in Figure 7-1 represents theoretical relative frequency, then the proportion of the total area that lies between 110 and $+ \infty$ (indicated in the figure by the shading) represents the theoretical relative frequency of applicants having sociability scores greater than 110. Suppose the relative frequency of this area is .27 or 27% of the total. Therefore, the probability that the next applicant has a sociability score above 110 is .27. To formalize:

The proportion of the total area under the curve of a theoretical relative frequency distribution that exists between any two points represents the probability of obtaining the events contained within the interval delimited by those two points.

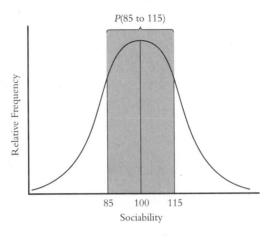

Figure 7-2 The proportion of the total area under the curve that is shaded represents the probability that the next applicant has a sociability score between 85 and 115.

Consider another example: What is the probability that the next applicant will have a sociability score between 85 and 115? This probability is given by the proportion of the total area that exists between scores of 85 and 115, as illustrated in Figure 7-2. In this case, this proportion is 65% of the total, so the theoretical relative frequency and the probability that the next applicant has a sociability score between 85 and 115 is .65.

The Standard Normal Distribution and Probability

The standard normal distribution is a theoretical relative frequency distribution. It can be used in the ways described in Chapter 4 to determine theoretical relative frequency or probability for problems like the two examples just given. The percentiles or relative frequencies for the standard normal distribution are presented in Table A of Appendix 2. In the pages that follow, sampling distributions, which reflect the amount of sampling error in a statistic, will be related to a known theoretical relative frequency distribution,[1] such as the standard normal, and this will permit probabilities to be assigned to the likelihood that one or another circumstance is true. This is the fundamental strategy of inferential statistics.

---

[1] The equation for a theoretical relative frequency distribution, such as the standard normal and several other distributions to be considered, is known as a **probability** (or **density**) **function.** To determine the probability of the occurrence of events located between any two points on the dimension, this function is integrated between these points by the methods of calculus. The values in Appendix 2, Table A, and several other tables in this book represent the results of such a process.

ESTIMATION

One of the most important purposes of statistical inference is to use sample statistics to estimate their corresponding population parameters. Of course, a statistic must be a "good" estimator of its respective parameter, and you recall from Chapter 3 that the denominator of the formula for the sample variance is $N - 1$ rather than just N to make the sample variance a better estimator of the population variance.

Characteristics of a Good Estimator

A "good" estimator is defined by several criteria, including unbiasedness, consistency, relative efficiency, and sufficiency.

Unbiasedness   A good estimator should be unbiased.

> An **unbiased estimator** of a population parameter is one whose average over all possible random samples of a given size equals the value of the parameter.

In actuality, statisticians determine whether a statistic is an unbiased estimator by using special mathematical procedures called **expectation theory,** which are beyond the scope of this text. But we can get an idea of what the criterion of unbiasedness implies when applied to the sample mean as an unbiased estimator of the population mean by looking at Table 7-3.

Suppose a population has a mean of 90 on a given test. If a single sample of size N were drawn, it is unlikely that the mean of the sample would be exactly 90. Perhaps, for example, it is 73, a difference of -17 between sample and population means. This information is presented in the first line of Table 7-3. Now suppose you were to create a small empirical sampling distribution of the mean by taking five independent samples, each of size N, calculating the mean of each sample, and then computing the mean of those five means (that is, determining the mean of this empirical sampling distribution of the mean). Suppose that mean is 96, which represents a difference of $+6$ between the sample estimate and the population value. This information is presented in the second line of Table 7-3. Now suppose empirical sampling distributions were created that were composed of 10, 50, 100, 1000, and then an infinite number of samples, each of size N. Notice in Table 7-3 that as the number of samples each of size N increases, the difference between the average sample estimate and the population value tends to become smaller and smaller. Therefore, as the number of samples increases, the average value of the sample estimate gets closer and closer to the population value until in the end, with an infinite number of samples, the average value of the estimator equals the value of the population parameter. If it does, as it does in this case, the sample statistic is defined to be an unbiased estimator of this population parameter. The sample mean, then, is an unbiased estimator of the population mean,

	Average of Hypothetical Sample Means for Different Numbers of Samples	
7-3		

Number of Samples	Average of Sample Means	Difference between Average Sample Mean and Population Mean (90)
1	73	− 17
5	96	+6
10	95	+5
50	87	−3
100	89	−1
1000	90.03	+ .03
∞	90.00	0

although this fact is actually demonstrated by expectation theory, not by empirical sampling distributions, which are given here for purposes of illustration only.

It might seem that all sample statistics would be unbiased estimators of their corresponding parameters, but this is not the case. Recall that if the sample variance is defined with N in the denominator, it would not be an unbiased estimator of the population variance. In fact, through expectation theory, it can be shown that such a sample variance tends to underestimate the population variance. But if the denominator of the sample variance is composed of $N - 1$ rather than N, as it is throughout this text, then this sample value is an unbiased estimator of the population variance. Ironically, even though s^2 is an unbiased estimator of σ^2, its square root, the sample standard deviation, s, is not an unbiased estimator of the population standard deviation. However, the bias is quite small, especially if large samples are involved, and corrections are rarely used.

Consistency   Another criterion of a good estimator is consistency.

A **consistent estimator** tends to get closer to the value of the population parameter as the size of the sample increases.

The mean and variance of a sample are likely to be closer to their corresponding population values for larger than for smaller samples. Therefore, the sample mean and variance are consistent estimators of the population mean and variance, respectively.

But are not all estimators consistent—the larger the sample the closer the sample value is likely to be to the population value? The answer is no. Suppose the first score in a sample is used to estimate the population mean. It, like the sample mean, is an unbiased estimator. But the value of the first score does not tend to converge on the value of the population mean as the sample size increases, because all the remaining cases sampled after the first are irrelevant to its value. Therefore, the first score selected in a distribution is an unbiased but not consistent estimate of the population mean.

Relative Efficiency   A third criterion for a good estimator is relative efficiency.

> A **relatively efficient estimator** is one whose sampling distribution has a smaller standard error than another estimator for samples of any particular size.

For example, the sample mean and median are both unbiased estimators of the population mean in normal distributions (but not necessarily in other types of distributions). But the variability of sample means (that is, the standard error of the mean) is less than the variability of sample medians (that is, than the standard error of the median). Therefore, the mean is a relatively more efficient estimator of the population mean in normal distributions than is the sample median. Because the normal distribution is relatively common and because many more advanced statistical procedures require normal distributions, the mean is often preferred over the median because of its relative efficiency. Note, however, that this may not be the case for certain distributions that are not normal.

Relative efficiency should not be confused with the concept of resistance introduced in Chapter 4 in association with the median and fourth-spread as indicators of distributions. It was stated there that the median was a more resistant indicator of central tendency than the mean, because its value was less influenced than that of the mean by changes in a few scores, especially atypical deviant scores. While a few atypical scores have less influence on the median than the mean, the variability of the value of the median from one sample of size N to the next—that is, its standard error—is somewhat larger than the variability of the mean. Therefore, the median is more resistant but the mean is more efficient as an estimator of central tendency.

Sufficiency   A final criterion is sufficiency.

> A **sufficient estimator** is one that cannot be improved as an estimator by using any aspects of the sample data that are not already involved in its definition.

The sample proportion is a sufficient estimator of the population proportion because its accuracy cannot be improved by considering any aspects of the data not already involved in its definition. The sufficiency of many other statistics, including the sample mean and variance, are more complicated.

Interval Estimation

Suppose that a psychologist in a school system wants to know the average IQ of students in a given high school. It is rather expensive to give an IQ test to each student, so a random sample of 25 students is tested. Suppose the sample mean is 109.

If the psychologist were required to use this information to estimate with one value the mean IQ of the population of all students in the high school, the estimate would be 109. After all, the sample mean is an unbiased estimator of the population mean, and if the sample was indeed random, one would feel somewhat confident that the sample mean of 109 was near the population mean. This is called **point estimation** because a single value is used as the estimator.

However, if one asked the psychologist whether the population mean actually was *exactly* 109, the answer certainly would be no. An alternative approach is to give a range of values such that one is reasonably confident that the interval limited by these values includes the population mean. For example, the psychologist might say that the interval of 103–115 is likely to contain the population mean. This is called **interval estimation** because the estimator is an interval, not a single value.

The interval to be constructed is called a **confidence interval** and the values describing the boundaries of such an interval are called **confidence limits.** The degree of confidence in the proposition that such an interval actually contains the population mean is indicated by a probability value. Of course, one would expect that a very large interval would be more likely to contain the population value than a very small one (everything else being equal). There are potentially any number of confidence intervals, each having a particular probability associated with it. The most commonly used confidence intervals are the "95% confidence interval" and the "99% confidence interval." The 95% confidence interval, for example, is a range of values that is likely to contain the population mean 95% of the time—that is, the probability is .95 that the interval contains the population mean.

To understand how a 95% confidence interval for a sample mean is determined, consider the data from the above example (and see Figure 7-3). The sample mean was 109 for a sample of $N = 25$. Suppose it is known that the population standard deviation is 15. Now consider the sampling distribution of the mean. Given the data at hand, the sample mean of 109 is a good estimate of the mean of the sampling distribution of the mean (that is, the population mean), and $\sigma_x/\sqrt{N} = 15/\sqrt{25} = 3$ is the standard deviation of this distribution of means (that is, the standard error of the mean, $\sigma_{\bar{x}}$). Now, 95% of the sample means will fall between $P_{.025}$ and $P_{.975}$ of the sampling distribution of the mean. The standard normal or z distribution presented in Table A of Appendix 2 can be used to determine these percentile points. To find $P_{.975}$, look down the third column in each set of three columns in Table A. This column gives the proportion of area under the standard normal to the right of a given z value or the proportion of area "in the tail" of the distribution. $P_{.975}$ is the point such that $2\frac{1}{2}\%$ or .0250 of the area falls to its right. Look down the column until you find .0250. It corresponds to the z value of 1.96. Since the standard normal distribution is symmetrical, $z = -1.96$ corresponds to $P_{.025}$. Therefore, 95% of the area of the standard normal distribution falls between the z values of -1.96 and $+1.96$. Recall, now, that the mean of the standard normal is 0 and the standard deviation is 1.00. As a result, a z value of a point corresponds to the number of standard deviations that point is away

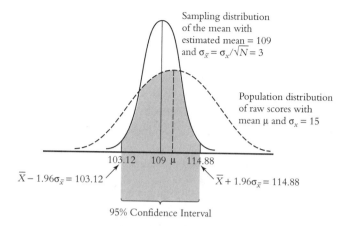

Figure 7-3 Illustration of a 95% confidence level for a sample mean. The population of raw scores has a mean μ (dashed vertical line) and a standard deviation of 15. A sample of 25 is drawn, with $\overline{X} = 109$. The sampling distribution of the mean is estimated to have a mean of 109 and it has a standard deviation of $\sigma_x/\sqrt{N} = 15/\sqrt{25} = 3$. The percentiles of the normal distribution indicate that 95% of such sample means would fall between $\overline{X} \pm 1.96\sigma_{\overline{x}}$, or $109 \pm 1.96(3)$ or 103.12 to 114.88. This is the 95% confidence interval for the mean; it indicates that in 95% of such samples of size 25, the population mean would fall within the interval $\overline{X} \pm 1.96\sigma_{\overline{x}}$.

from the mean. Therefore, 95% of the area of a normal distribution—*any* normal distribution—falls between -1.96 standard deviations and $+1.96$ standard deviations of the mean. In the present case, the normal distribution is the sampling distribution of the mean. It has a mean estimated to be $\overline{X}$ and a population standard deviation (that is, a standard error of the mean) of $\sigma_{\overline{x}}$. So, formally stated,

95% confidence limits for the mean of a normal distribution are $\overline{X} - 1.96\sigma_{\overline{x}}$ and $\overline{X} + 1.96\sigma_{\overline{x}}$. In words, the 95% confidence limits for the mean of a normal distribution equal the sample mean plus and minus 1.96 times the standard error of the mean.

In this example, $\overline{X} = 109$ and $\sigma_{\overline{x}} = \sigma_x/\sqrt{N} = 15/\sqrt{25} = 15/5 = 3$. So the 95% confidence limits are

$$\overline{X} - 1.96\sigma_{\overline{x}} = 109 - 1.96(3) = 109 - 5.88 = 103.12$$

$$\overline{X} + 1.96\sigma_{\overline{x}} = 109 + 1.96(3) = 109 + 5.88 = 114.88$$

Thus, we are 95% confident that such an interval, in this case 103.12 to 114.88, contains the population mean. In other words, if a 95% confidence interval were computed on each of the unlimited number of samples of $N = 25$ drawn from this population, on the average 95% of such intervals would include the population mean value within their limits.

Note that the probability statement applies to the interval and not to the population mean. The population mean is a fixed value, whereas the sample mean and the confidence interval are different from sample to sample. Therefore, the statement "the probability is p that the population mean falls within the interval" is technically bad form, because it implies that the value of μ varies and might or might not happen to land in the stated interval. Actually, it is the interval that is variable, and thus a more correct statement is "the probability is p that the interval includes the population value."

Suppose one wants to be especially cautious and construct an interval that 99 times out of 100 would include the population mean. Following the same logic as above, one would go to Table A to find the z values corresponding to $P_{.995}$ and $P_{.005}$. These points are such that $\frac{1}{2}$% or .0050 of the area of the standard normal falls beyond them. Again, looking down the third column in Table A for .0050, we find .0051 and .0049, so .0050 falls halfway between. It corresponds to a z value of 2.575. Thus, 99% confidence limits for the mean are 2.575 standard errors above and below the mean. Formally,

> 99% confidence limits for the mean of a normal distribution are $\overline{X} - 2.575\sigma_{\bar{x}}$ and $\overline{X} + 2.575\sigma_{\bar{x}}$. In words, the 99% confidence limits for the mean of a normal distribution equal the sample mean plus and minus 2.575 times the standard error of the mean.

In the present example, these limits are

$$\overline{X} - 2.575\sigma_{\bar{x}} = 109 - 2.575(3) = 109 - 7.725 = 101.275$$
$$\overline{X} + 2.575\sigma_{\bar{x}} = 109 + 2.575(3) = 109 + 7.725 = 116.725$$

Therefore, the probability is .99 that such an interval, in this case 101.275 to 116.725, contains the population mean.

Two things should be observed. First, recall that the sampling distribution of the mean will be normal in form if the distribution of raw scores is normal or if the sample size is large even if the distribution of raw scores is not normal. So, in most cases, the assumption of a normal distribution will be met, and the values from the standard normal distribution can be used as above. Second, the population standard error is required. When it is not available and it must be estimated from sample data, the z distribution is no longer appropriate and another theoretical distribution, the Student's t distribution, is needed. This will be described in Chapter 8. Similarly, confidence limits can be determined for statistics other than the mean. Often this requires different formulas and approaches than above, and these procedures will be presented later in this text.

SUMMARY

Statistical inference consists of using probability to make decisions about a population on the basis of a sample of observations. A sample may be obtained by simple random sampling in which each element of the population has an equal probability of being selected for the sample of N observations. Usually the likelihood that any element is selected is independent of the likelihood that any other element will be selected. Since a sample is a subset of a population, a statistic calculated on a sample will not necessarily be the same value from sample to sample. This is sampling error. The distribution of a statistic determined on separate independent samples of size N drawn from a given population is called a sampling distribution, and the standard deviation of this distribution, called the standard error of that statistic, is a numerical index of this sampling error. Given random sampling, the sampling distribution of the mean, for example, is normal in form if the population of raw scores is normal, or it approaches normality as the size of the sample increases. This implies that the standard normal distribution can be used to determine the probability that the sample mean takes on certain values. Statistics are often used to estimate population parameters, either as point estimates or to calculate interval estimates, such as confidence intervals for a parameter. "Good" estimators are unbiased, consistent, relatively efficient, and sufficient.

Formulas

1. **Standard score form**

$$z = \frac{X - \mu}{\sigma_x} \qquad z = \frac{\overline{X} - \mu}{\sigma_{\bar{x}}}$$

2. **Standard error of the mean**

$$\sigma_{\bar{x}} = \frac{\sigma_x}{\sqrt{N}} \quad \text{(population or theoretical)}$$

$$s_{\bar{x}} = \frac{s_x}{\sqrt{N}} \quad \text{(sample or empirical)}$$

3. **Confidence limits for the mean (population $\sigma_{\bar{x}}$ known)**

   95% limits:

   $$\overline{X} - 1.96\sigma_{\bar{x}} \quad \text{and} \quad \overline{X} + 1.96\sigma_{\bar{x}}$$

   99% limits:

   $$\overline{X} - 2.575\sigma_{\bar{x}} \quad \text{and} \quad \overline{X} + 2.575\sigma_{\bar{x}}$$

Exercises

For Conceptual Understanding

1. Discuss the appropriateness of the following samples for the stated populations:

   a. A researcher in education wants to test the effectiveness of two different teaching methods on college students in a university, using one method with an 8 A.M. class and the other with a 2 P.M. class.

   b. A social psychologist is investigating patterns of group dynamics in the context of a jury-simulation experiment to identify the factors involved in the change of attitudes among young adults that occurs during jury debates. Students taking Introduction to Psychology may volunteer to participate in an experiment. Eighty students select the jury-simulation experiment from the 20 possible research projects available to them.

   c. An advertiser wants to know how the public is responding to the new Wonderease Soap Powder package. The package had just been displayed on television, so the advertiser quickly sets up a telephone survey in a major city to inquire whether the people who had seen the commercial liked the new package.

2. A population of 10 scores is listed below:

   $$X_i = 2, 3, 3, 4, 5, 5, 7, 8, 8, 9$$

   Write each score on a piece of paper and place it in a hat. Randomly select 10 samples each of size 4 from this population, recording the values and computing the mean for each sample. (Replace the selected numbers after each drawing of 4.) Create an *empirical* sampling distribution of the mean from these data by calculating the mean of the means and the standard deviation of the means. Then estimate the *theoretical* standard error of the mean by using the first sample you collect and the appropriate formula. What does the standard error of the mean tell you about the sample mean as an estimator of the population mean?

3. The sampling distribution of the mean is nor-

   mal in form if either one of two conditions is met. What are those two conditions?

4. Under what circumstances are the mean and variance independent?

5. What is the relationship between an idealized experiment, probability, and theoretical relative frequency?

6. Describe four characteristics of a good estimator.

For Solving Problems

7. For the population standard deviations given below, what is the standard error of the sampling distributions of the mean for the samples of size N?

   a. If $\sigma_x = 16$, $N = 16$

   b. If $\sigma_x = 16$, $N = 64$

   c. If $\sigma_x = 50$, $N = 25$

   d. If $\sigma_x = 50$, $N = 100$

   e. If $\sigma_x = 35$, $N = 49$

   f. If $\sigma_x = 1.26$, $N = 36$

   g. If $N = 25$ and $\sigma_{\bar{x}} = 10$, what is σ_x?

8. In a normal distribution with $\mu = 55$ and $\sigma = 10$, what is the probability of randomly sampling a subject who scores

   a. 62 or higher?

   b. 70 or higher?

   c. 40 or higher? 40 or lower?

   d. between 45 and 58?

9. Suppose IQ is distributed normally in the population, with a mean of 100 and a standard deviation of 16. What is the probability of randomly sampling a person with an IQ that is

   a. 100 or higher?

   b. 100 or lower?

   c. between 84 and 116?

   d. higher than 120?

   e. between 92 and 124?

10. Assuming a normal population distribution with $\mu = 85$, $\sigma = 20$, what is the probability of obtaining the following means for groups of subjects chosen by random sampling?

a. $\overline{X} = 96$ or higher, $N = 4$?

b. $\overline{X} = 92$ or higher, $N = 16$?

c. $\overline{X} = 78$ or higher, $N = 25$? $X = 78$ or lower, $N = 25$?

d. $\overline{X}$ between 74 and 83, $N = 9$?

11. Suppose it is known from previous studies that high school seniors in the country average a score of 140 with a standard deviation of 20 on a national test of basic skills. If a counselor gives the test to all seniors in her school ($N = 169$), what is the probability that they should obtain a mean

a. higher than 142?

b. between 140 and 143?

c. lower than 139?

d. between 138 and 141?

e. between 141 and 144?

12. Suppose we assume that scores on a national achievement test are normally distributed. If a counselor gives the test to all children in the school, determine 95% and 99% confidence intervals for the mean for the following grades:

a. Third grade: $\overline{X} = 70$, $\sigma_{\bar{x}} = 9$

b. Sixth grade: $\overline{X} = 110$, $\sigma_{\bar{x}} = 12$

c. Tenth grade: $\overline{X} = 180$, $\sigma_{\bar{x}} = 27$

13. Suppose that a formula is created, based upon a regression determined on a very large number of cases, that predicts adult height from knowledge of a child's height at 12 months of age, with a standard error of 1.33 inches. The Grahams, proud parents of one-year-old Keith, use the formula and find that Keith's predicted adult height is 71 inches. Assuming normal distributions, determine a 90% interval estimate for Keith's adult height.

For Homework

14. A politician wants information on how the 2500 residents of the district feel about a tax hike. However, the survey is to include only 200 families. Comment on each of the following ways of obtaining the sample.

a. Mail questionnaires to 200 people randomly selected from the previous year's tax return files.

b. Place questionnaires in local shopping centers with self-addressed, stamped envelopes.

c. Randomly select names from the phone book and call between the hours of 10:00 A.M. and 2:00 P.M.

d. Stand at a major intersection in the principal city and question people on the street.

e. Send letters home from school with children.

15. What is the standard error of the sampling distribution of the mean for the following?

a. $\sigma_x = 84$, $N = 49$

b. $\sigma_x = 100$, $N = 100$

c. $\sigma_x = 72$, $N = 81$

d. $\sigma_x = 27$, $N = 9$

16. In a normal distribution with $\mu = 50$ and $\sigma = 10$, what is the probability of randomly obtaining a score

a. greater than 60?

b. greater than 45?

c. less than 42?

d. less than 53?

e. between 46 and 57?

17. In a normal distribution with $\mu = 50$ and $\sigma = 12$, what is the probability in a sample of 36 that $\overline{X}$ will be

a. greater than or equal to 46?

b. between 48 and 52?

c. less than or equal to 46 or greater than or equal to 56?

d. less than or equal to 53?

e. greater than 52 but less than 54?

18. To examine the effect of sample size in a normal distribution with $\mu = 25$ and $\sigma = 5$, determine the probability of finding $\overline{X} = 25.5$ or lower with

a. $N = 4$

b. $N = 25$

c. $N = 100$

19. One of the first signs of hypothyroidism in newborn babies is high birth weight. Suppose at a given hospital the average newborn birth weight for males is 121 ounces with a standard deviation of 10 ounces. A doctor knows that infants with hypothyroidism are always above the 90th percentile in birth weight. At what birth weight would the doctor become concerned about the possibility of a male baby having hypothyroidism?

20. Support beams are manufactured and rated according to their ability to withstand weight. To ensure good quality, random checks are made during which increasing amounts of weight are placed on a beam until the beam no longer sup-

ports the load. One classification of beams will buckle at an average load of 8 tons with a standard deviation of 1 ton.

a. One beam is tested and is found to be able to hold 7.6 tons before buckling. Is it likely that this beam is a member of the above-mentioned classification? Why?

b. If 100 beams are tested and found, on the average, to hold 7.6 tons before buckling, is it likely that this sample of beams is within the classification?

Introduction to Hypothesis Testing: Terminology and Theory

n the previous chapter, sampling distributions and the concept of probability as relative frequency were introduced. Now we present a simple example using a sampling distribution and the concept of probability as relative frequency in which the learning performance of a group of people who had been given a drug is compared with the performance of a nondrugged population. The example is presented informally and phrased in common language. Then, it is recast into more traditional statistical language to introduce the formal terminology, theory, and procedures of hypothesis testing. The chapter also introduces another theoretical relative frequency distribution often used to determine probabilities in behavioral science research: the Student's *t* distribution. It is used in place of the standard normal when certain population parameters are unknown and must be estimated from quantities calculated on samples.

STATISTICAL TERMINOLOGY

Although the specific techniques of statistical inference vary depending upon the research question being asked, the logic of the general approach is quite similar in diverse cases. To illustrate this logic and the terminology that accompanies it, a relatively simple example will be used.[1] In subsequent chapters, more complex problems closer to those actually encountered in behavioral science will be presented.

Hypothesis Testing: Informal Examples

The idea of taking a pill to improve one's ability to learn or remember may not be just a fanciful idea. To explore this possibility, scientists start by trying to improve a very simple and specific memory task before advancing to more complex skills. For example, suppose it is known that adults asked to learn a list of 15 nouns typically remember 7 with a standard deviation of 2 after an 80-minute retention interval. Symbolically, the population learning performance for this task is $\mu = 7$ and $\sigma = 2$. Now suppose that a psychologist and a physician have a scientific hunch that injecting the drug physostigmine between a learning session and a memory-testing session might alter performance, but they don't know whether it will improve or retard memory. The scientists decide to make some relatively informal observations, first on one person and then on a small group to which they will administer the drug. Later, they plan to conduct a formal experiment comparing a large group of drugged subjects with a large group of nondrugged subjects.

A Single Case   First the scientists randomly select one person from the population of available subjects. They teach that person the 15 nouns, administer the drug, wait 80 minutes, and then test the person's recall of the nouns. If this one drugged person

---

[1] Based upon, but not identical to, research reported by K. L. Davis et al., "Physostigmine: Improvement of Long-Term Memory Processes in Normal Humans," *Science* 201: 272–74.

performs quite differently from the mean of the nondrugged population, it is possible that the drug has had some effect on memory performance in this situation.

Suppose the drugged subject remembers 11 nouns correctly. Although this is clearly more than the population mean of 7, is it enough to warrant pursuing this line of research? How differently would the person have to perform to be sufficiently different from what would be expected of nondrugged subjects?

The logical strategy the scientists employ is quite simple. Suppose the drug has no effect. Then, the experimental subject should perform like a randomly selected member of the population of nondrugged subjects, which has a mean $\mu_x = 7$ and a standard deviation $\sigma_x = 2$. But if the experimental subject scores so well or so poorly that one would expect few (fewer than 5% by custom) of the nondrugged subjects to behave so extremely, then the scientists will be encouraged in the belief that the drug might have an effect.

The task, then, is to determine whether the subject's score falls into the most extreme 5% of the distribution of nondrugged subjects. But a subject could score extremely high or extremely low, so the "extreme 5%" must be divided into two parts: the lowest $2\frac{1}{2}\%$ and the highest $2\frac{1}{2}\%$, which together constitute the most extreme 5% of the distribution. The scientists, then, assume that the distribution of nondrugged scores is normal in form, and that the lowest-scoring $2\frac{1}{2}\%$ will fall below $P_{.025}$ and the highest-scoring $2\frac{1}{2}\%$ will fall above $P_{.975}$ in that normal distribution. What, then, is the percentile rank of the observed score of 11 in a normal distribution with a mean of $\mu = 7$ and a standard deviation of $\sigma = 2$? Is that percentile rank lower than $P_{.025}$ or higher than $P_{.975}$? If it is, then the scientists will conclude that the observed score is "extreme," which implies that the drug may have had some effect.

The situation is illustrated in Figure 8-1, which is similar to the graphs drawn to solve the relative frequency problems in Chapter 4. At the top of Figure 8-1 is a normal population distribution for scores X on the memory test with $\mu = 7$ and $\sigma_x = 2$. The axis for the standard normal distribution (z) is presented below the axis for X. From Table A in Appendix 2 we can determine that $P_{.025}$ is at $z = -1.96$ and $P_{.975}$ is at $z = +1.96$, and values more extreme than these points would be in the most extreme 5% of the population distribution.

To determine the percentile rank for $X = 11$, we need to translate it into a z value. Recall from Chapter 4 that

$$z_i = \frac{X_i - \mu_x}{\sigma_x}$$

translates any score in a normal distribution into a z score (also called a **standard normal deviate**) that can be related to the percentiles of the standard normal distribution. Assuming the drug had no effect, the experimental subject thus reacted like a member of the normal population of nondrugged subjects and $X_i = 11$, $\mu_x = 7$, and $\sigma_x = 2$. Then

$$z = \frac{X_i - \mu_x}{\sigma_x} = \frac{11 - 7}{2} = \frac{4}{2} = 2.00$$

Therefore, $X = 11$ is equivalent to $z = 2.00$ or 2.00 standard deviations above the mean, which is more extreme than $z = 1.96$, which corresponds to $P_{.975}$. Specifically,

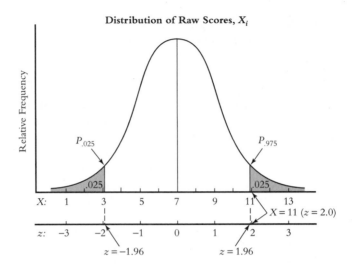

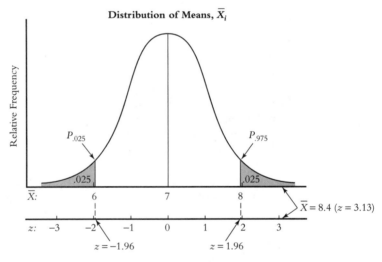

Figure 8-1 At the top is the population distribution of raw scores for nondrugged subjects. At the bottom is the sampling distribution of means from that population ($N = 20$). In each case, $P_{.025}$ and $P_{.975}$ are marked, and the shaded area represents the extreme 5% of the distribution.

looking at Table A in Appendix 2, and locating $z = 2.00$ in the left-hand column, we see that .0228 of the distribution falls to the right of $z = 2.00$, which means that $1 - .0228 = .9772$ or 97.72% of the distribution represents scores lower than 2.00. Therefore, $z = 2.00$ is at $P_{.9772}$. This is more extreme than $P_{.975}$, so the scientists conclude the drug may have had some effect on this subject's memory performance.

Because theoretical relative frequency (in this case, percentile ranks in the standard normal distribution) may also be interpreted in terms of probability, this logic also can be phrased in terms of probability. The scientists want to know whether the probability is very small (less than .05) that a nondrugged subject would score as extremely as did the experimental subject. Because the subject scored more extremely than $P_{.025}$ or $P_{.975}$, the probability is less than .05, or one in 20, that a nondrugged subject would score this extremely. Perhaps the drug has an effect.

A Single Group   Encouraged by this result, the scientists decide to randomly select 20 people, not just one, and determine their performance after receiving the drug. The same logic as before is used, except that now a mean, rather than a single score, is involved. If the mean of the group of drugged subjects is sufficiently different from what would be expected of nondrugged subjects, the scientists will conclude that perhaps the drug does influence learning performance.

The scientists tentatively assume that the drug has no effect—that is, that the group of 20 subjects will have a mean similar to the population mean of 7 for non-drugged subjects. If the mean performance of the drugged sample is more extreme than $P_{.025}$ or $P_{.975}$ in the sample distribution of means, the conclusion will be that it is not very likely that they were comparable to nondrugged people—their performance was too atypical—and perhaps the drug had some effect. If not, the experiment provides no support for the drug's effectiveness.

Graphically, this situation is presented at the bottom of Figure 8-1. It is essentially the same as that for a single score (top of Figure 8-1), except that the abscissa is now values of the mean $\overline{X}$, not individual raw scores. Therefore, the graph at the bottom of Figure 8-1 is a theoretical sampling distribution of the means from samples of size 20 from the nondrugged population having a population mean of 7. The question, then, is whether the observed mean is more extreme than $P_{.025}$ or $P_{.975}$ in this sampling distribution.

Suppose now that the sample of 20 subjects has a mean of 8.4. To determine its percentile rank, this value must be translated into a standard normal deviate. The task looks just like the previous case, except that now a mean, rather than a raw score, must be transformed into standard normal units (z). The translation formula derives from the fact that the expression

$$z = \frac{X_i - \mu_x}{\sigma_x}$$

is quite general. It can be read: The quantity of, any value minus the population mean of such values, divided by the population standard deviation of such values, will equal z. This could be written:

$$z, \text{ a standard normal deviate } = \frac{(\text{a value}) - (\text{the population mean of such values})}{(\text{the population standard deviation of such values})}$$

In the first example, the observation was a single score, X_i, and the population of values was the distribution of all nondrugged subjects. Now, a group mean ($\overline{X}$) is the value, and the theoretical distribution of means (the sampling distribution of the mean) is the population of values. Hence, the value will be $\overline{X} = 8.4$, the mean of such observations will be $\mu_{\bar{x}} = 7$ (since $\mu_{\bar{x}} = \mu_x = \mu$), and the standard deviation of such observations will be the standard deviation of the sampling distribution of means (that is, the standard error of the mean), which will equal $\sigma_x/\sqrt{N} = 2/\sqrt{20}$. Thus, the observed mean for a sample of 20 subjects may be translated into standard deviation units by

$$z = \frac{\overline{X} - \mu_{\bar{x}}}{\sigma_{\bar{x}}} = \frac{\overline{X} - \mu}{\sigma_x/\sqrt{N}}$$

$$z = \frac{8.4 - 7}{2/\sqrt{20}} = 3.13$$

Looking again at Table A in Appendix 2, we see that only .0009 of the cases in a normal distribution would be expected to fall above such a value. This mean of 8.4 is at $P_{.9991}$ of the sampling distribution of means based upon samples of $N = 20$ taken from the population of nondrugged subjects. Hence, the probability is indeed less than .05 that a group of 20 nondrugged subjects would score so extremely, and one may conclude that their performance may have been influenced by the drug.

Notice that although the formulas that were required to translate the observed value (X_i or $\overline{X}$) into standard normal units differ slightly in the two preceding examples, the logic behind the procedure is the same. Randomly select a subject or a group of subjects and tentatively assume that, despite the fact that they receive the drug, the drug does not influence their performance, which should be no different from that of nondrugged subjects. Arbitrarily select some percentiles (or probabilities) that define a result that would provoke the interpretation that these experimental subjects might be different from nondrugged subjects ($P_{.025}$ and $P_{.975}$). Translate a summary statistic of their performance (for example, X_i or $\overline{X}$) into a form compatible with a theoretical relative frequency distribution and determine its percentile rank. If that rank exceeds the limits set above, suspect that the tentative assumption that these people were just like nondrugged subjects may have been wrong, since their performance was very atypical or improbable. We now examine the details of this logic and the statistical vocabulary used to describe it.

Statistical Assumptions and Hypotheses

The kind of statistical inference described above requires that certain statistical assumptions be made.

> **Assumptions** are statements of circumstances in the population that the logic of the statistical process requires to be true but that will not be proved or decided to be true.

In the example, two assumptions are made:

1. The 20 subjects were randomly and independently selected from a nondrugged population; and

2. The sampling distribution of the mean is normal in form.

Both of these statements must be true to use the percentiles of the standard normal distribution in the way that the statistical process describes, but their truth or validity will not be examined during the statistical process. Of course, the researchers attempted to sample randomly and independently, but these characteristics are really simply assumed to be true. Also, we know from Chapter 7 that the sampling distribution of the mean will be normal if the distribution of raw scores is normal or if the sample size is large. It may be a reasonable guess that this condition is met, but it is not really examined here.

Next some hypotheses are stated.

> **Hypotheses** are statements of circumstances in the population that the statistical process will examine and decide their likely truth or validity.

The scientists have two hypotheses about what is actually the case: either the drug has an effect or the drug does not have an effect. First, notice that the hypotheses are contradictory or **mutually exclusive**—they both cannot be true. Indeed, the purpose of the statistical process is to decide between the two. They are also **exhaustive**—together they cover all the possible circumstances regarding the effectiveness of the drug. Finally, one of the hypotheses is tentatively held to be true. In this case, the scientists temporarily proceed as if the drug did *not* have any influence on memory.

> The hypothesis that is tentatively held to be true is called the **null hypothesis** and is customarily symbolized by H_0. The **alternative hypothesis** is represented by H_1.

In the present example the null and alternative hypotheses may be stated:

> H_0: The observed mean is computed on a sample drawn from a population with $\mu = 7$ (that is, the drug has no effect).
>
> H_1: The observed mean is computed on a sample drawn from a population with $\mu \neq 7$ (that is, the drug has some effect).

Assumptions and hypotheses are similar in certain ways. For example, they are both statements about what exists in the population and they are both crucial to the statistical logic. Specifically, if the null hypothesis that the drug has no effect is tentatively held to be true, then the random sample of 20 people should behave like a typical sample of 20 nondrugged subjects. The assumption that the population distribution is normal in form means that the sampling distribution of the mean will also be normal, and that the percentiles of the standard normal distribution can therefore be used to determine the probability that a random sample of 20 should have a mean as deviant from $\mu = 7$ as is the observed mean of 8.4.

But there are also important differences between assumptions and hypotheses. **Hypotheses** represent a set of two or more contradictory and often exhaustive possibilities, only one of which can actually be the case. Moreover, one of the hypotheses (the null hypothesis) is temporarily held to be true, and it is the reasonableness of this hypothesis that is being tested by these statistical procedures. The logic is as follows: tentatively hold the null hypothesis to be true (that is, assume the drug has no effect). If the results of the analysis could have reasonably occurred given that circumstance (that is, if the mean of the drugged group could probably have been scored by a nondrugged group), there is no reason to question the validity of the hypothesis of no drug effect. But if the analysis indicates that the likelihood is very small that a nondrugged group could have produced this mean, the null hypothesis is rejected in favor of the alternative hypothesis. In contrast, the **assumptions** of random sampling and normality remain constant and are assumed true regardless of the outcome of the experiment; only the null hypothesis is tentatively held and then either rejected or not rejected.

Significance Level

Recall part of the statistical logic described above: if the probability is very small that the observed sample mean could be produced by a nondrugged group, then one will reject the null hypothesis. How small does that probability need to be before the null hypothesis is rejected? That probability value is the significance level of the test of the null hypothesis:

> The **significance level** (or **critical level**), symbolized by α (alpha), is the probability value that forms the boundary between rejecting and not rejecting the null hypothesis.

In behavioral science research, the precise value of α is customarily taken to be .05. In the current example, the significance level of .05 indicates that unless the observed mean is so different from the population mean, $\mu = 7$, that it would be expected of nondrugged subjects less than 5% of the time, the researchers will not challenge the validity of the null hypothesis, H_0. However, if the probability is less than .05 that the observed mean for the drugged group would be obtained by nondrugged subjects, the null hypothesis that the drug has no effect will be rejected and the conclusion will be that the drug probably did exert some influence. This result is then said to be "significant at the .05 level," which may be written "$p < .05$," in which p stands for the probability of the result given the validity of the null hypothesis. The significance level, in this case .05, represents the probability value that separates a decision to reject from a decision not to reject the null hypothesis.

Decision Rules

Once the significance level has been established, it is used to formulate decision rules.

> **Decision rules** are statements, phrased in terms of the statistics to be calculated, that dictate precisely when the null hypothesis will be rejected and when it will not.

In the example above, the researchers selected the .05 level of significance, meaning that the null hypothesis would be rejected only if the observed sample mean deviated from the mean of the nondrugged population to an extent likely to occur in non-drugged samples less than 5% of the time. The percentile points $P_{.025}$ and $P_{.975}$ of the sampling distribution of the mean mark the endpoints of a range of values that would include 95% of the means of nondrugged samples but exclude 5% of such means, $2\frac{1}{2}\%$ because they are extremely low and $2\frac{1}{2}\%$ because they are extremely high. Since the sampling distribution of the mean is normal in form, the standard normal distribution may be used to translate these percentile ranks into percentile points (that is, z values). $P_{.975}$ corresponds to a z value such that .025 of the distribution falls to its right (above it). Scanning the rightmost column of Table A in Appendix 2 for .0250, one finds the corresponding z to be 1.96. Since the standard normal is a symmetrical distribution, we also know that .025 of the distribution falls to the left of (below) a z of -1.96. Thus, $P_{.025}$ and $P_{.975}$ correspond to the points $z = -1.96$ and $z = 1.96$.

This situation is illustrated at the bottom of Figure 8-1, which presents the theoretical sampling distribution of the means for samples of size 20 from the nondrugged population. The $P_{.025}$ at $z = -1.96$ and $P_{.975}$ at $z = 1.96$ are marked, and the shaded area represents the extreme 5% of the sampling distribution. An observed mean in this shaded region would result in rejection of the null hypothesis. Since the observed sample mean ($\overline{X} = 8.4$) will be translated into an equivalent z value, the formal statement of the decision rules is:

If z is between -1.96 and $+1.96$, do not reject H_0.

If z is less than or equal to -1.96 or greater than or equal to $+1.96$, reject H_0.

The same statements can also be expressed by using the symbols $<$ (less than), $>$ (greater than), $\leq$ (less than or equal to), and $\geq$ (greater than or equal to):

If $-1.96 < z < +1.96$, do not reject H_0.

If $z \leq -1.96$ or $z \geq +1.96$, reject H_0.

The z values of ± 1.96 in this example are called **critical values** because they separate z values that will result in a decision to reject or not to reject the null hypothesis. Also, the shaded areas in the sampling distribution at the bottom of Figure 8-1 define the **critical region** or **region of rejection** of the null hypothesis.

Computation

The statistical procedures take advantage of theoretical relative frequency distributions for which percentiles are already available. In the example we have been considering, the standard normal was used, but there will be other theoretical distributions employed in different contexts later in the text. The computational procedures essentially convert the observed statistic (for example, the mean) into a standard normal deviate so that this value can be compared with the critical values previously established for the standard normal distribution. In this case, the formula

$$z = \frac{\overline{X} - \mu_{\bar{x}}}{\sigma_{\bar{x}}}$$

was used to convert $\overline{X} = 8.4$ to $z = 3.13$.

To discriminate observed from critical values of the standardized variable, subscripts are often used. In our example,

$$z_{\text{obs}} = 3.13$$

represents the value corresponding to the observed data, whereas

$$z_{\text{crit}} = \pm 1.96$$

indicates the critical values of z.

Decision

In the example, $z = 3.13$ falls within the critical region (Figure 8-1), which implies that the obtained mean of 8.4 ($z = 3.13$) would be expected to occur in less than 5% of samples of 20 nondrugged subjects. Since it is unlikely that such a value would occur merely as a function of sampling error associated with selecting certain people as subjects and not others, the null hypothesis of no drug effect is rejected. When the null hypothesis is rejected, the observed difference, in this case between the sample mean and the population mean, is said to be **statistically significant** or just **significant**.

It is important to understand the relationship between the logic of this statistical process and the concept of sampling error as introduced in the previous chapter. The sampling distribution of the mean reflects the extent to which the mean of one sample will differ from the mean of another sample, even if these samples are treated exactly alike. If the observed mean is very extreme, so extreme that it is not a likely value in this sampling distribution, we reject the null hypothesis, because it is unlikely that a mean of this magnitude would be obtained merely because of chance variation from sample to sample—that is, it would be unlikely to occur because of sampling error alone. On the other hand, if the observed value is not sufficiently extreme to reject the null hypothesis, one can conclude that the mean of the drugged group could have occurred even if the drug had had no effect, since it could reasonably be expected to deviate from the population average on the basis of sampling error alone (random variation between samples, chance, and so on).

This interpretation can be understood by examining the formula for translating the observed sample mean into a standard normal deviate of the theoretical sampling distribution of the mean:

$$z = \frac{\overline{X} - \mu}{\sigma_{\bar{x}}}$$

The formula states that z is the ratio of the difference between the sample mean and the mean of the nondrugged population ($\overline{X} - \mu$) relative to the amount of expected sampling error as expressed by its numerical index, the standard error of the mean ($\sigma_{\bar{x}}$). Because this formula yields the values that define the critical region, it is sometimes called a **critical ratio.**

In a sense, the statistical test simply asks how different the observed mean was from the population mean, relative to the sampling error one might expect. As this ratio increases in size, one begins to suspect that the difference between observed and population means is probably not simply a function of random variation between samples, chance, or—more precisely—of sampling error.

Note that the decision to reject the null hypothesis (when this is appropriate) is *not* equivalent to accepting the alternative hypothesis. One rejects H_0; one does *not* accept H_1. The reason for taking such care in stating the decision is that only one hypothesis, the null hypothesis, was tested. It was tentatively held to be true, and statistical procedures were carried out to determine whether the data were reasonable under these conditions. Since they were not, H_0 was rejected as probably not a valid characterization of what really exists. Since it was H_0 that was tested, the decision concerns H_0, not H_1.

Similarly, if the results provoke the decision not to reject the null hypothesis, one does not accept it, since the results demand only that one not reject H_0. It may seem like quibbling to insist on a distinction between "not rejecting" and "accepting" the null hypothesis, but the failure to understand the difference often reflects a failure to understand the logic of this statistical procedure. Suppose the observed value of z is not in the critical region and thus is not large enough for the researchers to reject the null hypothesis. This does not necessarily say that the drug has no effect. It may influence learning performance, but not very much. Thus, the difference it produces is not large enough, relative to the amount of sampling error, to produce a z value sufficient to reject the null hypothesis in a sample of size N. But the same difference might lead to rejecting H_0 in a larger sample. Therefore, one does not prove or accept the null hypothesis—one simply has insufficient cause to reject it.

Decision Errors

Generally, the purpose of statistical inference is to make an educated guess about what exists in a population when only a small sample of cases from that population has been studied. Since the decision is based on an estimate, it may be wrong. In fact, if the significance level is .05, then in 5% of the cases in which the drug actually has no effect, the researchers will decide to reject H_0, but they will be wrong.

This is part of the statistical logic. You observe a mean score, and if that mean should occur less than 5% of the time if H_0 were true (that is, if the drug has no effect), you will decide to reject H_0. So if H_0 *really is true,* you will decide to reject H_0 5% of the time, and you will be wrong on those occasions. The drug actually had no effect, but you will have decided that it does. This is called a type I error.

A **type I error** occurs when the decision is to reject the null hypothesis when it is actually true. Given the validity of the null hypothesis, the probability that it is erroneously rejected by these procedures equals α, the significance level.

Another type of error, a type II error, is made when the null hypothesis is *actually wrong*—for example, as when the drug does indeed have an effect—but the *decision* is made not to reject it.

> A **type II error** occurs when the decision is not to reject the null hypothesis when it is actually false. The probability of this type of error is symbolized by β (beta).

The decision-making process is illustrated in Figure 8-2. The distribution at the left, labeled H_0, represents the sampling distribution of means (for samples with $N = 20$) for nondrugged subjects. The shaded areas in the tails of this distribution constitute the region of rejection. If the null hypothesis is actually true and the drugged sample behaves no differently from a nondrugged sample, then this area represents the probability of a type I error, since an observed mean in this region would lead to the erroneous decision to reject the null hypothesis when it is actually true. Now shift your attention to the right-hand distribution, labeled H_1, a theoretical sampling distribution of possible means for drugged subjects. Assume now that H_1 is, in fact, true and the drug has an effect. Because of the overlap in the distributions, it is possible that some sample means in the H_1 distribution will fall close enough to the means expected for the nondrugged population (i.e., the H_0 sampling distribution) so that the decision will be not to reject the null hypothesis when it is actually false—a type II error. An observed mean corresponding to point *a* would be such a case. The shaded area labeled β in Figure 8-2 corresponds to means for drugged samples for which the statistical process would erroneously decide that the drug probably had no effect. It represents the probability of a type II error. Unfortunately, the distribution for drugged samples when the drug actually has an effect (labeled H_1) is rarely available, and hence it is usually impossible to estimate the precise size of β. However, as the likelihood of a type I error decreases, the probability of a type II error increases.

There are two other areas to notice in Figure 8-2. The portion of the H_0 distribution that is not in the region of rejection represents the probability of correctly refusing to reject H_0 (probability $= 1 - \alpha$). The portion of H_1 that is not part of β represents the probability of correctly rejecting H_0 (probability $= 1 - \beta$). The latter probability is called the **power** of a test. Table 8-1 summarizes these points.

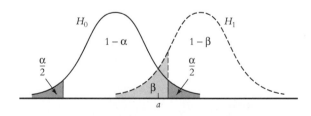

Figure 8-2 Diagram of the relationship between type I (α) and type II (β) errors and power $(1 - \beta)$.

8-1	**The Four Possible Outcomes of a Simple Decision Process and Their Associated Probabilities**

Decision	Actual Situation	
	H_0 Is True	H_0 Is False
Reject H_0	Type I error probability $= \alpha$	Correct decision probability $= 1 - \beta$ (power)
Do not reject H_0	Correct decision probability $= 1 - \alpha$	Type II error probability $= \beta$

Power of a Statistical Test

The power of a statistical test can be defined as follows:

> The **power** of a statistical test is the probability that the test will lead to a decision to reject H_0 when H_0 is indeed false.

High power is a major requirement of a good statistical test. If H_0 is actually not true and should be rejected, we want to know that the statistical test we are using will lead to that conclusion. Since type II errors, in which the test concludes not to reject H_0 when H_0 is actually false, occur with a probability of β, the probability of correctly rejecting H_0 is $1 - \beta$. The actual value of this probability is difficult to obtain, since it requires one to know the H_1 distribution. We rarely have this information—indeed, that is why we are performing the statistical test in the first place. Obviously, researchers would like their statistical procedures to have as much power as possible: if an effect of a treatment actually exists, they want their statistical procedures to be powerful enough to detect it. Despite the fact that one usually cannot precisely calculate the power of a statistical test without making several crucial assumptions, there are some ways of increasing this probability.

1. **The power of a statistical test increases as the significance level, α, increases.** Thus, a test performed at a significance level of .05 has more power than one at .01. Obviously, the less stringent the significance level, the more likely the null hypothesis will be rejected. But both correct and incorrect rejections will be more likely; simply having a larger value of α will increase the power but will also increase the probability of a type I error.

2. **The power of a test increases with an increase in sample size.** Thus, the likelihood of correctly rejecting H_0 is higher for a sample of 30 than a sample of 20. This is because increasing the N reduces the size of the standard error of a statistic, increasing the accuracy of estimates of parameters.

3. **Some statistical tests are more powerful than others.** For example, the more specific the assumptions that can be made about the population, the more powerful will be the statistical test. The techniques presented in Chapters 9, 10, and 13 are called **parametric tests,** because they make specific assumptions about the population parameters, such as, that the dependent variable is measured with an interval or ratio scale, that it is normally distributed, and that variances are similar from group to group. **Nonparametric tests,** presented in Chapter 14, make less specific assumptions. As a result, parametric tests tend to be more powerful than nonparametric tests.

Selecting the Significance Level

Recall that the significance level is the probability of erroneously rejecting the null hypothesis when it is, in fact, true (type I error). How does one go about setting this level? The value of α reflects the investigator's feeling of how much error is tolerable in making a decision to reject H_0 on the one hand, balanced against how much power is wanted in the statistical test on the other. Ordinarily, social and behavioral scientists accept the .05 level of significance as the value of α, but this is merely a convention that has grown up over the years, and there is nothing absolute about it. When testing some hypotheses, other levels of significance might be adopted depending upon how crucial it is to avoid being wrong in rejecting the null hypothesis. For example, if you were a brain surgeon and were giving a test to determine whether a patient needed a very delicate type of surgery, it might be that you could not afford to risk an operation if the patient really didn't need one (that is, if your diagnosis was wrong). Therefore, you might operate only if the probability was less than .01 or .001 that the test result was just a sampling error. However, the more you lower the probability of incorrectly rejecting the null hypothesis and performing the operation (type I error), the more you raise the probability that a patient who needs the operation will be incorrectly diagnosed as being able to do without it (incorrectly failing to reject H_0; a type II error). In any situation, it is important to remember that the level of significance is arbitrarily established on the basis of the researcher's tolerance for error in decision making and the desire for statistical power. However, depending upon the circumstances, the reader may feel differently from the researcher about the most appropriate relative sizes of error and power. Therefore, a reasonable procedure is for the researcher to adopt a significance level but also to report the actual probability found as a result of testing H_0, regardless of whether H_0 is rejected or not. Also, some statisticians dislike the combination of arbitrary significance levels and dichotomous reject/do not reject decisions and prefer to report probabilities without decisions and to determine confidence limits for parameters.

Directional Tests

The test of the null hypothesis described in the drug example was a **nondirectional test.** It was nondirectional because the hypotheses did not specify the direction of the possible influence of the drug. Would it help or hinder learning? The alternative hypothesis H_1 suggested only that the drug would influence performance, one way or the other. Consequently, the critical region was established so that extreme values in either direction would lead to a decision to reject H_0.

However, suppose the same drug had been found by other scientists to improve memory in rats, and the scientists then wondered whether its administration would also improve memory in humans. Since it is very unlikely that the drug would actually hinder memory, the researchers would establish the alternative hypothesis as

> H_1: The observed mean is computed on a sample drawn from a population with $\mu > 7$ (that is, the drug facilitates memory).

This is a **directional** hypothesis, because it prescribes that the mean will be greater than, not just different from, 7.

If the alternative hypothesis is made directional, then the null hypothesis is also made directional so that it includes all the other possibilities—namely, that μ is not just equal to 7 but that it is less than or equal to 7. This strategy makes the two hypotheses, H_1 and H_0, exhaustive of the possibilities. The H_0 to accompany the directional alternative given above is

> H_0: The observed mean is computed on a sample drawn from a population with $\mu \leq 7$.

When directional hypotheses are appropriate, it would not make sense to divide the critical region into two parts so that extreme means in either direction lead to rejecting the null hypothesis. Suppose the previous research information had suggested that the probability was almost zero that drugged subjects would exhibit an extremely low mean. Therefore, to maintain the significance level (that is, the probability of erroneously rejecting the null hypothesis) at .05, the region of rejection would have to include 5% of the scores in the right-hand tail of the distribution. Thus, when looking in Appendix 2, Table A (the standard normal), one would search for the z value that has .0500 of the area lying to its right. The critical value is $z = 1.645$, so the decision rules would be:

If $z_{obs} < 1.645$, do not reject H_0.

If $z_{obs} \geq 1.645$, reject H_0.

A comparison of the regions of rejection for these nondirectional and directional tests is presented in Figure 8-3.

Since a nondirectional test locates the critical region in both tails of the theoretical distribution, it is frequently called a **two-tailed test.** Conversely, since a directional test locates the critical region in only one end of the distribution, it is called a **one-tailed test.**

In determining whether a test should be directional or nondirectional, one considers whether there is other evidence or a theory that might predict the result. There needs to be some basis for presuming that a sample mean will deviate in only one direction from the population value. Then H_1 and the critical values are established to agree with that prediction. However, *this choice of a directional or nondirectional test must*

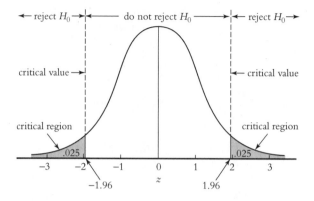

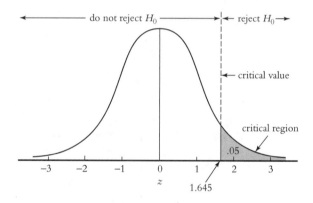

Figure 8-3 Decision process for a nondirectional (two-tailed) test (top) and a directional (one-tailed) test (bottom).

be established before one knows the experimental result. Otherwise, if one waits to know the data and then decides to make a directional instead of a nondirectional test, the theoretical probabilities involved in the significance level (for example, $\alpha = .05$) are no longer appropriate, because one will be tempted to apply the one-tailed test only in cases where z_{obs} is between the critical values for a directional and a nondirectional test. Therefore, the statistical hypotheses must be stated before the experimental observations are made.

Students frequently have difficulty understanding the rationale behind the size of the critical region in directional tests. One might look at Figure 8-3 and wonder, "If you have some reason for predicting that the sample mean will be greater than the mean of the population of nondrugged subjects, why do you select a *lower* critical value, which seems to make it easier to reject the null hypothesis?" It seems that it should be the other way around: if you know that the difference between means will likely be positive (to the right of $z = 0$), then the critical value of z should be more stringent (that is, farther to the right) to balance out the advantage gained by being able to predict the direction of a difference.

The explanation rests in the purpose of α, the significance level. Most researchers are interested in rejecting the null hypothesis. We want to have significant results. But, the significance level refers to the probability of rejecting the null hypothesis *by mistake*—that is, to the probability of deciding there is a difference when no difference actually exists. The problem, then, is to have a directional and a nondirectional test at the .05 level of significance comparable to each other in terms of the probability of falsely rejecting H_0—that is, the probability of a type I error. A nondirectional, or two-tailed, test must allow for error in rejecting the null hypothesis as a function of extreme differences in either direction due to sampling variation. However, if there is some reason to think that the sample mean will be larger than the population mean, because either empirical evidence or a theory suggests this will be the case, there is very little chance that an extremely large negative difference will be anything but sampling variation. To use a two-tailed, nondirectional alternative in this instance would really constitute a test at the .025 level of significance, because it is known that extreme negative differences will be sampling error and not drug effects. Therefore, to ensure that the probability of a type I error is indeed the value of α, it is necessary to place all of the critical region in the right-hand tail.

HYPOTHESIS TESTING WHEN $\sigma_{\bar{x}}$ IS ESTIMATED BY $s_{\bar{x}}$

A major step in hypothesis testing is to compare the statistic of interest (for example, $\overline{X}$) to its theoretical sampling distribution to obtain the relative frequency—the probability—that such a sample value could be obtained under the null hypothesis. To do this, a formula is used, and heretofore the formula requires population parameters, such as $\sigma_{\bar{x}}$, for its calculation. But parameters are typically not available and must be estimated with statistics. How can this be done?

Specifically, in the expression that translates a sample mean into a standard normal deviate,

$$z = \frac{\overline{X} - \mu}{\sigma_{\bar{x}}}$$

z is a standard normal deviate (if the sampling distribution of the mean is normal), μ is the population mean of the sampling distribution of $\overline{X}$'s determined for all possible samples of size N from the population, and $\sigma_{\bar{x}}$ is the theoretical standard error of such $\overline{X}$'s. Regrettably, the value of this parameter $\sigma_{\bar{x}}$ is usually not available, so it must be estimated.

The sample standard error of the sampling distribution of the mean, $s_{\bar{x}}$, which can be calculated on the basis of a single sample of scores ($s_{\bar{x}} = s_x/\sqrt{N}$), may be used for this estimation. However, when $s_{\bar{x}}$ is invoked to estimate $\sigma_{\bar{x}}$, the accuracy of $s_{\bar{x}}$ as an estimator and thus the accuracy of the probability value derived from the standard normal distribution, depends upon the size of the sample, N. If the sample is quite large ($N > 50$ or 100), the probabilities derived by using the standard normal distribution as described above are fairly accurate. But most samples in behavioral science are not so large, and it happens that when $\sigma_{\bar{x}}$ is estimated by $s_{\bar{x}}$ calculated on a small sample, the standard normal distribution does not provide sufficiently accurate probabilities. Fortunately, another theoretical sampling distribution exists that is appropriate for these situations and can be used instead of the standard normal distribution.

Student's *t* Distribution

The **Student's *t* distribution** is a theoretical relative frequency distribution developed by W. S. Gosset, who wrote under the name Student. When scores are transformed in the following way, the Student's *t* distribution, rather than the standard normal, is appropriate:

$$t = \frac{\overline{X} - \mu}{s_{\bar{x}}} = \frac{\overline{X} - \mu}{s_x/\sqrt{N}}$$

Notice that except for substituting t for z and $s_{\bar{x}}$ for $\sigma_{\bar{x}}$, the formula is the same as that for the standard normal. The t distribution, however, differs from the standard normal. In addition to permitting the parameter $\sigma_{\bar{x}}$ to be estimated by the sample statistic $s_{\bar{x}}$, there is a different t distribution for each size of sample upon which $s_{\bar{x}}$ is computed. The appropriate t distribution is determined not by the sample size, N, but by its **degrees of freedom** (df), which is $N - 1$ in this case. Figure 8-4 shows the t distribution for several degrees of freedom. It also depicts the standard normal distribution ($df = \infty$). Notice that the t and z distributions are quite different when the sample size (and thus the number of degrees of freedom) is small, but as the number of degrees of freedom increases, the t distribution becomes more and more like the standard normal distribution. In fact, when the sample size is allowed to be theoretically infinite, the t and standard normal distributions are identical.[2]

Degrees of Freedom Undoubtedly one of the most difficult concepts for students to grasp, but also one of the most necessary, is that of degrees of freedom.

> The number of **degrees of freedom,** abbreviated **df**, for any statistic is the number of components in its calculation that are free to vary.

The meaning of this definition is best explained by considering some examples. In a sense, every statistic in a specific context has a certain number of degrees of freedom

---

[2] Some texts suggest that if the sample size is 30 or more, the normal and t distributions are similar enough that the standard normal distribution (z) may be used even though $\sigma_{\bar{x}}$ has been estimated. While an examination of the tables for z and t in Appendix 2 certainly indicates a great degree of correspondence between the standard normal and t for samples of 30 or more, it will be the policy of this book to suggest that in hypothesis testing the t distribution always be used when a parameter is being estimated, regardless of sample size. This policy results in greater accuracy in determining probabilities and in somewhat less confusion.

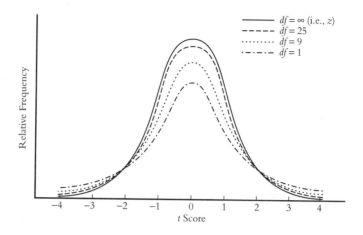

Figure 8-4 The *t* distribution for various degrees of freedom. With an infinite number of degrees of freedom, the *t* and the standard normal (*z*) distributions are identical.

associated with it. For example, the sample mean, $\overline{X}$, has N degrees of freedom. How does one know that? Begin by examining the formula for the mean:

$$\overline{X} = \frac{\Sigma X_i}{N}$$

The first question to ask is which components of this formula can vary at all and which cannot. When a mean is calculated from any specified sample of scores, the N does not vary. It is fixed in value for that sample. In contrast, the scores, X_i, which compose the ΣX_i, do take on different values—they are free to vary. That is, each of the scores could assume any possible value, and knowing the values of all but one score does not tell us the value of the last score. If there are five scores, and we know four of them, (3, 7, 5, 2, ?), there is no way for us to determine the value of "?". Thus, we must know all N scores to calculate the mean. It is in this sense that all N scores are free to vary (that is, to take on any values), and thus the number of degrees of freedom for the statistic $\overline{X}$ is simply N.

The task becomes more complicated when we consider the number of degrees of freedom for the variance. Again, begin with the formula:

$$s_x^2 = \frac{\Sigma(X_i - \overline{X})^2}{N - 1}$$

Neither the sample size, N, nor the mean, $\overline{X}$, is free to vary. For any specified sample, they are constants. However, the deviations between each score and the mean ($X_i - \overline{X}$) take on different values depending upon the scores in the distribution. How many of these N deviations are free to vary? It happens that all but one of these deviations are free to vary, because the sum of the deviations of scores about their own mean is always zero, $\Sigma(X_i - \overline{X}) = 0$. Knowing $N - 1$ of these deviations, you can always determine the last one because its value is such that the sum of all the deviations is zero. For example, consider the distribution (5, 7, 9). The mean is 7. Note what happens if any two of the deviations are computed:

X_i	$\overline{X}$	$(X_i - \overline{X})$
5	7	-2
7	7	0
9	7	$?$
		$\Sigma(X_i - \overline{X}) = 0$

Since the sum of the deviations from the mean is always zero, the third deviation must be $+2$. It is in this sense that the last deviation to be computed, regardless of which particular score in the distribution is remaining, is not free to vary but is determined. Knowing $N - 1$ of the deviations, you can determine the value of the last deviation. Therefore, only $N - 1$ deviations are free to take on any value. Thus, the number of degrees of freedom for the sample variance (and sample standard deviation) is $df = N - 1$.

The degrees of freedom for t are also $N - 1$. The formula for t is

$$t = \frac{\overline{X} - \mu}{s_x / \sqrt{N}}$$

The population mean μ, the mean $\overline{X}$ for this sample, and N are all constants. Thus, the degrees of freedom for t are the degrees of freedom for s_x, which we have just seen is $N - 1$. When selecting the proper critical values for a statistical test involving the t distribution, one must know the degrees of freedom for t. The present application of the t distribution dictates that t have $N - 1$ degrees of freedom. Other applications of t discussed in the next chapter may have different numbers of df associated with them because they involve different formulas for t.

Application of the *t* Distribution   Suppose the researchers decide to randomly select a sample of 20 subjects and give each of them the drug prior to the recall test, but suppose also that the standard deviation of the population of nondrugged people is not available. The population distribution is still regarded as normal in form with $\mu = 7$, and the researchers wish to test the hypothesis that the observed mean is computed on a sample drawn from this population having $\mu = 7$. However, since the population standard deviation is not known, $s_{\overline{x}} = s_x / \sqrt{N}$ is used to estimate $\sigma_{\overline{x}}$.

The expression

$$t = \frac{\overline{X} - \mu}{s_x / \sqrt{N}}$$

changes the observed sample mean into standardized t units. With certain assumptions similar to those invoked to use the standard normal, the t distribution may be employed in the same manner as the standard normal distribution.

The t distribution is found in Table B of Appendix 2 and deserves a brief examination. The table gives the critical values of t for various significance levels at each of several degrees of freedom for directional and nondirectional tests. The degrees of freedom are listed in the leftmost column. To their right, one finds various values of t for different significance levels for directional and nondirectional tests. Since the t distribution (like the z) is symmetrical, the values shown in the table for nondirectional tests hold not only for the right-hand (positive) half of the distribution but also, with the addition of a minus sign, for the left-hand portion. For example, in a two-tailed test at the .05 level at $N - 1 = 20 - 1 = 19$ degrees of freedom, the table shows that $2\frac{1}{2}\%$ of the distribution falls to the left of -2.093 and $2\frac{1}{2}\%$ falls to the right of 2.093. If a one-tailed test at 19 df at the .01 level of significance was appropriate, the entire 1% of the distribution would fall to the right of 2.539 (or possibly to the left of -2.539). Thus, the t table gives the critical values of t for various degrees of freedom and levels of significance for both directional and nondirectional tests.

One should observe two additional things about t. First, for any given level of significance, the value of t that is required to reject H_0 (critical value of t) becomes smaller as the number of degrees of freedom increases. Consider the third column for a two-tailed test at the .05 level of significance. With only one degree of freedom, $t = 12.706$ is required to reject H_0, whereas if $df = 2$, $t = 4.303$ is needed. If there is an infinite number of degrees of freedom, $t = 1.96$ is necessary, which is precisely the value of the standard normal deviate z for this situation. Thus, as the number of degrees of freedom increases, the critical value of t decreases, until it reaches the same value that the standard normal would dictate. So, the bigger the sample size and df, the greater the likelihood of correctly rejecting H_0 (i.e., the greater the power).

Second, the smaller the standard error, the more likely the null hypothesis will be rejected. This cannot be seen from the t table, but it can be appreciated by examining the expression for t:

$$t = \frac{\overline{X} - \mu}{s_{\bar{x}}}$$

Since the ratio of two quantities becomes larger as the denominator becomes smaller, it can be seen that the value of t grows larger as the standard error of the mean ($s_{\bar{x}}$) becomes smaller. Thus, the smaller the standard error the more likely the null hypothesis will be rejected when it is false.

Hypothesis Testing: A Formal Example

A formal presentation of the solution to this hypothesis-testing problem follows. Its format may be used as a model to follow when formal hypothesis-testing solutions are required.

Research Question   If the mean number of correct trials of the population of non-drugged subjects is 7, does the mean $\overline{X}$ of a random sample of 20 subjects who are given a drug differ sufficiently from that population to warrant the conclusion that the drug has an effect on memory performance?

Statistical Hypotheses

H_0: $\overline{X}$ is computed on a sample from a population with $\mu = 7$.

H_1: $\overline{X}$ is computed on a sample from a population with $\mu \neq 7$.

Assumptions and Conditions   It is assumed that (1) the subjects are randomly and independently sampled, and (2) the population of nondrugged subjects is normal with $\mu = 7$.

Significance Level   Adopt the .05 level.

Critical Values   Since the alternative hypothesis H_1 is nondirectional and $\alpha = .05$, and since s_x must be used to estimate σ_x, t values must be selected so that $2\frac{1}{2}\%$ of the t distribution falls to the left of one critical value and $2\frac{1}{2}\%$ falls to the right of the other. Looking in Appendix 2, Table B, with $N - 1 = 20 - 1 = 19$ *df*, one finds these critical values are -2.093 and $+2.093$.

Decision Rules   If t is between -2.093 and $+2.093$, do not reject H_0. If t is less than or equal to -2.093 or greater than or equal to $+2.093$, reject H_0. In symbols,

If $-2.093 < t_{obs} < +2.093$, do not reject H_0.

If $t_{obs} \leq -2.093$ or $t_{obs} \geq +2.093$, reject H_0.

Computation   Suppose the experiment is conducted, the observed mean is 8.4, and the standard deviation is 2.3. Placing these values into the expression for t, one obtains

$$t_{obs} = \frac{\overline{X} - \mu}{s_x/\sqrt{N}} = \frac{8.4 - 7}{2.3/\sqrt{20}} = 2.72$$

Decision and Interpretation   The observed t of 2.72 exceeds the critical value of $+2.093$ and falls within the province of the second decision rule to reject H_0. It is concluded that, given the expected amount of variability in means of samples of size 20, the mean of 8.4 is too deviant from $\mu = 7$ to be attributed simply to sampling error.

Once the statistical decision is made, the result must be interpreted with respect to the original research question. The simplest interpretation is to rephrase the decision in terms of the hypotheses. In this case, the result can be interpreted as showing that the drug may have an effect on memory performance; presumably the drug improves it.

Usually, however, interpreting a research study requires much more than a simple rephrasing of the statistical result. Often it consists of discussing the possibility that various other factors might have produced the result, not the one that was of focal attention. For example, the current study showed that the mean for the drugged group was higher than the mean of the population of nondrugged people, but this difference might not be associated with the drug. Perhaps it was associated with having an injection—the drug itself might not have improved performance, but the fact that a needle was stuck in the subjects' arms alerted them to the importance of remembering what they had learned. This possibility could have been eliminated if two groups of subjects had been sampled, one given the drug and one an injection of physiological saline solution that is known to have no effect on memory performance. In this case, if a difference still exists between these two groups, it is not associated simply with the injection, since both groups had that. Presumably, it must be caused by the drug (see Chapter 9).

Interpreting research depends upon a knowledge of research design, and a brief introduction to research design is presented in Chapter 11. It also depends upon a thorough knowledge of the previously reported research pertinent to the study in question and a very critical attitude. Statistics is only one part of the entire research enterprise—it is only one among many tools that a scientist uses to describe and interpret the results of a study. Hereafter, we will concentrate on the statistical decision; but the research is not complete until the result is interpreted in the context of the research design and the research literature relevant to the topic of the study.

Confidence Intervals for the Mean

In the previous chapter, confidence limits for the mean were described when the population standard error of the mean was known. In practice, it is unlikely that $\sigma_{\bar{x}}$ is available, so it must be estimated by calculating $s_{\bar{x}}$ from sample data.

In general, confidence limits for the mean are given by

$$\overline{X} \pm t_{\alpha}\,(s_{\bar{x}})$$

in which

$\overline{X}$ = the sample mean
α = 1 − level of confidence desired
t_{α} = the critical t value corresponding to a nondirectional test at significance level α
$s_{\bar{x}}$ = the estimated standard error of the mean or $s_x/\sqrt{N}$

Specifically, in the illustrative case above, determining 95% confidence limits for the mean requires that:

$\overline{X}$ = 8.4
α = 1 − .95 = .05
t_{α} = $t_{.05}$ = 2.093 (Table B, $df = N - 1 = 20 - 1 = 19$, nondirectional)
$s_{\bar{x}}$ = $s_x/\sqrt{N} = 2.3/\sqrt{20} = .51$

Therefore, 95% confidence limits are given by

$\overline{X} \pm t_{\alpha}(s_{\bar{x}})$
8.4 − (2.093)(.51) and 8.4 + (2.093)(.51)
8.4 − 1.07   8.4 + 1.07
7.33   9.47

Thus, the probability is .95 that such an interval, in this case 7.33 to 9.47, contains the population mean.

Limits also can be constructed at levels of confidence other than 95%. To do so, use the above formula but determine the critical values of t corresponding to a non-directional test at whatever level of confidence is desired (such as, for example, α = .01).

SUMMARY

Hypothesis testing is a procedure to determine the probability that a given observed numerical result could occur by sampling error under certain conditions. Typically, a research question is posed and is translated into specific statistical hypotheses stating two or more mutually exclusive and often exhaustive possible circumstances (for example, that the sample mean is from a population with $\mu = 7$ or from a population having a different mean). The null hypothesis is the one being tested. Assumptions are made about sampling, the nature of distributions, and characteristics of the population. A significance level is adopted, which is the probability value that separates a decision to reject or not reject the null hypothesis. It is used to determine critical values and decision rules for making that decision, keeping in mind the two types of decision errors that could be made. Then the observed statistical values and decision rules for making that decision are stated, and the observed statistical values are translated into values that can be compared to a theoretical sampling distribution. Which theoretical sampling distribution is selected depends on the nature of the statistical information available. This comparison produces the probability that the values of the observed statistic could occur by sampling error given the truth of the null hypothesis. If this probability is very small, then it is unlikely that the null hypothesis is true, and it is rejected. Otherwise, it is not rejected, and it is concluded that the results could have occurred because of sampling error alone.

Formulas

1. Standard form for a score

$$z = \frac{X - \mu}{\sigma_x}$$

2. Standard form for a mean

$$z = \frac{\overline{X} - \mu}{\sigma_{\bar{x}}} = \frac{\overline{X} - \mu}{\sigma_x / \sqrt{N}}$$

in which N = number of cases in the sample for $\overline{X}$

3. Standard form for a mean, parameter estimated

$$t = \frac{\overline{X} - \mu}{s_{\bar{x}}} = \frac{\overline{X} - \mu}{s_x / \sqrt{N}}$$

in which N = number of cases in the sample for $\overline{X}$

$$df = N - 1$$

4. Confidence intervals for the mean

$$\overline{X} - t_\alpha(s_{\bar{x}})$$

$$\overline{X} + t_\alpha(s_{\bar{x}})$$

in which α = 1 − the desired level of confidence

t_α = the critical t value corresponding to a nondirectional test at significance level α

Exercises

For Conceptual Understanding

1. What is the difference between statistical assumptions and hypotheses?

2. Some people say that a statistical test helps decide whether an observed experimental outcome was just a chance result. Explain what they mean. Why is it that a statistical test never proves with absolute certainty that an observed outcome was not a chance result?

3. Why does a statistician "reject H_0" rather than "accept H_1"?

4. Distinguish between the two types of decision errors.

5. What is meant by the power of a statistical test?

6. Consider the experiment with the drug that may enhance memory. If, before conducting the experiment, the researchers decided to draw a sample of 40 rather than of 20, what effect would the increase in sample size have on the relative sizes of the following?

    a. power

    b. β

    c. α

7. How would the relative sizes of β and power be changed if the researchers decided to use a smaller α (.01 rather than .05)?

8. What considerations go into selecting the value of α?

9. What is the basis of deciding between a directional and a nondirectional set of hypotheses?

10. When do you use the standard normal (that is, z) distribution and when do you use the t distribution as the theoretical relative frequency distribution for making a statistical test?

For Solving Problems

11. List the assumptions, hypotheses, formulas, significance level (adopt the .05 level), critical values, decision rules, computation, and decision for the following tests of hypotheses.

    a. Suppose a test for thought disorders had a mean of 81 and a standard deviation of 10 in a population of normal adults. A group of 25 adults between 20 and 30 years of age, each of whom has one parent who has been diagnosed as schizophrenic, take the test and score a mean of 85 (higher scores imply greater thought disorder). Determine whether the children of schizophrenics have more or less thought disorders than the normal population.

    b. What difference would it make in 11a above if you do not know the population standard deviation but you do know that the sample standard deviation is $s_x = 10$?

    c. Suppose males in a given area of the world average 62 inches in height. One tribe in the same area and of comparable genetic stock pierces the lips and molds the heads of its young children. The natives believe this piercing and molding make their males grow stronger. A psychological theory also suggests that a moderate amount of stress in infancy produces skeletally larger adults. The average height of a sample of 36 males in this special tribe is 64.5 inches, and the standard deviation is 7. Is this information consistent with the theory and tribal beliefs?[3] Why do these data not constitute proof that the piercing and molding cause this height advantage?

    d. The average IQ of male students at State University is 113. A random sample of 49 male students who were the firstborn children in their families was found to have a mean IQ of 117 with a standard deviation of 15. Is it reasonable to conclude that firstborn male students have a different IQ from that of the general population of males at State?

----

[3] Adapted from research reported by T. K. Landauer and J. W. Whiting, "Infantile Stimulation and Adult Stature of Human Males," *American Anthropologist* 66: 1007–28.

12. Determine 95% and 99% confidence intervals for the means in the following cases:

    a. $\overline{X} = 40$, $s_x = 7$, $N = 25$

    b. $\overline{X} = 46$, $s_{\bar{x}} = 10$, $N = 16$

For Homework

13. A marine biologist has evidence from other studies that the number of fish is declining in certain lakes because of pollution. The biologist decides to compare the fish population of these lakes to the number of fish in the population of lakes of comparable size. A sample of 25 lakes has a mean of 53,000 fish with a standard deviation of 15,000. The population mean is 58,000. List the hypotheses, t, df, decision rules, critical values, computation, and interpretation of the test. Adopt the .05 level of significance.

14. A researcher wants to determine whether a sample mean of 60 with a standard deviation of 22 cases on 121 subjects is significantly different from a population mean of 56.2. Determine t_{obs}, df, t_{crit}, and whether the sample is from the original population ($\alpha = .05$).

15. A sample of 16 scores has a mean 12 and a standard deviation 3. Estimate .95 and .99 confidence intervals for the population mean.

16. A drug company uses a .05 significance level to reject any lot of medication that does not contain an average of 400 units of a certain chemical. Ten ampuls are selected to test the quality of six lots. Below are listed the resulting mean numbers of units of the chemical for six lots of the medication. Which lots should be rejected as having failed to meet quality standards?

    Lot A: $\overline{X} = 402.3$, $s = 2.3$

    Lot B: $\overline{X} = 402.1$, $s = 2.6$

    Lot C: $\overline{X} = 399.8$, $s = .8$

    Lot D: $\overline{X} = 398.2$, $s = 1.2$

    Lot E: $\overline{X} = 400.5$, $s = .9$

    Lot F: $\overline{X} = 400.9$, $s = .2$

Elementary Techniques of Hypothesis Testing

The general logic of hypothesis testing was discussed in Chapters 7 and 8. Briefly, a hypothesis about a characteristic of the population is tentatively held to be true, data are collected, and statistical techniques are used to determine whether the results are likely to occur within the expected range of sampling error. If the results do not deviate markedly from what would be expected on the basis of sampling variation, there is no reason to doubt the validity of the hypothesis previously held to be true. If the results do deviate markedly from what would be expected on this basis, perhaps the hypothesis is not true.

The illustration of this procedure compared the mean number of nouns recalled by a group of subjects who were given a drug before testing with the known mean number recalled correctly by a population of nondrugged adults. The question was whether the mean of the drugged sample differed significantly from the mean of the nondrugged population. If the observed mean differed by an amount that could not be attributed to sampling error, then perhaps the drug had an effect.

INFERENCES ABOUT THE DIFFERENCE BETWEEN MEANS

The illustration described above required that the sample mean for the drugged group be compared to a known population mean for nondrugged individuals. However, in practice, population values are rarely available. Moreover, even if a very large group of nondrugged subjects had been tested and its mean determined, that group might have been assessed at a different time or under different circumstances than the drugged group. For example, perhaps people's memories are sharper in the morning after a good night's sleep, or maybe the injection, not the drug, affected performance. If the drugged group is not tested at the same time and under the same conditions as the nondrugged population, such procedural factors—not the drug—may produce differences in the performance of the two groups.

A scientifically better and more common procedure would be to select two random samples of 20 subjects each. Before the testing session, the people in one group would receive the drug, while those in the other group would be given an injection of physiological saline solution—a neutral, harmless, and ineffectual substance. None of the participants would know which substance, the drug or the saline solution, they received. Thus, each group would be under whatever stress is associated with being given an injection, and each group would have the same expectation for the consequences of that injection. The null hypothesis is that the drug and saline solution have equivalent effects on their respective population means. Thus, $H_0: \mu_1 = \mu_2$. Conversely, the drug might affect memory differently than does saline, in which case the population means would not be identical. Thus, $H_1: \mu_1 \neq \mu_2$. But the statistical logic demands that we tentatively hold the null hypothesis to be true: We will suppose that we have two samples drawn from identical populations and that whatever difference we observe in their mean values is simply sampling error. Given that the experiment has been well designed and rigorously executed, how does one determine whether

H_0 is true when there are two samples to be compared and no population parameters available?

Specifically, the point of the two-sample experiment just described is to ask whether the difference between the mean of the drugged group (call it $\overline{X}_1$) and the mean of the saline group (label it $\overline{X}_2$) could occur through sampling error alone. If the two groups performed exactly the same, then the difference between their means would be zero: $(\overline{X}_1 - \overline{X}_2) = 0$. However, this difference is not likely to be precisely 0 but instead will vary with sampling error. Thus, one needs to know the probability that the difference between the two groups, $(\overline{X}_1 - \overline{X}_2)$, could be just sampling error.

To obtain a formula for this question, consider again the formula:

$$t = \frac{\overline{X} - \mu}{s_{\bar{x}}}$$

As mentioned previously, this expression is quite general, and it can be read "a value minus the population mean of such values divided by an estimate of the standard error of such values is distributed as t, a theoretical sampling distribution." It could be written:

$$\frac{t, \text{ a theoretical}}{\text{sampling distribution}} = \frac{(\text{a value}) - (\text{the population mean of such values})}{(\text{an estimate of the standard error of such values})}$$

If "the difference between two sample means" is substituted for "a value" in the above expression, then the difference between two sample means minus the difference between the means of the two populations from which the samples are drawn, divided by an estimate of the standard error of such differences between sample means, will be distributed as t. In symbols,

$$t = \frac{(\overline{X}_1 - \overline{X}_2) - (\mu_1 - \mu_2)}{s_{\bar{x}_1 - \bar{x}_2}}$$

in which $(\overline{X}_1 - \overline{X}_2)$ is the difference between two sample means, μ_1 is the mean of the population from which the first sample was drawn, μ_2 is the mean of the population from which the second sample was drawn, and $s_{\bar{x}_1 - \bar{x}_2}$ is an estimate of the standard error of the difference between two sample means.

The computational task requires that we obtain values for each of the three terms in this formula. The first term, $(\overline{X}_1 - \overline{X}_2)$, will be provided by the observed means of the two groups in the experiment.

To obtain a value for the second term, $(\mu_1 - \mu_2)$, recall that the null hypothesis states that the drug has no effect and thus in the population the mean of the drugged group equals the mean of the nondrugged group. If $\mu_1 = \mu_2$, then $\mu_1 - \mu_2 = 0$. Since the logic of these statistical procedures demands that we tentatively hold the null hypothesis to be true, the term $\mu_1 - \mu_2$ in the above formula will be zero under H_0:

$$t = \frac{(\overline{X}_1 - \overline{X}_2) - \overbrace{(\mu_1 - \mu_2)}^{\text{equals 0 by } H_0}}{s_{\bar{x}_1 - \bar{x}_2}} = \frac{(\overline{X}_1 - \overline{X}_2)}{s_{\bar{x}_1 - \bar{x}_2}}$$

The third term—the denominator of this formula—is an estimate of the standard error of the difference between two sample means drawn from the same population. Under the null hypothesis, the populations of drugged and saline-injected people are tentatively presumed to be identical—that is, the two sample means are presumed to be drawn from the same population under H_0. But the mean of one sample will not precisely equal the mean of another sample, even when the two samples are drawn from the same or identical populations. They will differ somewhat from each other because of sampling error. One can imagine selecting two samples from a single population, computing the difference between their means, collecting another two samples, and so on. These differences between pairs of means would form a distribution of differences between pairs of means. As a distribution of a statistic, $(\overline{X}_1 - \overline{X}_2)$, this would constitute an empirical sampling distribution, and its standard deviation would be the standard error of the difference between pairs of means drawn from a single population. This population standard error may be estimated by $s_{\bar{x}_1 - \bar{x}_2}$, which value reflects the extent to which two sample means from identical populations will differ from one another on the basis of sampling error alone.

The task remaining is to determine an expression for $s_{\bar{x}_1 - \bar{x}_2}$. It happens that the standard error of the difference between means varies, depending upon whether the scores in the two groups are **independent** or **correlated.** If different subjects compose the two groups, the scores in these groups will be independent. However, if the same subjects are measured twice—such as before a treatment is administered and again afterward—the scores in the two groups will be correlated. This happens because some factors that influence an individual subject's performance in one circumstance (e.g., on the pretest) are likely to influence that subject's performance in another circumstance (e.g., on the posttest). For example, if children are given a special mathematics training program, it is likely that individual pupils who score high relative to the group before training, perhaps because they have better mathematics ability or are more motivated, will also score high after it. Thus, the before-and-after scores will be correlated. In this event, the estimate of $s_{\bar{x}_1 - \bar{x}_2}$ based upon independent groups is not appropriate and special techniques are required. We will consider both cases below.

The above general formula, then, expresses the ratio of the observed difference between sample means to a measure of the expected sampling variation in such differences. This ratio represents a value in the *t* distribution. Using Table B of Appendix 2, which lists critical values for the *t* distribution, we can determine the probability that the observed difference between means is simply sampling error. If this probability is small, suggesting that such a difference could not be reasonably expected from sampling error alone, the null hypothesis that these samples come from identical populations will be rejected and we will conclude that the drug has some differential effect over saline. The general logic is the same as in the previous two chapters; only the formula is different.

The Difference between Means: Independent Groups

Research Question A developmental psychologist was interested in how divorce affects the nursery school behavior of young boys.[1] It was possible that boys four to

---

[1] Based on, but not identical to, a study by E. M. Hetherington, M. Cox, and R. Cox, "Play and Social Interaction in Children Following Divorce," paper presented at National Institute of Mental Health Divorce Conference, 1978.

six years old might become more or less aggressive as a response to the divorce of their parents. To test this, the psychologist had research assistants observe over a two-week period the number of aggressive acts of 17 boys whose parents had recently divorced (for our purposes, the "divorced group") and 15 boys whose parents had not divorced (the "nondivorced group"). A formal summary of the statistical treatment of this experiment is given in Table 9-1, and the details are discussed below.

Statistical Hypotheses   The hypotheses are as follows:

$$H_0: \mu_1 = \mu_2$$
$$H_1: \mu_1 \neq \mu_2$$

in which μ_1 is the population mean for the divorced group and μ_2 is the mean for the nondivorced group.

Notice two things. First, the statistical hypotheses are stated in terms of population parameters, not statistics. This is because the decision to be made involves what is true about the populations—boys of divorced and nondivorced parents—not just about the two specific samples observed in this experiment. Second, the alternative hypothesis, H_1, does not specify whether μ_1 is greater than or less than μ_2. It just states that the population means are different—boys of divorced parents may be more *or* less aggressive than boys of nondivorced parents. Thus, the statistical test should be sensitive to differences in either direction and a nondirectional, or two-tailed, test is required.

Assumptions and Conditions   First, the subjects must be **randomly** and **independently sampled.** This means that each subject in the population has an equal opportunity of being selected and that the inclusion of one subject in the sample does not influence the probability of selecting any other member of the population into the sample.

Second, these procedures apply to two **independent groups.** That is, the divorced and nondivorced groups are composed of different boys who independently qualified for these two groups. The next section will describe procedures to be used with correlated groups.

Third, the population variances of the two groups are presumed to be equal, a characteristic called **homogeneity of variance.** Unfortunately, it is often difficult to decide whether this condition is met by the data to be analyzed. Since the variance of one sample from a population is not likely to equal the variance of another sample from that population because of sampling error, the question is, how different can the two sample variances be before one suspects that their population parameters are not equal? While precise statistical tests are available that can be used to make this decision, many statisticians think their application to this problem is not very worthwhile. In addition, moderate violations of this assumption can occur—especially if the number of subjects is approximately the same in each group—without seriously biasing the results. If the number of observations in one group is the same as in another and there are at least five subjects in each group, one sample variance may be as much as twice the size of the other without markedly altering the results of the statistical analysis.[2]

---

[2] G. E. P. Box, "Some Theorems on Quadratic Forms Applied in the Study of Analysis of Variance Problems: II. Effect of Inequality of Variance and of Correlations of Errors in the Two-Way Classification," *The Annals of Mathematical Statistics,* 25: 484–98.

9-1 | **Summary of the Test of Difference Between Means for Independent Groups (Divorced vs. Nondivorced Example)**

Hypotheses

H_0: $\mu_1 = \mu_2$
H_1: $\mu_1 \neq \mu_2$ (nondirectional)

Assumptions and Conditions

1. The subjects in each group are **randomly** and **independently sampled**.

2. The groups are **independent**.

3. The population variances are **homogeneous**.

4. The population distribution of $(\overline{X}_1 - \overline{X}_2)$ is **normal** in form (i.e., the X_1 and X_2 distributions are normal or N_1 and N_2 are not small).

Decision Rules (from Table B)

Given: .05 significance level, a nondirectional test, and

$$N_1 + N_2 - 2 = 17 + 15 - 2 = 30 \; df$$

If $-2.042 < t_{obs} < 2.042$, do not reject H_0.

If $t_{obs} \leq -2.042$ or if $t_{obs} \geq 2.042$, reject H_0.

Computation

	Divorced (Group 1)	Nondivorced (Group 2)
	$\overline{X}_1 = 40$	$\overline{X}_2 = 35$
	$s_1^2 = 16$	$s_2^2 = 18$
	$N_1 = 17$	$N_2 = 15$

$$t_{obs} = \frac{\overline{X}_1 - \overline{X}_2}{\sqrt{\left[\dfrac{(N_1 - 1)s_1^2 + (N_2 - 1)s_2^2}{N_1 + N_2 - 2}\right] \cdot \left[\dfrac{1}{N_1} + \dfrac{1}{N_2}\right]}}$$

$$= \frac{40 - 35}{\sqrt{\left[\dfrac{(17 - 1)16 + (15 - 1)18}{17 + 15 - 2}\right] \cdot \left[\dfrac{1}{17} + \dfrac{1}{15}\right]}}$$

$$t_{obs} = 3.43$$

Decision

Reject H_0.

Thus, it is best to have approximately equal numbers of subjects in each group and adequate-sized samples; then homogeneity of variance probably will not be a problem.

Finally, it must be assumed that the **sampling distribution of the differences between pairs of sample means $(\overline{X}_1 - \overline{X}_2)$ is normal in form.** This is necessary for the percentiles of the t distribution to be appropriate and accurate. This condition will be satisfied if the populations of raw scores are normal or if the sample sizes are not too small.

Formula   The formula relating the difference between sample means for independent groups to its standard error is

$$t = \frac{\overline{X}_1 - \overline{X}_2}{s_{\bar{x}_1 - \bar{x}_2}} = \frac{\overline{X}_1 - \overline{X}_2}{\sqrt{\left[\dfrac{(N_1 - 1)s_1^2 + (N_2 - 1)s_2^2}{N_1 + N_2 - 2}\right] \cdot \left[\dfrac{1}{N_1} + \dfrac{1}{N_2}\right]}}$$

The denominator of this formula is the standard error of the difference between means from independent groups. It is estimated on the basis of the variances of the two samples, s_1^2 and s_2^2. The degrees of freedom for t are the sum of the df for each sample variance:

$$df = \overbrace{(N_1 - 1)}^{df \text{ for } s_1^2} + \overbrace{(N_2 - 1)}^{df \text{ for } s_2^2} = N_1 + N_2 - 2$$

Significance Level   Use the .05 level of significance.

Critical Values   The critical values will depend upon the degrees of freedom, the nature of the alternative hypothesis, and the level of significance. A glance at Table 9-1 shows that there are 17 and 15 subjects in the groups, and thus the degrees of freedom are $(N_1 + N_2 - 2) = (17 + 15 - 2) = 30$. The alternative hypothesis is nondirectional and the .05 level has been set. Entering Table B in Appendix 2 with this information, one finds the critical values to be -2.042 and 2.042.

Decision Rules   If the observed (computed) value of t is between -2.042 and $+2.042$, the null hypothesis will not be rejected, and one will conclude that the difference between means is within the range of sampling error. In this event the results would not provide evidence supporting an association between divorce and aggressive behavior of young boys in nursery school. Conversely, if the observed value of t is greater than or equal to 2.042 or less than or equal to -2.042, the null hypothesis will be rejected. The reasoning is that the probability is too remote that such an observed difference in means reflects merely sampling error, and that it is likely that the divorce experience is associated with aggressive social behavior. In symbols,

If $-2.042 < t_{obs} < 2.042$, do not reject H_0.

If $t_{obs} \leq -2.042$ or if $t_{obs} \geq 2.042$, reject H_0.

Computation   To find the value of the observed t, one proceeds by listing the relevant known information and the formula for t, calculating the components of the formula, and computing the result. This work is presented in Table 9-1. The observed t is $t_{obs} = 3.43$.

Decision   The observed value of t conforms to the second decision rule, reject H_0. The probability that such a difference between means would occur merely as a function of sampling error is so small that it is likely that the two samples have been drawn from different populations. If the experiment has been well conceived, the implication is that parental divorce is associated with aggressive social behavior in the young boys of those couples. Notice that the interpretation is not that divorce *produced* the aggressive behavior. It may have, but it is also possible that the aggressive behavior of the young boys was present before divorce; in fact, it actually might have helped cause the parents' divorce rather than the reverse. One would need to demonstrate that the children in the divorced and nondivorced groups were equal on aggressiveness before the divorce. Even then other factors could cause this result. Only random assignment of children to families would come close to assuring the interpretation that divorce causes aggressiveness, but, of course, it is not possible to randomly assign children to divorcing and nondivorcing parents.

The Difference between Means: Correlated Groups

Sometimes the two means to be compared are obtained from the same subjects or from matched pairs of subjects (such as identical twins; pairs of individuals matched for age, gender, social class, IQ; and so on). In either case, some of the factors that influence scores in one group will also influence them in the other group (especially the factors associated with individual subjects), and thus the scores in the two groups will be correlated to some extent. In that event, the assumption of independent groups cannot be made, and alternative procedures must be used.

Research Question   Suppose it is important to ascertain whether stimulation of one area of the brain is more reinforcing to a rat than stimulation of another area. Electrodes are sunk into the two areas in each of 10 rats. After an initial training session, the rats are allowed to press two bars, one stimulating the first area of the brain and the other stimulating the second area. The number of presses on each bar is recorded; the mean number of presses for stimulation in area 1 will be compared with the mean for area 2. In addition, suppose a set of experiments by other investigators and a current theory strongly suggest that area 1 should be more effective in producing responses than area 2. Therefore, the research question is whether area 1 is a more reinforcing site for brain stimulation than is area 2. A summary of the statistical procedure testing this question is presented in Table 9-2.

Statistical Hypotheses   As indicated above, theory and empirical evidence suggest that the stimulation of area 1 may be more reinforcing than stimulation of area 2. Therefore, the hypotheses are directional rather than nondirectional. That is, the null hypothesis states that μ_1 either equals or is less than μ_2, while the alternative hypothesis specifies that μ_1 is larger than μ_2. In symbols,

$H_0: \mu_1 \leq \mu_2$

$H_1: \mu_1 > \mu_2$ (directional)

Recall that hypotheses must be exhaustive—that is, they must cover all the possibilities. It is for this reason that the null hypothesis states that the population means are equal

9-2 Summary of the Test of Difference Between Means for Correlated Groups (Brain Stimulation Example)

Hypotheses

$H_0: \mu_1 \leq \mu_2$

$H_1: \mu_1 > \mu_2$ (directional)

Assumptions and Conditions

1. The subjects are **randomly** and **independently sampled.**
2. The scores of the two groups are **correlated.**
3. The population distribution of the D_i is **normal** in form.

Decision Rules (from Table B)

Given: .01 significance level, a directional test, and

$N - 1 = 10 - 1 = 9 \ df$

If $t_{obs} < 2.821$, do not reject H_0.

If $t_{obs} \geq 2.821$, reject H_0.

Computation

Subject	X_1 (Area 1)	X_2 (Area 2)	$X_{1_i} - X_{2_i} = D_i$	D_i^2
a	58	42	16	256
b	45	50	−5	25
c	61	23	38	1444
d	55	50	5	25
e	58	45	13	169
f	90	85	5	25
g	26	30	−4	16
h	35	20	15	225
i	42	50	−8	64
j	48	60	−12	144

$N = 10 \quad \Sigma X_1 = 518 \quad \Sigma X_2 = 455 \quad \Sigma D_i = 63 \quad \Sigma D_i^2 = 2393$

$\bar{X}_1 = 51.8 \quad \bar{X}_2 = 45.5 \quad \bar{D} = 6.3$

$$t_{obs} = \frac{\bar{D}}{s_D / \sqrt{N}} = \frac{\Sigma D_i}{\sqrt{\dfrac{N\Sigma D_i^2 - (\Sigma D_i)^2}{N - 1}}} = \frac{63}{\sqrt{\dfrac{10(2393) - (63)^2}{10 - 1}}}$$

$t_{obs} = 1.34$

Decision

Do not reject H_0.

or the first is less than the second. If the result of the study is that the sample mean of the first group is very much less than the second—enough to reject a null hypothesis if a two-tailed test were conducted—the conclusion in the present case nevertheless would be not to reject this directional H_0. (The researcher, however, might seriously consider the nature of the study and the theory and previous literature that led to the use of a directional test.)

Formula As can be seen from Table 9-2, the basic data are two scores per subject, one for area 1 and one for area 2. The computational routine for this case differs from the preceding example and rests upon the following fact:

> The difference between two means equals the mean difference between pairs of scores.

This states that if the difference between each subject's two scores is computed, the mean of such differences will equal the mean over all subjects of scores for area 1 minus the mean of the scores for area 2. For example, look at the computation section of Table 9-2. The table lists for each subject the number of bar presses that stimulated area 1 (X_1), the number of bar presses that stimulated area 2 (X_2), and the difference between these two values $(X_{1_i} - X_{2_i} = D_i)$. Notice in the rows under the data that the mean for area 1 was $\overline{X}_1 = 51.8$, the mean for area 2 was $\overline{X}_2 = 45.5$, and the mean for the differences between pairs of scores was $\overline{D} = 6.3$. One can see that the difference between the two means $\overline{X}_1 - \overline{X}_2 = 51.8 - 45.5 = 6.3$ is identical to the mean of the differences between pairs of scores, $\overline{D} = 6.3$.

Applied to the current situation, this fact implies that if there is absolutely no difference between the two sample means $(\overline{X}_1 - \overline{X}_2 = 0)$, $\overline{D}$ will also equal zero. Because of sampling error, $\overline{D}$ rarely will be exactly zero even if the two populations are identical. However, if the means of the two samples are both estimates of the same population mean value, as the null hypothesis states, the positive and negative differences should cancel out and the sampling distribution of $\overline{D}$ will have a mean of zero. Consequently, the statistical question reduces to the probability that $\overline{D}$ should deviate from zero relative to the sampling error of $\overline{D}$.

The formula for this translation derives from the basic formula for t in which a sample mean minus the population mean is divided by an estimate of the population standard error of that mean. Now, however, the "mean" is not $\overline{X}$ or $(\overline{X}_1 - \overline{X}_2)$ but $\overline{D}$:

$$t_{obs} = \frac{\overline{D} - \mu_D}{s_{\overline{D}}}$$

But, as noted above, μ_D will equal 0 under the null hypothesis of no difference between the two population means, and substituting the expression for the standard error, we have

$$t_{obs} = \frac{\overline{D}}{s_D/\sqrt{N}}$$

If you are using a hand calculator, this formula will be convenient to use if your calculator will automatically compute the mean and standard deviation of the D_i. If not, you will find the following equivalent formula easier to compute:

$$t_{obs} = \frac{\Sigma D_i}{\sqrt{\dfrac{N\Sigma D_i^2 - (\Sigma D_i)^2}{N-1}}}$$

Note that the N in both these formulas is the number of subjects or *pairs* of scores (i.e., $N = 10$ in this example).

Assumptions and Conditions   The data are pairs of scores and it is assumed that the two sets of scores are **correlated.** Often this results from having the same or closely matched subjects contribute both scores. It is assumed that the *pairs* are **randomly** and **independently sampled** and that in the population the D_i are **normally** distributed. Normality is assumed to allow use of the percentiles of the t distribution. The D_i will be normal if the X_1 and X_2 distributions are normal.

Significance Level   Suppose that the implications of this study are such that a type I error—incorrectly rejecting the null hypothesis—would be exceptionally undesirable. Therefore, the .01 level of significance will be selected.

Critical Values   The test to be performed will involve a directional alternative. This means that the test is one-tailed: the critical region lies in only one of the two tails of the t distribution. Since additional information dictates that area 1 should be more reinforcing than area 2, the converse possibility, that area 1 might be poorer in reinforcing bar pressing, is so unlikely that the probability that this should occur is nearly zero. Under these conditions it is no longer appropriate to expect that a negative value of t might occur, and so the left-hand tail of the theoretical distribution is not included in the critical region. As a result, the critical value appropriate for this test is such that 1% of the area under the curve falls to its right. Looking in the table of t values with $N - 1 = 10 - 1 = 9$ degrees of freedom, one finds the .01 level, one-tailed critical value is 2.821.

Decision Rules   The decision rules are as follows:

If $t_{obs} < 2.821$, do not reject H_0.

If $t_{obs} \geq 2.821$, reject H_0.

Computation   The raw data and computation are illustrated in Table 9-2.

Decision   The observed value of t ($t_{obs} = 1.34$) is less than the critical value of 2.821, and according to the decision rules one does not reject the null hypothesis. An observed $\bar{D}$ of 6.3 is within the realm of sampling error for correlated samples of size 10 drawn from populations having the same mean. That is, if the two areas of the brain were identical in their reinforcing potential, with the amount of sampling variability involved it would be quite possible to obtain a sample having a $\bar{D} = 6.3$. The researcher must conclude that the data provide insufficient evidence that stimulation of area 1 is more reinforcing than stimulation of area 2.

What if area 1 actually had produced substantially *fewer* responses than area 2, despite the prediction to the contrary? As discussed above, if a directional test had been selected before the experiment was conducted (and this decision must be made beforehand), then a large negative t would conform to the first decision rule, which dictates that one not reject the null hypothesis. This would be the decision even if t were very large, such as -4.00, that is, large enough to be "significant" if a two-tailed test had been conducted.

Independent and Correlated Groups Compared

The two types of comparisons described above, independent groups and correlated groups, are different in more ways than just the formulas used to calculate t. They represent two different approaches to conducting research.

One approach is to make a **between-subjects comparison.** For example, one treatment might be given to one group of subjects and another treatment (or no special treatment) might be given to another, separate, independent group of subjects. If only two groups are involved, the means of the two groups might be tested statistically with the independent-groups t test described above.

Another approach is to make a **within-subject comparison.** In this case, for example, the two treatments might be given to the same subject or subjects that are closely matched in some way (e.g., siblings; pairs of subjects matched for gender, IQ, family background, etc.). Alternatively, the same subjects might be observed under two conditions, such as before and after a treatment or particular experience. These two groups of scores might be tested statistically with the correlated-groups t test described above.

Generally, between-subjects and within-subjects approaches are not simply two equivalent strategies aimed at answering the same question. They differ logically, statistically, and practically from each other. For example, suppose a researcher wanted to evaluate the effectiveness of a new preschool mathematics program designed to teach four- and five-year-old children basic quantitative concepts (e.g., counting, class inclusion and exclusion, etc.) in preparation for school. The first question the researcher might want to address is whether the children learned the concepts taught by the program and improved in their general quantitative readiness. To do this, the researcher might administer a test of quantitative readiness to the children before they started the training program and one after they finished the program. This particular within-subjects approach is sometimes called a **before-and-after** or a **pretest–posttest** strategy. If the program is successful, one would expect that the children would score better on the posttest than on the pretest. The logic of this strategy is clear: If the intent of the program is to produce *changes within individuals,* then one needs to observe or test the same subjects at two points in time to document such *changes within individuals.* Presumably, one would use a correlated t test to assess such change in the preschool children.

But while it seems necessary to show that children indeed improved from pretest to posttest, such improvement may or may not have been produced by the new training program per se. If the pretest was given when the children were four years old and the posttest given when they were five years old, perhaps the children improved simply because they were older and had acquired quantitative concepts at home, from television, or by natural experience and development, not because the program taught them. To determine that the program per se produced the gains, a between-subjects strategy might be used in which children are randomly assigned either to the special program or to a similar preschool but without the special quantitative program. An independent-groups *t* test might be used to compare the two groups at five years of age after the program was completed. If the program produced benefits, the children experiencing the special program should score better.

Would not the latter between-subjects comparison be sufficient? By demonstrating the effectiveness of the special program versus no program, does not the between-subjects comparison also show that the treatment children learned the program content and improved their quantitative skills? Not necessarily. Special training programs, for example, are often expensive and administered to small groups of subjects. As we have seen, two groups of randomly assigned subjects, especially small groups, could differ "significantly" by sampling error alone—that is, a type I error might be made. Perhaps, for example, the children assigned to the program group were more skilled quantitatively to begin with, not because the program improved them. Also, it is possible that the no-program comparison children *declined* in performance between four and five years of age (which is common among low-income children), whereas the program children did not. In this case, the program did not produce *gains* in the treated children but only prevented them from declining. The program children may be better than the no-program children at posttest, but a between-groups strategy would not detect that they did not improve or learn the specific content the program taught. The between-groups and the within-groups approaches answer different questions and both would seem to be useful.

Unfortunately, from a practical standpoint, it is not always possible to conduct both approaches. For example, between-groups studies are required when the treatments will have a more or less permanent effect on the participants, making it impossible to administer another treatment to the same people. A pregnancy prevention program for adolescents or the effectiveness of a drug or medical procedure in preventing or treating an illness usually can be evaluated only between groups, because the participants will be changed by the treatment or cannot receive simultaneously or sequentially both treatment and no treatment or several different treatments. On the other hand, to chart developmental change within individuals, document the effects of training or learning experiences, or demonstrate preferences for different stimuli (e.g., which candidate do you prefer? Which advertising scheme do you prefer?), within-groups approaches are usually required.

The two types of *t* tests are also different statistically. Generally, sampling error is produced by variability associated with individual differences between subjects, measurement error, and unreliability (see Chapter 1). Within-subjects comparisons are made, as the name implies, within subjects, so that much of the variability produced by individual differences between subjects is eliminated using the within-subjects approach. As a result, within-subjects comparisons tend to have smaller sampling errors and are more likely to detect group differences.

For example, look at the data in Table 9-2. Subject **f** had very high scores (i.e., 90 and 85) relative to the other subjects. From a between-subjects perspective, the deviations of these scores from the group means of 51.8 and 45.5, respectively, contribute to the variances of these groups, which are used to estimate the standard error of the mean in a between-subjects comparison. The larger these between-subjects variances, the larger the standard error of the mean, and the greater must be the difference between means to produce a large *t* and a significant between-subjects statistical result. Conversely, in the within-subjects comparison, the fact that subject **f** had extreme scores does not matter; only the *difference* between those scores, D_f, and the variability of those differences across subjects enter the calculation of *t*. Usually, the variability of differences between two identical measurements on the same subject is less than the variability between subjects of those same measurements. This tends to produce larger values of *t* for the within-subjects than the between-subjects comparisons. For example, the within-subjects *t* in Table 9-2 was 1.34, but if these data were treated as a between-subjects comparison, the independent-groups *t* would be only .77. It would be smaller because the individual differences between subjects would contribute to the estimate of sampling error in the between- but not the within-subjects comparison. In this sense, the within-subjects approach is often considered more sensitive or powerful. This advantage of the within-subjects approach exists only if the two distributions are at least moderately and positively correlated. If the correlation is negative, the within-subjects sampling error may be greater, not smaller, than if the distributions were treated as independent samples; and if the correlation is positive but small, the sampling error may not be reduced enough to compensate for the smaller degrees of freedom in the within-subjects approach. As discussed above, however, it is often not possible to simply pick either a within- or between-subjects approach, and the two approaches tend to provide answers to different questions.

INFERENCES ABOUT CORRELATION COEFFICIENTS

In the previous section, inferences about differences between means were considered. But tests of the values of other population parameters can also be made. For example, one might be interested in inferences about the degree of relationship between two variables. This section is concerned with two types of questions. First, does a linear relationship between two variables exist in the population? That is, if an *r* is computed on a sample, what is the probability that in the population the correlation is actually zero and that *r* deviates from $\rho = .00$ (ρ is "rho," the population correlation coefficient) by sampling error alone? Second, are two *r*'s drawn from independent groups of subjects significantly different from each other? That is, what is the probability that the two sample *r*'s differ from one another by sampling error alone and that in the population there is no difference between ρ_1 and ρ_2?

The Significance of *r*

Suppose a correlation of .74 is found between two variables. One would want to know whether an *r* of this magnitude is merely an imperfect reflection of a population in which the correlation is actually zero or whether an $r = .74$ faithfully mirrors a nonzero linear relationship in the population. Specifically, what is the probability that the sample $r = .74$ deviates from $\rho = .00$ by sampling error alone?

Research Question   Some tests of infant development contain items that are designed to assess the amount and nature of vocalizations made by the infant. For example, a bell is rung in front of the infant and the presence and extent of the baby's vocalization in response to this stimulus is measured. Although many items on such infant scales tend to assess motor development and do not seem to relate to later tested IQ, it might be reasonable to inquire whether there is a relationship between vocalization in 12-month-old female infants and verbal intelligence at 26 years of age.[3] A formal summary of the following procedure appears in Table 9-3.

9-3 | ### Summary of the Test of Significance of a Correlation Coefficient (Female Infant Vocalization–IQ Example)

Hypotheses

   $H_0: \rho = .00$

   $H_1: \rho \neq .00$ (nondirectional)

Assumptions and Conditions

   1. The subjects are **randomly** and **independently sampled.**

   2. The population distributions of both X and Y are **normal** in form.

Decision Rules (from Table C)

   Given: .05 significance level, a nondirectional test, and

   $N - 2 = 27 - 2 = 25\ df$

   If $-.3809 < r_{obs} < .3809$, do not reject H_0.

   If $r_{obs} \leq -.3809$ or $r_{obs} \geq .3809$, reject H_0.

Computation

   $r_{obs} = .74$

Decision

   Reject H_0.

---

[3] Inspired by J. Cameron, N. Livson, and Nancy Bayley, "Infant Vocalizations and Their Relationship to Mature Intelligence," *Science* 157 (1967): 331–33; R. B. McCall, P. S. Hogarty, and N. Hurlburt, "Transitions in Infant Sensorimotor Development and the Prediction of Childhood IQ," *American Psychologist* 27 (1972): 729–48; T. Moore, "Language and Intelligence: A Longitudinal Study of the First Eight Years," *Human Development* 10: 88–106.

Statistical Hypotheses The data reveal a correlation of .74 between infant vocalization and adult verbal intelligence for a group of 27 females. However, it is possible that no relationship actually exists in the population from which the infants were sampled, that is, $\rho = .00$. In such an event, the observed relationship of .74 is merely a function of sampling error. Theoretically, it would be possible to continue to select sample after sample of size 27 and compute a correlation on each. The distribution of such sample r's forms the sampling distribution of the correlation coefficient, and its standard deviation is the standard error of r. The null hypothesis states that the population correlation is actually .00. The alternative hypothesis dictates that a relationship of some nonzero magnitude exists in the population. Since there is no theory or empirical basis for predicting whether such a relationship, if it exists, is positive or negative, the alternative hypothesis is nondirectional. In symbols,

$$H_0: \rho = .00$$
$$H_1: \rho \neq .00$$

Formula The formula relating a sample correlation coefficient to a theoretical sampling distribution of r's follows the same general pattern as the previous formulas:

$$t = \frac{r_{obs} - \rho}{s_r}$$

This states that a sample correlation, r, minus the population correlation, ρ, divided by an estimate of the standard error of sample correlations, s_r, is distributed as t, but this time with $N - 2$ degrees of freedom. Since the null hypothesis dictates that $\rho = 0$, this formula reduces to $t = r/s_r$. Substituting the computational expression for s_r and then making a few algebraic simplifications, one has

$$t = \frac{r_{obs}}{s_r} = \frac{r_{obs}}{\sqrt{\dfrac{1 - r_{obs}^2}{N - 2}}} = \frac{r_{obs}}{\dfrac{\sqrt{1 - r_{obs}^2}}{\sqrt{N - 2}}}$$

$$t = \frac{r_{obs}\sqrt{N - 2}}{\sqrt{1 - r_{obs}^2}} \text{ with } df = N - 2$$

It is possible to substitute the observed value of r (in this case $r_{obs} = .74$) into the above formula, compute t_{obs} (which equals 5.50), and determine its probability under H_0 by looking in the table of t values as before. However, since this conversion is required so often and since the value of t depends only upon r_{obs} and upon its degrees of freedom, critical values of r for various df have been computed and are presented in Table C of Appendix 2. Once r_{obs} is calculated, no further computations are required. To use this table, decide whether the alternative hypothesis requires a directional or nondirectional test and select the significance level. Then locate the appropriate column and find the row corresponding to the degrees of freedom ($df = N - 2$). The critical value r_{crit} is given at the intersection. For the present data, a nondirectional test at the .05 level with $df = 27 - 2 = 25$ has a critical value of $r_{crit} = .3809$. Since the sampling distribution is symmetrical and this is a nondirectional test, there are two critical values, $-.3809$ and $+.3809$.

Assumptions and Conditions   It is assumed that the subjects are **randomly** and **independently sampled** and that the population distributions of both X and Y are **normal** in form.[4] Notice that one can *compute* a correlation on two variables regardless of the nature of their distributions, but normality is required to ensure the accuracy of this *statistical test* of the significance of r.

Significance Level   Adopt the .05 level.

Critical Values   In this case, the critical values can be expressed in terms of r rather than t by using Appendix 2, Table C. For a nondirectional test at .05 with $df = N - 2 = 27 - 2 = 25$, the critical values of r are $-.3809$ and $+.3809$. Therefore,

If $-.3809 < r_{obs} < .3809$, do not reject H_0.

If $r_{obs} \leq -.3809$ or $r_{obs} \geq .3809$, reject H_0.

Computation   Once r_{obs} is determined, no further computation is required.

Decision   Since the observed r ($r_{obs} = .74$) exceeds the critical value .3809, the second decision rule is used and H_0 is rejected. The probability is very low that the observed correlation of .74 could be drawn from a population in which the correlation between these two variables is actually zero. The implication is that there is a relationship in the female population between vocalization at age 1 and verbal intelligence at age 26.

Confidence Limits for the Correlation Coefficient

The hypothesis-testing procedures described immediately above ask a very limited question: What is the probability that the observed correlation could differ from a population value of $\rho = .00$ by sampling error alone? If this hypothesis is rejected, then we conclude that the population correlation is some value other than .00. But what value is it?

The observed correlation of $r = .74$ is one estimate of that population value. Recall from Chapter 7 that whenever a statistic—the sample r or $\overline{X}$, for example—is used to estimate its corresponding population parameter, it is called a **point estimate,** because a single value or point is used to estimate the population value. While a point estimate is helpful, we know it is only approximate. That is, the observed statistic or point estimate—r, for example—will still deviate from the population value, now presumed to be some value other than .00, by sampling error. So it is reasonable to specify a range of values that is likely to contain the population correlation. This range or interval of values constitutes an **interval estimation** of the parameter. When this interval is specified to have a particular numerical probability (e.g., .95 or .99) that the interval contains the population value, we call it a **confidence interval,** which we have already learned to calculate for the mean for the case in which the population standard error of the mean is known (Chapter 7, pages 200–202) and when it must

---

[4] Technically, the joint distribution of X and Y is bivariate normal.

be estimated from the data (Chapter 8, page 229). The task now is to determine confidence intervals for the correlation coefficient.

Confidence intervals for the correlation coefficient are computed in essentially the same general way that confidence intervals for the mean are calculated, except that the sample r and the standard error of the correlation coefficient are used instead of the sample $\overline{X}$ and the standard error of the mean, respectively. One other difference exists: The sampling distribution of the correlation coefficient r approximates a normal distribution (if N is moderately large) only when the null hypothesis that $\rho = .00$ is true. That is, the correlation coefficient can range only between -1.00 and $+1.00$, so when the population value is .00, sample values can range equally above or below .00, producing a symmetrical distribution. But when the population value is not .00, and especially when it is very high or very low, the sampling distribution of r is not symmetrical or normal because r's cannot be greater than $+1.00$ or less than -1.00. Therefore, while the t distribution can be used in the previous section to determine whether r differs from a population $\rho = .00$, it cannot be used to construct confidence intervals when $\rho \neq .00$. For the same reason, the standard error of the correlation coefficient used in the previous section is appropriate when $\rho = .00$, as is assumed in testing that null hypothesis, but another estimate of the standard error must be used if the population correlation is a value other than .00, which it will be if the null hypothesis is rejected.

This problem is solved by transforming the sample r into standard score form such that the standardized r's do have approximately a normal distribution for samples of moderate size. This transformation of r to z_r is given by

$$z_r = \tfrac{1}{2}\log_e(1 + r) = \tfrac{1}{2}\log_e(1 - r)$$

Fortunately, this formula has been computed for many values of r and the results have been tabled. This r-to-z transformation is given in Table D of Appendix 2. To use this table, simply locate the value of r in the left-hand column of the pair of columns and, immediately to its right on the same line, read the corresponding transformed value, z_r. Similarly, to transform a value of z_r back to r, locate the value of z_r in the right-hand column and read the corresponding r on the same line in the left-hand column.

Now, if we had an empirical sampling distribution of r's that had all been transformed in this way into z_r's, that distribution of z_r's would be approximately normal in form, so we could use the percentiles of the standard normal distribution. Further, the standard deviation of that sampling distribution of z_r's, that is, the standard error of the z_r, would be

$$\sqrt{\frac{1}{N - 3}}$$

Recall from the construction of confidence intervals for the mean that such intervals were simply the sample mean plus and minus the number of standard errors that would define an interval that would cover 95% or 99% of the sampling distribution of means. The same approach is used for constructing confidence intervals for the correlation coefficient, except that the r must first be transformed to z_r, and then the percentiles of the standard normal distribution can be used to determine the number of standard errors [i.e., $\sqrt{1/(N - 3)}$] needed to define the desired interval. Then these limits must be transformed back from z_r's to r's.

The endpoints of confidence intervals for the correlation coefficient are given by

$$z_r - z_\alpha \sqrt{\frac{1}{N - 3}} \quad \text{and} \quad z_r + z_\alpha \sqrt{\frac{1}{N - 3}}$$

in which

- r = the observed sample correlation
- z_r = the z transformed value for the observed sample correlation r (Table D)
- α = 1 − the desired level of confidence
- z_α = the critical value of z for a nondirectional test at significance level α (Table A)

The z_r's obtained above must be transformed back to r's using Table D.

Intervals can be constructed at any level of confidence, such as 90%, 95%, or 99% confidence. To do so, first determine α to be 1 − confidence level (e.g., $\alpha = 1 - .95 = .05$), then look in Table A of the standard normal distribution to locate the value of z_α corresponding to a nondirectional test at significance level α (i.e., $\alpha = .05$). A nondirectional test requires that half the critical region be above the mean and half below it, so one wants a z_α that has an area of half of α above it. For 95% confidence intervals, $\alpha = 1 - .95 = .05$, so $z_{\alpha = .05}$ must have $.05/2 = .025$ of the area above it. Table A shows this to be $z_{\alpha = .05} = 1.96$. For 99% confidence intervals, $\alpha = 1 - .99 = .01$, so $z_{\alpha = .01}$ must have $.01/2 = .005$ of the area above it. The required $z_{\alpha = .01}$ falls between 2.57 and 2.58, so we take $z_{\alpha = .01} = 2.575$. Although confidence intervals can be constructed for any level of confidence in this manner, the 95% and 99% intervals are most common, so the formulas for them can be stated simply:

The endpoints of 95% confidence intervals for the correlation coefficient are

$$z_r - 1.96 \sqrt{\frac{1}{N - 3}} \quad \text{and} \quad z_r + 1.96 \sqrt{\frac{1}{N - 3}}$$

The endpoints of 99% confidence intervals for the correlation coefficient are

$$z_r - 2.575 \sqrt{\frac{1}{N - 3}} \quad \text{and} \quad z_r + 2.575 \sqrt{\frac{1}{N - 3}}$$

For the example given above, the correlation between vocalizations during infancy and later IQ, the observed correlation is $r = .74$ for a sample of $N = 27$. To transform $r = .74$ into a z_r, turn to Table D in Appendix 2, locate $r = .74$, and find its corresponding value of $z_r = .950$. To find a 95% confidence interval, substitute $z_r = .950$ and $N = 27$ into the formula to obtain

$$z_r - 1.96 \sqrt{\frac{1}{N-3}} \quad \text{and} \quad z_r + 1.96 \sqrt{\frac{1}{N-3}}$$

$$.950 - 1.96 \sqrt{\frac{1}{27-3}} \quad \text{and} \quad .950 + 1.96 \sqrt{\frac{1}{27-3}}$$

$$.950 - .400 \quad \text{and} \quad .950 + .400$$

$$z_r = .550 \quad \text{and} \quad z_r = 1.350$$

Notice that these are values of z_r, not r, so we must return to Table D to transform them back from z_r to r. A $z_r = .550$ corresponds to an $r = .500$ (the closest z_r is .549) and $z_r = 1.350$ corresponds to an $r = .875$ ($z_r = 1.35$ is slightly smaller than 1.354). So the 95% confidence interval for the observed $r = .74$ is .50 to .88. The endpoints of 99% confidence intervals would be given by

$$z_r - 2.575 \sqrt{\frac{1}{N-3}} \quad \text{and} \quad z_r + 2.575 \sqrt{\frac{1}{N-3}}$$

$$.950 - 2.575 \sqrt{\frac{1}{27-3}} \quad \text{and} \quad .950 + 2.575 \sqrt{\frac{1}{27-3}}$$

$$.950 - .526 \quad \text{and} \quad .950 + .526$$

$$z_r = .424 \quad \text{and} \quad z_r = 1.476$$

$$r = .400 \quad \text{and} \quad r = .900$$

Thus, we are 99% confident that the interval (.40, .90) contains ρ. Therefore, in the previous section, we tested the hypothesis that the observed $r = .74$ was within sampling error of a population with $\rho = .00$. This hypothesis was rejected, implying that the population ρ is a value other than .00. The observed r is a point estimate and the confidence intervals are interval estimates of that population value. Specifically, we can conclude that the population correlation is very approximately .74, that the probability is .95 that the interval .50 to .87 contains the population value, and that the probability is .99 that the interval .40 and .90 contains the population value.

The Difference between Two Correlations: Independent Samples

Research Question The preceding $r = .74$ applies to females. It might be the case that the relationship between vocalization at age 1 and verbal IQ at age 26 is stronger or weaker for females than for males. To examine this possibility, a sample of 22 males was measured at 12 months and at 26 years of age in precisely the same manner as described for females. Whereas the correlation for females was .74, the correlation for males was .09. What is the probability that such a difference in observed correlations would occur merely as a function of sampling error when, in fact, in the population there is no difference between males and females in the degree of this relationship?

Note that the two correlations being considered, .74 and .09, have been computed on independent (that is, different) groups of subjects. The following procedures apply only to such cases. Somewhat different techniques are required to compare two correlations computed on the same group of subjects.[5] A summary of the procedures for independent groups is given in Table 9-4.

Statistical Hypotheses If there is really no difference in the degree of correlation for the two groups (genders), then the correlation coefficient in the population for the males should equal that parameter for females ($\rho_1 = \rho_2$). On the other hand, if there is a difference in the magnitude of the relationship for the two groups, their population values should not be equal. The alternative in this case is nondirectional (two-tailed) and the hypotheses are as follows:

$H_0: \rho_1 = \rho_2$

$H_1: \rho_1 \neq \rho_2$

Formula In the test of the null hypothesis that $\rho = .00$, it was reasonable to assume that the sampling distribution of r's is approximately normal in form, because with a population mean of zero, sample r's could vary equally above and below this value. As noted in the previous section on confidence intervals, however, if the population correlation is not zero, and especially if it is large, sample r's are not likely to be distributed normally because they cannot be higher than $+1.00$. Similarly, the distribution of the differences between two r's will not likely be normal in form. However, as above, if the r's are transformed to standard score form, z_r, then the sampling distribution of the difference between two transformed correlation coefficients is approximately normal for samples of moderate size. The r-to-z_r transformation is given in Table D in Appendix 2. We have seen in the previous section that for $r_1 = .74$, $z_{r_1} = .950$; similarly, for $r_2 = .09$, $z_{r_2} = .090$.

Once the r-to-z transformations have been made on each of the two values of r, the difference between the transformed correlation coefficients (z_r) relative to the standard error of such differences is given by the usual standard score formula of a value minus its population mean divided by the population standard error of such values. In the present case, a "value" is the transformed difference between two sample correlation coefficients; that difference in the population is zero under the null hypothesis, and the standard error reduces to a simple form, leaving the following formula:

$$z_{obs} = \frac{z_{r_1} - z_{r_2}}{\sqrt{\dfrac{1}{N_1 - 3} + \dfrac{1}{N_2 - 3}}}$$

Note that the distribution employed as the theoretical distribution is the standard normal (observe that z_{obs} refers to the standard normal deviate and z_r refers to the transformed correlations). Because the standard normal is being used and not the t, it

---

[5] See G. A. Ferguson, *Statistical Analysis in Psychology and Education* (New York: McGraw-Hill, 1981).

9-4 **Summary of the Test of Difference Between Two Independent Correlation Coefficients (Infant Vocalization–IQ Example)**

Hypotheses

H_0: $\rho_1 = \rho_2$

H_1: $\rho_1 \neq \rho_2$ (nondirectional)

Assumptions and Conditions

1. The subjects in each group are **randomly** and **independently sampled.**

2. The groups are **independent.**

3. The population distributions of X and Y for both groups are **normal** in form.

4. Both N_1 and N_2 are **greater than 20.**

Decision Rules (from Table A)

Given: .05 significance level and a nondirectional test

If $-1.96 < z_{obs} < 1.96$, do not reject H_0.

If $z_{obs} \leq -1.96$ or $z_{obs} \geq 1.96$, reject H_0.

Computation

Group 1 (females): $N_1 = 27$, $r_1 = .74$, $z_{r_1} = .950$ (from Table D)

Group 2 (males): $N_2 = 22$, $r_2 = .09$, $z_{r_2} = .090$ (from Table D)

$$z_{obs} = \frac{z_{r_1} - z_{r_2}}{\sqrt{\dfrac{1}{N_1 - 3} + \dfrac{1}{N_2 - 3}}}$$

$$= \frac{.950 - .090}{\sqrt{\dfrac{1}{27 - 3} + \dfrac{1}{22 - 3}}}$$

$z_{obs} = 2.80$

Decision

Reject H_0.

is necessary to have an adequate number of cases in the two samples. It is probably best to have N_1 and N_2 each greater than 20.

Assumptions and Conditions There are several assumptions required for performing this test. First, the sample r's are assumed to be computed on **randomly** and

independently sampled subjects. Second, it is necessary to have two **independent groups,** involving different or unmatched subjects. If this condition is not met, the formula above for the sampling error of the difference between correlation coefficients will not be appropriate, for reasons similar to those discussed under the testing of the difference in means for independent versus correlated groups. Third, the X and Y distributions for both groups of subjects must be **normal** in form. Fourth, the two **sample sizes must be greater than 20** for the statistical test to be accurate.

Significance Level   Assume the .05 level.

Critical Values   A nondirectional test at the .05 level using the standard normal distribution requires an observed z_{obs} in excess of ± 1.96 (Appendix 2, Table A).

Decision Rules

If $-1.96 < z_{obs} < 1.96$, do not reject H_0.

If $z_{obs} \leq -1.96$ or $z_{obs} \geq 1.96$, reject H_0.

Computation   The computation is summarized in Table 9-4. Note that the information given in the problem is stated first. Then the r's are transformed to their corresponding z_r values with the use of Appendix 2, Table D. The values are substituted into the formula and a $z_{obs} = 2.80$ is obtained.

Decision   The observed value of z ($z_{obs} = 2.80$) exceeds one of the critical values of ± 1.96 and thus conforms to the second decision rule, reject H_0. This means that the probability is very remote that two such correlation coefficients could be drawn from a common population purely on the basis of sampling error. Therefore, it is likely that these r's represent two populations that have different magnitudes of relationship between infant vocalization and later verbal intelligence. Specifically, it would appear that such a relationship is stronger for females than for males.

Comparison of $H_{0:}$ $\rho = .00$ and $H_{0:}$ $\rho_1 = \rho_2$

It is important to distinguish between the null hypothesis that a given correlation is zero (H_0: $\rho = .00$) and the null hypothesis that two correlations are equal (H_0: $\rho_1 = \rho_2$). Suppose the correlation for females is .74 but the correlation for males is .35. As illustrated, the $r = .74$ is "significant"; that is, it is unlikely that such a value has been computed on a sample drawn from a population that possesses no correlation at all ($\rho = .00$). More informally stated, the significant correlation implies that there is a relationship between infant vocalization and later verbal IQ for females. However, if the correlation $r = .35$ calculated on a sample of 22 males is tested for the null hypothesis $\rho = .00$, H_0 will not be rejected, implying that the evidence is not sufficient to conclude that a relationship exists between infant vocalization and verbal IQ for males. Now, when the two correlations (.74 and .35) are compared to determine whether they are significantly different from each other (H_0: $\rho_1 = \rho_2$), the test fails to reject the null hypothesis, suggesting that the relationship is not different for males and females. But this combination of results does not seem to make sense. How is it possible that there is a relationship for females but not for males, even though females and males do not differ significantly in the amount of the relationship?

This anomaly stems from the fact that tests of hypotheses often are interpreted in terms of a yes–no decision. Either there is or there is not a relationship. Although sometimes it is useful to think in this manner, attempting to make a dichotomous decision is a somewhat artificial procedure when the tool employed to make that decision is a probability value that may range from .00 to 1.00 and ordinarily does not fall neatly into one of two distinct classifications. To illustrate more clearly by citing an extreme example, suppose that with a sample of 32 girls and 32 boys, the correlations are .35 and .34, respectively. With a nondirectional test at the .05 level, the relationship would be judged significant for girls but not for boys, yet the two correlations certainly are not significantly different from each other. The lesson to be learned from this example is that one should refrain from making simple dichotomous decisions, especially when not rejecting the null hypothesis. Remember, not being able to reject H_0 does not entitle you to say that "there is no relationship." Rather, the evidence at hand is not sufficient to conclude that there is a relationship. In this case, the subtle difference between not rejecting H_0, which is the appropriate conclusion, and accepting or (worse yet) "proving" H_0, which are not appropriate conclusions, becomes important.

Another point illustrated by this example is that finding a relationship for females but finding no evidence that supports a relationship for males does not necessarily mean that the two genders are different in the relationship being examined. Performing a test of H_0: $\rho = .00$ separately on both males and females does not address the same statistical question as making a test of H_0: $\rho_1 = \rho_2$, which is necessary to draw a conclusion about whether the two genders are different in this regard.

How does one interpret a situation like the one described in which the combination of results is ambiguous? The correlation of .74 for girls is substantial enough to warrant the conclusion that there is probably a relationship between vocalization and verbal IQ for females. However, the data may be interpreted as inconclusive for males (assuming $r = .35$), in that while the correlation is not substantial enough to warrant a conclusion that a nonzero relationship exists for males, the observed correlation was not so different from the substantial correlation of .74 for females.

A COMPARISON OF THE DIFFERENCE BETWEEN MEANS AND CORRELATION

It will be instructive to compare the implications of a test of the difference between means and a test of the significance of a correlation. Fortunately, a very interesting comparison between testing for mean differences and using correlational procedures exists.[6] Researchers have been concerned with whether heredity or environment is

---

[6] This example was suggested by results reported in M. Skodak and H. M. Skeels, "A Final Follow-Up Study of One Hundred Adopted Children," *Journal of Genetic Psychology* 75: 85–125; M. P. Honzik, "Developmental Studies of Parent–Child Resemblance in Intelligence," *Child Development* 28: 215–28.

responsible for intelligence. One way to approach this issue is to investigate the IQs of adopted children and their biological and foster mothers. In general, the results have shown a correlation of approximately .38 between a measure of intellectual performance for the biological mothers and one for their children, but a correlation of almost zero between estimates of the IQs of the foster mothers based upon their years of education and the IQs of the children they reared. By itself, this information would appear to suggest that heredity may contribute more to IQ than environment. However, the mean IQ of the children was approximately 21 points higher than the mean IQ of their biological mothers but quite close to the estimated mean IQ of their rearing mothers. Thus, the children had an average IQ that was much more like the mean IQ of their foster mothers than that of their biological mothers, evidence that seems to suggest that environment may contribute more to IQ than heredity. How is this pattern of results possible, and what does it say about the procedures for testing a difference between means in contrast to a correlation?

Consider the hypothetical data presented in Figure 9-1. These data have been exaggerated somewhat to illustrate the statistical point more clearly. Notice first that the scores are plotted according to their value, with higher IQ scores at the top, and that they are clustered into scores for biological mothers, children, and foster mothers. The lines connecting pairs of scores designate which biological mother goes with which child and which child goes with which foster mother. A correlation is high if the lines linking corresponding scores do not cross excessively. Notice that the children tend to line up within their group in much the same order as their biological mothers, and thus the correlation between biological mother and child in this hypothetical example is fairly high. However, the extensive crossing of lines relating child's IQ and

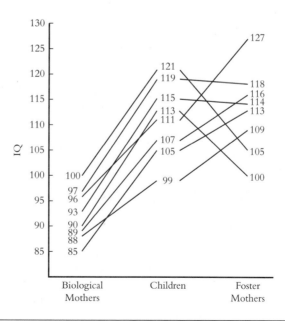

Figure 9-1 Diagram contrasting correlation and difference between means.

foster mother's IQ suggests that the ordering of children and foster mothers is quite unpredictable and thus the correlation is low. In contrast, the general height (vertical distance from the horizontal axis) of a group on this graph represents the mean of the group, and it is plain that in this respect the children are more like their foster mothers and less like their biological mothers. Therefore, depending upon certain factors, the possibilities are that two groups of scores may have a correlation but no mean difference, or a correlation and a mean difference. Further, the correlations may be either positive or negative. In short, for most distributions, *the correlation between two sets of measurements is independent of the means of those two sets of measurements.*

Correlations and mean differences often present two distinct types of information, both of which are valuable to the interpretation of the general results. Very often, researchers perform only one of these two types of data analyses when, in fact, both could be performed and would provide different types of information on the processes at work in the experiment or observation. For example, sometimes a test is given before a certain experimental treatment is introduced and then a posttest is given after the experimental treatment. Usually, the researcher looks for a difference in the means of the two groups of scores. Perhaps an experimental teaching method has been used and the investigator wants to know whether the performance of the group taught by this method has improved significantly over its performance prior to the experimental program. Suppose that some improvement is noted and the test of the difference between pretest and posttest means is significant. If there is also a correlation between the pretest and posttest, then one can also presume that the program affected each student to approximately the same extent: the better students were still at the top of the group on the posttest, and so on. However, if there is no significant correlation, such a result would suggest that although the treatment did raise the mean of the group, it seemed to influence some students more than others, changing their relative ranking within their group. As a consequence, a researcher might wish to investigate the characteristics of students who tended to show marked improvement and the characteristics of those who did not.

Therefore, the difference between the means of two sets of scores and the correlation between those sets are independent types of statistics, and they convey different messages about the events and processes being studied.

THE INTERPRETATION OF SIGNIFICANCE

Probability and Significance

There are several problems associated with declaring a difference simply as statistically significant or nonsignificant. First, the critical value used to make such a decision is based upon the significance level, which the researcher selects arbitrarily. Thus, whether or not a statistical test produces a significant result depends upon the arbitrary selection of the probability of a type I error. Fortunately (or unfortunately) there is some agreement among behavioral scientists as to what level that probability should be. Usually it is set at .05, and sometimes (though rarely) at .01 or .001. In fact, many of the tables in Appendix 2 of this book list critical values corresponding to only these significance levels.

Furthermore, as we saw above, the significant–nonsignificant dichotomy is rather artificial. A correlation of .35 may be significant but a correlation of .34 may not be significant. Moreover, a single pair of scores in a small sample can sometimes determine whether the relationship is significant or nonsignificant. But, despite the apparently artificial dichotomy between results labeled significant and nonsignificant, techniques of statistical inference were developed largely for just this purpose—to decide whether an observed event deviates from a specific population value because of sampling error alone—yes or no, not maybe. To decide after a statistical evaluation that the observation may be or may not be a chance result is merely to leave us where we started—with subjective decisions. Therefore, the custom of dichotomous decisions remains.

Significance and *N*

Another way that the significance of a statistical test is rather arbitrary is that the statistical result depends on *N*, the number of subjects in the sample. If the null hypothesis is actually false, it is more likely to be rejected as *N* increases (if H_0 is true, the size of *N* does not influence the result). This is because the standard error of the mean and the critical values of *t* and *r*, for example, depend on *N*.

To illustrate first with correlations, consider Table C of Appendix 2, which gives the critical values for *r* for various degrees of freedom (i.e., *N* − 2). Look down any column in the table: As you go down the column, the number of degrees of freedom increases and the size of *r* required to be statistically significant decreases. Notice, in the extreme, that if 100 degrees of freedom are available (i.e., *N* = 102), a correlation of .195 is significant at the .05 level (two-tailed). But a much larger correlation of .75 is not significant if only 5 degrees of freedom (*N* = 7) are available. In fact, *a correlation coefficient of any size can be statistically significant if it comes from a large enough sample.*

The same is true for differences between means: *any difference between means can be statistically significant if enough cases are sampled.* For example, suppose a sexual education program is developed for junior and senior high school students with the goal of encouraging these youth to delay the start of sexual activity. Suppose students are randomly assigned either to the program or to a no-program comparison group. Suppose further that the average age of first intercourse was 16.8 years for the program group ($s_1^2 = 1.20$) and 16.2 years for the comparison group ($s_2^2 = 1.30$). Now, if the research was conducted on two groups with only 10 youth in each, the independent-groups *t* would be 1.20, which would not be sufficient to reject the null hypothesis. If there were 20 youth in each group, the *t* would equal 1.70, again not significant. But if there were 30 in each group, *t* would equal 2.08, which would exceed the two-tailed critical value of 2.00 for 58 degrees of freedom at the .05 level. Viewed another way, if 1000 youth participate in each group, a difference of five weeks in the average age of initial sexual activity between the groups would be significant, and if 10,000 youth were in each group, a difference of 11 days would be significant. In short, whether a given result is statistically significant depends, somewhat arbitrarily, on the size of the sample.

From one standpoint, this is as it should be. The purpose of a statistical significance test is to demonstrate that an observed difference is or is not likely to be due to sampling error or chance alone. The more cases that are available, the more confident we feel that a difference is not just chance. If you flip a coin five times and it comes up heads three times, that is not sufficient evidence that the coin is biased. If you flip it ten times and six are heads, still no problem. But if you flip it 100 times and 60 are heads, it looks suspicious; and if you flip it 10,000 times and 6,000 are heads, you decide the coin is biased. Even though the percentage of flips that were heads was always 60%, confidence that the result is not simply chance increases with sample size. Therefore, deciding that a relationship or difference between means is not chance should depend in part on the sample size, but that strategy also implies that the conclusion to reject or not reject the null hypothesis may be somewhat arbitrarily influenced by whether the researcher has the time, research money, and access to a large sample.

Size of Relationship

The discussion above suggests that declaring a result—a correlation coefficient or a difference between means—to be "statistically significant" is not saying much. It says that *some* relationship or difference likely exists; that is, the observed relationship or difference is probably not sampling error, and this is important information. But it does not say *how much* of a relationship or difference exists, and even very small relationships or differences may be significant if the sample is large enough.

Consequently, it would be useful to have some idea of the *size* of a relationship or difference between means to supplement the knowledge of whether it is statistically significant. In the case of correlations, we can obtain one such index by squaring the sample r, which reflects the proportion of the total variability in one set of measurements, Y, that is associated with variability in another set of measurements, X. If the correlation between infant vocalization and later IQ is .74 for females, then $r^2 = (.74)^2 = .55$ or 55% of the variability in female adult IQ is associated with individual differences in infant vocalizations. If $r = .74$ in different samples, this estimate of the size of the relationship will be the same regardless of how many subjects are in those samples. The significance test tells the researcher that some relationship exists that is not likely to be just sampling error, whereas the proportion of variance based upon the observed correlation gives a point estimate of the size of the relationship (estimates of the proportion of variance based upon the confidence limits would give a range of possible sizes of the relationship).

Such indices also exist for the case of the difference between means, and the general logic of these estimates is similar to the logic that underlies determining the square of the correlation coefficient. Recall from Chapter 6 that the variance associated with Y's relationship to X was divided by the total variance in Y to produce r^2, the proportion of the total variability associated with the relationship between Y and X. This can be seen at the left of Figure 9-2, which is similar to Figure 6-1 from Chapter 6 on correlation. The *total* variance in the Y_i's is represented by the deviation between a score Y_i and the mean $\overline{Y}$, that is, $(Y_i - \overline{Y})$. The part of the variance in Y that is associated with Y's *relationship* to X is represented by the deviation between the value predicted by that relationship, $\hat{Y}_i$, and the mean of the Y, that is, $(\hat{Y}_i - \overline{Y})$. The deviation between a score and the regression line, that is, $(Y_i - \hat{Y}_i)$, represents *error*—that part of the total variation that is not associated with the relationship and remains after that relationship is used to predict Y.

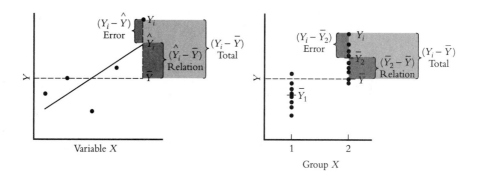

Figure 9.2 The contributors to total variance, variance associated with the relationship, and error variance for the case of correlation (left) and difference between two means (right).

The situation for the difference between means depicted at the right of Figure 9-2 is essentially the same except that the scores are clustered into only two groups along the X-axis and the means of those two groups, symbolized by $\overline{Y}_1$ and $\overline{Y}_2$, are substituted for the regression line, $\hat{Y}$. In this case, the *total* variation of the Y_i's is the same as for correlation—it is represented by the deviation of a score about the grand mean of the scores, symbolized by the unsubscripted $\overline{Y}$, ignoring group membership—that is, by $(Y_i - \overline{Y})$. That part of the total variation in Y that is associated with Y's *relationship* to X is represented by the deviation between a group mean and the grand mean, for example, $(\overline{Y}_1 - \overline{Y})$. This is based on the notion that if you did not know a subject's group, you would predict that subject's score to be the grand mean; but if you knew the subject's group, you would predict that group mean. So, the difference between the group and the grand mean represents that part of the total variation associated with group membership. Finally, the deviations between each score Y and its respective group mean, for example, $(Y_i - \overline{Y}_j)$, represent the variation in Y still remaining after knowing group membership. This within-group variation is *error* from the standpoint of predicting a subject's score by using its group mean.

So, an estimate of the proportion of variance in Y associated with X in both the correlational and difference between means situations is given by

$$\frac{\text{variance in } Y \text{ associated with } Y\text{'s relationship to } X}{\text{total variance in } Y}$$

A statistic based on the above strategy that provides an estimate of the proportion of variance associated with membership in two independent groups is called **omega squared,** and its population value is symbolized by the Greek letter omega squared, ω^2. A computational formula that provides a rough estimate of omega squared is

$$\text{est. } \omega^2 = \frac{t_{\text{obs}}^2 - 1}{t_{\text{obs}}^2 + N_1 + N_2 - 1}$$

in which t_{obs} is the observed t from the t test for independent groups and N_1 and N_2 are the sample sizes of the two groups.

Applied to the data in Table 9-1 regarding aggressive social behavior among young boys of divorced or nondivorced parents, $t_{obs} = 3.43$, $N_1 = 17$, and $N_2 = 15$, so

$$\text{est. } \omega^2 = \frac{t_{obs}^2 - 1}{t_{obs}^2 + N_1 + N_2 - 1} = \frac{(3.43)^2 - 1}{(3.43)^2 + 17 + 15 - 1} = .25$$

This implies that approximately 25% of the variability of aggressive social behavior of young boys in this study was associated with whether the parents of the boys were divorced or not. Notice that despite the fact that the *t* was clearly "significant," the amount of variance associated with divorce in this example was not very large. Indeed, most results in the social and behavioral sciences do not represent large portions of the total variability in the behavior being studied.

With respect to the example of the sex education program with 30 subjects in each group, $t_{obs} = 2.08$, $N_1 = 30$, and $N_2 = 30$, so

$$\text{est. } \omega^2 = \frac{t_{obs}^2 - 1}{t_{obs}^2 + N_1 + N_2 - 1} = \frac{(2.08)^2 - 1}{(2.08)^2 + 30 + 30 - 1} = .0525$$

This result suggests that the sex education program accounted for only 5.25% of the variance in age of first intercourse among these youth. The fact that the program produced a delay of 7 months or 5.25% of the variability in age of initiation of sexual activity may be more useful to policy makers who must decide whether to fund such a program across a state or across the country, perhaps for millions of taxpayer dollars, than simply the knowledge that the program produces a "statistically significant" difference.

Limitations   While the proportion of variance associated with group membership expressed by these estimates is useful for communicating the size of the relationship between group membership and the measured variable of interest, it has limitations.

For one thing, statisticians have developed more than one method of characterizing the size of a relationship between group membership and the measured variable, and each has advantages and disadvantages. Omega squared is one of the more commonly used indices, and its general logic is similar to that used in correlation. But the formula given here for its computation is approximate. For example, it can produce a negative estimate of the proportion of variance if t_{obs} is less than 1.00, in which case the estimate for ω^2 is set to 0%. Further, while the value of omega squared does not depend on *N* as much as the significance of *t*, omega squared nevertheless is influenced by *N*. This is because t_{obs} is part of the formula for omega squared, and t_{obs} is computed with the standard error of the mean, which contains *N* in its denominator. For example, in the sex education example in which a difference of .6 year (7 months) was tested for groups of different sizes, omega squared was 2.2% for groups of size 10, 4.5% for groups of 20, and 5.3% for groups of 30.

Another problem is that the generality of both these estimates is very limited, more limited than the proportion of variance estimates obtained for the square of the Pearson product moment correlation, r^2. In the case of *r*, presumably a random sample of subjects is selected and the values of both *X* and *Y* are free to vary and are representative of that sample. Thus, r^2, the proportion of variance in *Y* associated with *X*, can be generalized to the population for which this sample is representative. But in the case of the difference between means, the particular treatments or conditions that define the groups are not usually randomly selected or free to vary, and therefore they are not representative of some population of treatments or conditions. Moreover, the nature of those treatments or conditions may well influence the size of the total variability of Y in unusual and unknown ways. Different treatments or conditions can influence the amount of variability as well as the mean of a measure. Perhaps boys of divorced parents are more variable as well as having a higher average amount of social aggressiveness. While boys of divorced parents constituted more than half of the research sample (17 of 32 subjects or 53%), they might compose only 25% of the population of preschool boys. Therefore, the research sample would overestimate the *total variability* of aggressive behavior in the population, which would make the estimate of the proportion of variance associated with divorce inaccurate. This situation arises often, because researchers often select for study two groups that are approximately equal in size regardless of their proportional representation in the population. And when the two groups represent rather artificial conditions (e.g., infants playing with different types of toys, learning under different reinforcement conditions), what is the total variability in the population? Indeed, what is the population of such treatments or conditions?

In short, in the case of most studies of two (or more) groups, one has essentially no guarantee that the total variation in Y is typical even of the population from which the subjects were drawn, which in turn limits the generality of the estimate of the proportion of variance associated with group membership. Thus, such estimates usually apply only to the specific treatments, circumstances, and subjects actually studied. Some statisticians find these limitations so severe that they do not encourage the use of omega squared or other measures of the size of relationship in such situations. Others believe that, although limited and approximate, these estimates communicate something about the size of the relationship, and something, if appropriately interpreted, is better than nothing.

Statistical versus Scientific or Practical Significance

It should be apparent that the interpretation of scientific results is a complex matter. Of paramount importance is the fact that statistical procedures provide the researcher with a set of tools that can help interpret the results of experiments, but statistical procedures must be properly and carefully interpreted and no amount of statistics can replace the adequate collection of data under well-controlled conditions.

This proposition implies two important points. First, if the data are sloppily collected or inappropriate to the research question being posed, all the statistical significance in the world will not make the experiment worthwhile. Statistics do not add to or change the meaning of the data; they merely help reveal whatever conclusions lie hidden within the collection of measurements.

The second point is that **statistical significance** and **scientific** or **practical significance** are not the same. Scientific and practical significance, although difficult to define, reflect the extent to which a result influences thought (for example, scientific theory) or action (for example, whether the government provides Head Start with more funds to enroll more children). While statistical significance is typically required as evidence that some phenomenon exists, it alone does not also guarantee scientific or practical significance. Estimates of the size of the relationship or difference may help. But even though an experimenter may find very large differences between two groups, this knowledge may be completely useless—no one really cares (for example, a lack of scientific and practical significance). Politicians who desire to cut appropriations for science are fond of publicizing research that is statistically significant but appears to be of trivial practical interest (although sometimes the research actually may have considerable scientific or even practical application that was not explained).

STATISTICS IN THE JOURNALS

Reports of research are usually filled with statistical information, often presented in symbols. Unfortunately, no single system of symbols is used by every journal or textbook. The symbols used in this text are similar to those used in several more advanced texts, but they differ from some of the symbols used in other texts and some scientific journals. All the journals of the American Psychological Association use the same symbol system, which is given here:

M = sample mean ($\overline{X}$ in this text)

μ = population mean

SD = sample standard deviation (s in this text)

σ = population standard deviation

r = sample correlation coefficient

ρ = population correlation coefficient

a = sample intercept

b = sample slope

SE = sample standard error, usually with a subscript indicating the statistic involved (SE_M = standard error of the mean; $s_{\bar{x}}$ in this text)

N = total number of cases

n = number of cases in a subgroup (usually written with a subscript indicating the group)

p = probability value

df = degrees of freedom

t = value of the t distribution, sometimes written with a subscript t_{obs} or t_{crit} to indicate whether it is an observed or critical value

The results of tests of statistical inference are usually reported by giving in symbols the statistic calculated, any degrees of freedom, and the probability level. For example, "the girls' mean, $M = 4.79$, exceeded that of the boys, $M = 2.54$, $t(40) = 2.31$, $p < .05, \ldots$," meaning that $t_{\text{obs}} = 2.31$, $df = 40$, and the test of the difference between two independent means rejected the null hypothesis with a probability of less than .05. Similarly, one might write that "the correlation for girls was significant, $r(45) = .39$, $p < .01$, and was greater than for boys, $r(39) = -.04$, $z_{\text{obs}} = 2.04$, $p < .05$." This says that the correlation for girls of .39 with $df = 45$ was significantly different from .00 with probability less than .01. The correlation for boys of $-.04$ with $df = 39$ was not significantly different from .00, and the test of the difference between these two correlations produced an observed z of 2.04, which led to rejecting the null hypothesis at a probability of less than .05.

SUMMARY

The basic logic of many techniques of hypothesis testing is the same. They differ in the way the statistical information is translated into a theoretical relative frequency distribution that will provide the probability that the observed results could occur if the null hypothesis were true. Techniques are described for testing the difference between two means from independent groups of subjects and the difference between two means from correlated groups (groups having the same or matched subjects). Methods are also described to test whether a correlation is significantly different from zero and whether two correlations are different from each other. Depending upon certain factors, inferences about the differences between means for two correlated groups and about their correlation are independent. While statistical significance implies that a difference between means exists in the population beyond sampling error, it does not indicate the size of that difference relative to the variability within groups. The proportion of the total variability that is associated with this difference between group means may be estimated with ω^2. Statistical significance does not necessarily imply that the result is important scientifically or from a practical standpoint.

Formulas

1. Difference between means

a. Independent groups

$$t_{obs} = \frac{\bar{X}_1 - \bar{X}_2}{\sqrt{\left[\frac{(N_1 - 1)\, s_1^2 + (N_2 - 1)s_2^2}{N_1 + N_2 - 2}\right] \cdot \left[\frac{1}{N_1} + \frac{1}{N_2}\right]}}$$

$$df = N_1 + N_2 - 2$$

$$\text{est. } \omega^2 = \frac{t_{obs}^2 - 1}{t_{obs}^2 + N_1 + N_2 - 1}$$

b. Correlated groups

$$t_{obs} = \frac{\bar{D}}{s_D / \sqrt{N}} = \frac{\Sigma D_i}{\sqrt{\frac{N\Sigma D_i^2 - (\Sigma D_i)^2}{N - 1}}}$$

$$df = N - 1$$

2. Correlations

a. Test that the correlation is zero

Look for the critical values of r in Appendix 2, Table C.

b. Confidence limits for population correlation

$$z_r - z_\alpha \sqrt{\frac{1}{N-3}} \text{ and } z_r + z_\alpha \sqrt{\frac{1}{N-3}}$$

in which

r = the observed sample correlation

z_r = the z transformed value for the observed sample correlation r (Table D)

α = 1 − the desired level of confidence

z_α = the critical value of the standard normal distribution (z) for a nondirectional test at significance level α (Table A; 1.96 for $\alpha = .05$; 2.575 for $\alpha = .01$)

The z_r's obtained above must be transformed back to r's using Table D.

c. Difference between two independent correlations

$$z_{obs} = \frac{z_{r_1} - z_{r_2}}{\sqrt{\dfrac{1}{N_1-3} + \dfrac{1}{N_2-3}}}$$

(Consult Appendix 2, Table D, for r-to-z_r transformations.)

Exercises

For Conceptual Understanding

1. Describe the basic logic of hypothesis testing.

2. What is the general formula for translating any value into units of the t distribution?

3. In the test of the difference between means for independent groups, why must it be assumed that the population distribution of the difference between means is normal, and under what conditions can such an assumption be made?

4. Researchers sometimes have a choice between a correlated-groups design and an independent-groups design in planning an experiment. What considerations would be involved in making this choice?

5. Construct a numerical example in which two groups of eight scores are significantly different if the same eight subjects each contributed a score to both groups (that is, for correlated groups), but at the same time, the groups are not significantly different if the same scores are considered to be from two independent groups of subjects.

For Solving Problems

6. A theory suggests that when people experience cognitive dissonance (for example, conflicting motives or thoughts), they will attempt to reduce this dissonance by altering their perceptions of the circumstances. In one experiment, college students were asked to perform a boring task for a long period of time. Some of the students were paid $20 and others were paid $1 to tell the next student that the task was actually interesting and fun. Later, in private, the students were asked to rate their real feelings about the task. The theory predicts that the group that was paid $20 would not experience as much dissonance when lying about the task because $20 is a fair wage for that kind of fib. Therefore, they would most likely rate the task as being quite boring. However, the group that was paid $1 would probably experience more dissonance about lying and as a consequence would tend to view the task more favorably to justify their lie. The ratings (the higher the number, the more interesting the subject rated the task) for

the two groups are presented below. In view of what the theory predicts, evaluate these data using the general form of the tables in the text. Assume a directional alternative hypothesis and that $\alpha = 0.5$.[7]

$1	$1	$20	$20
4	7	2	3
4	8	6	4
6	7	3	3
9	9	3	2
10	6	1	

7. An industrial psychologist sought to evaluate the effects of a new profit-sharing incentive plan on productivity in her firm, a computer assembly plant. The plant had 10 assembly teams. The number of units produced by each team during the eight weeks prior to the beginning of the new incentives and for eight weeks afterward are presented below. Test the hypothesis that the new incentive plan had no effect on productivity using the formal organization illustrated in this chapter.

Team	Before	After
1	85	80
2	79	82
3	78	79
4	80	78
5	81	85
6	108	102
7	76	68
8	69	66
9	78	71
10	74	76

8. To compare the effectiveness of two types of psychotherapy in the treatment of anxiety, two groups of 12 college students were selected from a large group of students in a freshman speech course. The 24 students were chosen following a series of interviews, questionnaires, and physiological measurements, because they had the highest composite scores for anxiety about giving a speech. Each student was then randomly assigned to receive ten sessions of therapy that followed one of two forms. One type, *behavioral therapy,* attempted to teach the subject to relax and to mentally associate this state of relaxation with the thought of giving a speech. The other group received *insight therapy,* in which the therapist attempted to discern the causes of each person's anxiety and to provide that person with insight into his or her problem. Later, each student was asked to give another speech and the same type of composite score of anxiety was measured. The *after − before differences* between the first and second scores for each subject are given below. (Note that high scores indicate a great deal of improvement.) Evaluate these data in terms of the possible difference in the effectiveness of the two methods of therapy.[8]

Behavior Therapy	Insight Therapy
4	2
−2	3
3	−4
7	0
6	−1
5	5
9	2
2	1
4	4
5	0
7	1
1	3

[7] Inspired by L. C. Festinger and J. M. Carlsmith, "Cognitive Consequences of Forced Compliance," *Journal of Abnormal and Social Psychology* 58 (1959): 203–10.

[8] Inspired by G. L. Paul, *Insight vs. Desensitization in Psychotherapy: An Experiment in Anxiety Reduction* (Stanford: Stanford University Press, 1966).

9. Compute the correlation for the data in Exercise 7 above and determine whether the r is significant. What implications do the results have regarding the stability of differences between assembly teams?

10. Suppose the correlation between the IQs of 35 sets of identical twins is .85 and the correlation between 30 sets of fraternal twins is .58. Test the obvious theoretical prediction that the correlation is higher for identical twins than for fraternal twins.[9]

11. A new program is developed to enrich the kindergarten experience of children in preparation for first grade. Hillmont school system tries out the new curriculum in one classroom and compares it with another classroom using the old curriculum. Pupils in each classroom are tested at the beginning of the school year (the pretest) and again at the end of the school year (the posttest). The test gives a score of 10 if the pupil performs at grade level. The scores for the two classes are given at right.

   a. Determine separately for each group whether performance improved from pretest to posttest. Assume $\alpha = .05$ and a nondirectional alternative hypothesis.

   b. Determine separately for each group whether the posttest scores are significantly higher than grade-level performance, that is, $\mu = 10$. Assume $\alpha = .05$ and a two-tailed test. (Note: One-sample tests; see Chapter 8.)

   c. Calculate the correlation between pretest and posttest scores separately for each group and test its significance. Assume $\alpha = .05$ and directional hypotheses (that r is positive).

   d. Determine whether the correlations in part c are different from one another. Assume $\alpha = .05$ and a nondirectional alternative hypothesis.

---

[9] Inspired by L. Erlenmeyer-Kimling and Lissy F. Jarvik, "Genetics and Intelligence: A Review," *Science* 142: 1477–78.

New Curriculum

Pupil	Pretest	Posttest
a	9	16
b	6	11
c	14	14
d	12	10
e	9	14
f	8	12
g	12	15
h	8	11
i	11	14

Old Curriculum

Pupil	Pretest	Posttest
j	12	13
k	7	9
l	7	7
m	11	10
n	11	10
o	14	15
p	6	8
q	10	9
r	11	12
s	12	11

12. In Exercise 7, suppose the correlation was .22 between years of experience and units produced before the incentive program for a larger sample of 62 employees. Determine 95% confidence limits for the population correlation based on this r. What do these limits tell you about the significance of this r?

13. In Exercise 8, determine the proportion of variance in difference scores associated with the two treatments (assume $t = 2.63$).

For Homework

14. A team of researchers has developed a test that attempts to predict the level of physical punish-

ment, neglect, and abuse parents will use against their children. A series of observations on the parents with their babies in the newborn nursery, a questionnaire, and interviews are used to determine a "proneness to violence" score. When the children of these parents are six years old, additional observations, questionnaires, and interviews are conducted and a score is determined that reflects the frequency and extent of violence reportedly used by a parent on the child. This score is called "violence." Suppose these procedures are used with a group of 10 mothers and a separate group of 10 fathers (*not* the husbands of the mothers sampled). Given the data provided, answer the questions below. (Use $\alpha = .05$ and ignore the small N used to make computations easier.) In formulating your answer to each question, list the hypotheses, assumptions, decision rules (including the critical values, directionality of the test, and degrees of freedom if appropriate), computation, and decision.

Father	Proneness to Violence (X)	Violence (Y)
k	9	19
l	11	17
m	9	16
n	6	14
o	12	20
p	11	18
q	5	12
r	8	18
s	8	15
t	7	16

Mother	Proneness to Violence (X)	Violence (Y)
a	10	7
b	13	9
c	11	9
d	7	3
e	16	11
f	12	6
g	6	2
h	12	8
i	10	5
j	8	6

a. Test the hypothesis that there is no relationship between the proneness to violence score and the violence score for mothers.

b. Test the hypothesis that there is no relationship between the proneness to violence score and the violence score for fathers.

c. Test the hypothesis that the two sets of parents do not differ in the magnitude of the relationships assessed in parts **a** and **b**.

d. Suppose that the correlations computed in parts **a** and **b** were both significant and that part **c** showed that the two correlations were significantly different from one another. Is this possible? If so, what does it mean?

e. Test the hypothesis that the mothers and fathers sampled do not differ in their proneness to violence score; in their violence score.

f. Suppose the data came from 10 mother–father pairs (mother **a** is the wife of father **k,** etc.). Again test the hypothesis that mothers and fathers do not differ in their proneness to violence scores; in their violence scores.

g. Twice you have tested the hypothesis that mothers and fathers do not differ with respect to their proneness to violence scores, once using an independent-groups test (**e**) and once using a correlated-groups test (**f**).

Suppose, as is quite possible, the correlated-groups analysis gave a significant result (H_0 rejected) but the independent-groups analysis did not (H_0 not rejected). Explain how this would be possible, and explain what advantage correlated-groups research designs might have over independent-groups designs (if one has a choice).

h. Determine 95% confidence intervals for the two population correlations in parts **a** and **b** above.

i. Determine the proportion of variance associated with the differences between means in part **e**.

Simple Analysis of Variance

echniques that compare the difference between two means were examined in the previous chapter. The usual null hypothesis for those tests is that the two samples are drawn from populations having the same mean (H_0: $= \mu_1 = \mu_2$). However, very often a researcher wants to compare means from more than two samples and asks what the probability is that these several samples are all drawn from populations having the same mean (H_0: $\mu_1 = \mu_2 = \mu_3 = \cdots = \mu_p$). A statistical procedure that addresses this broader question is called the **analysis of variance** (often abbreviated **ANOVA**).

To illustrate, consider an experiment on the need of normal adults to dream. Some years ago it was discovered that an observer can tell whether a person is dreaming by monitoring his or her brain waves, EEG, heart rate, respiration rate, eye movements, and so on.[1] Among other things, this discovery made it possible to study the effects of disturbing people while they are dreaming, thus depriving them of the opportunity to dream. For example, the researcher might ask whether deprivation of dreaming has any effects on a person's temperament during the waking day. If normal adults need to dream, perhaps they become anxious and irritable if the opportunity to dream is curtailed. Alternatively, dreaming itself may be disturbing, and people may be more relaxed if their dreaming is minimized or prevented.[2] Suppose one group of subjects has their sleep interrupted five times during the night, but never during a dream. A second group is aroused the same number of times, but on two occasions during each night this wakening is during a dream. A third group is awakened an equal number of times, but each of those times is during a dream. Therefore, the three groups of subjects can be characterized as undergoing either no, some, or much dream interruption. These procedures are carried out in a special sleep laboratory for six consecutive nights. During each day the subjects are interviewed and given tests to evaluate how anxious and irritable they are. Each subject is assigned a total score over the entire six-day period, with a high score indicating a very irritable individual.

Suppose the means for the three groups (no, some, and much dream interruption) are 4.00, 8.00, and 17.75, respectively. This seems to indicate that the prevention of dreaming produces increased irritability in normal adults. But the observed differences between means could be a function of sampling error and no real difference exists between these groups in the population. The null hypothesis is that the population means are equal to one another and, therefore, to a common value—H_0: $\mu_1 = \mu_2 = \mu_3 = \mu$. The statistical problem is to determine the probability that such results could occur if H_0 is true.

In the previous chapter, techniques were presented for analyzing the difference between two means. Although it would be possible to use the t test on each pair of means (AB, AC, BC) that can be formed from the three samples (A, B, C), this approach is inappropriate for two main reasons. First, when the .05 level of significance is adopted, one expects to reject the null hypothesis incorrectly by chance or sampling error alone (a type I error) in 5 of every 100 *independently* sampled pairs. But when several t tests are used, the pairs of means are not all independent: out of k groups, each mean will be part of $k - 1$ pairs, and an extreme mean obtained by chance for a single group can lead to an incorrect decision in $k - 1$ cases. Therefore, one does

---

[1] See N. Kleitman, "Patterns of Dreaming," *Scientific American* 203: 82–88.
[2] Based on, but not identical to, W. Dement, "The Effect of Dream Deprivation," *Science* 131: 1705–707.

not know just what the probability of a type I error would be if t tests are performed on all pairs of means.

Second, the researcher frequently wants to ask a broader question than whether specific pairs of means are different. For example, the scientist might be less interested in whether no versus two interruptions or two versus five interruptions makes a difference than in the general question of whether the extent of dream interruption—from no to much interruption—influences temperament. If pairs are tested, some pairs may be significantly different while others are not. Although this specific information might be important and useful, it leaves ambiguous the answer to the general question of whether a relationship exists between dream deprivation and temperament, because the differences may be significant for some but not all pairs. The analysis of variance gives a single answer to this general question.

LOGIC OF THE ANALYSIS OF VARIANCE

The purpose of simple analysis of variance is to determine the probability that the means of several groups of scores deviate from one another merely by sampling error.[3] The analysis of variance partitions the variability in the total sample in much the same way as was illustrated in the discussion of correlation and regression that opened Chapter 6. In that context, the total variability of the Y_i is divided into a portion that is attributable to X and a portion that is not associated with X. Similarly, in Chapter 9 in the discussion of estimation of the proportion of variance associated with membership in two groups, the total variability was divided into a portion associated with group membership and a portion that was not (i.e., error). In the case of the analysis of variance, the total variability in the scores is partitioned into a portion that reflects differences between the means of the several groups and a portion that is not influenced by the differences between those means.

The partitioning of variability is performed in such a way that two estimates of the variance of the scores in the population are computed. One of these estimates is based upon the deviation of the group means about the *grand mean* (the mean over all scores in the total analysis; symbolized by μ). The size of this estimate is influenced both by the variability of individual subjects (since their scores are involved in both the group and grand means) and by differences between group means. Because this variance estimate is based upon the deviation of group means about the grand mean, it is called the **between-groups estimate** of the population variance. The second estimate of the variance of the scores in the population is based upon the deviation of scores about their respective group means. This estimate is influenced not by differences between group means but only by the random variability of individual subjects within each group about the mean of that group. Therefore, it is known as the **within-groups estimate** of the population variance.

---

[3] The discussion of the analysis of variance in this chapter and Chapter 13 assumes that the factors in the research design are fixed rather than random. This assumption is explained in Chapter 13.

Conceptually, the two variance estimates differ only in that the between-groups estimate is sensitive to differences between the group means whereas the within-groups estimate is not. As the group means become increasingly different from one another, the between-groups variance estimate grows larger. However, the null hypothesis being tested is that in the population all the group means are equal (H_0: $\mu_1 = \mu_2 = \mu_3$). Since the null hypothesis is tentatively held to be true while its validity is being tested, the between-groups estimate would not be influenced by population differences between means because under H_0 the means are assumed to be equal to one another. Thus, given the null hypothesis, the between-groups estimate is influenced only by the same random variation in scores that influences the within-groups estimate.

Therefore, the analysis of variance provides two estimates of the population variance. Given the null hypothesis, the estimates should be identical except for sampling error. The probability that two independent variance estimates differ from one another only by sampling error can be determined by taking the ratio of the two sample variances, in this case,

$$\frac{s^2_{between}}{s^2_{within}}$$

Under the null hypothesis that these variances estimate the same population value, this ratio has a known theoretical distribution called *F*, the percentile points of which are listed in Table E of Appendix 2. If the ratio is so large that the probability is exceedingly small that $s^2_{between}$ and s^2_{within} estimate the same population variance, then one may assume that the additional influence of differences between group means has inflated the value of $s^2_{between}$, causing the *F* ratio to be unusually large. In that case, the null hypothesis should be rejected.

Notice that although this technique is called the analysis of *variance,* its purpose is really to assess differences in group *means.* It accomplishes this goal by comparing two variance estimates, one of which can be influenced by differences in group means and one of which cannot. The remainder of this section is devoted to expanding and clarifying the rationale behind the technique.

Notation and Terminology

Notation   It will help to provide some notation that will serve as a convenient language with which to discuss the analysis of variance. Table 10-1 presents such notation for the general case. Any score, X_{ij}, is written with two subscripts. The first, or i, subscript denotes the subject's number within its group, and the second, or j, subscript indicates the group to which that subject belongs. So X_{ij} denotes the ith subject in the jth group, and X_{24} is the second subject in the fourth group. Table 10-1 is arranged to show several subjects in each of the groups. The i subscript runs up to n_1 for the first group, where n_1 indicates the total number of subjects in the first group. Similarly, n_2 indicates the number of subjects in the second group, and n_j denotes the number of subjects in the jth group, where j signifies "some" group in the set. There are p groups, so the number of subjects in the last group is n_p.

The total of all scores in group j is symbolized by T_j and the mean of the jth group is $\overline{X}_j = T_j/n_j$. Thus, $\overline{X}_1$ is the mean of the first group and $\overline{X}_p$ is the mean of the last or pth group. In general, $\overline{X}_j$ is the mean of "some" group (the jth group). The

10-1 General Notation for Simple Analysis of Variance

	Group 1	Group 2	$\cdots$	Group p	
	X_{11}	X_{12}	$\cdots$	X_{1p}	
	X_{21}	X_{22}	$\cdots$	X_{2p}	
	X_{31}	X_{32}	$\cdots$	X_{3p}	
	X_{41}	X_{42}	$\cdots$	X_{4p}	
	$\vdots$	$\vdots$	$\vdots$	$\vdots$	
	$X_{n_1 1}$	$X_{n_2 2}$	$\cdots$	$X_{n_p p}$	
Group Totals	T_1	T_2	$\cdots$	T_p	**Grand Total** T
Group Means	$\overline{X}_1 = \dfrac{T_1}{n_1}$	$\overline{X}_2 = \dfrac{T_2}{n_2}$	$\cdots$	$\overline{X}_p = \dfrac{T_p}{n_p}$	**Grand Mean** $\overline{X}$

X_{ij} = score for the ith subject in the jth group

p = number of groups

n_j = number of subjects in the jth group

$N = n_1 + n_2 + \cdots + n_p =$

$\quad \displaystyle\sum_{j=1}^{p} n_j$ = total number of subjects over all groups

T_j = total of all scores in group j

$\overline{X}_j$ = mean for group $j = \dfrac{T_j}{n_j}$

T = grand total over all subjects

$\overline{X}$ = grand mean $= T/N$

grand mean, symbolized by $\overline{X}$ without any subscripts, is the mean of all the scores from all the groups. It can be expressed as

$$\overline{X} = \frac{T}{N}$$

in which T is the total of all scores and N is the total number of subjects in the design. One should note that the grand mean equals the average of the group means *only* if all n_j are equal.

Variance Terminology The logic of the analysis of variance as briefly sketched above requires two estimates of the population variance, the between-groups and the within-groups variance estimates. It will help to refine some terminology and notation about variances at this point. Earlier in the text the sample variance (variance estimate) was defined to be

$$s^2 = \frac{\sum_{i=1}^{N}(X_i - \overline{X})^2}{N-1}$$

The numerator of this fraction is the sum of squared deviations about the mean, and thus it is sometimes called the **sum of squares** and is represented by **SS.** The denominator is the **degrees of freedom** associated with the sum of squares, and its symbol is *df.* In these terms the variance, s^2, is a type of average sum of squares, and in the context of the analysis of variance it is known as a **mean square** or **MS.** Consequently, a variance estimate can be symbolized by

$$s^2 = \frac{\sum_{i=1}^{N}(X_i - \overline{X})^2}{N-1} = \frac{SS}{df} = MS$$

Although this expression uses a different vocabulary from that used earlier, there is no important conceptual difference. The new terms, sum of squares (*SS*) and mean square (*MS*), are customarily associated with the analysis of variance. Therefore, a variance estimate in the context of the analysis of variance is called a mean square, or *MS*.

Hypotheses   The purpose of the analysis of variance is to estimate the probability that the means of several groups differ from one another merely by sampling error. The null hypothesis is that in the population the group means ($\mu_1 = \mu_2 = \cdots = \mu_p$) are equal to one another and thus to the grand mean over all the groups (μ). The null hypothesis is symbolized by

$$H_0: \mu_1 = \mu_2 = \cdots = \mu_p = \mu$$

The alternative hypothesis states that in the population the group means are not all equal, which is best expressed as follows:

$$H_1: \text{not } H_0$$

Partitioning of Variability

Derivation of MS   Determining the two estimates of the population variance involves partitioning the total variability into two portions in a manner quite similar to what was done in regression and correlation. Recall that (as shown in Figure 6–1) the total variability in scores was divided into a portion associated with the dependent variable's relationship to the predicting variable and a portion not associated with that relationship (that is, error). Then, in Chapter 9 (illustrated in Figure 9–2) total variability was partitioned into a portion associated with membership in one of two groups and a portion not associated with group membership (i.e., error). The same strategy is used in the analysis of variance except more than two groups are involved: The total variability in scores is divided into a portion associated with the group membership of the

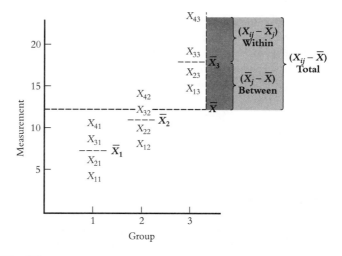

Figure 10-1 Partitioning of deviations in the analysis of variance. When squared and summed over all subjects, the deviations $(\overline{X}_j - \overline{\overline{X}})$ contribute to the between-groups estimate and the deviations $(X_{ij} - \overline{X}_j)$ contribute to the within-groups estimate of the population variance.

subjects and a portion not associated with that affiliation. If one thinks of group membership as the predicting variable, the partitioning in regression and correlation and in the analysis of variance are essentially the same.

The partitioning of deviations for a single score in the analysis of variance is illustrated in Figure 10-1. The analysis of variance deals with the relationship between two variables—group affiliation (for example, dream deprivation), which is marked off along the abscissa, and some measurement or dependent variable (for example, temperament), which is marked off along the ordinate. The scores for individual subjects are represented by X_{ij}, in which i signifies a specific person in the jth group. The height of the X_{ij} on the graph reflects the value of the dependent variable for that subject—the higher the score, the higher the X_{ij} is located on the graph. The means for each of the three groups are indicated with a light dashed line and the grand mean for all subjects is indicated with a heavy dashed line. The total variability of the scores is composed of their deviations from the grand mean of all scores. The total deviation for the particular score X_{43} in Figure 10-1 is the braced vertical distance between the score and the grand mean $(X_{ij} - \overline{\overline{X}})$, which is composed of two parts: the deviation of the group mean from the grand mean $(\overline{X}_j - \overline{\overline{X}})$ plus the deviation of the score from its own group mean $(X_{ij} - \overline{X}_j)$. In symbols,

$$(X_{ij} - \overline{\overline{X}}) \quad = \quad (\overline{X}_j - \overline{\overline{X}}) \quad + \quad (X_{ij} - \overline{X}_j)$$

$$\downarrow \qquad\qquad\qquad \downarrow \qquad\qquad\qquad \downarrow$$

(total deviation = (deviation of group + (deviation of score
from grand mean)   mean from   from group mean)
                    grand mean)

total variability = between-group source + within-group source

It now should be clear that scores differ from one another for two reasons. First, scores are members of different groups, and if the group means differ from one another, the scores within a group will, on the average, differ from the scores in another group. Thus, the expression $(\overline{X}_j - \overline{X})$ indicates the magnitude by which the mean of the jth group deviates from the grand mean; it also represents the average difference between scores in that group and the average of all the scores. This is the **between-group** source of variability. Second, the scores within a group differ from one another—they do not all equal their group mean. This fact is expressed in the quantity $(X_{ij} - \overline{X}_j)$, which is the difference between a score and its group mean. This is the **within-group** source of variability. Thus, scores differ from one another because they belong to different groups $(\overline{X}_j - \overline{X})$ and because of something unique to each subject that causes a subject's score to differ from the mean of its own group $(X_{ij} - \overline{X}_j)$.

The above deviations are for a single score. Variability is usually measured in squared deviations, and it needs to be determined for all subjects in the study. Therefore, to obtain a formula for the total variability, begin with the above expression for a single score

$$(X_{ij} - \overline{X}) = (\overline{X}_j - \overline{X}) + (X_{ij} - \overline{X}_j)$$

Then square each side, sum over the n_j scores within each group, and then over the p groups:

$$\sum_{j=1}^{p} \sum_{i=1}^{n_j} (X_{ij} - \overline{X})^2 = \sum_{j=1}^{p} \sum_{i=1}^{n_j} [(\overline{X}_j - \overline{X}) + (X_{ij} - \overline{X}_j)]^2$$

When these operations are carried out, the result is

$$\sum_{j=1}^{p} \sum_{i=1}^{n_j} (X_{ij} - \overline{X})^2 \quad = \quad \sum_{j=1}^{p} n_j(\overline{X}_j - \overline{X})^2 \quad + \quad \sum_{j=1}^{p} \sum_{i=1}^{n_j} (X_{ij} - \overline{X}_j)^2$$

$$\downarrow \qquad\qquad\qquad\qquad \downarrow \qquad\qquad\qquad\qquad \downarrow$$

$$\left(\begin{array}{c} \text{Total squared} \\ \text{deviations} \\ \text{in } X \end{array} \right) \quad = \quad \left(\begin{array}{c} \text{squared deviations} \\ \text{between group} \\ \text{means} \end{array} \right) \quad + \quad \left(\begin{array}{c} \text{squared deviations} \\ \text{within} \\ \text{groups} \end{array} \right)$$

Notice that each component of the above expression is a sum of squared deviations (SS). Thus, the expression can be written

$$SS_{\text{total}} = SS_{\text{between}} + SS_{\text{within}}$$

This statement indicates that the total sum of squares is composed of the sum of squared deviations of group means from their grand mean (SS_{between}) plus the sum of the squared deviations of scores from their respective group means (SS_{within}). Notice that the sum of squared deviations between a group mean and the grand mean is multiplied by n_j. The multiplication is done because this deviation is meant to apply to each subject in the group, and there are n_j subjects in the jth group.

Recall from the general rationale of the analysis of variance that rather than two *sums of squares,* one sensitive to group differences and the other not, two *mean squares* with these characteristics are required. A variance estimate is a sum of squares divided by its degrees of freedom. Therefore, the degrees of freedom for the sums of squares discussed above must be determined.

Recall from Chapter 8 that the degrees of freedom for a statistic equals the number of elements in its calculation that are free to vary. In the case of the sample variance,

$$s^2 = \sum \frac{(X_i - \bar{X})^2}{N - 1}$$

the degrees of freedom is one less than the total number of scores (that is, $N - 1$). In the above formula, N and $\bar{X}$ are constants, and the deviation of the last score about the mean is not free to vary because it must equal that number which, when added to the sum of the other deviations, will make the total of all deviations equal to zero. So the degrees of freedom for a variance estimate is one less than the number of cases involved in its sum of squares.

We now apply this principle to the notation and terminology of the analysis of variance. The total sum of squares is composed of the squared deviations of each score about the grand mean, so its degrees of freedom is one less than the total number of scores (N):

$$df_{total} = N - 1$$

The $SS_{between}$ is composed of the squared deviations of each group mean about the grand mean, so its degrees of freedom is one less than the total number of group means (p):

$$df_{between} = p - 1$$

The SS_{within} is composed of the squared deviations of each score about its own group mean, so its degrees of freedom is the sum of the $n_j - 1$ across all groups, which equals the total number of cases (N) less one for each of the group means (p):

$$df_{within} = (n_1 - 1) + (n_2 - 1) + \cdots + (n_p - 1)$$
$$= (n_1 + n_2 + \cdots + n_p) - (1 + 1 + \cdots + 1)$$
$$df_{within} = N - p$$

It happens that the df_{total} equals the sum of the degrees of freedom for $SS_{between}$ and SS_{within}:

$$df_{total} = df_{between} + df_{within}$$
$$df_{total} = (p - 1) + (N - p)$$
$$df_{total} = N - 1$$

Therefore, the degrees of freedom for the total sample, like the sum of squares, is the sum of a between- and a within-group component.

To obtain a mean square, one divides a sum of squares by its degrees of freedom. Therefore, the mean square sensitive to between-group differences is

$$MS_{between} = \frac{SS_{between}}{df_{between}}$$

and the mean square within groups is

$$MS_{within} = \frac{SS_{within}}{df_{within}}$$

One must be careful *not* to suppose that, because

$$SS_{total} = SS_{between} + SS_{within}$$

and

$$df_{total} = df_{between} + df_{within}$$

the total mean square equals the sum of the between and within mean squares. It does not:

$$MS_{total} \neq MS_{between} + MS_{within}$$

Comparison of $MS_{between}$ and MS_{within}   The logic of the analysis of variance rests on the fact that one of the variance estimates just discussed, $MS_{between}$, can be influenced by population differences between the means of the several groups, whereas the other, MS_{within}, cannot be so influenced. This section attempts to make that vital difference more obvious.

First, consider why two subjects from *different* groups do not have the same score. Recall the example discussed at the beginning of the chapter. Suppose the temperament score for the first subject in the some-deprivation group was 10 and that of the first subject in the much-deprivation group was 16. Why are these two scores different? There are two major possibilities. First, one subject had only some dream deprivation while the other had much. Perhaps the difference in temperament between them is related to their group affiliation—that is, to the amount of dream deprivation. This contribution to differences between scores stems from possible differences between the groups, and often those groups represent different treatments.

A **treatment effect** occurs when the population mean corresponding to a given group in the analysis differs from the average of the population means for all groups (for example, $\mu_j \neq \mu$). The treatment effect is estimated by the deviation between a group mean and the grand mean, $(\overline{X}_j - \overline{X})$.

If the null hypothesis is true, however, there are no treatment effects, because the population means of all groups are identical. But if H_1 is the case, population group means will differ from one another and treatment effects are said to exist. Therefore, one possible reason two subjects do not have the same score is treatment effects.

In addition to potential treatment effects, two people assessed under identical circumstances usually do not respond the same way because of a variety of uncontrolled factors. For example, some subjects are happier than others because they tend to cope well with almost any circumstance. The entire collection of such potential causes of differences, presumably unrelated to differences in treatments, is called **error.** Therefore, differences or variability in scores from *different* groups (i.e., between groups) may stem from treatment effects and error.

In contrast, consider why two scores from subjects in the *same* group are not identical. Both subjects have the same amount of dream deprivation, so differences in treatment condition cannot be a determinant of variability within a group. However, all the influences subsumed under the concept of error may still be operative. Therefore, in contrast to variability between groups, variability within groups, as reflected in MS_{within}, is a function of error but not of treatment effects. It is for this reason that MS_{within} is sometimes labeled MS_{error}.

The fact that $MS_{between}$ is sensitive to treatment effects whereas MS_{within} is not may be illustrated with a numerical example. Suppose one consults a table of random numbers and selects three groups of five numbers each. Since these are random numbers, the three groups are analogous to three samples of size five drawn from a common population (sampling under the null hypothesis). Thus, no treatment effects exist. These numbers are displayed in the top half of Table 10-2. The group means are 4, 6, and 5; the grand mean is 5. The between-groups variance estimate is computed below the listing of the groups: $MS_{between} = 5.0$. The SS_{within} for each group is presented and the within-groups estimate is at the right: $MS_{within} = 7.33$.

Now suppose that there are group differences or treatment effects. To implement this condition, 3 is added to each score in group 2 and 12 is added to each score in group 3. Consider the two variance estimates presented in the lower half of Table 10-2. Notice that the MS_{within} is precisely the same as it was before the effects were introduced. This is a reflection of the fact that the variability of scores within a group is not influenced by adding a constant to each score in the group. However, the value of $MS_{between}$ has indeed changed as a function of the treatment effects. Now $MS_{between} = 215.0$. This is because a different constant was added to each of the three groups and thus to the three group means, and $MS_{between}$ is sensitive to these differences. It should be clear from this numerical example that the within-group variance estimator is not influenced by differences in group means (that is, treatment effects), whereas the between-group estimate is influenced by treatment differences. In brief,

$MS_{between}$ reflects treatment effects + error

MS_{within} reflects error

The *F* Test

The final step in the rationale of the analysis of variance is to compare the two variance estimates under the assumptions of the null hypothesis. Recall that the null hypothesis states that the populations means are equal (there are no treatment effects), so, given the above, both $MS_{between}$ and MS_{within} reflect only error and both should estimate the same parameter. Since they are estimates, they will probably not precisely equal each other but will vary as a function of sampling error. But in the long run we would expect them to tend toward the same value. Therefore, their ratio, which has a known theoretical distribution called *F*,

$$F = \frac{MS_{between}}{MS_{within}}$$

10-2

Numerical Illustration: $MS_{between}$ Is Influenced by Group Differences and MS_{within} Is Not

Group 1	Group 2	Group 3	
$SS_1 = \Sigma(X_{i1} - \bar{X}_1)^2$	$SS_2 = \Sigma(X_{i2} - \bar{X}_2)^2$	$SS_3 = \Sigma(X_{i3} - \bar{X}_3)^2$	
7	2	9	
5	6	3	
3	9	5	$MS_{within} = \dfrac{SS_1 + SS_2 + SS_3}{df_1 + df_2 + df_3}$
4	9	6	
1	4	2	$MS_{within} = \dfrac{20 + 38 + 30}{4 + 4 + 4} = 7.33$
$SS_1 = 20$	$SS_2 = 38$	$SS_3 = 30$	
$df_1 = 4$	$df_2 = 4$	$df_3 = 4$	
$\bar{X}_1 = 4.0$	$\bar{X}_2 = 6.0$	$\bar{X}_3 = 5.0$	$\bar{\bar{X}} = 5.0$

$$MS_{between} = \frac{\displaystyle\sum_{j=1}^{p} n_j(\bar{X}_j - \bar{X})^2}{p - 1}$$

$$MS_{between} = \frac{5(4 - 5)^2 + 5(6 - 5)^2 + 5(5 - 5)^2}{3 - 1} = 5.0$$

Treatment effect (added to each score)

0	+3	+12
Group 1	**Group 2**	**Group 3**

Group 1		Group 2		Group 3	
7	$SS_1 = 20$	5	$SS_2 = 38$	21	$SS_3 = 30$
5		9		15	
3	$df_1 = 4$	12	$df_2 = 4$	17	$df_3 = 4$
4		12		18	
1		7		14	
$\bar{X}_1 = 4.0$		$\bar{X}_2 = 9.0$		$\bar{X}_3 = 17.0$	

$$MS_{within} = \frac{SS_1 + SS_2 + SS_3}{df_1 + df_2 + df_3}$$

$$MS_{within} = \frac{20 + 38 + 30}{4 + 4 + 4} = 7.33$$

$$\bar{X} = 10.0$$

$$MS_{between} = \frac{\sum_{j=1}^{p} n_j(\bar{X}_j - \bar{X})^2}{p - 1}$$

$$MS_{between} = \frac{5(4 - 10)^2 + 5(9 - 10)^2 + 5(17 - 10)^2}{3 - 1} = 215.0$$

should vary from a value of 1.00 only because of sampling error. The F distribution is another theoretical sampling distribution, named F after Sir Ronald Fisher, who first described the analysis of variance and this sampling distribution. The percentiles of the F distribution may be used to determine the probability of obtaining an F ratio of a specified size purely by sampling error. If the probability is very small that an observed F of this size is merely a function of sampling error, then perhaps the null hypothesis of no treatment effects is wrong. Presumably, since $MS_{between}$ but not MS_{within} is influenced by treatment effects, a high F ratio may mean that $MS_{between}$ is large relative to MS_{within} because of treatment effects.

The percentile points for the theoretical sampling distribution of the ratio of two independent variance estimates (that is, F) are listed in Table E of Appendix 2. However, just as there is a different t distribution for each number of degrees of freedom, there is a separate F distribution for each *combination* of degrees of freedom for the mean square in the numerator ($MS_{between}$) and the mean square in the denominator (MS_{within}) of the F ratio, that is, $df_{between} = p - 1$ and $df_{within} = N - p$.

Table E in Appendix 2 lists the .05 (roman type) and .01 (**boldface** type) critical values for F's of various degrees of freedom. To locate a critical value, find the column of the table corresponding to the degrees of freedom for the variance in the numerator ($df_{between}$). Then locate the row corresponding to the degrees of freedom for the variance in the denominator (df_{within}) of the F ratio. Suppose the degrees of freedom for the numerator and denominator are 2 and 30, respectively. Locate the intersection of the column labeled 2 and the row labeled 30 degrees of freedom. There are two values at that point. The first (roman type) is 3.32; that is the critical value for a test at the .05 level. Below this figure you will find the value **5.39** (boldface type), which represents the critical value for a test at the .01 level.

Independence   The use of the F distribution in the analysis of variance requires that the sample variances in the ratio be independent. As discussed previously, estimates of the population mean and variance are independent if they are based upon samples from populations that are normally distributed. Thus, if the population of scores X_{ij} is normally distributed, $\overline{X}_j$ and s_j^2 are independent for any group j, and the collection of all the group means ($\overline{X}_j$) is independent of the collection of all the group variances (s_j^2). $MS_{between}$ is based upon the $\overline{X}_j$ while MS_{within} involves the deviations of scores about their group mean (s_j^2). Therefore, $MS_{between}$ and MS_{within} are two independent estimates of the population variance if the population is normally distributed. In this case the F distribution may be used.

> The ratio of two independent variance estimates is distributed as F with two parameters: the degrees of freedom of the variance estimate in the numerator and the degrees of freedom of the variance estimate in the denominator.

Logic of the F Test   The logic of the F test can now be summarized as follows:

1. MS_{within} is an estimate of the population variance based upon the deviations of scores about their respective group means. It is not influenced by mean differences between groups (treatment effects).

2. $MS_{between}$ is also an estimate of the population variance if the null hypothesis is true. It is based upon the deviations of group means about the grand mean. Since it is influenced by any treatment effects that exist in the population, it estimates the same population variance as MS_{within} only if those treatment effects are zero, that is, if the null hypothesis is true.

3. Since the two variance estimates are independent if the population distribution is normal and since the logic of hypothesis testing demands that the null hypothesis be tentatively held to be true, the ratio of the two independent variance estimates is distributed as F:

$$F = \frac{MS_{between}}{MS_{within}}$$

4. Under conditions of the null hypothesis, the two MS's estimate the same population value, so the ratio should approach 1.0 in the long run.[4] The observed value of F is compared to the theoretical sampling distribution of such ratios to determine the probability that such an F value could be obtained merely by sampling error.

5. As the observed F ratio becomes larger, the probability becomes smaller that an F of this size should be obtained merely by sampling error. If there are treatment effects, $MS_{between}$ will be sensitive to them and MS_{within} will not. Therefore, an improbably large F value suggests that the null hypothesis (no treatment effects) was not appropriate and that treatment effects probably do exist in the population. Thus, the null hypothesis should be rejected.

Assumptions and Conditions Underlying the Simple Analysis of Variance

Five assumptions must be made to perform the statistical manipulations necessary for the analysis of variance. These assumptions are essentially the same as those required for the t test for means of two independent groups (Chapter 9), but they are now applied to several groups.

First, it is assumed that the groups involved in the analysis are composed of **randomly** and **independently sampled** subjects. Second, the groups of scores being analyzed must be **independent;** this will be the case if the subjects in one group are not the same individuals and are not matched with individuals who are in the other groups.

The procedures for the analysis of variance outlined in this book all require independent groups. However, occasionally it is desirable not to have independent groups but to measure the same subjects under each of several different conditions and to have the analysis determine whether these correlated groups of scores have different means. The procedures for this type of analysis are called **repeated-measures analyses of variance,** because subjects are measured under more than one condition. The difference between the analysis of variance for independent groups and repeated-

---

[4] Actually, the ratio approaches 1.00 in the long run only as the number of subjects becomes very large.

measures analyses of variance is analogous to the difference between the *t* test for means of independent groups and the *t* test for means of correlated groups. Repeated-measures analyses of variance are presented in more advanced texts.[5]

The third assumption required for the analysis of variance is **homogeneity of within-group variances.** That is, it is assumed that the populations from which the groups are drawn have equal variances. In symbols,

$$\sigma_1^2 = \sigma_2^2 = \cdots = \sigma_p^2$$

One reason for this assumption is that it enables one to pool the variability about each group mean into a single estimate of the population variance, MS_{within}. Without homogeneity of variance, one group with a very large variance might contribute disproportionately to this single estimate, and MS_{within} would not be representative of the variability within each group.

There are procedures for testing the homogeneity of variances.[6] However, some statisticians have argued that these procedures do not provide an entirely appropriate test.[7] Further, if a sufficient number of cases are sampled and the number of subjects in each group is the same, moderate violations of this assumption do not alter the result of the analysis of variance very much.

Fourth, it is assumed that each sample is drawn from a population of scores that is **normal** in form. This condition should be reflected in each of the groups sampled: Each should have a relatively normal distribution.

As discussed previously, the population distribution must be normal so that the two variance estimates will be independent, permitting the *F* test to be used. If the means are independent from the variability of scores about these group means (as they are in a normal distribution), then the between-groups and within-groups variance estimates will be independent. Moreover, the probability levels for the *F* statistic are accurate only if the two variance estimates are based on normal distributions.

Violations of the assumption of normality are not terribly damaging if a sufficient number of cases are sampled and if the departure from normality is not severe. However, when the distributions are decidedly not normal or there are not many cases in each group, a nonparametric analysis may be performed. Some nonparametric techniques are presented in Chapter 14.[8]

Fifth, the set of groups in the analysis are said to be fixed rather than random. A **fixed set** of groups is one in which the nature of the groups (for example, the amount of dream interruption) is established and "fixed" by the researcher. Consequently, the results of the analysis apply only to the groups evaluated. In contrast, a **random set** of groups is selected at random from a much larger potential set and the results are intended to apply to that larger set. To clarify this distinction, suppose an educational

---

[5] B. J. Winer, D. R. Brown, and K. M. Michels, *Statistical Principles in Experimental Design,* 3rd ed. (New York: McGraw-Hill, 1991). For a simplified discussion of the problems and assumptions in performing these analyses, see R. B. McCall and M. I. Appelbaum, "Bias in the Analysis of Repeated-Measures Designs: Some Alternative Approaches," *Child Development* 44: 401–15.

[6] Winer et al., *Statistical Principles.*

[7] W. L. Hays, *Statistics,* 5th ed. (Fort Worth: Harcourt Brace, 1994).

[8] S. Siegel and N. J. Castellan, *Nonparametric Statistics for the Behavioral Sciences* (New York: McGraw-Hill, 1988).

researcher wanted to evaluate the effects of introducing a prekindergarten program. However, the educator suspects that the results of the program will depend on the specific teacher, so three teachers are selected to have a preschool class. How the teachers are selected determines whether the set of three groups is fixed or random. Suppose the researcher put the names of all the qualified preschool teachers in the school system into a hat, randomly drew three names, and then assigned each of the three teachers the task of teaching a preschool class. These three teachers would be a random set, representing all the preschool teachers in the school system, and the results could be generalized to the group of all preschool teachers in the system. Conversely, if the researcher simply picked three teachers who happened to be available and interested in the project, this set of teachers would be fixed and the results would apply only to those specific teachers. Similarly, in the example presented in this chapter, the three levels of dream deprivation—no interruptions, two interruptions, and five interruptions—are fixed because they were selected deliberately, not selected randomly from a larger population of conditions. All the analyses of variance presented in this text assume fixed sets of groups. Procedures for analyzing random sets are presented in more advanced texts.[9]

COMPUTATIONAL PROCEDURES

General Format

Computations for the analysis of variance are more complicated than for other statistics presented in this text. Further, few hand calculators will compute the necessary statistics with a simple push of a few buttons. You may, however, be able to accumulate the n_j, ΣX, and ΣX^2 for each group with one button, which will help. Computers, though, will perform all the calculations automatically once the data for each group have been entered. Again you will need to read the instructions for your particular machine and program (see the *Study Guide* for this text for instructions to use StataQuest, MINITAB, and SPSS). The computational procedures presented here assume that you have either no computational aids or a simple hand calculator. Even if you expect to use a computer program to perform the calculations, you must read this section to understand the terminology, general procedure, and format for results of an analysis of variance, and you should compute one or more analyses by hand to get a feel for the technique before relying on the computer.

Table 10-3 gives a general computational scheme for the simple analysis of variance. Section A lists the scores in the p groups and several quantities needed for the computations. Under each group are the statistics for that group: the total of all the scores (T_j), the n_j, the mean $(\overline{X}_j)$, the sum of squared scores $\left(\sum_{i=1}^{n_j} X_{ij}^2\right)$, and the squared

[9] Hays, *Statistics*.

10-3 General Computational Procedures for the Simple Analysis of Variance

A. Data	Group 1	Group 2	$\cdots$	Group p	Total Sample
	X_{11}	X_{12}	$\cdots$	X_{1p}	
	X_{21}	X_{22}	$\cdots$	X_{2p}	
	X_{31}	X_{32}	$\cdots$	X_{3p}	
	$\cdots$	$\cdots$	$\cdots$	$\cdots$	
	$X_{n_1 1}$	$X_{n_2 2}$	$\cdots$	$X_{n_p p}$	
Totals	$T_1 = \sum\limits_{i=1}^{n_1} X_{i1}$	$T_2 = \sum\limits_{i=1}^{n_2} X_{i2}$	$\cdots$	$T_p = \sum\limits_{i=1}^{n_p} X_{ip}$	$T = \sum\limits_{j=1}^{p} T_j$
n_j	n_1	n_2	$\cdots$	n_p	$N = \sum\limits_{j=1}^{p} n_j$
Means	$\overline{X}_1 = \dfrac{T_1}{n_1}$	$\overline{X}_2 = \dfrac{T_2}{n_2}$	$\cdots$	$\overline{X}_p = \dfrac{T_p}{n_p}$	
Sum of squared scores	$\sum\limits_{i=1}^{n_1} X_{i1}^2$	$\sum\limits_{i=1}^{n_2} X_{i2}^2$	$\cdots$	$\sum\limits_{i=1}^{n_p} X_{ip}^2$	$\sum\limits_{j=1}^{p}\left(\sum\limits_{i=1}^{n_i} X_{ij}^2\right)$
Squared sum of scores divided by n_j	$\dfrac{T_1^2}{n_1}$	$\dfrac{T_2^2}{n_2}$	$\cdots$	$\dfrac{T_p^2}{n_p}$	$\sum\limits_{j=1}^{p}\left(\dfrac{T_j^2}{n_j}\right)$

B. Intermediate Quantities

$$(\text{I}) = \frac{T^2}{N} \qquad (\text{II}) = \sum_{j=1}^{p} \left(\sum_{i=1}^{n_j} X_{ij}^2 \right) \qquad (\text{III}) = \sum_{j=1}^{p} \left(\frac{T_j^2}{n_j} \right)$$

C. Basic Formulas

$$SS_{\text{between}} = (\text{III}) - (\text{I}) \qquad df_{\text{between}} = p - 1 \qquad MS_{\text{between}} = \frac{SS_{\text{between}}}{df_{\text{between}}}$$

$$SS_{\text{within}} = (\text{II}) - (\text{III}) \qquad df_{\text{within}} = N - p \qquad MS_{\text{within}} = \frac{SS_{\text{within}}}{df_{\text{within}}}$$

$$SS_{\text{total}} = (\text{II}) - (\text{I}) \qquad df_{\text{total}} = N - 1$$

D. Summary Table

Source	df	SS	MS	F
Between groups	$p - 1$	SS_{between}	MS_{between}	$\dfrac{MS_{\text{between}}}{MS_{\text{within}}}$
Within groups	$N - p$	SS_{within}	MS_{within}	
Total	$N - 1$	SS_{total}		

sum of the scores divided by n_j $\left(\dfrac{T_j^2}{n_j}\right)$. Remember that the sum of squared scores (ΣX^2) is determined by squaring each score and then summing. The squared sum of the scores $(\Sigma X)^2$ divided by n_j is calculated by first summing the scores, then squaring the sum, and finally dividing by n_j. To the right of these group quantities are the totals for the entire sample: the total of all scores $(T, \text{ sum the } T_j)$, the total number of cases $(N, \text{ sum the } n_j)$, the total sum of the squared scores

$$\sum_{j=1}^{p}\left(\sum_{i=1}^{n_j} X_{ij}^2\right), \text{ sum the } \sum_{i=1}^{n_j} X_{ij}^2$$

and the squared sum of all the scores divided by n_j

$$\sum_{j=1}^{p}\left(\frac{T_j^2}{n_j}\right), \text{ sum all the } \frac{T_j^2}{n_j}$$

These are the basic quantities needed to calculate the sums of squares for the analysis of variance.

Section B lists three intermediate quantities that facilitate the computation of the *SS*. Notice that these nonsubscripted quantities [i.e., **(I)**, **(II)**, and **(III)**] are defined differently than those used in the computation of regression and correlation. Quantity **(I)** is the squared total sum of scores divided by N; that is, square T and divide by N:

$$\textbf{(I)} = \frac{T^2}{N}$$

Quantity **(II)** is simply the total sum of the squared scores as found previously:

$$\textbf{(II)} = \sum_{j=1}^{p}\left(\sum_{i=1}^{n_j} X_{ij}^2\right)$$

Quantity **(III)** is the total across groups of the squared group sum of scores (T_j^2) divided by its group n_j, also calculated previously:

$$\textbf{(III)} = \sum_{j=1}^{p}\left(\frac{T_j^2}{n_j}\right)$$

Section C of the table lists the formulas for the three sums of squares, the three degrees of freedom, and the two mean squares required for the analysis of variance. Notice that the formulas for the *SS* are expressed in terms of the three intermediate quantities [**(I)**, **(II)**, **(III)**] previously computed. The degrees of freedom are determined by using p (the number of different groups) and N (the total number of subjects in the sample).

Section D presents the traditional summary table for the simple analysis of variance. The first column denotes the source of the variance estimate. This is followed by the degrees of freedom, sums of squares, mean squares, and F ratio.

Numerical Example

Research Question   We can illustrate the statistical procedure by continuing with the dream example in which the question was whether deprivation of dreaming produces differences in the temperaments of normal adults. The hypothetical results are presented in Tables 10-4 and 10-5. The means for the three groups (no, some, and much deprivation) are 4.00, 8.00, and 17.75, respectively. While it appears that the prevention of dreaming produces increased irritability and anxiety in normal adults, the observed differences between means could be a function of sampling error, there being no real difference in means among the three groups. Given the null hypothesis of no differences in the population means, what is the probability that the observed difference between these three sample means is merely a function of sampling error? The details of the analysis are described below and summarized in Tables 10-4 and 10-5.

Statistical Hypotheses To answer the question above requires determining the probability that the observed group means could occur if the population means for the three groups are in fact equal. Symbolically stated, the null hypothesis is

$$H_0: \mu_1 = \mu_2 = \mu_3 = \mu$$

10-4   **Summary of the Simple Analysis of Variance (Dream Example)**

Hypotheses

$H_0: \mu_1 = \mu_2 = \mu_3 = \mu$

H_1: Not H_0

Assumptions and Conditions

1. The subjects in each group are **randomly** and **independently sampled.**
2. The groups are **independent.**
3. The population variances for the groups are **homogeneous** ($\sigma_1^2 = \sigma_2^2 = \sigma_3^2$).
4. The population distribution of scores is **normal** in form.
5. The set of groups is **fixed.**

Decision Rules (from Table E)

Given: .05 significance level, $df = 2, 12$

If $F_{obs} < 3.88$, do not reject H_0.

If $F_{obs} \geq 3.88$, reject H_0.

Computation (see Table 10-5)

$F_{obs} = 24.50$

Decision

Reject H_0.

10-5

Computational Example of the Simple Analysis of Variance (Dream Example)

A. Data	Group 1 (No Deprivation)	Group 2 (Some Deprivation)	Group 3 (Much Deprivation)	Total Sample
	7	5	21	
	5	9	15	
	3	12	17	
	4	12	18	
	1	7		
		3		
Totals	$T_1 = (7 + 5 + \cdots + 1)$ $T_1 = 20$	$T_2 = (5 + 9 + \cdots + 3)$ $T_2 = 48$	$T_3 = (21 + 15 + 17 + 18)$ $T_3 = 71$	$T = \sum_{j=1}^{p} T_j = 139$
n_j	$n_1 = 5$	$n_2 = 6$	$n_3 = 4$	$N = \sum_{j=1}^{p} n_j = 15$
Means	$\bar{X}_1 = \dfrac{20}{5} = 4.00$	$\bar{X}_2 = \dfrac{48}{6} = 8.00$	$\bar{X}_3 = \dfrac{71}{4} = 17.75$	
Sum of squared scores	$\sum X_{i1}^2 = 100$	$\sum X_{i2}^2 = 452$	$\sum X_{i3}^2 = 1279$	$\sum_{j=1}^{p} \left(\sum_{i=1}^{n_j} X_{ij}^2 \right) = 1831$

Squared sum of scores divided by n_j

$$\frac{T_1^2}{n_1} = \frac{(20)^2}{5} = 80 \qquad \frac{T_2^2}{n_2} = \frac{(48)^2}{6} = 384 \qquad \frac{T_3^2}{n_3} = \frac{(71)^2}{4} = 1260.25 \qquad \left| \sum_{j=1}^{p}\left(\frac{T_j^2}{n_j}\right) = 1724.25 \right.$$

B. $(\mathbf{I}) = \dfrac{T^2}{N} = \dfrac{(139)^2}{15} = 1288.07$

$(\mathbf{II}) = \displaystyle\sum_{j=1}^{p}\left(\sum_{i=1}^{n_j} X_{ij}^2\right) = 1831$

$(\mathbf{III}) = \displaystyle\sum_{j=1}^{p}\left(\frac{T_j^2}{n_j}\right) = 1724.25$

C. $SS_{\text{between}} = (\mathbf{III}) - (\mathbf{I}) = 1724.25 - 1288.07 = 436.18$

$SS_{\text{within}} = (\mathbf{II}) - (\mathbf{III}) = 1831 - 1724.25 = 106.75$

$SS_{\text{total}} = (\mathbf{II}) - (\mathbf{I}) = 1831 - 1288.07 = 542.93$

$df_{\text{between}} = p - 1 = 3 - 1 = 2$

$df_{\text{within}} = N - p = 15 - 3 = 12$

$df_{\text{total}} = N - 1 = 15 - 1 = 14$

$MS_{\text{between}} = \dfrac{SS_{\text{between}}}{df_{\text{between}}} = \dfrac{436.18}{2} = 218.09$

$MS_{\text{within}} = \dfrac{SS_{\text{within}}}{df_{\text{within}}} = \dfrac{106.75}{12} = 8.90$

$F = \dfrac{MS_{\text{between}}}{MS_{\text{within}}} = \dfrac{218.09}{8.90} = 24.50^{**}$

D. **Summary Table**

Source	df	SS	MS	F
Between groups	2	436.18	218.09	24.50**
Within groups	12	106.75	8.90	
Total	14	542.93		

Critical values $(df = 2, 12)$ $\quad *F_{.05} = 3.88, p < .05$

$**F_{.01} = 6.93, p < .01$

The alternative is

H_1: not H_0

Assumptions and Conditions   The assumptions underlying this test were explained above and are listed in Table 10-4.

Significance Level   Adopt $\alpha = .05$.

Critical Values   The computational work presented in Table 10-5 shows that in this case the numerator of the F ratio has 2 degrees of freedom and the denominator has 12. At $\alpha = .05$ with $df = 2, 12$, the critical value for F is 3.88 according to Table E of Appendix 2.

Decision Rules   The decision rules are:

If $F_{obs} < 3.88$, do not reject H_0.

If $F_{obs} \geq 3.88$, reject H_0.

The statement of the decision rules makes it look as if a directional test is being performed. In one sense, it is; in another sense, it is not. Since the mean square sensitive to treatment effects is always placed in the numerator of the F ratio, the F test in the analysis of variance rejects H_0 only when the value of F_{obs} is very large. In this sense, it is a one-tailed test because the critical region is in only one tail of the theoretical sampling distribution. However, the analysis of variance detects only whether the means differ by more than sampling error—which particular means are larger or smaller than other means is not considered. Therefore, the analysis of variance is making a nondirectional test because the direction of differences between means is not specified. It is possible to state alternative hypotheses that do specify the ordering of group means or the form of the relationship between the levels of the factor and the dependent variable, but these techniques are beyond the scope of this text.[10] Consequently, students should not be concerned about one- versus two-tailed, directional versus nondirectional, tests in the analysis of variance in this text. All F tests are nondirectional.

Computation   The calculations are presented in Table 10-5 and follow the general procedures set forth in Table 10-3.

Decision   The observed F is $F = 24.50$. Clearly this is greater than the critical value of 3.88 at $\alpha = .05$. In fact, as presented at the bottom of Table 10-5, the observed value exceeds the critical value when the significance level is .01. Customarily, the .05 level is used as the *minimum* significance level for stating that the data suggest the rejection of the null hypothesis. However, the researcher may indicate when the observed F exceeds the critical value at a higher level of significance. Frequently, the value of F is followed by ★ or ★★ or ★★★ if it exceeds the critical value for $\alpha = .05, .01$, or .001, respectively. Thus, since 24.50 is larger than the critical value for $\alpha = .01$, this fact is indicated by writing the F value in the summary table as 24.50★★.

---

[10] Winer et al., *Statistical Principles*.

(Actually, $F_{obs} = 24.50$ exceeds the .001 level, but our table gives only the .05 and .01 critical values.)

Typically, if these results were published in a journal, they might be written in the following form: "The means (M's = 4.00, 8.00, and 17.75 for the none, some, and much dream interruption groups, respectively) were significantly different, $F(2, 12) = 24.50$, $p < .01$."

The obtained F value results in a decision to reject the null hypothesis. This means that the probability is very small that the three means (4, 8, and 17.75) differ merely by sampling error. Therefore, it is likely that the between-groups variance estimate has been influenced by treatment effects and that the population group means probably do differ from one another. The conclusion can be drawn that it is likely that the interruption and prevention of dreaming in adults leads to increased anxiety and irritability.

COMPARISONS BETWEEN SPECIFIC MEANS

The significant F test in the analysis of variance indicates that the sample group means tested are probably not from populations with identical means. It also implies that the population means of at least two groups in the set are likely to be significantly different. But which pair? How many pairs? In the above example, it is possible that a large amount of dream interruption does produce significantly more irritability than does no dream interruption, but does a small amount of dream interruption also produce more irritability than no interruption? Sometimes, then, it is desirable to test the difference between two of the means in a set, or even between all possible pairs of means, to help describe or specify the general result of the analysis of variance.

These statistical tests between means are called comparisons or contrasts.

> A **comparison** or **contrast** is the difference between two or more means assessed in the context of a larger or more general analysis.[11]

Calculating the Comparison

The comparison between two means in a simple analysis of variance is determined by computing an F ratio. Specifically, the sum of squares for the comparison between the ith and jth means is:

$$SS_{comp} = \frac{(\overline{X}_i - \overline{X}_j)^2}{\dfrac{1}{n_i} + \dfrac{1}{n_j}}$$

---

[11] Contrasts involving weighted sets of means can also be tested but are beyond the scope of this text.

in which $\bar{X}_i$ and $\bar{X}_j$ are the two means and n_i and n_j are the sizes of their samples, respectively. Since the degrees of freedom for this comparison is one, the mean square is $MS_{comp} = SS_{comp}/1 = SS_{comp}$. The F test for the comparison is MS_{comp}/MS_{within}. To summarize:

A comparison between two means in a simple analysis of variance can be tested by using

$$F_{comp} = \frac{(\bar{X}_i - \bar{X}_j)^2}{\left(\frac{1}{n_i} + \frac{1}{n_j}\right) MS_{within}}$$

in which $\bar{X}_i$ and $\bar{X}_j$ are the two means, n_i and n_j are their respective sample sizes, and MS_{within} is the mean square within groups from the analysis of variance.

One of the advantages of using this expression rather than a simple t test is that the above procedure is more powerful. Recall from Chapter 8 that the power of a statistical test is the probability that the test will lead to a decision to reject the null hypothesis when the null hypothesis is truly false—that is, when it should be rejected. The above statistic is more powerful than a simple t test because the denominator is composed of an estimate of within-group variability based on the subjects in all the groups of the analysis of variance. In contrast, this estimate would be based only upon groups involved in the comparison if a simple t test were performed. Assuming that the variability within groups is similar from group to group, this broader estimate of within-group variability is more stable because it is based upon more subjects, and power, you may recall, increases with increases in sample size. You have seen this fact when looking for critical values in the tables for t or F—the greater the degrees of freedom, the smaller the critical value and the easier to reject the null hypothesis.

The Critical Value

The critical value for the comparison statistic given above depends upon the reason the comparison is being made and how many such comparisons are to be made.

First, consider the reason for making the comparison. In one situation, the researcher may plan before the experiment to make certain comparisons. These are called a priori, or before-the-fact, comparisons. Conversely, a researcher may conduct the experiment and the analysis of variance, find a significant difference between the means, notice that certain means appear larger than others, and as a consequence of seeing the data want to find out whether those means are indeed significantly different. These are a posteriori, or after-the-fact, comparisons.

Comparisons planned before the experiment is conducted are called **a priori** comparisons. Comparisons planned after the general analysis is conducted and decided upon partly as a consequence of seeing the results of that analysis are called **a posteriori** comparisons.

The critical value for the comparison test will depend in part on whether the comparisons are a priori or a posteriori. But why should the reason for making the comparisons influence the critical value? Recall that the critical level, α, is the probability of a type I error—the probability of incorrectly rejecting the null hypothesis. If the researcher plans the comparisons before conducting the study and analysis, the critical level for any single comparison is equal to α—that is, whatever critical level the researcher adopts. But after the researcher sees that the analysis of variance is significant and observes that one mean is substantially different from another, the test of that particular comparison is very likely to be significant. That is so because the researcher deliberately picked the pair to test on the basis of information already known—a significant result from the analysis of variance and the selection of certain means that "look" like they might be significantly different. That information strongly suggests that they will be statistically different. Therefore, when comparisons are a posteriori, this bias must be corrected by using a more stringent critical value that will make the probability of a type I error for such comparisons equal to the critical level, α.

Now consider why the critical value, α, may depend upon the number of comparisons to be tested. Recall that if $\alpha = .05$, then on the average 5 of 100 such tests will be significant by chance alone when the null hypothesis is true—that is, the null hypothesis will be incorrectly rejected (a type I error). So when just one pair of means is tested, the likelihood that the null hypothesis is incorrectly rejected at $\alpha = .05$ is .05, or 5 times in 100. But if two independent tests are conducted, the likelihood that this error will be made in one of the two tests is approximately $.05 + .05 = .10$. In three independent tests, it is approximately .15. So, as the number of comparisons conducted increases (although they are all not independent of each other), so does the likelihood of obtaining at least one significant difference by chance alone. Thus, some correction to the critical value is often used when several comparisons are to be made to keep the probability of a type I error more nearly equal to α for the *set* of comparisons, not just each *individual* comparison.

The need to adjust critical values as a function of the a priori or a posteriori nature of the comparisons and the number of comparisons to be performed creates several situations in which the critical values for the comparisons are determined in different ways. Only a few of these situations are described here.[12]

Explaining a Significant Result   First and most common, the researcher makes one or at most two comparisons if and only if the analysis of variance produces a significant result. For example, in the illustrative case used in this chapter, the researcher may test the difference between much and no dream interruption and between some and no dream interruption after the analysis of variance F test is found significant.

---

[12] The procedures recommended here provide the best blend of power and protection against type I error. As a result, teachers will notice that the commonly used Neuman–Keuls, Duncan multiple range, and Scheffé procedures are not described. For a more complete discussion of this topic see Hays, *Statistics*.

When one or two comparisons are made for the purpose of describing more specifically the nature of a significant analysis of variance result, the value of F for the comparison is evaluated against a critical value of F with degrees of freedom equal to 1 and the degrees of freedom associated with the mean square within. In symbols,

Test statistic: F_{comp}

Critical values: $F(df = 1, df_{within})$

This case requires that only one or two comparisons be conducted regardless of the number of groups in the analysis of variance, that the F test from the analysis of variance be significant, and that the primary purpose of the comparisons be to explain more specifically this general result. Notice that the critical value is the usual critical value for an F with degrees of freedom of 1 and df_{within}—that is, no correction to the critical value is made. No correction is necessary if only one or two comparisons are conducted to explain a significant analysis of variance result, even if the specific comparisons are determined after seeing the means, because the significant F implies that at least one pair of population means is likely to be significantly different. Such comparisons then simply specify the one or two pairs of means that the analysis of variance has already indicated are likely to be significantly different.

To illustrate numerically, the results of the sample problem given in this chapter are

No interruption: $\overline{X}_1 = 4.00$, $n_1 = 5$ $MS_{within} = 8.90$

Some interruption: $\overline{X}_2 = 8.00$, $n_2 = 6$ $df_{within} = 12$

Much interruption: $\overline{X}_3 = 17.75$, $n_3 = 4$

To test the difference between much versus no dream interruption, calculate the observed F for the comparison

$$F_{comp} = \frac{(\overline{X}_3 - \overline{X}_1)^2}{\left(\frac{1}{n_3} + \frac{1}{n_1}\right) MS_{within}} = \frac{(17.75 - 4.00)^2}{\left(\frac{1}{4} + \frac{1}{5}\right)(8.90)} = 47.21$$

and determine from Table E the critical value

$$F_{.05} (df = 1, df_{within} = 12) = 4.75$$

So the comparison of much versus no interruption is highly significant.

One might also wonder whether some interruption is different from none. In this case,

$$F_{comp} = \frac{(8.00 - 4.00)^2}{\left(\frac{1}{6} + \frac{1}{5}\right)(8.90)} = 4.90$$

Since the critical value is the same for this comparison as the first ($F_{crit} = 4.75$), this too is significant.

A Priori Comparisons Sometimes a researcher is interested in testing the general difference between a set of groups in an analysis of variance but also has a special interest even before the study is conducted in the comparison between two particular groups. For example, a market analyst might conduct a survey of owners of different brands of television sets to determine how satisfied these people are with the brand they selected. The researcher is interested both in whether satisfaction is different for the five leading brands and whether satisfaction is different for brand X, the set produced by her company, versus brand Y, X's nearest competitor in sales. This is an a priori comparison because the researcher decided to make the test before seeing the data. If just one a priori test is to be made and there is ample a priori theoretical or practical justification for making the test, then the procedures described above may be used. However, if several a priori tests are to be made, other procedures not presented here must be used to adjust the critical value. So:

> If one a priori comparison is to be made, then the test statistic is F_{comp} with degrees of freedom of 1 and df_{within}.
>
> Test statistic: F_{comp}
>
> Critical values: $F(df = 1, df_{within})$

All Possible A Posteriori Comparisons Sometimes all possible pairs of means are to be compared. Typically, this occurs after the analysis of variance has produced a significant result and the researcher wants to know which means are significantly different from which other means for the entire set of means in the analysis. Therefore, these are a posteriori comparisons because they are carried out only if the analysis of variance produces a significant result.

> When all possible a posteriori pairwise comparisons are to be made, the appropriate test statistic is the **Studentized range statistic, q,** which equals
>
> $$q_{obs} = \sqrt{2F_{comp}}$$
>
> in which F_{comp} is the F for the comparison as described above. The q_{obs} is tested with a critical value of q having two parameters—the number of groups p in the total analysis of variance and the degrees of freedom for df_{within}.
>
> Test statistic: $q_{obs} = \sqrt{2F_{comp}}$
>
> Critical values: $q(p, df_{within})$

Critical values of q are given in Table F of Appendix 2.

In this case, the comparisons are determined entirely a posteriori—after the analysis of variance has produced a significant result. Further, the procedure is designed for the case in which all possible pairs of means are to be compared. If fewer than all possible comparisons are actually tested, the result will be conservative—that is, the actual significance level will be smaller than the α selected for the test.

The approach recommended here is called **Tukey's honestly significant difference (HSD)** procedure. It keeps the probability of a type I error for all possible pairwise comparisons equal to or smaller than α.

Suppose after obtaining the significant analysis of variance result in the illustrative experiment, the researcher decides to test the difference between all three pairs of the three groups. In this situation, a new test statistic is required, but it is based on the F comparisons that have been used in the other cases described here. The F's for the three comparisons, two of which have been calculated above, and the corresponding q_{obs} calculated with the above formula are:

Much vs. no interruption:$\quad\quad F_{comp} = 47.21, \quad q_{obs} = 9.72$

Some vs. no interruption:$\quad\quad F_{comp} = 4.90, \quad\quad q_{obs} = 3.13$

Much vs. some interruption:$\quad F_{comp} = 25.64, \quad q_{obs} = 7.16$

For example, the first q_{obs} for much versus no interruption, which had an $F_{comp} = 47.21$, is obtained as follows:

$$q = \sqrt{2F_{comp}} = \sqrt{2(47.21)} = \sqrt{94.42} = 9.72$$

The critical value for q is given in Table F in Appendix 2, and it has two parameters. The first parameter, p, is the number of means in the entire analysis of variance. In the illustrative example, this is 3. Notice, this is *not* the number of such comparisons being performed (which also happens to be three in this example), *nor* is it the number of degrees of freedom for $MS_{between}$; it is the total number of means or groups in the analysis of variance—much, some, and no dream interruption. The other parameter is the degrees of freedom for the within-groups mean square from the analysis of variance—df_{within}—which is 12 in this case. Looking now at Table F in Appendix 2, find the column labeled Number of Groups in Design = 3 and look down until you find the row corresponding to $df_{within} = 12$. The critical value at the .05 level is 3.77, and at the .01 level it is 5.04. Therefore, while the much versus no interruption and the much versus some interruption comparisons are significant, the some versus no interruption comparison is not. Notice that the some versus no interruption comparison was significant when just two pairs of means were tested, but not in this case when all three pairs of means were compared. This shows that the tests become more stringent the more comparisons one makes.

SIZE OF RELATIONSHIP

The finding of a statistically significant effect in the simple analysis of variance indicates that the groups likely come from populations having different means. In short, it says *some* treatment effect probably exists, but that fact alone does not indicate the *size* of the treatment effect. Even the size of the F from the analysis of variance does not reflect the size of the treatment effect. Of course, the sample means provide point estimates of their respective population means, but such estimates do not take into consideration the amount of variability and sampling error one would expect for the circumstances at hand.

Confidence Intervals

One approach to describing the findings more informatively is to compute confidence intervals for each sample mean to specify an interval that likely includes the population mean. Such procedures were described for a single mean using the sample variance to estimate the standard error of the mean on page 229.

Consider now constructing confidence intervals for a set of two or more means.[13] For example, the dream interruption data in Table 10-5 include three groups of participants who received either zero, two, or five dream interruptions per night for six nights, and the measured variable was a score on a temperament test (higher scores represent greater irritability). Confidence intervals could be constructed separately for each of these three means, using the procedures presented on page 229. For example, recall that confidence intervals can be determined by using the formula $\overline{X} \pm t_{\alpha}s_{\bar{x}}$, in which $\overline{X}$ is the observed sample mean, t_{α} is the value of t for a nondirectional test at significance level α (i.e., $\alpha = 1$ minus the level of confidence; $\alpha = 1 - 95\% = .05$ for 95% confidence intervals), and $s_{\bar{x}}$ is the estimate of the standard error of the mean (that is, the expected sampling variability of the mean). This formula could be used to calculate intervals for each group based upon only the $\overline{X}$ and $s_{\bar{x}}$ observed for each specific group.

But the simple analysis of variance makes an assumption that the population variances of each of the groups in the analysis are equal (i.e., the homogeneity of variance assumption). If this assumption is true, the sample variances of the three groups all estimate the same population variance and could be combined or "pooled" into a single estimate of that population variance. The MS_{within} is that single estimate of the population variance based upon the variation within all three groups combined or pooled. Its square root, $\sqrt{MS_{within}}$, provides a single estimate of the population standard deviation for all three groups. Now to obtain an estimate of the standard error of the mean, recall that the standard deviation is divided by the square root of N in the manner of $s_{\bar{x}} = s_x/\sqrt{N}$. Applied to the present case in which $\sqrt{MS_{within}}$ estimates s_x and n_j is the number of cases in the jth group,

$$s_{\bar{x}_j} = \frac{s_x}{\sqrt{n_j}} = \frac{\sqrt{MS_{within}}}{\sqrt{n_j}} = \sqrt{\frac{MS_{within}}{n_j}}$$

---

[13] From G. Keppel, *Design and Analysis: A Researcher's Handbook* (Englewood Cliffs, NJ: Prentice Hall, 1991).

Therefore:

> The endpoints of confidence intervals for the mean of the jth group in a simple analysis of variance are given by
>
> $$\bar{X}_j - t_\alpha s_{\bar{X}_j} \quad \text{and} \quad \bar{X}_j + t_\alpha s_{\bar{X}_j}$$
>
> in which
>
> $\bar{X}_j$ is the observed sample mean of the jth group
>
> α is 1 minus the desired level of confidence
>
> t_α is the critical value of t for a nondirectional test at significance level α with df_{within}
>
> $s_{\bar{X}_j}$ is the estimated standard error of the mean of group j based upon the pooled MS_{within} from the analysis of variance and the size of group j (i.e., n_j):
>
> $$s_{\bar{X}_j} = \sqrt{\frac{MS_{\text{within}}}{n_j}}$$
>
> So the endpoints of confidence intervals for the mean of group j are given by
>
> $$\bar{X}_j - t_\alpha \sqrt{\frac{MS_{\text{within}}}{n_j}} \quad \text{and} \quad \bar{X}_j + t_\alpha \sqrt{\frac{MS_{\text{within}}}{n_j}}$$

Notice that the critical value of t is based upon the degrees of freedom for the pooled variance estimate, df_{within}, *not* $n_j - 1$, because MS_{within} is the variance used to estimate the standard error of the mean.

For the dream interruption data in Table 10-5, $MS_{\text{within}} = 8.90$ with $df_{\text{within}} = 12$. Group 1 (no deprivation) had a mean $\bar{X}_1 = 4.00$ and $n_1 = 5$, and a 95% confidence interval requires a t value for a nondirectional test at $\alpha = 1 - .95 = .05$ and $df_{\text{within}} = 12$, which Table B in Appendix 2 gives to be 2.179. So, the 95% confidence interval for group 1 is given by

$$\bar{X}_1 - t_{.05, df=12} \sqrt{\frac{MS_{\text{within}}}{n_1}} \quad \text{and} \quad \bar{X}_1 + t_{.05, df=12} \sqrt{\frac{MS_{\text{within}}}{n_1}}$$

in which $\bar{X}_1 = 4.00$, $t_{.05,\ df=12} = 2.179$, $MS_{\text{within}} = 8.90$, and $n_1 = 5$, so

$$4.00 - 2.179 \sqrt{\frac{8.90}{5}} \quad \text{and} \quad 4.00 + 2.179 \sqrt{\frac{8.90}{5}}$$

$$1.09 \quad \text{and} \quad 6.91$$

For group 2 (some deprivation), $\bar{X}_2 = 8.00$ and $n_2 = 6$ while $MS_{\text{within}} = 8.90$ and $t_{.05, df=12} = 2.179$ remain the same, so the 95% confidence limits are

$$8.00 - 2.179 \sqrt{\frac{8.90}{6}} \quad \text{and} \quad 8.00 + 2.179 \sqrt{\frac{8.90}{6}}$$

$$5.35 \quad \text{and} \quad 10.65$$

For group 3 (much deprivation), $\overline{X}_3 = 17.75$ and $n_3 = 4$, so the limits are

$$17.75 - 2.179 \sqrt{\frac{8.90}{4}} \quad \text{and} \quad 17.75 + 2.179 \sqrt{\frac{8.90}{4}}$$

$$14.50 \quad \text{and} \quad 21.00$$

A major advantage to using the pooled variance estimate from the analysis of variance rather than the individual group variances to determine the standard error of the group means when calculating confidence intervals is that the pooled variance estimate MS_{within} permits one to use df_{within} rather than the smaller n_j for determining the value of the t_α. Recall that the required value of t gets smaller as the degrees of freedom increase, so confidence intervals based upon the pooled estimate and its degrees of freedom as a set (but not necessarily individually) will be narrower and presumably more informative. These advantages, however, exist only if the assumption of homogeneous within-group population variances is true.

Now, a graph of these results is sometimes drawn in the manner of Figure 10-2. Each of the three groups is located on the abscissa. In this case, the groups can be conceived to represent unequal steps (0, 2, 5) along the scale of "number of dream interruptions." But, if one did not have confidence that this scale possessed meaningful equal intervals, the groups could have been conceived as representing "no," "some," and "much" dream deprivation and spaced evenly along a nominal scale forming the abscissa. Then points representing the means are placed over each group at a height corresponding to their location on the measured scale forming the ordinate, in this case, "temperament score." If the abscissa is at least an ordinal scale, the means may

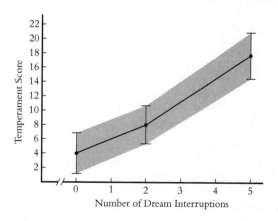

Figure 10-2 A plot of the means and confidence intervals for the three dream interruption groups in Table 10-5.

be connected with lines as has been done in Figure 10-2. Finally, the confidence interval for each mean is represented with horizontal tick marks placed at the limits, which are then connected with a vertical line passing through the group mean.

This picture of the set of confidence intervals communicates *approximate* information about the nature of the differences between groups, about the form of the possible relationship between groups and the measured variable, and about the size of the group differences relative to the variability within groups. For example, if the confidence interval for at least one group does not overlap with the confidence interval for at least one other group, it is likely that the analysis of variance will produce a significant result. Clearly, the five-interruption group's interval does not overlap with the other two, and the F from the analysis of variance was 24.50, which was highly significant. Further, the lack of overlap in confidence intervals between two specific groups reflects a possible significant difference between those two groups within the set. Again, the five-interruption group's interval does not overlap with the interval for either the zero- or two-interruption group. Indeed, the tests of specific comparisons conducted on pages 298–300 showed that the five-interruption group was significantly different from both the zero- and two-interruption groups, but the zero- and two-interruption groups were not different from each other. It is important to note that lack of overlap in the confidence intervals should not be taken as statistical evidence for a significant comparison. Rather, the specific comparison must be calculated, because it may adjust critical values depending upon the number of comparisons calculated and whether they are conducted a priori or a posteriori.

Finally, although the solid line in Figure 10-2 represents the best single estimate of the relationship between number of dream interruptions and temperament, the confidence limits can be thought of as representing a band (i.e., the shaded area in Figure 10-2) that likely includes the population relationship, which might take any one of an uncountable number of forms within the shaded band. The probability that this shaded band includes the population relationship (i.e., the probability that *all* the confidence intervals include their respective population means) is represented by the confidence level used to determine the intervals raised to the pth power, where p is the number of groups in the analysis of variance. In the present example, this probability is $(.95)^3 = .86$.

Proportion of Variance Interpretation

Another approach to describing the size of the difference between group means is to estimate the proportion of the total variance in the measured variable (i.e., temperament scores) that is associated with the differences in means between the groups (i.e., amount of dream interruption). This approach is very similar in rationale (but not computation) to determining in a regression and correlation situation the proportion of the variance in one measure that is associated with differences in the other measure. This rationale was applied to the case of the difference between two group means in Chapter 9, and it can be extended to cover p groups in the simple analysis of variance. The logic and interpretation of this strategy and its statistic, **omega squared,** symbolized by ω^2, were discussed on pages 260–263, which material should be reviewed.

In the case of p groups, omega squared may be estimated approximately with the formula

$$\text{est. } \omega^2 = \frac{SS_{\text{between}} - (p - 1)\, MS_{\text{within}}}{SS_{\text{total}} + MS_{\text{within}}}$$

For the data in Table 10-5 on dream interruption, the quantities required for this formula are

$$SS_{\text{between}} = 436.18$$
$$p = 3$$
$$MS_{\text{within}} = 8.90$$
$$SS_{\text{total}} = 542.93$$

So an estimate of omega squared is given by

$$\text{est. } \omega^2 = \frac{436.18 - (3 - 1)\, 8.90}{542.93 + 8.90}$$

$$\text{est. } \omega^2 = .76$$

This indicates that approximately 76% of the variance in temperament scores of the subjects in this study was associated with differences between the three dream interruption groups. The limitations on interpreting this estimate, especially its restricted generality, were discussed in Chapter 9 and should be kept in mind.

SUMMARY

The analysis of variance is a method of testing the significance of the difference between two or more means—in the present case, from independent groups of subjects. Two estimates of the population variance are computed. One is calculated on the difference between the group means and the grand mean. It is the mean square between groups, and it is sensitive to whatever differences may exist between group means. The other estimate of the population variance is calculated within each of the groups and is pooled across groups. It is called the mean square within groups, and it is not sensitive to whatever differences may exist between group means. If the null hypothesis is true—that is, if the several groups all have the same population mean—then the ratio of the between-groups to the within-groups mean squares, known as the F ratio, should deviate around the value of 1.00 by sampling error. But if the F ratio is large, it probably reflects the influence of group differences on the numerator (the between-groups mean square), and the null hypothesis should be rejected. The size of the differences between group means relative to the variability within groups may be seen by constructing and plotting confidence intervals for the group means and by estimating ω^2, the proportion of the total variability that is associated with differences between the group means.

Exercises

For Conceptual Understanding

1. Suppose a researcher wants to test the efficacy of four different programs designed to help people stop smoking. Volunteers are randomly assigned to each of the four programs or to a fifth group that receives no help. If *t* tests are used to compare each group to every other group, how many statistical tests will be conducted? Why is the analysis of variance preferred over this approach?

2. The four sets of numbers given at the end of this exercise were obtained from a random number table.

   a. Compute the analysis of variance, following the format outlined in Table 10-5.

   b. Add 5 to each score in group 3. This process is analogous to the existence of what in the population? Now, compute the analysis of variance again and explain any difference between this result and that found in your answer in part **a.**

   c. Start with the data as revised in part **b,** and add 20 to the last score in each group. Recompute the analysis of variance and explain any differences you observe.

Group 1	Group 2	Group 3	Group 4
9	3	2	5
3	8	3	8
6	7	6	1
4	3	8	7
4	7	3	0
	3	8	6

3. Explain how the partition of variability in the analysis of variance is analogous to the partition of variability in regression and correlation.

4. Summarize the logic of the *F* test in the analysis of variance, and explain how an analysis of *var-*

iance can determine whether differences between *means* exist in the population.

5. Why are critical values for paired comparisons adjusted as a function of the number of such comparisons to be made and whether they are a priori or a posteriori?

For Solving Problems

6. According to the theory of the *inoculation effect,* people are more resistant to changing their attitudes toward something if they have been mildly attacked for those attitudes in the past than if they have never been challenged. Suppose college students who agree with the statement "Most forms of mental illness are not contagious" are divided into three groups. The "support" group receives information that is mildly supportive of the statement. The "inoculation" group receives information that is mildly critical of the attitude. The control group receives no information about the attitude. Subsequently, all subjects receive a message that vigorously attacks the belief, and they are later tested for attitude change.[14] The results are presented below. A larger score indicates greater attitude change. Determine whether the groups are significantly different.

Support Group	Inoculation Group	Control Group
7	1	5
5	4	7
3	3	4
6	3	5
4	2	7
7	5	8

---

[14] Based upon, but not identical to, W. J. McQuire and D. Papageorgis, "The Relative Efficacy of Various Types of Prior Belief-Defense in Producing Immunity Against Persuasion," *Journal of Abnormal and Social Psychology* 62: 327–37.

7. A controversy exists over how best to teach preschool children or preschoolers. One approach is very nondirective. It relies on teachers to take advantage of circumstances in the children's play and other activities to teach language, mathematics, and other subjects. Another approach is directive. In it, children are taught academic subjects directly, with special lessons and practice for each subject. Suppose a researcher arranges for small preschool classes to be taught for one year in either a nondirective or a directive way, and wants to compare the results to a small control group that receives no particular preschool experience. After the year, all children go to the same public school. In addition to assessments of the amount of subject matter learned, the researcher also is interested in IQ differences, because the previous literature shows that preschool is associated with increases in IQ relative to controls while the children are in the program but not three years later.

Below are the IQ scores for the children in the three groups, first at the end of Year One and then at the end of Year Three.

Year One

Nondirective	Directive	Control
99	100	91
106	110	103
103	108	89
109	112	97
99	103	101
104	105	99
107		95
		94

Year Three

Nondirective	Directive	Control
103	97	98
106	105	107
109	107	95
112	103	106
105	95	109
110	101	107
107		101
		99

In separate analyses of variance, test for group differences for Year One and for Year Three.

8. Now that you have seen the ANOVA results, test the comparisons between all pairs of means in Exercise 6.

9. Perform the following comparisons on the data in Exercise 7.

a. Suppose, as a result of knowing that preschool programs improve IQ while the children are in the preschool programs, the researcher decides before the experiment is conducted to compare the Nondirective group with the Control group on the Year One data.

b. Suppose after seeing the Year Three analysis of variance, the researcher wishes to specify that it is the Nondirective group that is different from the Directive group.

c. Suppose after seeing the Year Three analysis of variance, the researcher wants to make all possible comparisons to determine which pairs of groups are significantly different from one another. Explain any disparities in the outcome of part b above and this test.

10. For the data in Exercise 6, determine .95 confidence intervals for the means, plot them and the means on a graph, and determine the percent variance associated with the treatments. Discuss the limitations of the latter estimate.

For Homework

11. From the data below, prepare an analysis of variance summary table and determine the significance of the result ($\alpha = .01$).

Group 1	Group 2	Group 3	Group 4
4	6	2	8
4	5	4	7
2	6	5	8
0	3	1	5

12. Five different physical fitness programs are tried by a random selection of men 35–40 years of age. After four weeks, the men are given a test with a maximum score of 20. Given the data below, is there any evidence that the five programs produced different results ($\alpha = .05$)?

		Program		
1	**2**	**3**	**4**	**5**
12	8	11	6	12
7	10	6	8	11
14	11	5	7	9
17	9	5	10	7
9	13	8	5	8
10	12	6	10	8
8	10	7	12	7
11	7	8	6	10

13. Conduct the following comparisons for the data in Exercise 12.

   a. To explain a significant overall result from the analysis of variance, compare the means for programs 1 vs. 3 and for programs 2 vs. 3.

   b. Suppose for theoretical reasons, the comparison between programs 1 and 5 was of particular interest regardless of the significance of the differences between the set of five programs. Test this difference.

   c. If all possible comparisons were made as a consequence of finding a significant overall result in the analysis of variance, which pairs of programs would be significantly different? Explain any differences you find here versus those found in parts **a** and **b**.

14. Estimate the proportion of variance associated with program mean differences in Exercise 12. Then determine 95% confidence intervals for each program and plot on a graph (note that the programs do not form a scale).

Three

SPECIAL TOPICS

The first two parts of this text have presented the

basic elements of descriptive and inferential

statistics. Typically, this material is the core of a

first course in applied statistics in the behavioral

sciences. But there is much more, and this part of

the text provides additional information that many

instructors add to the core topics, including a

summary of research design, topics in probability,

two-factor analysis of variance, and nonparametric

techniques.

Introduction to Research Design

cientific research, whether in physics or psychology, is an attempt to answer questions in a systematic, objective, and precise manner. Science is distinguished from other methods of answering questions by its reliance on empirical methods.

> **Empirical methods** use experimentation and especially direct, careful, and systematic observation.

Scientists rely only on information they can observe and measure as evidence that a relationship exists or that one event causes another.

A major component of empirical methods is research design.

> **Research design** consists of the methods and strategies scientists use to conduct experiments and produce empirical observations that help them determine the relationship between two or more things or that one event causes another.

Designing and conducting empirical research can be divided into five general steps:

1. **Formulate a scientific question** about a relationship between two or more things. For example, does viewing violent television programs by children produce aggressive behavior toward others?

2. **Operationalize the "things" of interest and their relationship.** What operations, for example, would be used to define which television programs would be considered "violent"; what operations would be used to measure "aggressive behavior" in children; and what operations would be conducted to observe the possible relationship between the two?

3. **Collect data;** that is, perform the operations decided above, make the systematic observations, and record the measurements.

4. **Analyze the data** in ways that answer the original question. This step usually requires the use of certain statistical techniques of the kind presented in this text and others.

5. **Draw conclusions and interpret the results** of the observations. The real skill in designing and carrying out research is to collect the observations in such a way as to obtain a clear and unequivocal answer to the original question that rules out all other possible explanations. This is not an easy task.

You can see from this description of research that statistics is involved in a major way in only one of the five steps—data analysis. Further, whether the research produces valid conclusions depends upon how the research was designed, not just upon how the statistics were carried out. Statistics is but one of several tools used by scientists in research design. Even the most brilliant statistical analysis cannot produce meaningful answers if the research was poorly designed. Valid scientific conclusions, then, come from well-designed and well-analyzed research. We now consider the five steps of research in greater detail.

SCIENTIFIC QUESTIONS

Because science is empirical, scientific questions must be about things that can be observed and measured objectively and precisely in a direct or indirect manner. Some topics simply cannot be studied scientifically because the main object of interest cannot be observed. The existence of God cannot be studied scientifically, because God is not directly observable. The existence of God is a matter of faith, not science. Nevertheless, some things that scientists study are not directly observable, but their influence is observable. No one has seen gravity, for example, but physicists certainly study it. They can do so because they can specify precisely what effects gravity has on a variety of objects and events, and they can observe and measure those effects objectively and precisely. Similarly, psychologists cannot see intelligence but they can define it to include certain behaviors but not others, and those behaviors can be observed and measured in objective and precise ways. Therefore, although gravity and intelligence cannot be observed directly, their effects *can* be specified and observed. God is said to influence events, but it is also said that "God works in strange and unpredictable ways," making it impossible to specify what is and what is not an act of God. So God is not open to scientific study, even indirectly.

Sometimes questions are potentially scientific but practical problems prevent them from being studied directly. Psychologists, for example, would like to know whether the play activity of infants and toddlers provides them with the experiences necessary for developing language. Picking up a ball and dropping it on the floor, for example, is a physical analogue of the grammatical form: subject (I)—verb (drop)—object (ball). Perhaps such exploration allows the toddler to experience the basic concepts underlying grammar, and without such experiences language might not develop at all, be delayed, or be of a different form than that of our current language. Potentially, this is a scientific question. We can certainly define such experiences precisely and in observable ways, we can measure language development and grammatical production and accuracy, and we can imagine a way to provide some but not other infants with exploratory experiences. But in reality we cannot permit some toddlers to play with objects and people and totally prevent other toddlers from having such experiences. So the question, at least in this form, cannot be answered directly by scientific means for practical reasons, although other questions on this general topic can be answered scientifically and the results interpreted to address this same issue.

Variables and Relationships

Besides being empirical, scientific questions usually involve a relationship of some sort. For example, is something associated with something else? Is there a correlation between watching violent television programs and aggressive social behavior? Is parental divorce associated with greater school problems in the children of these parents? Is the level of achievement motivation in young women related to the amount of encouragement they received from their fathers? Does Head Start improve the later school performance of low-income children?

Variables   The "things" of scientific relationships are called variables.

> A **variable** is the general characteristic being measured on a set of subjects, objects, or events, the members of which may take on different numerical values.

So in the question about the relationship between watching violent television and aggressive social behavior, violent television viewing is a variable because children will not all watch the same amount of violent television. Aggressive social behavior is another variable, because the children will display different amounts of aggressiveness.

Variables are of two kinds—dependent and independent.

> The **dependent variable** is the one whose values are thought to "depend" or be influenced by another variable, called the **independent variable.**

Social aggressiveness is presumed to depend in part on watching violent television. Therefore, in the above example, social aggressiveness is the dependent variable and watching violent television is the independent variable. Similarly, in an experiment, a researcher may expose different groups of children to violent television cartoons, action programs, dramatic programs, and no special television programming, and then observe the amount of aggressive social behavior displayed by the children on the playground. Type of television program would be the independent variable and the amount of aggressive social behavior would be the dependent variable.

Causal and Noncausal Relationships   Relationships may be causal or noncausal.

> In a **causal relationship** the independent variable actually influences in some way the value of the dependent variable.

For example, watching violent behavior on television makes some children more aggressive under some circumstances. It produces, in part, the aggressiveness. Sometimes an independent variable completely determines the value of the dependent variable. When the temperature of water at sea level is brought below 0° Celsius, the water freezes—essentially every time and for each sample of pure water. But such perfect causality rarely occurs in behavioral science. Research indicates that watching violent behavior on television probably makes *some* children (especially those who are already somewhat aggressive) a little more aggressive in *some* situations (but not necessarily all), and that watching violence on television is *not the only cause* of aggressiveness. The relationship is causal, but imperfectly so.

> A **noncausal relationship** is one in which the independent variable does not actually produce or influence the dependent variable, but the values of the independent and dependent variables are in some way related.

For example, some mothers give their six-month-old infants a great deal of intellectual stimulation (they provide mobiles for the infants, talk to them, play with them, and so on). The children grow up to be somewhat more intelligent than those of mothers who do not provide such stimulation. However, it is unlikely that the stimulation provided at six months of age *makes* the children brighter. Rather, the mothers who stimulate their infants at six months are the same mothers who provide a good language model, require their children to think and ask questions, and offer them books and diverse experiences when the children are older. It is the stimulation the same mothers provide in early childhood—not at six months of age—that likely produces somewhat higher intelligence. Therefore, the relationship between maternal stimulation at six months and child intelligence at six years may be noncausal.

Observational vs. Experimental Research

Two general types of research exist, depending upon the nature of the relationship the researcher wishes to demonstrate.

Observational Research   This is often the first step in research.

> In **observational research,** the researcher usually makes observations in a naturalistic or specially designed context, but does not deliberately attempt to manipulate the independent variable to produce changes in the dependent variable. The purpose is to determine whether a relationship— causal or noncausal—exists between two variables.

This was the first step in research on television and aggressiveness. Scientists simply observed how much violent television young boys watched at home and how much aggressive behavior they displayed at preschool. They did not deliberately show some boys violent programs and others not; that is, they did not deliberately manipulate the independent variable of watching violent television programs—they simply observed what the boys did at their own choosing. As a result, observational research cannot by itself determine whether a relationship is causal, because too many other factors might influence the result. So the purpose of observational research is to determine whether a relationship exists at all and if so, how strong it is. If no association had been found between watching violent behavior on television and displaying aggressive behavior in preschools, researchers might have abandoned the whole issue.

But these associations did occur, thus providing a second question: Is the association causal? The observed association by itself cannot answer this question. For example, the television networks claimed that youngsters who are aggressive in preschool watch more violence on television, and there is some truth to this statement. They argued that watching violence on television does not cause aggressiveness in school; rather, both these behaviors are caused by an aggressive personality.

Experimental Research   To determine whether an association is causal, it is usually necessary to conduct an experiment.

> In **experimental research,** the scientist controls and deliberately manipulates the independent variable in some way for the purpose of observing the effects of that manipulation on the dependent variable.

The crucial aspect of experimental research that distinguishes it from observational research is the deliberate manipulation by the scientist of the independent variable. For example, in a strictly observational study of violent behavior on television and aggressive social behavior, parents might be asked to keep a record of the shows their children watch over a two-week period. During the same period the number of aggressive acts by these children in preschool would be recorded by observers, so that the scientists could determine whether those children who watched more television violence were more aggressive. In such a case, the researcher does not control whether the children see many or few violent programs. The children and their parents, as well as other factors, determine how much and what kinds of television programming the children watch. In contrast, an experiment might be conducted in which the *researcher* showed some children a certain number of hours of programs with some violent behavior (such as "Roadrunner" cartoons and "Batman" cartoons or movies), while other children did not see these shows. In this case, who sees violence on television, what kind of violence, and how much violence are determined by the researcher. He or she is manipulating or experimenting with the independent variable, and such manipulation under careful conditions can permit the researcher to conclude that the independent variable actually causes or influences directly the dependent variable— that is, that viewing violent television programming makes children more aggressive.

Selecting the Appropriate Type   Scientific research typically addresses whether the relationship in question exists, to what degree it exists, and, if possible, whether that relationship is causal. It would seem, then, that experimental research would always be the method of choice, since it has greater potential to answer the question of causality.

In reality, observational methods are often used, either because experimental methods cannot be implemented or for a variety of other reasons. For example, the question of whether exploratory manipulation of objects is a necessary experience for the normal development of language cannot be answered directly by experimental methods because one could not deprive a group of children of all exploratory experience. Similarly, one cannot provide some girls with warm, caring, encouraging fathers and others with cold, uninvolved, nonencouraging fathers to see whether the first group would grow up to be more achievement oriented than the second group. However, one could conduct an observational study to determine whether a relationship existed between measures of warmth, caring, and encouragement of fathers and measures of their daughters' achievement later in life.

Sometimes the choice of observational or experimental research depends upon the nature of the question being asked. If one wants to know whether the viewing of violent television programs can cause increased aggressiveness among preschool children, then an experiment is probably required. But suppose one is interested in knowing whether the everyday aggressiveness typically displayed in preschools is caused by the everyday television viewing habits of children. In this case, the question itself stipulates that nothing be changed from the circumstances that actually exist now in American homes, in the television industry, and in preschools, and it therefore eliminates most forms of experimental manipulation. A naturalistic observational study is required. If no relationship between television viewing habits and aggressiveness were found, then despite experiments showing that watching violent behavior on television *can* cause aggressiveness, one would have to conclude that in current circumstances it *does not* cause the aggressiveness typically observed in preschools, at least not in any simple direct manner. If, on the other hand, the observational study did reveal a relationship, one still would not be perfectly certain that the relationship is causal. Watching violent behavior on television *could* produce aggressiveness, but perhaps everyday preschool fighting is caused by aggressive children who happen to watch more violence on television than do other youngsters.

In the long run, a combination of experimental and observational research like that just described, consisting of studies conducted under progressively more natural circumstances, has led scientists to believe that there probably is a causal connection: that some aggressiveness in society is produced by violent programs on television.

The Role of the Literature Review   While formulating a research question appears to require little more than the statement of a question about a relationship between two or more variables that can be studied empirically, the process is actually considerably more complicated and involved. Moreover, a well-conceived question is crucial to whether the research will contribute to knowledge and will be worth conducting.

Ideally, research questions are formed only after the scientist has read and studied all the published research and theory pertaining to the general topic of interest. Formally reviewing research literature is a valuable skill, and it takes practice to do it well and efficiently. For example, it helps to have a general familiarity with the field before starting the detailed analysis of each study. Then, each study is read and evaluated from a methodological standpoint. That is, judgments are made as to the scientific quality of the research and what conclusions can be drawn from the work, regardless of the conclusions the original authors drew. Then the results that are adequately supported by the research are catalogued according to the particular independent variable, dependent variable, characteristics of the subjects studied, and other factors. In this way, the reviewer attempts to decide on the basis of acceptable scientific evidence what factors influence the behavior in question, under what circumstances, for whom, and in what way. When several hundred studies have been published, each using a different procedure and sample, seeing common themes and explaining the exceptions to those themes can be very difficult, but that is the skill required in reviewing scientific literature.

The factors that are thought generally to influence the behavior are called parameters, the same word that has been used previously for a numerical characteristic of a population.

A **parameter** is a factor that is thought generally to influence a particular behavior, event, or characteristic. A specific meaning of the term *parameter* is a numerical characteristic of a population.

If viewing violent television programs does produce more aggressiveness in already aggressive children, then watching violent television is a parameter of aggressiveness. More specifically, it is a **situational parameter,** or **exogenous parameter,** because it occurs in the environmental situation external (or exogenous) to the child. But the personal characteristic of being an aggressive child is also a parameter of aggressiveness. It signals which children are likely to be adversely influenced by watching violent television. However, the personality characteristic of aggressiveness is a **subject parameter,** or **endogenous parameter,** because it is associated with the individual person.

The goal of a literature review is to identify situational and subject parameters of a given behavior and to describe the causal relationships that will explain and predict who will display the behavior, when, and how much of the behavior will occur under different circumstances. Inevitably, research literature provides only partial information, and questions about whether a factor is actually a parameter or about how one or more parameters work together to influence the behavior still remain. Those questions become the focus of future research—indeed, one of them may be the research question of the next study.

Literature reviews not only help researchers formulate important questions, they also provide information about other parameters that can be used in designing related research projects. For example, perhaps violent television influences boys more than girls, and such violence is more influential if the person shown committing the violent act is somehow rewarded for having done so (for example, if he or she achieves a goal through violence). Then, even though the researcher may want to study whether cartoon or real-life portrayals of violence are more influential in promoting aggressiveness, the study may focus only on boys and only on television shows in which the violent person is rewarded. Most behaviors are influenced by many parameters, not all of which can be studied in a single project, so choices of gender of subject, type of violent program, and so forth must be made. These choices should be made on the basis of the literature review.

The literature review can also help the researcher decide which operations might be used in the study, including different tests or methods of observing aggressiveness, different types of television programs and their association with aggressiveness, and so on (see the next section).

OPERATIONALIZING

The next step is to operationalize.

> To **operationalize** is to define in terms of specific acts or operations how the sample will be collected, how variables will be defined and measured, how manipulations will be conducted, and how subjects will be observed.

These operations should be so specific that another scientist could repeat the entire study in the same way as the original.

Sampling

Ideally, for the applications in this text, subjects are randomly sampled from a population. Sociologists, pollsters, and market researchers sometimes take truly random samples of certain populations, but psychologists rarely do. Instead, they use so-called "convenience samples," such as rats available in the colony or shipped from a supplier (which are not randomly selected); students enrolled in the university subject pool; parents of infants who answer a newspaper ad or agree to participate after answering a phone call; and so forth. Whatever the procedures, they must be specified in advance and written down for later publication.

Operational Definitions of Variables

The variables of interest must be defined in operational terms—that is, precisely how will the independent variable be defined and manipulated (if it is a true experiment), and how will the dependent variable be measured?

This seems to be a relatively straightforward matter, but it is not, and how the variables are operationally defined can greatly influence the results of the study. For example, suppose one were to operationalize the variable of social class. What might be a good index or measurement of social class? Often, educational level is used, but sometimes annual income or the status of the father's occupation are used. Years of education and income, however, are not the same; they are related but not identical. Some people, for example, hold very important jobs but do not have many years of education. Also, occupational status might be unfair to women in the sample who have elected to stay at home to rear their children or who made career sacrifices to have a family. Also, the several measures of social status might not relate equally to the other variables in the study. For example, if one were interested in the kinds of stimulation that mothers provided to their infants and toddlers, educational achievement might be more highly related to such stimulation than would the occupational status of the mother or her husband. So the operational definition of a variable can be a complex and very important matter. Indeed, one of the reasons that research literatures are often difficult to integrate and summarize is that each study may use different operational definitions of the major variables.

A related issue is determining how many and which dependent variables to measure. This text presents techniques designed to analyze only one dependent variable, but most research in the behavioral sciences involves many dependent variables. A thorough knowledge of the research literature is necessary to know which variables are likely to be affected by the independent variable. Even then, a treatment may have benefits on variables that were not the original focus of the study. For example, when early childhood enrichment programs were first designed and evaluated, the primary dependent variable used to assess their effects was IQ. It was shown that the IQs of poor children did rise while they were in the program, but such an advantage over children who did not have the enrichment program disappeared over the subsequent years. It appeared that early childhood programs produced no lasting benefit, and governmental programs, such as Head Start, almost died. Later, it was discovered that such experiences did have effects, not on IQ as originally suspected and hoped, but on preventing later school failure, the need for remedial services, and even teenage pregnancy and delinquency. It is often wise in the social and behavioral sciences to measure more than one dependent variable.

Reliability and Validity   A good operational definition of a variable will produce measurements that are reliable and valid.

> **Reliability** refers to whether the measurement procedures assign the same value to a characteristic each time it is measured under essentially the same circumstances.

Some operational definitions produce perfectly reliable measurements. If birth certificates are available, age can be determined with perfect reliability. The same age is obtained for each individual each time the measurement is taken on the same day (age, of course, does change over time). But in behavioral research, perfect reliability is not always achievable. For example, suppose two observers watch children in preschool and then rate each child's aggressiveness on a ten-point scale, ten being the most aggressive. These two observers will not always agree on the ratings they give each child. The ratings will not be perfectly reliable.

Usually, in behavioral research, if the measurements are recorded automatically, interobserver reliability is assumed to be perfect or acceptable. If human observers must judge the behavior, however, interobserver reliability must be evaluated and statistical procedures used to determine the degree of agreement or reliability. Then the scientist must judge whether the level of interobserver reliability is sufficiently high for the measurements to be meaningful. If it is not, then the measurement procedures might need refinement and the observers more training and practice.

> **Validity** refers to the extent to which the measurement procedures assign values that accurately reflect the conceptual variable being measured.

Does the measurement actually measure what it is supposed to measure? Sometimes a measurement is obviously valid. Years since birth is a perfectly valid measure of age. But the validity of other measurements is less obvious. Is the number of times a child hits other children in the preschool a valid index of the general personality trait of aggressiveness? It might be for the preschool context, but it might not be a valid index of aggressiveness on the athletic field. Also, aggressive boys hit, but girls are more likely to express their aggressiveness verbally than physically, so hitting might not be as valid a measure of aggressiveness for girls as boys. This example shows that a particular operational measurement is not valid by itself—hitting is not a valid measurement, but rather it may be a valid measurement of one concept but not another, for some children and not others, at some ages and not others, under some circumstances and not others, and so on. Determining the validity of a particular operational measurement can be a very complicated, time-consuming process.

Operationalizing Procedures

Once the process for selecting a sample has been determined and major variables have been operationalized with appropriate measurement techniques, procedures for actually carrying out the research must be specified. Perhaps the most important such procedure in experimental research is defining the procedures for manipulating the independent variable. These procedures determine in large part whether the researcher can conclude after the study is completed that the independent variable indeed influenced the dependent variable.

Experimental and Control Groups   In its simplest form, an experiment might have two groups of subjects, an experimental group and a control group.

> The **experimental group** receives the experimental manipulation or treatment of interest, while the **control** or **comparison group** is treated in the same way as the experimental group but without the manipulation or treatment of interest.

In the example, the experimental group would be shown violent television programs, for example, "Batman" (with parental permission), for a specific period of time (for example, 30 minutes each day for two weeks). In contrast, the control group would be shown nonviolent, but equally interesting and action-filled, television programs according to the same schedule. The ideal is for the experimental and control groups to be identical in every way *except* with respect to the specific circumstance—the independent variable—under study. If it were not for the control group, one could not conclude that watching violent programs produces social aggressiveness. Even if the experimental group showed a high degree of aggressiveness, perhaps it is simply the amount of aggressiveness common to children of that age, to that preschool, or to that time of year. The control group "controls" for these and other possible explanations of the results.

The more similar the experience of the control group is to that of the experimental group, the more of these other explanations will be eliminated. For example, suppose the control group saw no television programming at all at preschool. Now if the experimental group is more aggressive, is it because of the violent television viewing? Not necessarily. Perhaps it was due to watching television in general, not to the particular content of the programs.

To control for this possibility, one might have the control group watch the same amount of television at the same time and in the same place, but the programs would be about nature instead of violence. Now, if the experimental group is more aggressive, is it not because of the violent programming? Maybe, but violent programs differ from nature programs in more ways than simply their portrayal of violence. They have more action, for example, the scenes change more frequently, the music is louder and faster, the emotions expressed are more extreme, and so forth. Perhaps it is this package of factors associated with action and excitement that gets the children riled up and makes them more aggressive, not the violent acts per se that are depicted in the violent programs. To control for this possibility, a special television program might be created involving an athletic contest consisting of a chase over wild terrain. The same music, the same pattern of scene lengths, and the same emotions are used, but no violence. Now, does the viewing of violent acts produce more aggressiveness?

Maybe. The answer is only "maybe," because one usually can think of other factors that were not quite the same between the experimental and control group experiences besides the violent acts in the television programs that the two groups watched. The art of research design lies in creating those two groups to be as similar as possible except for the particular factor under investigation. In this way, if a difference between the two groups is observed on the dependent variable, then that difference can only be attributed to the factor that was different between them. Sometimes several control groups must be used to control for all the possible factors that might explain the difference.

Random Assignment of Subjects to Groups   It is crucial in a true experiment for the subjects in the sample to be randomly assigned to the different groups in the research design. For example, if the experiment on television violence and aggressiveness used three groups—violent programs, action programs, and "Mister Rogers' Neighborhood" (which teaches kindness and helpfulness rather than aggressiveness)—the children would be randomly assigned to those three groups. In fact, the researcher might assign numbers to the children in the sample and then use a random number table to determine which children were placed in each of the three groups. In observational research, random assignment is usually not possible—indeed, the inability to randomly assign subjects to experimental treatments may be the reason why observational methods rather than a true experiment have been selected.

Randomly assigning subjects to an experimental treatment and control group is one way of minimizing the possible influence of subject parameters, such as aggressiveness. That is, if subjects are randomly assigned, then the aggressive children are likely to be assigned in approximately equal numbers to all three groups, thereby minimizing the possibility that the results could be attributed to the aggressive personalities of the children and not to watching violent programs.

Extraneous Variables   In the above paragraphs, we have seen that variables other than the independent variable may cause or account for differences in the dependent variable in both observational and experimental research. Such variables are called extraneous variables.

> An **extraneous variable** is a factor other than the independent variable that could influence the dependent variable in such a way that it could explain the relationship observed between independent and dependent variables.

If the research question concerns the relationship between watching violent television and aggressiveness in preschool, any variable other than watching violent television (for example, an aggressive personality, the action rather than the violence component of the programs) that might possibly account for differences in aggressiveness between the groups would be extraneous.

Confounding   Extraneous variables produce what scientists call confounding.

> **Confounding** occurs in research when extraneous variables could possibly influence the dependent variable in such a way that they, and not the independent variable, could potentially explain the observed relationship. In this case, the independent variable and the extraneous variable are said to be **confounded.**

If the aggressive personalities of subjects are unevenly distributed among the groups, then the extraneous variable of aggressive personality is said to be confounded with the amount of violent television programming watched.

Confounding can occur in experimental or observational studies, and it can involve extraneous variables associated with the naturalistic or experimental situation, the subjects, or even the observers who are collecting the data. Several examples have been given above of confounding involving extraneous variables associated with situational factors (such as, the action and emotional content of violent programs rather than the violence per se) and extraneous variables associated with subject factors (for example, that aggressive children tend to watch more violent television). Minimizing situational confounding requires skill in designing the research, and minimizing subject factors can sometimes be accomplished in experimental studies by taking a large sample and randomly assigning subjects to groups.

Even then, other kinds of confounding can occur in both experimental and observational research. For example, the subjects can expect certain things to happen in the study and as a result do things that make those expectations become a reality.

> **Subject bias** occurs when the subjects in the research influence its outcome instead of or in addition to the independent variable, because they know, or think they know, something about the nature of the study.

This can occur even in an experiment in which subjects are randomly assigned to the experimental and control groups. For example, if subjects know they are receiving a drug that is supposed to make them tense or nervous, they may actually become tense or nervous because they *expect* they should feel that way, not because the drug makes them so. Some people believe that the effects of marijuana are enhanced by expectations of feeling relaxed and cheerful, especially when a group of people smoke together. To prevent this kind of subject bias in research on drugs, subjects who are not to be given the drug under study may be administered a **placebo,** such as a cigarette that tastes and smells like marijuana but has no corresponding physiological effect. The subjects given the placebo are led to believe that they are taking the drug so that their expectations will match those of the subjects who actually receive the drug.

In a similar way, the researcher and the observers can influence the result.

> **Observer bias** occurs when the researcher or the observers influence the outcome of a study instead of or in addition to the independent variable.

Observers who know the purpose of an experiment, or who know which groups are supposed to do "better" than other groups, may treat the subjects differently and may actually produce behavioral differences between the groups. Researchers may handle more gently those rats who are expected to learn very quickly; they may be more courteous and encouraging to people who are expected to perform better; they may even perceive and score the performance of some subjects in accord with their expectations. If you are observing the aggressive behavior of children in a preschool, you may be more prone to decide that a slight nudge between pupils is an aggressive act if you know that the nudger saw violent television programs. To guard against observer bias, many researchers attempt to use observers and research assistants who do not know the purpose of the study or who do not know whether subjects belong to the experimental or control group. Such observers (and subjects) are called *blind,* because they are blind to which subjects are in which groups, the purpose of the experiment, and so forth.

> When subjects and observers are both blind to the research and subject conditions, the study is said to constitute a **double-blind** research design. If either subjects or observers are blind, but not both, the study is called a **single-blind** design.

DATA COLLECTION AND DATA ANALYSIS

The researcher must make many decisions when specifying the procedures for data collection. Each of these decisions might influence the results. In the examples, the researcher must decide which television programs to use, how many times to show them, how much time to spend counting aggressive acts, what the definition of an aggressive act shall be, and so on. Ideally, each decision should be made on the basis of previous research (for example, which programs have been found to be effective in the past, how much viewing is necessary to produce a difference, and so forth). This is why a thorough knowledge of the research literature is absolutely essential. But rarely does that literature provide information on all or even most of the procedural issues that must be decided. Therefore, many researchers make tentative decisions and then try them out on just a few subjects to see how well the procedures work. These are called **pilot subjects,** and sometimes an entire **pilot experiment** is conducted, especially if many of the procedures are new. In either case, the scientist must keep a log describing in detail all the procedural decisions made and why, because they must be included in the written report of the research.

Once all the decisions are made and recorded and pilot subjects are seen and evaluated, it is necessary to design a system for recording the data. Usually it is necessary to obtain some general information on each subject, such as age, gender, race, education (of the subject or parents of the subject), and any other information pertinent to the variables in the study, and this information is included in the written report of the research to describe the nature of the sample that was used.

In addition, the dependent variable or variables must be recorded. Data may be automatically recorded by electronics designed to detect the behavior of interest (for example, a machine that detects when a baby's head turns more than 20 degrees from center), or data are recorded by hand and then entered on score sheets along with all the other information gathered on each subject.

Usually, behavioral researchers collect substantial amounts of data and conduct complicated statistical analyses that often require computers. Then the recording system must match the needs of the computer programs to identify, store, retrieve, and analyze the data efficiently.

Once all the data are collected, data analysis begins. This text describes many simple statistical procedures, and more advanced texts present a wide variety of more complicated techniques. Two major requirements for the use of any statistical procedure are that the particular statistical test selected be one that will appropriately address the research question at issue and that the assumptions required by the particular statistical technique can be met in the context in which it will be applied.

CONCLUSIONS AND INTERPRETATIONS

The final step is to decide what conclusions can be drawn from the results and what implications those conclusions have for what is generally known about the topic being studied. Specifically, what statistically significant results were found? Can such results be attributed to the independent variables? What extraneous variables possibly could have produced these results instead? How do these results relate to information already reported in the literature? What general conclusions about the topic can now be drawn?

THE RESEARCH REPORT

If the study was successful—that is, if the results can be validly interpreted and if they contribute an important piece of new knowledge about the topic—then the scientist will write a report of the study and submit it for publication in a journal that specializes in that field.

Writing a Research Report

A research report follows a format that is roughly similar to the steps of research design described above. It consists of several sections:

1. **Introduction.** In this section (which usually has no heading) the past literature on the topic is reviewed briefly, often in approximately four to eight double-spaced manuscript pages. This review emphasizes the theoretical and practical importance of the topic of the study, what previous research has discovered about the topic, and why it is important to ask the questions that are posed in this research.

2. **Method.** This section, which begins with this heading, usually has two or more subsections. One subsection might include a description of the subjects, how they were selected, and their characteristics (such as gender, race, age, and so on). The operationalizing procedures are described in the next subsection—the operational definitions of the variables, the procedures for implementing the treatments of different groups, the methods used to assess reliability, and so forth.

3. **Results.** Here the results of all data analyses are presented, accompanied by the statistical details. While summary statements of the conclusions that follow from each analysis are appropriate, usually these conclusions are simply verbal statements similar to those presented in the chapters of this text after a statistical test has been conducted.

4. **Discussion.** Here broader conclusions are drawn and interpretations are made, limitations or possible confounds are considered, and a discussion of how the study contributes to the general body of knowledge on the topic is presented. It is only here in this final section that the researcher is permitted to depart from a strict description of what was actually done and observed to speculate on what the results might mean in broader terms. It is here, then, that generalizations to populations and speculations about possible causal mechanisms that might explain the results are made.

The details of this format can be found in the *Publication Manual of the American Psychological Association* for publishing in psychological journals, and in the manuals of other professional organizations for journals in those disciplines.

Evaluating a Research Report

Whether you actually conduct your own research or read the research reports of others, you must be able to evaluate the quality of research and to judge for yourself what can be accurately concluded. Typically it takes both practice and considerable knowledge of a field to become a shrewd critic, but below are a few guidelines to help you evaluate a research report.

1. **Is the research question important?** Will it influence theory, practical policies, or the understanding of a given behavior in a substantial way?

2. **Is the sample appropriate?** Does the sample faithfully reflect the characteristics of the intended population? Are certain groups or characteristics of individuals disproportionately represented? Would the results likely be different if the sample contained other groups or types of persons?

3. **Are the operational definitions of the major variables appropriate?** Would the results be different if those variables were defined differently, perhaps in ways more similar to the definitions used in other studies?

4. **Are the procedures appropriate?** Do they minimize the roles of possible extraneous variables, or do they introduce extraneous variables and permit confounding? What extraneous variables might account for the results?

5. **Are the statistics appropriate?** Do they address the questions they were intended to answer, and are the assumptions for each test met in these data?

6. **Are the conclusions appropriate?** Do they follow from the results of the study and the statistical analyses that were conducted? Are the implications and generalizations reasonable?

ETHICAL CONSIDERATIONS

Most research primarily benefits the researcher and ultimately, one hopes, society. With some notable exceptions, many research studies provide relatively little benefit to the individuals who cooperate in the study as "subjects," or the more contemporary label, "participants." As a result, the potential exists for participants to be unfairly and even unethically treated in the course of being observed or participating in a study in which the benefits to them personally may be minimal.

The ethics of conducting research is a broad topic ranging from how participants are treated during the observations to honesty and forthrightness in analyzing and reporting results. One of the more important and tangible ethical concerns involves submitting a plan for conducting the research to an **Internal Review Board,** a committee of scientific peers and representatives of potential participants established by the institution employing the researcher to review the research procedures for conformity to ethical principles. The purpose of the review, which is conducted before the research is begun, is to help the investigator conduct research in an ethical manner, protect participants from potential harm and abuse, and avoid misunderstanding, conflict, and lawsuits between the researcher, the institution, and the participants.

Generally, the researcher must submit a fairly detailed description of the procedures to be followed in the proposed research to the Internal Review Board. On the basis of that document, the Board must make several judgments, including whether the potential gain of the research outweighs its risks, whether subjects are adequately informed about the study to give knowledgeable consent to participate, and whether the rights of subjects, especially confidentiality, are protected.

Risk/Benefit

Most social and behavioral research contains essentially no risks for the participants that are associated specifically with the research procedures. The study may involve observing infants playing with different toys, asking adults their opinions of political issues or advertising layouts, or requesting college students to learn material under various circumstances. Except for the typical risks of everyday existence (e.g., driving to the research site, playing with toys), such studies are essentially free of potential dangers uniquely associated with the research procedures.

However, showing children violent television programs, for example, may not be risk free. Scientists may judge that, in contrast to the state of our knowledge several decades ago when such studies were first conducted, we now know as a result of those studies that viewing violent television programs promotes aggressive behavior, at least to some extent in some children. The Internal Review Board may decide that researchers cannot show young children vivid, lifelike, violent television programs, even though we also know that many children watch such programs at home when they are broadcast on commercial television. The fear may be that if a child happens to commit a severe aggressive act after the experiment, such as seriously harming another child, it could be alleged and substantiated that the experimental experience caused the child to commit that act and the researcher and the employing institution would be responsible for its consequences. Thus, some of the examples presented in this chapter, depending upon the procedures actually proposed, might be judged to be too risky and not approved. To minimize such claims, researchers often propose to use procedures that are very common in society, even though potentially dangerous. For example, most cartoons aimed at child audiences are quite violent, more so than most live dramas created for adult viewers, and young children watch a good number of such cartoons. With appropriate parental permission (see below), showing children cartoons may be judged to have some risk, but it is such a common experience in the lives of children that it does not constitute an unusual risk. Even showing violent lifelike programs that are aimed at children (e.g., "Batman") may be judged to represent a "common" risk and may be approved.

Such risks must be judged by the Board in the context of the potential gain to society of the research results. At issue may be the fact that such programs are watched by millions of children, and research information is needed to document their potential harm or benefit (it was once argued that watching such programs provided therapeutic release of aggressive tendencies) for the purpose of recommending and substantiating changes in television broadcasting regulations and to provide sound advice to parents. Weighing the possible benefit of a specific research project against its potential risks is usually a complex, difficult, apples–and–oranges decision.

Informed Consent

A major component of the research proposal to the Internal Review Board is a description of how informed consent will be obtained. This usually consists of creating a form to be signed by the participant and/or parent of a participant, called a **signed consent** or **signed release,** that briefly states the purpose of the research, what the participant will experience or be asked to do under what conditions, how long the procedures will take, what will be measured or recorded, and what will be done with the information after the observations are completed. The subject is usually assured that he or she may decline to participate in any part of the observation, withhold any information that is requested, and withdraw completely at any time.

For most studies, such information is benign and unobjectionable. But some purposes and procedures common in social and behavioral research may be resisted by potential participants. Some parents, for example, may not want their children to see cartoons because of their violent content; some members of minority groups may not want to contribute to studies of intelligence, teenage pregnancy, or delinquency, for example, that they fear will portray members of their group in an unfavorable light; and some potential participants may not want to be "randomly assigned" to one or another treatment (e.g., comprehensive services for low-income families versus no special services) when they perceive that one treatment is substantially more desirable. The requirement for informed consent has also all but eliminated the use of deception in social and behavioral research.

The signed release also must contain a statement that accurately and completely—but briefly—states the known and suspected benefits and risks of participating. Typically, there are no unusual risks or benefits, or the possible risks and benefits of a new treatment or program, for example, are not known or are presumed beneficial. In the case of the television example, however, the researcher would be required to inform potential participants that viewing cartoons may be associated with increased aggressive behavior in some children. Some parents may refuse to participate, and this may alter the representativeness of the sample actually obtained.

Confidentiality

Another major component of the signed consent form deals with protecting the participants' privacy. A standard assurance is that no identifying information (e.g., the subject's name) will be collected or, if obtained, such identifying information will be separated from the other information acquired during the study and the identification key (i.e., a list identifying each subject number with the name of the participant) will be known only to the researcher, stored in a locked file, and released to no one except by court order. Publications and presentations of results will communicate information on groups of participants, not individuals, or presentation of case studies will not contain identifying information.

Most research involves no information of a delicate nature—that is, information that people would not want someone else to know about them. However, studies of certain therapeutic treatments, for example, typically involve information about personal problems (e.g., drug and alcohol abuse, psychopathology, child abuse) that must be kept in strictest confidence and known by as few people as possible. Also, information about educational history and income levels may be considered personal and confidential, especially by potential participants who have especially low or high levels of these factors. And the frequent use of videotape, which is explicitly identifying, has raised additional concerns about who sees the tapes, under what circumstances, for what purposes, and for what period of time.

Honest and Forthright Reporting

Although not covered in the signed consent, another major ethical domain concerns the requirement for honest and forthright reporting of research procedures, including statistical procedures and results. Most research is conducted quite privately, so the researcher and possibly an assistant might be the only ones who would know about the fabrication or alteration of data, the selection of subjects that favor a given result, the judicious selection of statistical procedures that produce desired outcomes, and the deliberate omission of procedures or statistical tests that produced evidence that conflicted with the main result.

The vast majority of scientists are honest. But science is nothing if it cannot be trusted, so every research university and the federal government have established procedures to investigate allegations of dishonesty and to recommend sanctions and punishments. These **scientific integrity procedures,** which are conducted after an allegation of dishonesty is made, can be extraordinarily thorough. They may involve a committee of scientific peers that reviews the justification for each of the numerous major decisions that a researcher makes during any study, checks the raw data, statistically reanalyzes all the data, and studies the faithfulness of the reporting of procedures and results. If a judgment of **scientific misconduct** is rendered, the sanctions can destroy the trust and confidence of the scientific community in the guilty scientist, and punishments can include being barred from obtaining grants to support research and termination of employment. It can be science's "death penalty."

SUMMARY

Research design consists of the methods scientists use to make observations that will produce empirical information, usually about the relationship between two or more things. Five basic steps characterize research design: (1) formulating a scientific question, (2) operationalizing the "things" of interest and their relationship, (3) collecting data, (4) analyzing data, and (5) interpreting the results. The "things" of research are variables, specifically a dependent variable whose values are thought to "depend" on the independent variable. Relationships between variables may be causal or noncausal, and observational research tends to discover whether a relationship exists, whereas experimental research seeks to determine whether a given relationship is causal. In a simple experiment, an experimental group is given the treatment of interest, while a control group is given the same experience except for the treatment of interest. The skill in designing research is to minimize confounding caused by the influence of extraneous variables (rather than of the independent variable) that might produce the observed results. In this vein, it is sometimes helpful to conduct a double-blind experimental design in which neither the subjects nor the observers know to which group an individual subject belongs. Research must be conducted in an ethical manner, and Internal Review Boards decide before research is conducted that the benefits outweigh any risks and that participants will be given truly informed consent and have their privacy and confidentiality protected. Scientific integrity procedures are institutional methods that guide the investigation of allegations of scientific misconduct.

Exercises

For Conceptual Understanding

1. What are the five basic steps of research design?

2. Distinguish between the independent and dependent variables.

3. What is the association between causal and noncausal relationships on the one hand, and observational and experimental research on the other?

4. Give several reasons why the literature review is an important part of research design.

5. Distinguish between situational and subject parameters.

6. What are operational definitions and why are they important?

7. Distinguish between reliability and validity.

8. Why is random assignment of subjects to experimental and control groups crucial to interpreting the results of experiments?

9. Give an example of confounding.

10. Define a double-blind experiment, and explain why it is important in determining whether an observed relationship is causal.

11. Explain the value of conducting a pilot experiment.

12. What are the sections of a research report?

13. Discuss some of the ethical concerns and procedures in conducting and reporting research.

For Homework

14. Distinguish between a variable and its operational definition, and give three examples of each.

15. Distinguish between an independent and a dependent variable, and give three examples of each.

16. Define and distinguish between the reliability and the validity of a measurement. Illustrate with a concrete example.

17. Distinguish between causal and noncausal relationships, and give three concrete examples of each.

18. Compare and contrast the observational and experimental approaches to a research issue. Cite the advantages, disadvantages, and limitations of each strategy. Illustrate with concrete examples.

19. What is a control group? Give three concrete examples.

20. What are two attributes of a good sample?

21. Why is random assignment of subjects to groups important in designing and interpreting experimental research?

22. Discuss and give examples of several kinds of possible confounding.

23. Discuss why showing children typical television cartoons may or may not be considered unethical research.

Topics in Probability

he concept of probability is basic to statistical inference. This chapter presents an introduction to the formal study of probability.

SET THEORY

Sets and Relations among Sets

Probability is best understood in terms of set theory. Therefore, some elementary definitions and operations of set theory will be presented first.

> A **set** is a well-defined collection of things, typically objects or events.

The ordinary concept of a set, such as a set of drinking glasses or a set in tennis, is quite similar to the mathematical notion of a set. The crucial factor in the definition is that there be some quality that the objects in the set possess or some rule that they follow which defines them as members of the set and all other objects or events as not members of the set. For example, husbands are a set. Unmarried males, married females, and giraffes, for example, are not members of the set of husbands. If he is not married, if he is a she, if he is a giraffe, then he is not a member of the set of husbands. Thus, a set is a collection of objects or events that are distinguishable from all other objects on the basis of some particular characteristic(s) or rule. It is customary to label a set by some capital letter, such as A, B, or C. If the set is all positive even numbers less than 7, then 2, 4, and 6 are the elements of this set, which may be written by listing the elements within brackets, such as $\{2, 4, 6\}$.

> An **element** of a set is any one of the set's members.

For example, if Jim Simpson, a male, is married, then he is an element of the set H of husbands.

A very important concept for the study of probability is that of subset.

> If every element in set A is also an element of set B, then A is a **subset** of B.

Since the phrase *is a subset of* is rather long and cumbersome, it is customary to write it with the symbol $\subseteq$. A subset, like a set, is represented by a capital letter. So $A \subseteq B$ symbolizes the idea that A is a subset of B. For example, married fathers, which we might label M, is a subset of husbands, H, which can be represented by $M \subseteq H$.

There are two special sets that should be mentioned, the universal set and the empty set.

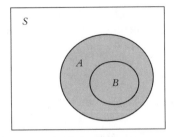

Figure 12-1 Set–subset relationship. *S* is all college students (universal set), *A* is college students at State (set), and *B* is a sample of the students at State (subset).

The **universal set** includes all things to be considered in any one discussion. It is symbolized by **S.**

The **empty set (null set)** contains no elements. It is symbolized by ∅.

The universal set is simply an inclusive set that defines the general type of objects or events being discussed. A set is a subset of this universal set; that set in turn may have subsets and elements. For example, one of the primary applications to statistics of the concepts of set and subset is their analogy to the concepts of **population** and **sample** (see pages 82–84). A sample is a subset of the population, or universal set. In Figure 12-1, one might consider the entire space, *S* (the universal set), as the population composed of all college students. Perhaps the sample is composed only of college students attending State University and is symbolized by *A*. Further, the researcher randomly selects only a few college students at State, symbolized by *B*. Thus, $A \subseteq S$, $B \subseteq S$, and $B \subseteq A$.

If *S* is the entire space and *A* is a set in *S*, then the symbol **A',** read "not A," denotes the set of all elements in *S* that are not in *A*. *A'* is called the **complement**[1] of *A*.

Graphically, *A'* in Figure 12-2 is the complement of *A*. In another example, if *S* includes all heterosexual married individuals, then all husbands, *H*, is a subset of *S*, and the complement of *H*, symbolized by *H'*, includes all wives. The complement of the set of all husbands and wives in *S* is the empty set, ∅.

---

[1] Other symbols that are sometimes used for the complement of *A* include $\bar{A}$, $\tilde{A}$, A^c, and $\sim A$.

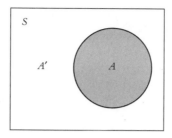

Figure 12-2 Complementation. *S* is the universal set, *A* is a set in *S*, and *A'* (not *A*) is the complement of *A* since it contains all the elements in *S* that are not in *A*.

If every element in *A* is also an element of *B* and if every element in *B* is also an element of *A*, then *A* and *B* are **equal sets** and ***A* equals *B*:**

If $A \subseteq B$ and $B \subseteq A$, then $A = B$.

If $A = \{1, 2, 5, 9\}$ and $B = \{5, 1, 9, 2\}$, then $A = B$. Note that the order in which the elements of *A* and *B* are listed within a set is irrelevant.

Operations

An important operation is the **union** of two sets.

Given two sets, *A* and *B*, the **union** of *A* and *B* is the set of all elements that are in *A*, in *B*, or in both *A* and *B*, which is written ***A* ∪ *B*.**

The word *union* is symbolized by ∪, so that the union of *A* and *B* is written $A \cup B$ and read "*A* union *B*." The union of all husbands (*H*) and wives (*W*) is all married persons (*M*), or $H \cup W = M$.

In Figure 12-3, the gray-shaded area represents the union of *A* and *B*, symbolized $A \cup B$. The crucial word to remember about the concept of union is *or*. The criterion for including any element in the union of *A* and *B* is whether that element is contained in *either* set *A* *or* set *B*. For example, if $A = \{1, 2, 3, 4, 5\}$ and $B = \{3, 4, 5, 6, 7, 8\}$, then $A \cup B = \{1, 2, 3, 4, 5, 6, 7, 8\}$. Notice that numbers contained in both *A* and *B* are not represented twice in $A \cup B$. To give another example, if a warning on a bottle of medicine says to call a doctor if a fever or vomiting occurs, then one should call the doctor if the union of fever (*F*) and vomiting (*V*) occurs, that is, $F \cup V$.

Another operation on sets is that of **intersection.**

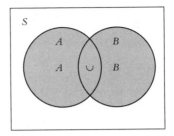

Figure 12-3 Union. *A* union *B* (*A* ∪ *B*) is represented by the gray-shaded area. *A* ∪ *B* contains all elements in *A*, in *B*, or in both *A* and *B*.

> Given two sets, *A* and *B*, the **intersection** of *A* and *B* contains all elements that are in both *A* and *B*, but not in *A* or *B* exclusively, which is written **A ∩ B.**

The symbol for intersection is ∩, and "*A* intersection *B*" is written $A \cap B$. The intersection of all husbands younger than 30 years and all husbands older than 20 years is all husbands between 20 and 30 years of age. The intersection of all husbands and wives includes no one—that is, it is the empty set. In Figure 12-4, which illustrates the intersection of *A* and *B*, $A \cap B$ is the gray-shaded portion of the diagram. Thus, if $A = \{1, 2, 3, 4, 5\}$ and $B = \{3, 4, 5, 6, 7, 8\}$, then $A \cap B = \{3, 4, 5\}$.

Here the emphasis is on the word *and*, because for an element to be a member of the intersection of *A* and *B*, it must be contained in *both A and B.* The intersection represents the common portion of two sets, or the elements shared by two sets. Winning a four-team, single elimination tournament, for example, requires that a team win both the first game (*F*) and the second game (*S*), that is, $F \cap S$.

It is useful to be more explicit about the criteria for union (∪) and intersection

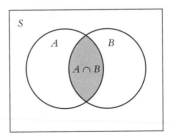

Figure 12-4 Intersection. *A* ∩ *B* is represented by the gray-shaded area, which includes all elements that are members of both *A* and *B*.

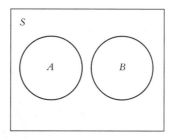

Figure 12-5 Disjoint sets.
$A \cap B = \varnothing$, that is, A and B
have no common elements.

($\cap$). An element belongs to the **union** of A and B if it qualifies under *any one* of the following criteria: (1) it is a member of A, (2) it is a member of B, or (3) it is a member of both A and B. However, in determining whether an element qualifies for inclusion in the **intersection** of A and B, (3) is the only criterion: An element must be a member of *both* A and B. The difference between the important words in the definitions also highlights the distinction between union and intersection. $A \cup B$ contains elements that are in A *or* B *or* in both A and B, whereas $A \cap B$ contains only those elements that are in both A *and* B simultaneously. Notice that elements in the intersection of A and B are a subset of the elements in the union of A and B.

There is a special case of intersection that defines a particular relationship between two sets. Suppose $A \cap B$ contains no elements, that is, $A \cap B = \varnothing$. Such a condition says that A and B share no common elements. No element in A is also in B and no element in B is also in A.

> Two sets, A and B, are said to be **disjoint sets** if no element in A is also in B and if no element in B is also in A. In symbols, if $A \cap B = \varnothing$, then A and B are disjoint.

Husbands and wives are two disjoint sets because no element is common to both. For most purposes, plants and animals are disjoint and so are fruits and vegetables (although biologists might suggest there exist some forms that have many properties of both).

Figure 12-5 illustrates a pair of disjoint sets. Another pair of disjoint sets is A and its complement, A'. Since A' contains all elements not in A, it is clear that for any set A, A and A' are disjoint.

A summary of the terms and concepts discussed in this section is presented in Table 12-1.

SIMPLE CLASSICAL PROBABILITY

Probability usually involves thinking about an **idealized experiment** in which a given phenomenon is repeatedly observed an indefinite number of times under ideal con-

12-1   Terminology of Set Theory

1. **Set:** A well-defined collection of things.
2. **Element:** A member of a set.
3. **Universal set, *S*:** The set of all things under discussion.
4. **Empty set, $\emptyset$:** A set containing no elements.
5. **Subset, $A \subseteq B$:** If every element in A is also in B, then A is a subset of B.
6. **Complement, A':** Set A' is the complement of set A if it contains every element in S that is not in A.
7. **Equality, $A = B$:** Set A equals set B if every element in A is also in B and every element in B is also in A. That is, if $A \subseteq B$ and $B \subseteq A$, then $A = B$.
8. **Union, $A \cup B$:** The union of sets A and B is the set of all elements that are in A, in B, or in both A and B.
9. **Intersection, $A \cap B$:** The intersection of A and B is the set of all elements that are in both A and B.
10. **Disjoint sets:** Two sets are disjoint if they share no common elements, that is, $A \cap B = \emptyset$.

ditions. For example, when someone says the chance of flipping a coin and obtaining a head is $\frac{1}{2}$, this means that over an uncountable number of flips of that coin under fair and ideal conditions, heads will turn up half the time. This imagined flipping of a coin over and over again is the idealized experiment. Similarly, when the weather forecaster says the chance of rain today is 65%, this means that it will rain on 65% of an uncountable number of days with circumstances just like today. The "idealized experiment" is the uncountable set of days like today.

In the study of probability, the concepts of set theory summarized above are given special names. Specifically, the collection of all the elements of an idealized experiment is called the **sample space,** which corresponds to the universal set and is symbolized by *S*. The elements of this set are known as **elementary events,** or **outcomes.** An **event** is any subset of elements in *S*. Therefore, an event may be an elementary event, a subset of elementary events, or all elementary events (that is, *S*). Thus, in the idealized experiment of rolling a single die, the sample space consists of the set of six elementary events (outcomes): {1, 2, 3, 4, 5, 6}. Examples of an event in this idealized experiment include a roll of a deuce {2}, a roll of three or less {1, 2, 3}, and a roll of an even number {2, 4, 6}. In the case of the weather forecast, the sample space may consist of only two elements, rain and no rain.

Simple classical probability involves making probabilistic statements about a sample space whose elementary events are all equally likely to occur. That is why discussions of probability frequently employ the flipping of coins, drawing of cards, and rolling of dice, because each of these experiments consists of a set of equally likely elementary events. This is often *not* the case for common, everyday events. For example, while people often think that the likelihood is equal of giving birth to a boy or girl, this is not so (that is, boys are slightly more likely than girls). Further, depending upon a couple's sexual behavior, especially the frequency and timing of intercourse during the woman's menstrual cycle, the likelihood of a boy may be much higher or lower for a specific couple than the general average. Similarly, horses in the Kentucky Derby are not equally likely to win, and the betting odds are one expression of that inequality. Nevertheless, much of probability theory is based upon equally likely events (although advanced procedures exist to deal with unequally likely events).

> In a sample space containing equally likely elementary events, the **probability** of a given event A is defined to be the number of outcomes in A divided by the total number of outcomes in the sample space, S. If #(A) signifies the number of outcomes in A and #(S) symbolizes the number of outcomes in S, then $P(A)$, the "probability of event A," is given by
>
> $$P(A) = \frac{\#(A)}{\#(S)}$$

In the experiment of tossing a coin, the sample space consists of the two equally likely outcomes, heads and tails, and the probability of a head is

$$P(\text{head}) = \frac{\#(\text{head})}{\#(\text{possible outcomes})} = \frac{1}{2}$$

In the idealized experiment of rolling a die, the sample space consists of the equally likely alternatives $\{1, 2, 3, 4, 5, 6\}$, and the probability of rolling a two is $\frac{1}{6}$. The probability of five is also $\frac{1}{6}$. The probability of a four or less is $\frac{4}{6}$. Similarly, when a medicine bottle says that 5% of the people who take this medicine experience nausea, it means that the probability is $\frac{5}{100} = \frac{1}{20} = .05$ that one will become nauseated as a result of taking this medicine. This statement assumes that every person has an equal likelihood of developing such side effects.

It is important to remember that in any situation, the probability of an event can be neither less than .00 nor greater than 1.00. A probability value always equals .00 or 1.00 or falls between those two values.

Conditional Probability

Conditional probability refers to a situation in which the probability of a given event depends upon the occurrence or nonoccurrence of another event.

Consider the following situation. Suppose one takes the 12 picture cards (4 jacks, 4 queens, and 4 kings) from a deck of playing cards and shuffles them. Define event A as drawing a king from this deck. Since there are four kings in the deck of 12 cards, the probability of event A is

$$P(A) = \frac{\#(A)}{\#(S)} = \frac{4}{12} = \frac{1}{3}$$

Suppose that whatever card is drawn on this trial is *not* put back into the deck before making a second draw. Such a procedure is known as **sampling without replacement**, because cards are not replaced into the deck after their selection. Now consider the probability of an event *B*, obtaining a king on the second draw. This probability depends upon the result of the first draw and thus is an instance of conditional probability. If a king was drawn first and not replaced, then for the second draw there would be three kings left in the deck, which now consists of 11 cards. Thus, $P(B) = \frac{3}{11}$. However, if a card other than a king was drawn on the first trial, all four kings would exist in the remaining deck of 11. Then, $P(B) = \frac{4}{11}$.

> **Conditional probability** is the probability that an event will occur given that some other event has already occurred. The probability that event *B* will occur given that event *A* has already occurred is written
>
> $P(B|A)$
>
> which is read "the probability of *B* given *A*."

In the example above, the probability of *B* (drawing a king on the second draw without replacement) is conditional upon the occurrence of event *A* (drawing a king on the first draw). Given that *A* has indeed occurred, the probability of *B* is written

$$P(B|A) = \frac{3}{11}$$

since three kings are left in the deck of 11 remaining cards after one king has been selected on the first trial. A more commonplace example is the likelihood of whether a married couple tries to have a second or especially a third child. This event is conditional in part on the gender of the existing children. A couple is more likely to try for a third child, for example, if their first two children are girls than if they are both boys or one of each gender. Whether a couple decides to have another child is "conditional" on the number and gender of the children they already have.

Dependent and Independent Events

The above example also illustrates the concept of dependent events.

> Two events are said to be **dependent** if the occurrence of one event alters the probability of the occurrence of the other.

Since the probability of obtaining a king from the special set of 12 cards on the second draw (event *B*) is either $\frac{3}{11}$ if a king was drawn on the first trial (event *A*) or $\frac{4}{11}$ if a nonking was picked first, events *A* and *B* are said to be dependent: The probability of *B* depends on the occurrence or nonoccurrence of event *A*. Similarly, the gender of the first two children and the decision to have a third child are dependent events.

> Two events are said to be **independent** if the occurrence of one event does not alter the probability that the other event will occur. This can be expressed in terms of conditional probability as
>
> $$P(B|A) = P(B)$$

When $P(B|A) = P(B)$, this implies that the probability of B is not affected by (that is, is not conditional upon) the occurrence of A.

In the example given above for conditional probability, cards are drawn from a 12-card deck without replacement. But suppose cards are replaced after a selection is made, a circumstance called **sampling with replacement.** Now consider the probability of event B, obtaining a king on the second draw, as a function of the occurrence of event A, obtaining a king on the first draw. If a king is selected first and put back into the deck before the second draw, the probability of getting a king on the second draw is $\frac{4}{12}$, since there will be four kings in the complete deck of 12 cards. If a king is not found on the first draw and the card is replaced, there are again four kings in 12 cards. Thus, the probability of getting a king on the second draw is independent of (that is, is not conditional upon) the outcome of the first event. In symbols,

$$P(B|A) = P(B) = \frac{4}{12},$$

which means that the probability of B given that A has occurred is simply the probability of B, ignoring what has happened with respect to event A.

An interesting application (or lack of application) of the concept of independence is known as the **gambler's fallacy.** The gambler's fallacy is the failure to appreciate the independence of some sequential events. Often, people who play games of chance, such as the lottery, believe that they have a better than usual probability of winning after a long losing streak. Surely, they "reason," their luck should change and that having lost so often in the past means that their chance of winning is better on the next game. People also sometimes feel after winning a few times that they are on a "hot streak" or that they "can't lose," as if their chance of winning is somehow greater on the next game because of their past good luck streak. But these beliefs are fallacies. If the game is truly chance, such as the lottery and honest gambling games, the probability of winning is the same on each occasion and is not dependent on previous performance. That is, each game is independent of the next.

But, while games of chance are likely to be composed of independent events, many other commonplace events are not. For example, if a basketball team loses 10 games in a row before facing its major rival, a school of approximately equal ability, the probability of winning the game may be more than .50 if the team is high-spirited and determined to salvage an otherwise poor season: "We have nothing to lose; let's give it all we've got." Of course, the probability of winning could be less than .50 if the 10 losses have so demoralized the team that it can hardly face the humiliation of another game. This is partly what many athletes mean when they say their sport is "mostly mental," or when a football announcer declares that "the momentum has shifted to Pittsburgh." Thus, although the independence of events is an important characteristic in assessing probabilities, in practice it is sometimes difficult to determine whether events are actually independent or not.

Mutually Exclusive Events

Two events may be mutually exclusive.

> Two events, *A* and *B*, of the same idealized experiment are considered to be **mutually exclusive** if they share no elementary events, that is, if *A* and *B* are disjoint sets ($A \cap B = \emptyset$).

Two events are mutually exclusive if their sets are disjoint. If *A* and *B* are mutually exclusive, that is, if $A \cap B = \emptyset$, then if *A* has occurred, *B* has not. Consider the event *A* of drawing a king on a single draw from that special deck composed only of four jacks, four queens, and four kings. Abbreviate the cards by giving them two initials, the first for the denomination and the second for the suit (for example, *KS* for the King of Spades). Since event *A* is drawing a king, set *A* contains four elements (all the kings):

$A = \{KS, KC, KH, KD\}$

Let event *B* be drawing a queen in a single draw. Set *B* also contains four elements:

$B = \{QS, QC, QH, QD\}$

Note first that there is no element in *A* that is also in *B*. Therefore, *A* and *B* are disjoint sets so $A \cap B = \emptyset$. Note also that if on a single draw *A* occurs, *B* has not occurred; if you draw a king, you do not also draw a queen. In this case, drawing a king and drawing a queen are mutually exclusive events.

In contrast, consider now the events of drawing a king (event *A*) and drawing a red card (event *B*) from this special deck:

$A = \{KH, KD, KS, KC\}$
$B = \{KH, KD, QH, QD, JH, JD\}$

These events are not mutually exclusive because they have two common elements, $\{KH, KD\}$; therefore $\#(A \cap B) = 2$, not 0, and *A* and *B* are not disjoint. Further, it is not necessarily true that if *A* occurs, *B* does not occur. If the king of hearts or king of diamonds is drawn, then both event *A* and event *B* have simultaneously occurred. More commonly, winning or not winning a tennis tournament, having a boy or a girl baby, and catching a fish today or not catching a fish are three examples of pairs of mutually exclusive events.

Although mutually exclusive events may appear to be very similar to independent events, the two concepts should not be confused. The difference between them is that *mutually exclusive* refers to not sharing *elements of sets,* whereas *independence* is defined in terms of the *probabilities* of two events. Therefore:

Mutually exclusive events means $A \cap B = \emptyset$
Independent events means $P(B|A) = P(B)$

For example, if you are on a photographic safari in Africa, seeing a lion and seeing a leopard are mutually exclusive events (i.e., seeing one animal type never constitutes a case of seeing the other type), but these two events may not be independent if, for example, lions and leopards tend to inhabit the same areas of the savanna.

Independent events are important in sampling and hypothesis testing. Many statistical tests require that the subjects in the research study are independent—that is, that each subject in the population has an equal likelihood of being sampled and that the probability of any subject being sampled is not dependent (or conditional) on any other subject being sampled. The concept of mutually exclusive events is used when forming the hypotheses to be tested statistically. The possible outcomes allowed under one hypothesis (for example, the null hypothesis) must be mutually exclusive of the possible outcomes allowed under the other hypothesis (such as, the alternative hypothesis). This is required so that the outcome will conform to only one of the two hypotheses, because no outcome is included in both hypotheses.

PROBABILITY OF COMPLEX EVENTS

Probability of Intersection, $A \cap B$

What is the probability of $A \cap B$?

> The probability that two events, A and B, will both occur is
>
> $$P(A \cap B) = P(A)P(B|A)$$

This means that the probability that both A and B will occur is the probability of A times the probability of B given that A has occurred.

In the special case when A and B are independent events, $P(B|A) = P(B)$. This is the definition of independent events. In such a case, substituting $P(B)$ for $P(B|A)$ in the above expression, we have the following:

> The probability that two *independent* events, A and B, will both occur is
>
> $$P(A \cap B) = P(A)P(B)$$

For example, recall the deck of 12 face cards used in the previous example and consider determining the probability of drawing 2 kings in succession without replacement. Restated, this amounts to finding the probability of obtaining a king on the first draw (event A) *and* obtaining a king on the second draw (event B). Thus, the task is to compute the probability that both A and B will occur, which can be symbolically stated in terms of their intersection:

$$P(A \cap B)$$

The formula given above states that the required probability will be the product of the probability of A times the conditional probability of B given that A has occurred. The probability of drawing a king on the first trial is $P(A) = \frac{4}{12}$, since there are 4 kings in the deck of 12 cards. The conditional probability of drawing a king on the second trial given that a king has already been drawn and not replaced is $P(B|A) = \frac{3}{11}$ (three kings would remain in the deck, which now contains 11 cards). Thus, the required probability is

$$P(A \cap B) = P(A)P(B|A)$$
$$P(A \cap B) = \frac{4}{12} \cdot \frac{3}{11} = \frac{12}{132} = .09$$

Thus, on the average one could expect to draw two consecutive kings without replacement only 9 times in 100 such two-card draws. The same logic and strategy apply to real card games (for example, the likelihood of being dealt four straight aces or drawing two aces in a row from the deck in a poker game), but the calculations are more involved because the deck is bigger and the events more complex.

Why are the probabilities multiplied? Consider the solution of the above problem in more detail. The task is to determine the probability of $A \cap B$, which by the laws of classical probability should be given by the number of ways $A \cap B$ can occur divided by the total number of two-card sequences in the sample space. In symbols,

$$P(A \cap B) = \frac{\#(A \cap B)}{\#(S)}$$

First, how many ways can $A \cap B$ occur? Simply stated, there are four ways a king could be obtained on the first draw (one way for each king in the deck), and *for each one of those four ways* there exist three possible ways a king could be selected on the second draw given that one was picked on the first draw. Therefore, there are four threes, or 4×3 ways of getting two consecutive kings without replacement. The 4 is the $\#(A)$ and the 3 is the $\#(B|A)$, so

$$\#(A \cap B) = \#(A) \times \#(B|A) = 4 \times 3 = 12$$

These 12 ways of obtaining two kings are listed in Table 12-2.

Second, how many total possible outcomes are there in this situation? The logic is the same: There are 12 possible outcomes on the first draw (one for each card in the deck), and *for each one of these 12 outcomes* there are 11 possible results on the second draw (one for each of the remaining 11 cards). Thus,

$$\#(S) = \#(S_A) \times \#(S_{B|A}) = 12 \times 11 = 132$$

where S_A and $S_{B|A}$ are the sample spaces for events A and $B|A$, respectively.

The probability in question is found by dividing the number of specified outcomes by the total number of possible outcomes in the idealized experiment. As determined above, this amounts to

$$P(A \cap B) = \frac{\#(A \cap B)}{\#(S)} = \frac{\#(A) \times \#(B|A)}{\#(S_A) \times \#(S_{B|A})} = \frac{4 \times 3}{12 \times 11} = \frac{12}{132} = .09$$

$$P(A \cap B) = P(A) \times P(B|A) = \frac{4}{12} \times \frac{3}{11} = \frac{12}{132} = .09$$

To summarize, the first part of the preceding expression is the probability of event A, $P(A)$, and the second part is the conditional probability of B given A, $P(B|A)$. The probability of the intersection of two events equals the probability of A times the conditional probability of B given A. In the case being illustrated, the probability of drawing two consecutive kings is $\frac{4}{12} \times \frac{3}{11} = \frac{12}{132} = .09$. This means that if you were repeatedly to draw two consecutive cards from the special deck, on the average you might expect that 9 of every 100 such pairs would be pairs of kings.

Turning to a different example, suppose you arrange to play a game of racquetball with a classmate. Your partner warns that arranging a 5 P.M. game is chancy, since her afternoon laboratory classes run late about 50% of the time. Moreover, the probability is only .60 that a court will be available at 5 P.M. What is the probability that the two of you play your game at 5 P.M.?

Define event A to be "your partner is able to play at 5 P.M.," and event B to be

12-2 | **Number of Ways of Drawing Two Consecutive Kings ($A \cap B$)**

First Draw	Second Draw	Number of Ways
King of spades	King of clubs King of hearts King of diamonds	3
King of clubs	King of spades King of hearts King of diamonds	3
King of hearts	King of spades King of clubs King of diamonds	3
King of diamonds	King of spades King of clubs King of hearts	3
[First draw, (A)]	× [Second draw, ($B|A$)]	= [Total, ($A \cap B$)]
4	× 3	= 12

"a court is available." Thus, the probability of the complex event, having a partner *and* a court free, is

$$P(A \cap B) = P(A)P(B|A)$$

Since the probability of a free court is not altered by whether your partner can play, A and B are independent, and $P(B|A) = P(B)$, which in this case is .60. $P(A)$ is the probability that your partner can play, which is .50, so

$$P(A \cap B) = P(A)P(B|A) = P(A)P(B) = (.50)(.60) = .30$$

In words, the likelihood that you can play a racquetball game with your partner at 5 P.M. is .30, or on 3 of every 10 such afternoons.

Note that one needs to remember only that

$$P(A \cap B) = P(A)P(B|A)$$

This formula allows for the independence of A and B, in which case $P(B|A) = P(B)$. The real problem in using this expression is translating the verbal statement of the problem into events A and B, and then determining whether the intersection of A and B is required. Most often, the intersection of A and B is needed if the problem states (or could be restated to say) that *both A and B* must occur to satisfy the complex condition. The key word is *and:*

1. You must draw a king on the first try *and* on the second.
2. Your partner must finish in time *and* the court must be free.

The word *and* signifies that the joint occurrence of A and B is required, and thus $A \cap B$ is an appropriate description of the problem.

The formula

$$P(A \cap B) = P(A)P(B|A)$$

also provides an expression for the conditional probability of B given A. Dividing by $P(A)$ and transposing, we have

$$P(B|A) = \frac{P(A \cap B)}{P(A)}$$

Therefore, the **conditional probability** of B given A equals the probability that A and B will both occur divided by the probability of A. In symbols,

$$P(B|A) = \frac{P(A \cap B)}{P(A)}$$

If the probability that your team will win a two-game basketball tournament is .15 [$P(A \cap B) = .15$] and the probability of winning your first game is .50 [$P(A) = .50$], then before the tournament begins you can speculate that the probability that you win the tournament given that you win the first game is

$$P(B|A) = \frac{P(A \cap B)}{P(A)} = \frac{.15}{.50} = .30$$

Probability of Union, *A* ∪ *B*

> The probability of the occurrence of either one of two events, *A* or *B*, that is, *A* ∪ *B*, is
>
> $$P(A \cup B) = P(A) + P(B) - P(A \cap B)$$

This states that the probability that either *A* or *B* will occur equals the probability of *A* plus the probability of *B* minus the probability of both *A* and *B*. However, the probability of *A* ∩ *B* is zero if *A* and *B* are mutually exclusive. (This is because *A* ∩ *B* = ∅ and *P*(∅) = .00.) Therefore, if *A* and *B* are mutually exclusive events, the preceding formula is simplified:

> If *A* and *B* are mutually exclusive, the probability that either *A* or *B* will occur is
>
> $$P(A \cup B) = P(A) + P(B)$$

This states that if *A* and *B* are mutually exclusive the probability that either *A* or *B* will occur is the sum of the probability of *A* with the probability of *B*.

Return to the special deck of 12 face cards. What is the probability of selecting either a king or a red card in a single draw? If event *A* is drawing a king and event *B* is drawing a red card, then the question is: what is the probability of *A* or *B*, which can be represented as the probability of the union of these events, $P(A \cup B)$.

The longer formula given above requires three probabilities. The probability of getting 1 of the 4 kings in a deck of 12 cards is $P(A) = \frac{4}{12}$. Since half the cards are red, $P(B) = \frac{6}{12}$. The probability of *A* ∩ *B* is simply the likelihood that both a king and a red card turn up, which will happen if either the king of hearts or king of diamonds is selected—2 of the 12 cards. Thus, $P(A \cap B) = \frac{2}{12}$. Therefore, the required probability is

$$P(A \cup B) = P(A) + P(B) - P(A \cap B)$$
$$= \frac{4}{12} + \frac{6}{12} - \frac{2}{12} = \frac{8}{12}$$
$$= .67$$

But what is the rationale of this formula? Why does one add the probabilities of *A* and *B*, and why must the probability of their intersection be subtracted? Consider the example in more detail. The problem is to determine the probability of the union of *A* and *B*, and according to the laws of classical probability this should equal the number of outcomes in *A* ∪ *B* divided by the total number of outcomes on a single draw:

$$P(A \cup B) = \frac{\#(A \cup B)}{\#(S)}$$

Using the same abbreviations as before, the possible outcomes for events A (a king) and B (a red card) are

$$\#(A) = \#(KH,\ KD,\ KS,\ KC) = 4$$
$$\#(B) = \#(KH,\ KD,\ QH,\ QD,\ JH,\ JD) = 6$$

For an outcome to qualify for $A \cup B$ it must be either in A or in B. If one simply adds the outcomes in A to those in B, some outcomes will be counted twice—namely, those that are in both A and B, that is, in $A \cap B$. In this example, the outcomes in the intersection of A and B are KH and KD. If these shared outcomes are subtracted from the sum, the correct number of elements in $A \cup B$ is produced. Specifically,

$$\underbrace{\#(A)}_{(KH + KD + KS + KC)} + \underbrace{\#(B)}_{(\cancel{KH} + \cancel{KD} + QH + QD + JH + JD)} - \underbrace{\#(A \cap B)}_{(\cancel{KH} + \cancel{KD})}$$

Simplifying, one finds that

$$KH + KD + KS + KC + QH + QD + JH + JD$$

are the eight outcomes that are either a king or a red card. Thus,

$$\#(A \cup B) = \#(A) + \#(B) - \#(A \cap B) = 4 + 6 - 2 = 8$$

The probability of $A \cup B$ is

$$P(A \cup B) = \frac{\#(A \cup B)}{\#(S)}$$
$$= \frac{\#(A) + \#(B) - \#(A \cap B)}{\#(S)}$$
$$= \frac{\#(A)}{\#(S)} + \frac{\#(B)}{\#(S)} - \frac{\#(A \cap B)}{\#(S)} = \frac{4}{12} + \frac{6}{12} - \frac{2}{12} = \frac{8}{12} = .67$$
$$P(A \cup B) = P(A) + P(B) - P(A \cap B)$$

Consider some additional examples. First, suppose a student applies to two graduate schools, A and B, and assesses the chances of getting into them as $P(A) = .10$ and $P(B) = .25$, respectively. What is the probability of getting into at least one of the schools? Since either A or B satisfies the required outcome, the probability is given by

$$P(A \cup B) = P(A) + P(B) - P(A \cap B) = .10 + .25 - (.10)(.25) = .325$$

Second, what is the probability of getting an odd number or a six in a single roll of a die? If A is the event of getting an odd number and B the event of getting a six, then the required probability is

$$P(A \cup B) = P(A) + P(B) - P(A \cap B) = \frac{3}{6} + \frac{1}{6} - 0 = \frac{4}{6} = .67$$

Note that $P(A \cap B) = 0$ because A and B are mutually exclusive. As before, it is not necessary to change the formula for cases in which A and B are mutually exclusive. Instead, one merely sets $P(A \cap B)$ equal to zero.

The word *or* in a verbal statement of a problem tends to imply that the union of two or more events is required, just as the word *and* signals the intersection of two or more events. Thus, the above discussion on the probability of the union of two events dealt with (1) either a king *or* a red card, (2) getting accepted at either school *A or* school *B*, and (3) rolling an odd number *or* a six.

Although the two probabilistic laws described above have been stated only in terms of two events, the formulas can be generalized to cases involving more than two events. For example, if *A*, *B*, and *C* are *independent*, then

$$P(A \cap B \cap C) = P(A)P(B)P(C)$$

If *A*, *B*, and *C* are *mutually exclusive*, then

$$P(A \cup B \cup C) = P(A) + P(B) + P(C)$$

Consider a last example, which involves a combination of union and intersection. What is the probability of selecting either an ace or a king on the first draw and a red card or a face card (*J*, *Q*, *K*) on the second draw from an ordinary deck of 52 cards if selection is performed with replacement? Let

$A =$ an ace on the first draw
$B =$ a king on the first draw
$C =$ a red card on the second draw
$D =$ a face card on the second draw

Reducing the statement of the problem to symbols, the required complex event is [(*A* or *B*) and (*C* or *D*)], and thus the required probability is

$$P[(A \cup B) \cap (C \cup D)]$$

Because the selection is performed with replacement, the two events $(A \cup B)$ and $(C \cup D)$ are independent. Further, *A* and *B* are mutually exclusive, whereas *C* and *D* are not. Therefore, expanding the above expression,

$$P[(A \cup B) \cap (C \cup D)] = [P(A) + P(B)] \times [P(C) + P(D) - P(C \cap D)]$$

The necessary quantities are:

$$P(A) = \frac{4}{52} = \frac{1}{13}$$

$$P(B) = \frac{4}{52} = \frac{1}{13}$$

$$P(C) = \frac{26}{52} = \frac{1}{2}$$

$$P(D) = \frac{12}{52} = \frac{3}{13}, \text{ and}$$

$$P(C \cap D) = P(C)P(D) = \left(\frac{1}{2}\right)\left(\frac{3}{13}\right) = \frac{3}{26}$$

Hence, the result is

$$\left(\frac{1}{13} + \frac{1}{13}\right)\left(\frac{1}{2} + \frac{3}{13} - \frac{3}{26}\right) = \left(\frac{2}{13}\right)\left(\frac{8}{13}\right) = \frac{16}{169} = .095$$

METHODS OF COUNTING

The classical definition of probability rests on the principle of dividing the number of elementary events in *A* by the number of elementary events in *S*. However, determining how many different five-player basketball configurations can be put on the court if the team has 10 players who can play any position is very tedious if one has to write out all 252 five-player groups. Therefore, it is desirable to be able to assess the number of outcomes in a complex event in some other manner than by listing each of them. Certain counting methods are available to facilitate this task.

Permutations

> A **permutation** of a set of objects or events is an ordered sequence of the elements from that set. The number of ordered sequences of *r* objects, which can be selected from a total of *n* objects, is symbolized by
>
> $$_nP_r,$$
>
> which is read "the number of permutations of *n* things taken *r* at a time."

If one has four objects, *A, B, C,* and *D,* then *ABCD, ADBC,* and *ADCB* are some of the 24 possible permutations of the four objects taken four at a time. If these four objects are taken two at a time, then *AB, BA, AC, CA, AD, DA, BC,* and *CB* are some of the 12 permutations of four objects taken two at a time. Note that the definition states "ordered sequence." That means that *AB* and *BA* are two different permutations; that is, order makes a difference. Horse race fans know that some types of bets require that the bettor specify which horses will win, place, and show (that is, come in first, second, and third) in a single race. Since order makes a difference, the bettor must specify a permutation. Similarly, it is difficult to guess the combination to a combination lock because it is only one of many permutations.

> The number of permutations of *n* things taken *r* at a time, $_nP_r$, equals
>
> $$_nP_r = \frac{n!}{(n-r)!}$$
>
> If *r = n*, the number of permutations of *n* things (taken *n* at a time) equals
>
> $$_nP_n = \frac{n!}{(n-n)!} = \frac{n!}{0!} = n!$$
>
> Note that 0! = 1.

The symbol $n!$, read "n factorial," is defined as

$$n! = n(n - 1)(n - 2)(n - 3) \ldots \quad (1)$$

Thus, $5! = (5)(4)(3)(2)(1) = 120$.

Let us examine the logic behind these expressions for permutations more closely. Suppose one has five objects, A, B, C, D, and E. Consider first the number of permutations of these five objects taken five at a time, $_nP_n = {}_5P_5$. Think of the task as having to fill five positions. There are five possible objects with which to fill the first position. *For each one of those five selections,* the second position may be filled with any one of the four remaining objects, since an ordered sequence is comparable to sampling without replacement. This means that there are $(5)(4) = 20$ ways of filling the first two positions. Note that one multiplies because *for each one* of the first five possibilities, there exist four ways to fill the second. The third position may be filled with any one of the three remaining objects; there are two ways to fill the fourth; and only one object (or way) remains for the last position. Therefore, the total number of ordered sequences or permutations of five things is

$$(5)(4)(3)(2)(1) = 5! = 120$$

To generalize, there are $n!$ permutations of n things (taken n at a time),

$$_nP_n = n!$$

Now suppose you have five things but need to know the number of permutations of only three of the five elements at a time. Table 12-3 lists the 60 permutations of five things (A, B, C, D, and E) taken three at a time. The 60 permutations are arranged into 10 groups, each containing the 6 permutations of three elements.

The number of permutations of n things taken r at a time, $_nP_r$, in this case $_5P_3$, is determined with the following logic. There are five ways to fill the first position, four ways to fill the second position, and three ways to fill the third position, and that is all the positions required. Thus, one multiplies to arrive at the answer:

$$_nP_r = n(n - 1)(n - 2) \ldots (n - r + 1) = (5)(4)(3) = 60$$

The same result is obtained by dividing $n!$ by $(n - r)!$:

$$_nP_r = \frac{n!}{(n - r)!} = \frac{(5)(4)(3)(\cancel{2})(\cancel{1})}{(\cancel{2})(\cancel{1})} = (5)(4)(3) = 60$$

Thus, the formula for the number of permutations of n things taken r at a time is

$$_nP_r = n(n - 1)(n - 2) \ldots (n - r + 1) = \frac{n!}{(n - r)!}$$

12-3	**The Sixty Permutations and Ten Combinations (Underlined) of Five Things Taken at a Time**

<u>ABC</u>	<u>ABD</u>	<u>ABE</u>	<u>ACD</u>	<u>ACE</u>
ACB	ADB	AEB	ADC	AEC
BCA	BAD	BAE	CAD	CAE
BAC	BDA	BEA	CDA	CEA
CAB	DAB	EAB	DAC	EAC
CBA	DBA	EBA	DCA	ECA
<u>ADE</u>	<u>BCD</u>	<u>BCE</u>	<u>BDE</u>	<u>CDE</u>
AED	BDC	BEC	BED	CED
DAE	CBD	CBE	DBE	DCE
DEA	CDB	CEB	DEB	DEC
EAD	DBC	EBC	EBD	ECD
EDA	DCB	ECB	EDB	EDC

Consider the problem of picking the first-, second-, and third-place winners in a race involving seven horses of unknown, and thus presumably equal, speed. What is the probability of success at this task? There is only one way to pick the three horses and assign them to the proper places (event A), but there are $_7P_3$ ways of ordering seven horses in groups of three. The required probability is

$$P(A) = \frac{\#(A)}{\#(S)} = \frac{1}{_7P_3} = \frac{1}{\dfrac{7!}{(7-3)!}} = \frac{1}{\dfrac{(7)(6)(5)(\cancel{4})(\cancel{3})(\cancel{2})(\cancel{1})}{(\cancel{4})(\cancel{3})(\cancel{2})(\cancel{1})}} = \frac{1}{(7)(6)(5)}$$

$$P(A) = \frac{1}{210} = .0048$$

Therefore, in a crowd of 10,000 people, none of whom had any information on the horses and all of whom guessed randomly, on the average 48 people would pick correctly, in order, the first three horses to finish. Interpreted in another way, you are likely to win such a bet by chance alone 48/10,000 or less than five times for each 1000 times you bet it. There are many ordered sequences of even a few different events, which is why the payoff for winning such bets is so substantial. And people continue to bet because they believe the horses are not equally likely to win (true) and that *they know* which are most likely to win (may be true). If these premises are true, it improves the odds, but it is not clear by how much.

Combinations

> A **combination** is any set or subset of objects or events, regardless of their internal order. The number of groups of r objects that can be selected from n objects is symbolized[2] by $_nC_r$ and given by
>
> $$_nC_r = \frac{n!}{(n-r)!r!}$$

While AB and BA are two different *permutations,* they are the same *combination.* Permutations are ordered sequences, whereas combinations ignore the order of their elements. So, when the number of combinations of n things taken n at a time is desired, only one combination exists.

The initial hand one is dealt in most card games is a combination, because the sequence in which one receives the cards does not matter. In a twist on words, a combination will not open a combination lock, only a permutation will; but see how far you get asking for a "permutation lock" at the hardware store.

Consider the logic of determining $_nC_r$. Recall that the number of permutations of n objects taken r at a time is

$$_nP_r = \frac{n!}{(n-r)!}$$

Thus, if $n = 5$ and $r = 3$,

$$_5P_3 = \frac{5!}{2!} = 60$$

The number of combinations, however, is much less than 60, because many of the permutations are simple reorderings of a single combination. Table 12-3 presents the 60 permutations of five things taken three at a time, but they are arranged into groups corresponding to the 10 combinations. It can be seen that there are six permutations *for every one combination* (underlined in the table). In general, if objects are to be taken r at a time, there will be $r!$ permutations for every combination. Thus, since there are three elements in each group, there are $3! = (3)(2)(1) = 6$ permutations for each combination, as indicated by the groupings of six in Table 12-3. One divides the number of permutations by $r!$ to obtain the number of combinations, since there are $r!$ permutations for every combination. Therefore, the expression for combinations is

$$_nC_r = \frac{_nP_r}{r!} = \frac{\dfrac{n!}{(n-r)!}}{r!} = \frac{n!}{(n-r)!r!}$$

---

[2] Some other books use the symbol $\binom{n}{r}$.

To illustrate, suppose a football coach has seven guards on the team. Assuming that it makes no difference who plays the right and the left positions, how many different pairs of guards can be fielded? Since order is unimportant, this amounts to asking: how many combinations are there of seven people taken two at a time?

$$_7C_2 = \frac{7!}{5!2!} = \frac{(7)(\cancel{6})(\cancel{5})(\cancel{4})(\cancel{3})(\cancel{2})(\cancel{1})}{(\cancel{5})(\cancel{4})(\cancel{3})(\cancel{2})(\cancel{1})(\cancel{2})(\cancel{1})} = 21$$

What is the probability of being dealt a five-card poker hand containing all spades? If A is the event of being dealt all spades, any one of the many combinations of 13 cards in that suit of spades taken five at a time will do. The sample space S includes the number of five-card hands in a deck of 52 cards. The required probability is

$$P(A) = \frac{_{13}C_5}{_{52}C_5} = \frac{\dfrac{13!}{8!5!}}{\dfrac{52!}{47!5!}} = \frac{(13)(12)(11)(10)(9)}{(52)(51)(50)(49)(48)}$$

$$P(A) = .0005$$

Binomial Probability

A special application of determining the number of combinations occurs when there are only two possible outcomes—success/failure, heads/tails, rain/no rain, fish/no fish, win/lose—for each of several trials or occasions. For example, in six tosses of a coin, what is the probability of obtaining exactly two heads? It is important to notice that the question does not specify the order in which the heads and tails must appear. It requires only that exactly two of the six tosses be heads. Therefore, the solution to this problem may be achieved by first determining the probability of any one specific sequence of two heads and four tails and then determining how many such sequences are possible.

One potential sequence is (H, H, T, T, T, T). The probability of a head on any single flip is $\frac{1}{2}$ and the probability of a tail is also $\frac{1}{2}$. The probability of getting two heads in succession is $(\frac{1}{2})(\frac{1}{2})$, of two heads followed by a tail is $(\frac{1}{2})(\frac{1}{2})(\frac{1}{2})$, and of (H, H, T, T, T, T) is $(\frac{1}{2})(\frac{1}{2})(\frac{1}{2})(\frac{1}{2})(\frac{1}{2})(\frac{1}{2}) = \frac{1}{64}$.

Further, the probability of the sequence (T, T, T, H, H, T) is also $(\frac{1}{2})(\frac{1}{2})(\frac{1}{2})(\frac{1}{2})(\frac{1}{2})(\frac{1}{2}) = \frac{1}{64}$. Indeed, the probability of *any* particular sequence of two heads and four tails is $\frac{1}{64}$.

How many such sequences of two heads and four tails are there? The number of sequences of two heads in six tosses is the same as the number of combinations of six things (6 tosses) taken two (2 heads) at a time:

$$_6C_2 = \frac{n!}{r!(n-r)!} = \frac{6!}{2!(6-2)!} = 15$$

Since the event of two heads in six flips can occur in 15 different ways, each with a probability of $\frac{1}{64}$, the required answer is given by the sum of these 15 mutually exclusive probabilities, or

$$\frac{1}{64} + \frac{1}{64} + \cdots + \frac{1}{64} = 15\left(\frac{1}{64}\right) = \frac{15}{64} = .23$$

This problem is an example of **binomial probability,** and the process of finding its solution may be formalized into a general expression.

> In a sequence of n independent trials that have only two possible outcomes (arbitrarily call them "success" and "failure") with the probability p of a success and probability q of a failure (note that $q = 1 - p$), the probability of exactly r successes in n trials is
>
> $$_nC_r p^r q^{n-r}$$
>
> or
>
> $$P(r, n; p) = \frac{n!}{r!(n-r)!} p^r q^{n-r}$$

Suppose the likelihood of rain on any particular day of your one-week vacation is .30. What is the probability that it never rains during your vacation? In this case $n = 7$, $r = 7$, p (no rain) $= .70$, and q (rain) $= .30$, and

$$\frac{n!}{r!(n-r)!} p^r q^{n-r} = \frac{7!}{7!(7-7)!} (.70)^7 (.30)^0 = .08$$

Note that $0! = 1$ and $(.30)^0 = 1$. Pack an umbrella.

To illustrate further, suppose an urn contains one red and three green balls. If a ball is returned to the urn after each drawing, what is the probability of selecting three reds in five tries? A success is defined as the selection of a red ball. Since on any one trial only one of the four balls in the urn is red, the probability of a success is one-fourth ($p = \frac{1}{4}$). Since $q = 1 - p$ and since three of the four balls available on any one trial are failures, $q = \frac{3}{4}$. The event under consideration is to obtain three red balls ($r = 3$) in five draws ($n = 5$). Therefore, the desired probability is

$$P(r, n; p) = \frac{n!}{r!(n-r)!} p^r q^{n-r}$$

$$P(3, 5; \tfrac{1}{4}) = \frac{5!}{3!(5-3)!} (\tfrac{1}{4})^3 (\tfrac{3}{4})^{5-3}$$

$$= \frac{90}{1024}$$

$$= .088$$

As another example, suppose a school for severely disturbed youth knows that on the average only 40% of its graduates hold a full-time job, are in school, or are engaged in some combination of work and education one year after graduating. Dismayed with this figure, the school institutes a new program designed to improve such outcomes. Of the first class of 8 students in the new program, 7 are at work, in school, or both after a year. Given the previous track record, the school wants to know the probability that 7 or more students should be productively engaged as defined.

The desired answer requires the probability that 7 of 8 *or* 8 of 8 students are productively engaged, which is the sum of those two probabilities. The probability of a success is .40, and the probability of a failure is $1 - .40 = .60$. The probability of 7 of 8 is

$$P(r, n; p) = \frac{n!}{r!(n - r)!} p^r q^{n-r}$$

$$P(7, 8; .4) = \frac{8!}{7!(8 - 7)!} (.4)^7 (.6)^1$$

$$= (8)(.0016384)(.6)$$

$$= .00786432$$

The probability of 8 of 8 successes is

$$P(8, 8; .4) = \frac{8!}{8!(8 - 8)!} (.4)^8 (.6)^0$$

$$= (1)(.00065536)(1)$$

$$= .00065536$$

So the required probability of either 7 *or* 8 successes is the sum of these two probabilities, or

$$P(7 \text{ or } 8 \text{ of } 8) = .00786432 + .00065536 = .00851968$$

Since this observed result is unlikely using the probabilities from before the new program, perhaps the new program is more effective than the old.

SUMMARY

Probability is best understood in terms of set theory, so the definitions of a set, an element in a set, a subset, the universal set, the empty set, and the complement are presented. Then the operations of set equality, union, and intersection are described. The simple classical probability of an event A is given by the number of outcomes in A divided by the number of outcomes in the sample space (that is, the total possible outcomes). Then conditional probability, dependent and independent events, and mutually exclusive events are defined, and the probability of the intersection, the union, and the conditional probability of two events are discussed. Finally, methods of counting (for example, permutations and combinations) and binomial probability are presented.

Formulas

1. Classical probability

a. $P(A) = \dfrac{\#(A)}{\#(S)}$

b. $P(\varnothing) = \dfrac{\#(\varnothing)}{\#(S)} = \dfrac{0}{\#(S)} = 0$

c. $P(S) = \dfrac{\#(S)}{\#(S)} = 1.00$

2. Probability of intersection

a. $P(A \cap B) = P(A)P(B|A)$

b. If A and B are independent [that is, if $P(B|A) = P(B)$], then $P(A \cap B) = P(A)P(B)$.

3. Conditional probability

$$P(B|A) = \dfrac{P(A \cap B)}{P(A)}$$

4. Probability of union

a. $P(A \cup B) = P(A) + P(B) - P(A \cap B)$

b. If A and B are mutually exclusive (that is, if $A \cap B = \varnothing$), then $P(A \cup B) = P(A) + P(B)$.

5. Permutations

a. $_nP_r = \dfrac{n!}{(n - r)!}$

b. $_nP_n = n!$

6. Combinations

$$_nC_r = \dfrac{n!}{(n - r)!r!}$$

7. Binomial probability, the probability of r successes in n independent trials

$$P(r, n; p) = \dfrac{n!}{r!(n - r)!} p^r q^{n-r}$$

in which

$p =$ the probability of a success

$q =$ the probability of a failure ($q = 1 - p$)

Exercises

For Conceptual Understanding

1. In an idealized experiment, how many coin tosses would be required before the proportion of heads would equal exactly .50?

2. Determine which of the following events are mutually exclusive:

a. flipping a head, flipping a tail with one toss of a coin

b. rolling an even number, rolling less than a three with one roll of a die

c. selecting an unmarried male, selecting a married female in a single random drawing for a sweepstakes prize

d. catching a fish on a fishing trip and catching a trout on that trip

3. Determine which of the following events are independent:

a. flipping a head on the first toss and a tail on the second toss of a coin

b. flipping a head on one coin and simultaneously flipping a head on a second coin

c. having a boy as the first child and having a girl as the second child

d. selecting an ace and then selecting a red card when drawing is done with replacement, when drawing is done without replacement

e. winning the lottery one week and winning it again the next week

4. Why does one multiply the separate probabilities to obtain the probability of an intersection but add the separate probabilities to obtain the probability of a union of two independent and mutually exclusive events, respectively?

5. Which of the following complex events require

the intersection and which require the union of two elementary events?

a. The probability that one of four sections of a class you need to take to graduate has space available and that section meets in the morning

b. The probability that you go to the Spring Formal with either Harris or Mark

c. The probability that it will not rain either today or tomorrow so you can go on an all-day hike on one of the next two days

d. The probability that it does not rain on any of the next three days

6. Which is required, the number of combinations or permutations?

a. A nursery school has a total enrollment of 15, but only 12 children come on any given day. How many different sets of children could potentially attend on a given day?

b. How many different sets of teams could make the baseball playoffs if each of four divisions sent only one of its six teams to the playoffs?

c. How many different finishing sequences are there in an Olympic event if there are 30 entrants but only three medals awarded?

For Solving Problems

7. In a deck of 12 cards (4 jacks, 4 queens, 4 kings), determine the probability of drawing:

a. a queen.

b. a black jack or a black queen.

c. a red card or a king.

8. In a single roll of a die, what is the probability of obtaining:

a. at least a three?

b. less than a four or an odd number?

9. What is the probability of rolling two successive sixes with a die?

10. If an urn contains three red and four green balls, what is the probability of selecting:

a. a green ball given that a red ball has been drawn on the first selection and is not re-placed?

b. two successive green balls if the first selection is replaced?

c. two successive green balls if the first is not replaced?

11. Determine the following permutations:

a. $_6P_6$ **b.** $_4P_2$ **c.** $_5P_2$

12. If a room has four chairs, how many different seating arrangements are there if six people come in?

13. Determine the following combinations:

a. $_5C_5$ **b.** $_5C_3$ **c.** $_6C_2$

14. How many different relay teams, each composed of four swimmers, could a coach put together with eight swimmers to choose from? If the order in which the individuals swam made a difference, then how many teams could be put together?

15. What is the probability of naming the three top horses (regardless of specific position) in a field of nine? What is the probability of correctly designating first, second, and third positions for the field of nine? What is the probability of correctly assigning them to the top three positions, given that you have picked the top three horses?

16. What is the probability of correctly guessing the "combination" (actually, permutation) of a lock that has 35 numbers on its dial and uses a three-number sequence to open it? Assume that a number may not be repeated.

17. Suppose seven infants are familiarized with a given visual stimulus. Later they are permitted to look at that familiar pattern along with two other stimuli that represent two degrees of similarity to the familiar one. The discrepancy theory predicts that looking time should be an inverted-∪ function of magnitude of discrepancy. The probability of this pattern is $\frac{1}{3}$. What is the probability that exactly five of the seven infants should respond this way? Five or more of the infants?[3]

For Homework

18. What is the probability of drawing a three of diamonds or any five from a deck of 52 cards?

19. What is the probability of drawing anything but a two or a club from the 52-card deck?

---

[3] Inspired by R. B. McCall and J. Kagan, "Stimulus–Schema Discrepancy and Attention in the Infant," *Journal of Experimental Child Psychology* 5 (1967): 381–90.

20. A die is rolled once. What is the probability of rolling
    a. an even number?
    b. a 2, 3, or 6?
    c. more than a 3 and less than a 6?
    d. an odd number but not a 3?
    e. more than a 4 or less than a 6?

21. An urn contains 10 balls: 4 white, 1 red, 4 blue, and 1 black. If one ball is drawn, determine the probability that it is
    a. not white or blue
    b. blue or red
    c. black, white, or red
    d. orange, blue, or green

22. With the same urn and balls as in Exercise 21, what is the probability of drawing
    a. a red and then a black with no replacement?
    b. a red and then a blue with no replacement?
    c. a black and then a black with no replacement?
    d. a black and then a black with replacement?
    e. the sequence blue, red, white, blue, black with no replacement?

23. The weather bureau reports that the chance of precipitation tomorrow is 60%, while the chance that the temperature will be less than 32°F is 20%. What is the probability that tomorrow it will
    a. snow? (Assume that it snows at a temperature less than 32°F.)
    b. rain?
    c. be less than 32°F but not snow?
    d. be over 32°F but not rain?

24. Parents, hearing the forecast in Exercise 23, tell their family that they will go skiing if it does not rain. What is their chance of going skiing?

25. Certain diseases (e.g., sickle cell anemia, cystic fibrosis) are autosomal recessive. This means that each person has two relevant genes, and both genes must be for the disease before the person is actually afflicted. If the person has only one gene for the disease, he or she does not get the disease and is called a carrier. Suppose that two married adults are both carriers—that is,

each has one gene for the disease and one gene for normality. Each parent will pass on one gene to each child. What is the probability that
    a. the couple's first child has the disease?
    b. the couple's first child is a carrier?
    c. in a family of two children, neither will have a gene for the disease?
    d. in a family of three children, at least one child will have the disease?
    e. in a family of four, two will be perfectly normal (neither afflicted nor a carrier)?

26. If $P(A) = .3$, $P(B) = .4$, $P(C) = .5$, and A, B, C are independent but not mutually exclusive, determine the value of the following:
    a. $P(B \cap C)$
    b. $P(A \cup B)$
    c. $A \cap (B \cup C)$
    d. $A \cup (B \cup C)$
    e. $(A \cap B) \cup C$

27. Determine the following:
    a. $_5P_3$ c. $_4P_1$
    b. $_6P_2$ d. $_5P_5$

28. Determine the following:
    a. $_5C_3$ c. $_4C_1$
    b. $_6C_2$ d. $_5C_5$

29. A student has 5 independent assignments to do. These assignments can be done in how many orders if the student has time to do:
    a. all 5?
    b. only 3?

30. If the order in which the assignments in Exercise 29 are done is unimportant, how many different combinations of assignments can be done if the student has time to do:
    a. all 5?
    b. only 3?

31. Suppose on a fishing trip, a grandfather determines that the likelihood of his grandson catching a fish is .10 per cast of the line. What is the probability that the grandson will catch
    a. no fish in 5 casts?
    b. at least 1 fish in 5 casts?
    c. exactly 2 fish in 6 casts?

Two-Factor Analysis of Variance

he simple analysis of variance described in Chapter 10 tests the probability that the means from several sample groups estimate the same population mean and differ only because of sampling error. It was assumed that the groups were part of a classification scheme: different amounts of dream deprivation, different teaching programs, and so forth. These groups represent different levels or types of a single category of events, treatments, stimuli, and so on. But scientists more frequently design their experiments to include groups that simultaneously belong to more than one classification. Perhaps subjects are either males or females *and* some subjects of each gender have their dreams interrupted a few times, many times, or no times during each evening. Then groups belong to the classifications of both gender *and* dream deprivation. It is the purpose of this chapter to consider the extension of simple analysis of variance to those cases in which the groups fall into a two-way classification scheme.

TWO-FACTOR CLASSIFICATION

Children observe the behavior of others, and often they imitate that behavior, especially if they perceive it as bringing about positive rewards. Consider an experiment specifically designed to test this notion,[1] in which forty boys and forty girls are randomly selected from a given school and shown one of two movies. Both films feature an adult who hits, pounds, pushes, and otherwise assaults a large plastic balloon doll with a weighted base (called a Bobo doll). The adult in each film serves as a model, presumably for the children to imitate. Half of the children, including both boys and girls, see a film in which the model is rewarded by another adult with praise and congratulations for assaulting the doll; the other half of the children see a film in which the model is punished with verbal reprimands for demonstrating violent behavior. After viewing the films, both groups of children are brought into a room containing several age-appropriate toys, including a Bobo doll. During a 10-minute play period in which the children are closely observed, the investigators count the number of imitated aggressive behaviors directed at the Bobo doll. There are four groups in this experiment: (1) boys who saw the model being rewarded, (2) boys who saw the model being punished, (3) girls who saw the model being rewarded, and (4) girls who saw the model being punished. These four groups represent different levels of two classification schemes—one based on the gender of the subjects and the other based on the type of reinforcement the model received (reward vs. punishment). Psychologists call the reinforcement that a person sees being given to another person, as in the films, *vicarious reinforcement.*

Classification Terminology

A single classification scheme is called a **factor,** and each group within it represents a **level** of that factor. Notice that the levels do not necessarily represent different *amounts*

---

[1] Based upon, but not identical to, A. Bandura, D. Ross, and S. A. Ross (1963). "Imitation of Film-Mediated Aggressive Models," *Journal of Abnormal and Social Psychology* 66: 3–11.

13-1 **Example of a Two-Factor Research Design (Vicarious Reinforcement Study)**

		Factor B (Vicarious Reinforcement)	
		b_1 (reward)	b_2 (punishment)
Factor A (Gender of Subject)	a_1 (males)	Males who see the model rewarded	Males who see the model punished
	a_2 (females)	Females who see the model rewarded	Females who see the model punished

of the factor, as is demonstrated by the levels of the gender factor, male and female. The above study has a **two-factor design** with two levels of each factor. Designating a factor by a capital letter and each level by a subscripted lowercase letter of the same character, this design can be expressed in a tabular fashion as shown in Table 13-1.

Table 13-2 shows the average number of imitative aggressive responses for each of the four groups. In addition to the group means, the numbers at the ends of the rows and columns, called **marginals,** express the mean number of imitative responses for males (15) and for females (11) *collapsed over,* or *ignoring,* reinforcement conditions, and the mean number for subjects watching a rewarded (22) and a punished (4) model collapsed over gender of subject. The grand mean number of responses over all subjects is 13.

Main Effects and Interactions

Three types of questions could be asked about the data from such an experiment. First, is there a significant difference between the levels of **factor A?** In this case, on

13-2 **Means for the Vicarious Reinforcement Study**

		Factor B (Vicarious Reinforcement)		
		b_1 (reward)	b_2 (punishment)	
Factor A (Gender of Subject)	a_1 (males)	25	5	15
	a_2 (females)	19	3	11
		22	4	13

the average, do boys imitate aggressive behavior more often than girls do? Specifically, does the difference between the means of 15 and 11 likely exist in the population, or is this observed difference merely a function of sampling error? Second, is there a significant difference between the levels of **factor B?** In this case, does the vicarious experience of seeing the model rewarded or punished influence the extent to which children will imitate? Is the difference between the means of 22 and 4 due to sampling error? Third, is there an **interaction** between the gender of the child and the type of vicarious reinforcement? Specifically, is the effect of seeing the model rewarded or punished different for boys than for girls? The same question may be turned around—does the difference between boys and girls in their tendency to imitate depend on whether the model is rewarded or punished?

> An **effect** is a difference among population means. A **main effect** is a difference among population means for levels of a factor ignoring the other factors. An **interaction** occurs when the nature of the effect for one factor is not the same within all levels of another factor.

The meaning of these terms can be appreciated best by examining graphs of possible results of the two-factor example just described. In Figure 13-1, graph (A) depicts the case in which all the means are approximately the same. There are no main effects and no interaction. Graph (B) describes one main effect: males imitate more often than females do. Notice that the reward and punishment conditions do not appear to have influenced the amount of imitation. Graph (C) illustrates a main effect for vicarious reinforcement with no effects for gender and no interaction. Children of both genders imitated more if the model was rewarded than if the model was punished, and the amount of imitation was the same for both genders.

In graph (D), there are two main effects. Males imitated more than females, and the reward condition generated more imitation than the punishment condition. There is no interaction in graph (D) because the difference between males and females under both reinforcement conditions is about the same. Another way to say this is that the lines for the two genders are approximately parallel, indicating that the reinforcement conditions had a comparable effect on both genders.

Graphs (E) and (F) depict interactions between gender and vicarious reinforcement because the nature of the effects for one factor is not the same within both levels of the other factor. In graph (E), whether the model was rewarded or punished made a difference for males but not for females. In graph (F), the vicarious reinforcement had opposite effects for the two genders: a rewarded model produced more imitation in females than a punished model, but the reverse was true for males. Notice that in graph (F) there is no main effect for vicarious reinforcement or for gender. That is, if males and females are averaged separately within the reward and then within the punishment conditions, these means (indicated by the $\times$'s) are not very different from one another. Similarly, if the reward and punishment means are averaged separately within each gender, they are not different. Thus, graph (F) depicts an interaction with no main effects.

As can be seen in the above examples, any possible combination of results can occur. There may be one or two main effects with or without an interaction, or there

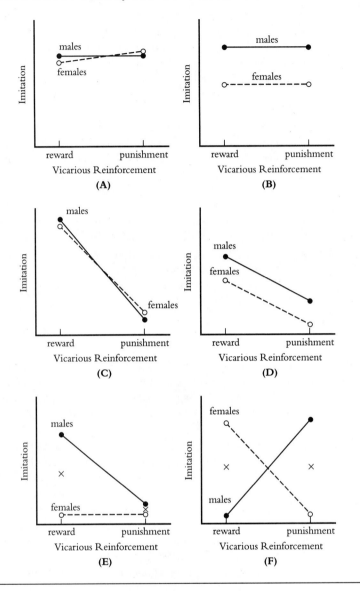

Figure 13-1 Examples of possible main effects and interactions.

may be an interaction with no main effects. However, when an interaction is found to be significant, considerable care must be taken when interpreting the significance or nonsignificance of main effects, and often main effects are ignored in this case. For example, if there is a significant interaction, that fact means that the influence of one factor differed depending upon the level of the other factor. Since a main effect represents a difference in means for one factor collapsed over the levels of the other factor, the occurrence of an interaction indicates that collapsing over the other factor is not meaningful: the effects for factor *A* are not the same within each level of factor *B*. For example, consider graph (E) in Figure 13-1. The $\times$'s indicate the means for the reward and punishment conditions ignoring the gender of the subject. It might be the case that the difference between the two means (the $\times$'s) is significant, and one would conclude that a rewarded model produces more imitation than a punished model. However, if the interaction depicted in that graph is also significant, it means that vicarious reinforcement makes a difference only for males and not for females, thus qualifying the general interpretation of the main effect for reinforcement. The reverse of this situation occurs in graph (F). Here the $\times$'s are not different, and it is likely there would not be a significant main effect for reinforcement. But looking at the graph, it would be wrong to conclude that reinforcement has had no influence on imitation—its influence depends on the gender of the subject, and these differences are masked when the two genders are averaged within reward and within punishment conditions. Therefore, *if an interaction is significant, considerable care must be taken in interpreting the nature of main effects for factors that are involved in that interaction, and, as a general rule, the main effects, whether significant or not, are ignored.*[2]

LOGIC OF TWO-FACTOR ANALYSIS OF VARIANCE

The logic of two-factor analysis of variance is a direct extension of the rationale underlying simple analysis of variance.[3] To review, in simple analysis of variance the total sum of squares is partitioned into two components, each of which provides an estimate of the variability in the population if the null hypothesis is true that the population mean of each group is the same value. However, one of these variance estimates is based upon the deviations of the group means about the grand mean **(between-groups estimate)**, whereas the other is based upon the deviations of the scores about their respective group means **(within-groups estimate)**. The between-groups estimate is influenced by any difference between groups (that is, treatment effects) that may exist, whereas the within-groups estimate is not. Consequently, if the between-groups estimate is very large relative to the within-groups estimate, then the tentative presumption that these groups are all samples from the same population (that is, the null hypothesis) may not be true. When the null hypothesis is true, the ratio of the between-groups estimate to the within-groups estimate is distributed as *F*; by determining the percentile rank of the observed ratio in this theoretical relative frequency distribution, one can find the probability of obtaining such a ratio by sam-

---

[2] This general "rule" does have an exception in the following event: an interaction exists, but all groups within a level of a factor are significantly higher than the corresponding groups within another level of that factor. The analysis would yield a main effect and an interaction. For example, suppose in graph (E) of Figure 13-1 the line for males was substantially higher than for females under each of the reinforcement conditions. If the males displayed more imitation than females under reward and under punishment conditions in addition to having the pattern difference displayed in graph (E), then the main effect for gender as well as the interaction would be meaningful. Determining whether a main effect should be interpreted when an interaction is also present is a complex issue and beyond the scope of this text.

[3] This discussion will be restricted to fixed factors, independent groups, and equal cell sizes, as discussed later.

pling error alone. If the probability is sufficiently small, the null hypothesis of no difference between the population means is rejected.

In two-factor analysis of variance, the total sum of squares is partitioned into four rather than just two components. As before, the four variance estimates are called **mean squares (MS)**, and they all estimate the same population variance if the null hypothesis is true.[4]

One estimate of the population variance (written MS_A) is based upon the deviations about the grand mean of the marginal means for the levels of factor A. This estimate is analogous to the between-groups estimate in a simple analysis of variance involving only the groups of factor A (that is, ignoring factor B). In the example, this estimate would be based upon the deviations of the means for males (15) and females (11) about the grand mean (13). MS_A is sensitive to differences between the means of the levels of factor A.

A second estimate of the variability in the population (written MS_B) is based upon the deviations about the grand mean of the marginal means for the levels of factor B. As such, it is analogous to MS_A, except the means for the levels of factor B (ignoring factor A) are used. In the example, this estimate would be based upon the deviations of the reward (22) and punishment (4) means about the grand mean (13). MS_B is sensitive to differences between the means of the levels of factor B.

A third estimate of the variability in the population (symbolized by MS_{AB}) is based upon the deviations of each group mean from what would be predicted on the basis of knowledge of the two main effects. This mean square is sensitive to the possible interaction between factors A and B.

Finally, the fourth estimate of the population variability is derived in the same manner as the within-groups estimate in simple analysis of variance (MS_{within}). It is based upon the deviations of each score from its respective group mean. In contrast to the other mean squares, it is not sensitive to differences between groups or between levels of factors; thus, it can be used as a standard against which the other estimates may be evaluated. For this reason, MS_{within} is sometimes called MS_{error}, because it represents the "error variability" for each of the other sources of variability.

The ratio of the mean square for factor A to the mean square within groups provides a test of the null hypothesis that the means of the levels of A differ from one another merely as a function of sampling error. Under the null hypothesis, the size of this ratio should not be very large because both variances presumably estimate the same value. However, since MS_A is sensitive to differences between the means for levels of factor A but MS_{within} is not, the ratio will be large to the extent that those means deviate from one another. If they are very different, the ratio will be so large that the probability that the deviations derive merely from sampling error is exceedingly small, and the null hypothesis will be rejected. The conclusion will be that the means for the levels of factor A are significantly different from one another.

Similarly, the ratio of the mean square for factor B to MS_{within} reflects the extent to which the means for the levels of factor B differ from one another. The general

---

[4] Technically, there are several null hypotheses, as discussed in the material that follows.

rationale for testing the null hypothesis that such differences are merely a function of sampling error is similar to that just described for factor A.

Last, the ratio of MS_{AB} to MS_{within} tests the proposition that any observed interaction between factors A and B represents sampling error and that no interaction actually exists in the population.

In short, the logic of two-factor analysis of variance is a direct extension of that employed for simple analysis of variance. The total sum of squares is partitioned into components, each sensitive to a different aspect of the classification scheme. Under the null hypothesis, all the groups in the design are randomly selected and have the same population mean. If this is true, the sample means of the groups and the means for the levels of the two factors should deviate from the grand mean only by sampling error. Since MS_A, MS_B, and MS_{AB} are sensitive to different possible effects within the design whereas MS_{within} is not, the ratio of each of these three mean squares to MS_{within} provides F ratios that test their respective null hypotheses.

Notation

The raw score notation follows the general plan described for simple analysis of variance except that an additional subscript is necessary for the second classification factor. This notation is summarized in Table 13-3. A single score is indicated by X_{ijk}, which represents the ith individual in the jth level of factor A and the kth level of factor B. There are a total of p levels of factor A and q levels of factor B. It will be assumed that there are an equal number of subjects in each of the jk groups[5] and that this number is n.

Any group in the experiment is called a **cell** of the design, and the location of any cell is determined by the intersection of a specific row and column as indicated in Table 13-3. Thus, the intersection of the jth row and the kth column defines the jkth cell. The total of the scores for all the subjects in that cell is represented by T_{jk} and the mean of that group is written $\bar{X}_{jk}$. Therefore, the total of all the scores in the cell represented by the intersection of the 2nd row and the 3rd column called cell 2, 3 (read "cell two-three") is T_{23} and the mean of that cell is $\bar{X}_{23} = T_{23}/n$. The total of all the scores in a given row is $T_{j.}$, in which j indicates the row and the dot (located where the column designation should be) signifies that the scores have been summed over all columns. Thus, the mean of the third row is written as $\bar{X}_{3.} = T_{3.}/nq$ since there are q groups, each containing n subjects, in each row. Similarly, the mean of the second column is defined by $\bar{X}_{.2} = T_{.2}/np$ or the mean of the kth column by $\bar{X}_{.k} = T_{.k}/np$. The dot now indicates scores have been summed over all rows. The grand total and grand mean (that is, over all rows and columns) are written $T_{..}$ and $\bar{X}_{..}$.

---

[5] The two-factor techniques described in this text demand that there be an equal number of subjects per cell and that n be greater than 1. Special procedures do exist for the case in which the groups do not have equal numbers of subjects and/or $n = 1$. See, for example, B. J. Winer et al. (1991). *Statistical Principles in Experimental Design,* 3rd ed. (McGraw-Hill).

13-3 Summary of Raw Score Notation for the Two-Factor Analysis of Variance

Factor A		Factor B b_1	b_2	$\ldots$	b_q	Row Means
a_1		X_{111}	X_{112}	$\vdots$	X_{11q}	
		X_{211}	X_{212}	$\vdots$	X_{21q}	
		X_{311}	X_{312}	$\vdots$	X_{31q}	
		$\vdots$	$\vdots$	$\ldots$	$\vdots$	
		$\underline{X_{n11}}$	$\underline{X_{n12}}$	$\vdots$	$\underline{X_{n1q}}$	
		$\bar{X}_{11} = \dfrac{T_{11}}{n}$	$\bar{X}_{12} = \dfrac{T_{12}}{n}$	$\ldots$	$\bar{X}_{1q} = \dfrac{T_{1q}}{n}$	$\bar{X}_{1.} = \dfrac{T_{1.}}{nq}$
a_2		X_{121}	X_{122}	$\vdots$	X_{12q}	
		X_{221}	X_{222}	$\vdots$	X_{22q}	
		X_{321}	X_{322}	$\vdots$	X_{32q}	
		$\vdots$	$\vdots$	$\ldots$	$\vdots$	
		$\underline{X_{n21}}$	$\underline{X_{n22}}$	$\vdots$	$\underline{X_{n2q}}$	
		$\bar{X}_{21} = \dfrac{T_{21}}{n}$	$\bar{X}_{22} = \dfrac{T_{22}}{n}$	$\ldots$	$\bar{X}_{2q} = \dfrac{T_{2q}}{n}$	$\bar{X}_{2.} = \dfrac{T_{2.}}{nq}$
$\ldots$		$\ldots$	$\ldots$		$\ldots$	$\ldots$

	b_1	b_2	$\cdots$	b_q	Grand Mean
a_p	X_{1p1} X_{2p1} X_{3p1} $\vdots$ X_{np1}	X_{1p2} X_{2p2} X_{3p2} $\vdots$ X_{np2}	$\cdots$ $\cdots$ $\cdots$ $\cdots$ $\cdots$	X_{1pq} X_{2pq} X_{3pq} $\vdots$ X_{npq}	
	$\bar{X}_{p1} = \dfrac{T_{p1}}{n}$	$\bar{X}_{p2} = \dfrac{T_{p2}}{n}$	$\cdots$	$\bar{X}_{pq} = \dfrac{T_{pq}}{n}$	$\bar{X}_{p.} = \dfrac{T_{p.}}{nq}$
Column Means	$\bar{X}_{.1} = \dfrac{T_{.1}}{np}$	$\bar{X}_{.2} = \dfrac{T_{.2}}{np}$	$\cdots$	$\bar{X}_{.q} = \dfrac{T_{.q}}{np}$	$\bar{X}_{..} = \dfrac{T_{..}}{npq}$

n = the number of subjects in each group

a_j = the jth level of factor A

b_k = the kth level of factor B

p = the number of levels of factor A

q = the number of levels of factor B

X_{ijk} = the ith score in the jkth group

T_{jk} = the total of all scores in the jkth group

$\bar{X}_{jk}$ = the mean of the jkth group

$T_{..}$ = the total of all scores

$\bar{X}_{i.}$ = the marginal mean for the ith row

$\bar{X}_{j}$ = the marginal mean for the jth column

$\bar{X}_{..}$ = the grand mean

Partitioning of Variability

In simple analysis of variance, the total variability of scores is partitioned into two parts, a between- and a within-groups component. In symbols,

$$SS_{total} = SS_{between} + SS_{within}$$

In the case of a two-factor design, subjects also differ in score value because of between-group differences, but the group differences are now more complicated because there are two classification schemes. A subject belongs to a level in factor A that may be associated with a higher or lower average score, and the subject belongs to a level of factor B that may also be associated with a certain average score. In addition, there might be something special about being in a specific cell that would not have been predicted solely on the basis of factors A and B. Finally, subjects vary within groups as a function of individual differences and measurement error. Therefore, in two-factor analysis of variance, subjects' scores differ from one another because of group differences associated with factor A (SS_A), factor B (SS_B), the interaction of factors A and B (SS_{AB}), and individual differences and measurement error (SS_{within}). In symbols,

$$SS_{total} = SS_A + SS_B + SS_{AB} + SS_{within}$$

Consider a numerical example as an illustration of these sources of variability in score value. Suppose Bill is a male in the experiment described above and has been shown the movie in which the adult's aggressive behavior was rewarded. Referring to Table 13-2, which shows the means for the cells in this study, one can see that Bill is a member of the male-reward cell, located in the top left corner. Suppose Bill's score is 28. Why did Bill score 28 and not the grand mean, 13? To what shall we attribute the difference between any score X_{ijk} and the grand mean $\overline{X}_{..}$? In Bill's case,

$$X_{ijk} - \overline{X}_{..} = 28 - 13 = 15$$

That is, Bill's score deviated 15 points from the grand mean. To what shall we attribute these 15 points?

First, Bill is male and the males in the study tended to score a little higher than average. Specifically, according to Table 13-2, males averaged 15 whereas the grand mean over all subjects was 13. Therefore, one source of variability resides in the difference between the mean of level a_1 of factor A ($\overline{X}_{1.}$) and the grand mean ($\overline{X}_{..}$). In Bill's case, males differed from the grand mean by

$$(\overline{X}_{1.} - \overline{X}_{..}) = 15 - 13 = 2,$$

which suggests that 2 points of the 15-point difference between Bill and the grand mean is because of factor A—that is, because Bill is male.

A second source of variability is the fact that Bill saw the model rewarded while some other children saw the same behavior punished. Notice in Table 13-2 that the reward condition averaged 22 relative to the grand mean of 13. Therefore, another source of variability is the difference between the mean of level b_1 of factor B ($\overline{X}_{.1}$) and the grand mean ($\overline{X}_{..}$). For Bill, the reward condition differed from the grand mean by

$$(\overline{X}_{.1} - \overline{X}_{..}) = 22 - 13 = 9,$$

which suggests that 9 points of the 15-point difference between Bill and the grand mean comes from factor B—that is, because Bill is in the reward condition.

Thus, being male and seeing the model rewarded would explain $2 + 9 = 11$ points of the 15-point difference between Bill's score and the grand mean. But perhaps there is something special about the particular combination of being male and seeing the model rewarded that contributes to the scores of individuals in this cell. According to the calculations just performed, the male–reward condition should have a mean that is 11 points above the grand mean, that is, a mean of $13 + 11 = 24$. But in fact, Table 13-2 indicates that the male–reward cell has a mean of 25, 1 point higher than what would be predicted on the basis of differences associated with factor A and factor B. This 1-point difference is the variability associated with the AB interaction. Generally, the interaction variability comes from the difference between the particular cell mean and the grand mean ($\overline{X}_{jk} - \overline{X}_{..}$) minus the effects for both factor A ($\overline{X}_{j.} - \overline{X}_{..}$) and factor B ($\overline{X}_{.k} - \overline{X}_{..}$). Algebraically, this reduces to

$$\overbrace{(\overline{X}_{jk} - \overline{X}_{..})}^{\text{cell}} - \overbrace{(\overline{X}_{j.} - \overline{X}_{..})}^{\text{factor } A} - \overbrace{(\overline{X}_{.k} - \overline{X}_{..})}^{\text{factor } B}$$

$$\overline{X}_{jk} - \overline{X}_{..} - \overline{X}_{j.} + \overline{X}_{..} - \overline{X}_{.k} + \overline{X}_{..}$$

$$\overline{X}_{jk} - \overline{X}_{j.} - \overline{X}_{.k} + \overline{X}_{..}$$

Applied to Bill's situation,

$$25 - 15 - 22 + 13 = 1,$$

which suggests that 1 point of the 15-point difference between Bill's score and the grand mean is associated with the specific combination of being a male who saw aggression rewarded—namely, the AB interaction.

The combination of the differences associated with factor A ($+2$), with factor B ($+9$), and with the AB interaction ($+1$) in conjunction with the grand mean (13) now accounts for the mean of 25 in Bill's male–reward cell:

$$2 + 9 + 1 + 13 = 25$$

But not everyone in that cell had a score of 25. Bill actually scored 28. Therefore, the final reason subjects score differently from one another resides in individual differences and measurement error. This source of variability is reflected in the difference between an individual subject's score and the cell mean ($X_{ijk} - \overline{X}_{jk}$), which in Bill's case was $28 - 25 = 3$.

Applying this discussion to all subjects, individuals vary in score value ($\overline{X}_{ijk} - \overline{X}_{..}$) as a function of their level of factor A ($\overline{X}_{j.} - \overline{X}_{..}$), level of factor B ($\overline{X}_{.k} - \overline{X}_{..}$), interaction or the combination of factors A and B ($\overline{X}_{jk} - \overline{X}_{j.} - \overline{X}_{.k} + \overline{X}_{..}$), and individual differences and measurement error that produce differences between individual subjects and their respective cell means ($X_{ijk} - \overline{X}_{jk}$). Algebraically, the sentence can be expressed as follows:

$$\underbrace{(X_{ijk} - \bar{X}_{..})}_{\text{total}} = \underbrace{(\bar{X}_{j.} - \bar{X}_{..})}_{A} + \underbrace{(\bar{X}_{.k} - \bar{X}_{..})}_{B} + \underbrace{(\bar{X}_{jk} - \bar{X}_{j.} - \bar{X}_{.k} + \bar{X}_{..})}_{AB} + \underbrace{(X_{ijk} - \bar{X}_{jk})}_{\text{within}}$$

In Bill's case,

$$\underbrace{\text{total}}_{28 - 13} = \underbrace{A}_{(15 - 13)} + \underbrace{B}_{(22 - 13)} + \underbrace{AB}_{(25 - 15 - 22 + 13)} + \underbrace{\text{within}}_{(28 - 25)}$$

$$15 = \quad 2 \quad + \quad 9 \quad + \quad 1 \quad + \quad 3$$
$$15 = \quad 15$$

Sums of Squares   Variability is expressed in terms of *squared* deviations summed over all subjects—that is, in terms of sums of squares. When the algebraic statement above is squared and summed over all subjects, the left side becomes $\sum_{i=1}^{n} \sum_{j=1}^{p} \sum_{k=1}^{q}$ $(X_{ijk} - \bar{X}_{..})^2$, which is the total sum of squares, SS_{total}. When the right side is squared and summed, all the cross-product terms cancel out and the expression reduces to

$$SS_{\text{total}} = SS_A + SS_B + SS_{AB} + SS_{\text{within}}$$

This is what is meant by partitioning. The total sum of squares is divided into separate components that sum to SS_{total} and represent different sources of variability (factor A, factor B, AB interaction, within cells or groups).

Degrees of Freedom   As before, the mean squares are obtained by dividing each sum of squares by its degrees of freedom. The total sum of squares involves the squared deviation of each score from the grand mean. Thus, the number of degrees of freedom for SS_{total} will be one less than the total number of observations in the entire design. If there are n subjects per group and pq groups, the total number of subjects is npq, which may be abbreviated by N. Therefore,

$$df_{\text{total}} = npq - 1 = N - 1$$

The sum of squares for factor A involves the squared deviation from the grand mean of each of the means for the levels of factor A. There are p such means, so the degrees of freedom for the factor A variance estimate is

$$df_A = p - 1$$

Similarly, factor B has q levels, so the degrees of freedom is

$$df_B = q - 1$$

Determining the degrees of freedom for interaction is less obvious. This sum of squares involves cell mean differences from a value determined by the grand mean and by the row and column means. Therefore, the task is to determine how many cell means in the table are free to vary given the grand mean and all marginal means. Consider, for example, a table with two rows and three columns (i.e., 2 × 3):

?	?		10
			20
10	15	20	15

A little thought indicates that once the values for any two cells in a row are established—for example, the ones with question marks—then the means of the remaining cells can be determined. The open cells must have a value that would make the average within that row (or column) equal to the given marginal average. Therefore, in the above example, there are two degrees of freedom for interaction, because once those two cell means are known, all the others can be determined. It happens that the number of cells free to vary is always the product of (the number of rows minus one) and (the number of columns minus one), because one vacant cell in any row or column can be determined if the other cells in that row or column are known. Consequently, the degrees of freedom for interaction is

$$df_{AB} = (p - 1)(q - 1)$$

The sum of squares within cells is composed of the squared deviations of the scores about their own cell mean. Within any one cell, $n - 1$ of those scores are free to vary. Since there are pq cells in the design, the degrees of freedom for SS_{within} is $pq(n - 1)$:

$$df_{within} = pq(n - 1)$$
$$= pqn - pq$$
$$df_{within} = N - pq$$

Notice that the degrees of freedom for the total equals the sum of the degrees of freedom for the component sources of variability:

$$df_{total} = df_A + df_B + df_{AB} + df_{within}$$

Mean Squares   The mean squares for the several sources of variability are given by the sums of squares divided by their respective degrees of freedom:

$$MS_A = \frac{SS_A}{df_A} \qquad MS_B = \frac{SS_B}{df_B} \qquad MS_{AB} = \frac{SS_{AB}}{df_{AB}} \qquad MS_{within} = \frac{SS_{within}}{df_{within}}$$

***F* Ratios**   Once the mean squares have been determined, F ratios can be constructed to test the significance of the several possible treatment effects. In each case, the mean square for factor A, B, or the AB interaction is divided by the mean square within to produce the F ratio for that treatment effect:

$$F_A = \frac{MS_A}{MS_{within}} \qquad F_B = \frac{MS_B}{MS_{within}} \qquad F_{AB} = \frac{MS_{AB}}{MS_{within}}$$

In each case, under the null hypothesis the numerator and denominator are both estimates of the same population value. However, the numerators (in contrast to MS_{within}) are sensitive to differences between particular sets of means. To the extent that these means differ from one another, the numerator of the F ratio will become large relative to its denominator. If the resulting ratio is so large that the probability is very small that an observed value of its size could be reasonably expected under the null hypothesis, then H_0 is rejected and one concludes that the differences between the means of the effect being considered are so great that they are not likely to be merely a result of sampling error.

Numerical Illustration of Treatment Effects

It will be helpful to consider a numerical example showing how the presence of treatment effects, that is, large differences between marginal or cell means and the grand mean, are reflected in the results of an analysis of variance.

Table 13-4 presents a 2×2 analysis of variance using randomly selected numbers between 0 and 9 as the scores. Since the numbers are random, it can be assumed that the null hypothesis is true. The several mean squares should all estimate the same population variance. These different estimates will probably not equal one another, but if H_0 is true, they should deviate only by sampling error. The lower portion of the table presents the form of the traditional summary of the analysis of variance. According to the F table given in Table E of Appendix 2, the critical value for F with 1 and 16 degrees of freedom is 4.49 at the .05 significance level. As shown by the F_{obs} values on the right side of Table 13-4, none of the effects in this example is significant.

Now consider the analysis presented in Table 13-5. The same numbers are used, except that 10 has been added to each score in both the b_2 cells. This is analogous to introducing a treatment effect for factor B. It raises the marginal mean for b_2 from 6 to 16. Observe how the several sums of squares respond to this specific manipulation by comparing them to those found before the treatment effect was introduced (shown in Table 13-4). First, as one might expect, the sum of squares and mean square for factor B increase considerably as a function of this change (and so does the total sum of squares). However, notice that the sums of squares and mean squares for A, AB, and within groups do not change. This example illustrates how a selective introduction of a specific main effect is reflected only in the appropriate sum of squares and mean square.

Next, turn to the analysis presented in Table 13-6. Here the same random numbers used in Table 13-4 (without treatment effects) have been modified by adding 10 to only the scores in the upper right-hand cell, ab_{12}. This treatment should produce an interaction between factors A and B, since the pattern of results for the levels of B depends upon or interacts with the level of A. As anticipated, the analysis shows an increase in the sum of squares for interaction, SS_{AB}. But observe that the sums of squares for A, B, and the total also increase. Only the SS_{within} remains uninfluenced by this change. The inflation of SS_A and SS_B in response to the increment of a single cell results from the fact that such a manipulation also changes the marginal means of the respective levels of A and B. The SS_{within} is not altered because adding a constant to any group of scores does not change the variability of those particular scores about

13-4 Two-Factor Analysis of Variance with No Effects

Factor B

	b_1	b_2	
a_1	5 4 3 4 2	8 9 4 2 5	$\bar{X}_{1.} = 4.6$
	$\bar{X}_{11} = 3.6$	$\bar{X}_{12} = 5.6$	
a_2	6 7 5 8 4	6 9 5 9 3	$\bar{X}_{2.} = 6.2$
	$\bar{X}_{21} = 6.0$	$\bar{X}_{22} = 6.4$	
	$\bar{X}_{.1} = 4.8$	$\bar{X}_{.2} = 6.0$	$\bar{X}_{..} = 5.4$

Factor A

$$SS_A = nq \sum_{j=1}^{p} (\bar{X}_{j.} - \bar{X}_{..})^2$$

$$= (5)(2)[(4.6 - 5.4)^2 + (6.2 - 5.4)^2]$$

$$SS_A = 12.80$$

$$SS_B = np \sum_{k=1}^{q} (\bar{X}_{.k} - \bar{X}_{..})^2$$

$$= (5)(2)[(4.8 - 5.4)^2 + (6.0 - 5.4)^2]$$

$$SS_B = 7.20$$

(continued)

13-4 Continued

Interaction

$$SS_{AB} = n \sum_{j=1}^{p} \sum_{k=1}^{q} (\bar{X}_{jk} - \bar{X}_{j.} - \bar{X}_{k} + \bar{X}_{.})^2$$

$$= 5[(3.6 - 4.6 - 4.8 + 5.4)^2 + (5.6 - 4.6 - 6.0 + 5.4)^2$$
$$+ (6.0 - 6.2 - 4.8 + 5.4)^2 + (6.4 - 6.2 - 6.0 + 5.4)^2]$$

$$SS_{AB} = 3.20$$

Within Groups

$$SS_{\text{within}} = \sum_{i=1}^{n} \sum_{j=1}^{p} \sum_{k=1}^{q} (X_{ijk} - \bar{X}_{jk})^2$$

$$SS_{\text{within}} = 75.60$$

Source	df			SS	$MS = \dfrac{SS}{df}$	$F = \dfrac{MS}{MS_{\text{within}}}$
A	$p - 1$	=	1	12.80	12.80	2.71
B	$q - 1$	=	1	7.20	7.20	1.52
AB	$(p - 1)(q - 1)$	=	1	3.20	3.20	.68
Within	$N - pq$	=	16	75.60	4.72	
Total	$N - 1$	=	19	98.80		

13-5 Two-Factor Design with a Treatment Effect Added to Level b_2

Factor B

	b_1	$b_2(+10)$	
a_1	5 4 3 4 2	18 19 14 12 15	
	$\bar{X}_{11} = 3.6$	$\bar{X}_{12} = 15.6$	$\bar{X}_{1.} = 9.6$
a_2	6 7 5 8 4	16 19 15 19 13	
	$\bar{X}_{21} = 6.0$	$\bar{X}_{22} = 16.4$	$\bar{X}_{2.} = 11.2$
	$\bar{X}_{.1} = 4.8$	$\bar{X}_{.2} = 16.0$	$\bar{X}_{..} = 10.4$

Factor A

$$SS_A = nq \sum_{j=1}^{p} (\bar{X}_{j.} - \bar{X}_{..})^2$$

$$= (5)(2)[(9.6 - 10.4)^2 + (11.2 - 10.4)^2]$$

$$SS_A = 12.80$$

$$SS_B = np \sum_{k=1}^{q} (\bar{X}_{.k} - \bar{X}_{..})^2$$

$$= (5)(2)[(4.8 - 10.4)^2 + (16.0 - 10.4)^2]$$

$$SS_B = 627.20$$

(continued)

13-5 Continued

Interaction

$$SS_{AB} = n \sum_{j=1}^{p} \sum_{k=1}^{q} (\overline{X}_{jk} - \overline{X}_{j\cdot} - \overline{X}_{\cdot k} + \overline{X}_{\cdot\cdot})^2$$

$$= 5[(3.6 - 9.6 - 4.8 + 10.4)^2 + (15.6 - 9.6 - 16.0 + 10.4)^2$$
$$+ (6.0 - 11.2 - 4.8 + 10.4)^2 + (16.4 - 11.2 - 16.0 + 10.4)^2]$$

$$SS_{AB} = 3.20$$

Within Groups

$$SS_{\text{within}} = \sum_{i=1}^{n} \sum_{j=1}^{p} \sum_{k=1}^{q} (X_{ijk} - \overline{X}_{jk})^2$$

$$SS_{\text{within}} = 75.60$$

Source	df			SS	$MS = \dfrac{SS}{df}$	$F = \dfrac{MS}{MS_{\text{within}}}$
A	$p - 1$	=	1	12.80	12.80	2.71
B	$q - 1$	=	1	627.20	627.20	132.88
AB	$(p-1)(q-1)$	=	1	3.20	3.20	.68
Within	$N - pq$	=	16	75.60	4.72	
Total	$N - 1$	=	19	718.80		

13-6 Two-Factor Design with a Treatment Effect Added to Cell ab_{12}

	Factor B		
	b_1	b_2	
a_1	5 4 3 4 2	(+10) 18 19 14 12 15	$\bar{X}_{1.} = 9.6$
	$\bar{X}_{11} = 3.6$	$\bar{X}_{12} = 15.6$	
a_2	6 7 5 8 4	6 9 5 9 3	$\bar{X}_{2.} = 6.2$
	$\bar{X}_{21} = 6.0$	$\bar{X}_{22} = 6.4$	
	$\bar{X}_{.1} = 4.8$	$\bar{X}_{.2} = 11.0$	$\bar{X}_{..} = 7.9$

(Factor A labels a_1, a_2 shown at left)

$$SS_A = nq \sum_{j=1}^{p} (\bar{X}_{j.} - \bar{X}_{..})^2$$

$$= (5)(2)[(9.6 - 7.9)^2 + (6.2 - 7.9)^2]$$

$$SS_A = 57.80$$

$$SS_B = np \sum_{k=1}^{q} (\bar{X}_{.k} - \bar{X}_{..})^2$$

$$= (5)(2)[(4.8 - 7.9)^2 + (11.0 - 7.9)^2]$$

$$SS_B = 192.20$$

(*continued*)

13-6

Continued

Interaction

$$SS_{AB} = n \sum_{j=1}^{p} \sum_{k=1}^{q} (\bar{X}_{jk} - \bar{X}_{j.} - \bar{X}_{.k} + \bar{X}_{..})^2$$

$$= 5[(3.6 - 9.6 - 4.8 + 7.9)^2 + (15.6 - 9.6 - 11.0 + 7.9)^2$$
$$+ (6.0 - 6.2 - 4.8 + 7.9)^2 + (6.4 - 6.2 - 11.0 + 7.9)^2]$$

$$SS_{AB} = 168.20$$

Within Groups

$$SS_{within} = \sum_{i=1}^{n} \sum_{j=1}^{p} \sum_{k=1}^{q} (X_{ijk} - \bar{X}_{jk})^2$$

$$SS_{within} = 75.60$$

Source	df		SS	$MS = \dfrac{SS}{df}$	$F = \dfrac{MS}{MS_{within}}$
A	$p - 1 =$	1	57.80	57.80	12.25
B	$q - 1 =$	1	192.20	192.20	40.72
AB	$(p - 1)(q - 1) =$	1	168.20	168.20	35.64
Within	$N - pq =$	16	75.60	4.72	
Total	$N - 1 =$	19	493.80		

their mean. This example also illustrates that significant main effects may not be very meaningful when the interaction is significant.

These comparisons demonstrate how the analysis of variance is sensitive to different types of treatment effects. The tables show how SS_A and SS_B reflect differences in the marginal means, how SS_{AB} is changed when some but not all cells within a single level of one factor are altered, and how SS_{within} remains uninfluenced by changes in cell or marginal means. Consequently, if the null hypothesis is true and the cells are all random samples from populations having the same mean, the several mean squares all estimate the same value. However, to the extent that treatment effects exist for *A, B,* and/or *AB,* their respective mean squares will be inflated. But the mean square within groups will not be altered because it is not sensitive to differences between cell or level means. Consequently, the three *F* ratios, $\frac{MS_A}{MS_{within}}$, $\frac{MS_B}{MS_{within}}$, and $\frac{MS_{AB}}{MS_{within}}$, will test the existence of treatment effects in the population.

Assumptions and Conditions Underlying Two-Factor Analysis of Variance

Two-factor analysis of variance requires essentially the same assumptions and conditions as a simple analysis of variance and for the same reasons.

First, the subjects must be **randomly** and **independently sampled.** Second, the groups in the design must be **independent** from one another. These assumptions are necessary so that the variability within one group is not correlated with the variability within another group. While all the analysis-of-variance procedures described in this text assume independent groups, techniques do exist for handling situations in which the groups are not independent, such as a before–after experiment in which the same subjects are measured before and then after a special treatment is administered.[6]

Third, the population distributions of scores for each group in the design are **normal** in form. The assumption of normality is made so that the variance estimates in each *F* ratio will be independent, and so that the probabilities of the *F* distribution will be appropriate. Fourth, it is assumed that the populations from which the groups are drawn have equal variances. **Homogeneity of variance** must be assumed so that MS_{within} will represent an appropriate pooling of the variability within each group in the design.

Fifth, the procedures require that the factors involved in the design be **fixed.** Recall from Chapter 10 that there are two types of factors, random and fixed. A *random factor* is one for which the levels are randomly selected, and the results are generalized to the entire set of levels of which the ones in the analysis are representatives. In contrast, a *fixed factor* is one in which the levels are determined by the researchers and the results apply only to that set of levels. It is possible to have a two-factor analysis of variance in which one factor is fixed and the other is random. This is called a **mixed model.** The procedures outlined in this text are applicable only to the **fixed model**

---

[6] See "Repeated Measures Designs," in Winer et al. (1991). *Statistical Principles;* R. B. McCall and M. I. Appelbaum, "Bias in the Analysis of Repeated-Measures Designs: Some Alternative Approaches," *Child Development* 44 (1973): 401–15.

(both factors are fixed), since this is the most common of the three designs. Techniques for analyzing the random and mixed models can be found elsewhere.[7]

Finally, the techniques outlined here assume that there are an **equal number of cases** in each group and that there is **more than one observation per cell.** Again, procedures do exist for designs having unequal *n*.[8]

COMPUTATIONAL PROCEDURES

General Format

Calculators and especially computers become more useful as the size and complexity of the analysis of variance increases. As with simple analysis of variance, hand calculators can help obtain certain quantities within cells; some scientific or statistical calculators will compute an entire simple analysis of variance, and a few will compute a two-factor analysis of variance. Computers, of course, will automatically perform all calculations of a simple or two-factor (and multiple-factor) analysis once the data are entered. Follow the manual for your machine and program or consult the *Study Guide* for this text for instructions on how to use StataQuest, MINITAB, and SPSS. Even if you have access to a computer, you will understand the analysis of variance, learn the procedures and format of such an analysis, and be able to interpret its results better if you first perform a few analyses by hand.

This section and Table 13-7 describe how to calculate a two-factor analysis of variance by hand or with a hand calculator. It is not difficult, but it can take some time if the design has many subjects or levels within factors. Recall that n = the number of subjects in each group, N = the total number of subjects in the design, p = the number of levels of factor A, q = the number of levels of factor B, and X_{ijk} = the score for the ith subject in the jth level of A and kth level of B.

The first task is to cast the data into a row and column table. Second, separately for each cell, determine the sum (T_{jk}) and the sum of the squares of all scores (ΣX_{ijk}^2), which can be done in one operation on some hand calculators. Third, put the cell totals you have just computed into a row and column table similar to that presented in section A of Table 13-7. Fourth, following the table, determine the row totals ($T_{j.}$), the column totals ($T_{.k}$), the grand total ($T_{..}$), and the squares of these totals ($T_{j.}^2$, $T_{.k}^2$, $T_{..}^2$).

Next, section B presents five intermediate quantities that facilitate the computation of the analysis of variance. Quantity **(I)** is the grand total squared and then divided by the total N in the design. Quantity **(II)** is simply the sum of all squared scores. These two quantities are defined and labeled the same as for a simple analysis of variance, but the remaining quantities are different. Specifically, quantity **(III)** is the sum of the squared totals for each level of factor A divided by nq, and quantity **(IV)**

---

[7] Winer et al., *Statistical Principles*.
[8] Ibid.

13-7 | **Computational Formulas for Two-Factor Analysis of Variance**

A. Table of Totals

		Factor B				Row Totals
		b_1	b_2	...	b_q	
Factor A	a_1	T_{11}	T_{12}	...	T_{1q}	$T_{1\cdot}$
	a_2	T_{21}	T_{22}	...	T_{2q}	$T_{2\cdot}$
	$\vdots$	$\vdots$	$\vdots$	$\vdots$	$\vdots$	$\vdots$
	a_p	T_{p1}	T_{p2}	...	T_{pq}	$T_{p\cdot}$
Column Totals		$T_{\cdot 1}$	$T_{\cdot 2}$	...	$T_{\cdot q}$	$T_{\cdot\cdot}$

B. Intermediate Quantities

$$\textbf{(I)} = \frac{T_{\cdot\cdot}^2}{N} \qquad \textbf{(II)} = \sum_{i=1}^{n}\sum_{j=1}^{p}\sum_{k=1}^{q} X_{ijk}^2 \qquad \textbf{(III)} = \frac{\sum_{j=1}^{p} T_{j\cdot}^2}{nq}$$

$$\textbf{(IV)} = \frac{\sum_{k=1}^{q} T_{\cdot k}^2}{np} \qquad \textbf{(V)} = \frac{\sum_{j=1}^{p}\sum_{k=1}^{q} T_{jk}^2}{n}$$

C. Formulas and Summary Table

Source	df	SS	MS	F
A	$p-1$	(III) − (I)	$\dfrac{SS_A}{df_A}$	$\dfrac{MS_A}{MS_{\text{within}}}$
B	$q-1$	(IV) − (I)	$\dfrac{SS_B}{df_B}$	$\dfrac{MS_B}{MS_{\text{within}}}$
AB	$(p-1)(q-1)$	(V) + (I) − (III) − (IV)	$\dfrac{SS_{AB}}{df_{AB}}$	$\dfrac{MS_{AB}}{MS_{\text{within}}}$
Within	$N-pq$	(II) − (V)	$\dfrac{SS_{\text{within}}}{df_{\text{within}}}$	
Total	$N-1$	(II) − (I)		

is the sum of the squared totals for each level of factor B divided by np. Quantity **(V)** is the sum of the squared totals for all the cells divided by n.

Section C presents the formulas for obtaining all the necessary entries in the analysis of variance summary table. Notice that the sums of squares are computed quite easily by adding and subtracting the intermediate quantities. The sum of the degrees of freedom and the sum of the *SS* should equal df_{total} and SS_{total}, respectively.

Formal Example

The relative contribution of heredity and environment to behavior has been a controversial topic for many years. The most common position is that heredity and environment work in concert. One cannot talk about the role of heredity without considering the environment and vice versa; both heredity and environment produce the behavior we observe.

One illustration of this idea is that the same environmental experience affects different species in different ways. For example, consider an experiment in which twelve dogs from each of four different breeds (basenji, shetland sheepdog, wire-haired fox terrier, beagle) were either indulged or disciplined between the third and eighth weeks of their lives.[9] The indulged animals were encouraged in play, aggression, and climbing on their caretaker. In contrast, the disciplined dogs were restrained on their handler's lap, taught to sit, stay, come on command, and so on. The indulged–disciplined treatment was inspired by reports that overindulged children often cannot inhibit their impulses in structured situations. Consequently, the test of the effects of these treatments was to take each animal into a room with the handler and a bowl of meat. The dog was hungry, but the handler prevented it from eating for three minutes by hitting the animal on the rump with a rolled newspaper and shouting "no." After this period of restraint, the handler left the room, and the length of time it took the dog to begin to eat the meat (latency) was recorded. Presumably, the indulged animals should go to the food more quickly (record shorter latencies) than the disciplined dogs.

Analysis A simplified set of hypothetical data is presented in section A of Table 13-8. Factor *A* (rearing condition) has two levels, indulged and disciplined. Factor *B* is breed and has four levels. Note that the levels of factor *B* constitute separate nominal categories, not an interval or higher scale. Consider some possible results. It may be that the disciplined dogs will refrain from eating the food while the indulged will feast quickly. In this case, one would expect that the analysis of variance would yield a significant effect for factor *A*. Another result might be that regardless of rearing condition, some breeds generally go to the food more rapidly than others, in which case there would be a significant effect for factor *B*. Last, the effect of indulged or disciplined rearing may depend on the breed. For one breed the rearing treatment might make a difference, whereas for another it might not (or it might even have an effect opposite to that for another breed). In that event, the analysis would yield a significant interaction between rearing and breed. If an interaction of this form turned out to be significant, then any main effects for rearing and breed would be reevaluated and perhaps disregarded. Section B of Table 13-8 gives the totals for each cell of the design and the marginal totals. Section C presents the computation of the intermediate quan-

---

[9] Inspired by, but not identical to, D. G. Freedman (1958). "Constitutional and Environmental Interactions in Rearing of Four Breeds of Dogs," *Science* 127: 585–86.

13-8 Computational Example of a Two-Factor Analysis of Variance (Rearing × Breed Example)

A. Raw Data

| | | Factor B (Breed) | | |
	b_1 (Basenjis)	b_2 (Shetlands)	b_3 (Terriers)	b_4 (Beagles)
a_1 (Indulged)	1	7	6	9
	4	10	9	7
	3	10	7	10
	1	9	8	10
	2	6	5	8
	2	8	10	9
a_2 (Disciplined)	5	9	1	2
	1	9	0	6
	4	8	3	3
	1	10	1	4
	2	5	2	5
	3	8	4	3

Factor A (Rearing)

(continued)

Continued

B. Table of Totals

	b_1	b_2	b_3	b_4	
a_1	$T_{11} = (1 + \cdots + 2)$ $= 13$	$T_{12} = (7 + \cdots + 8)$ $= 50$	$T_{13} = (6 + \cdots + 10)$ $= 45$	$T_{14} = (9 + \cdots + 9)$ $= 53$	$T_{1.} = 161$
a_2	$T_{21} = (5 + \cdots + 3)$ $= 16$	$T_{22} = (9 + \cdots + 8)$ $= 49$	$T_{23} = (1 + \cdots + 4)$ $= 11$	$T_{24} = (2 + \cdots + 3)$ $= 23$	$T_{2.} = 99$
	$T_{.1} = 29$	$T_{.2} = 99$	$T_{.3} = 56$	$T_{.4} = 76$	$T_{..} = 260$

C. Intermediate Quantities

(I) $\dfrac{T_{..}^2}{N} = \dfrac{(260)^2}{48} = 1408.3333$

(II) $\sum_{i=1}^{n} \sum_{j=1}^{p} \sum_{k=1}^{q} X_{ijk}^2 = (1^2 + 4^2 + 3^2 + \cdots + 5^2 + 3^2) = 1896.0000$

(III) $\dfrac{\sum_{j=1}^{p} T_{j.}^2}{nq} = \dfrac{(161^2 + 99^2)}{(6)(4)} = 1488.4167$

(IV) $\dfrac{\sum_{k=1}^{q} T_{.k}^2}{np} = \dfrac{(29^2 + \cdots + 76^2)}{(6)(2)} = 1629.5000$

(V) $\dfrac{\sum_{j=1}^{p} \sum_{k=1}^{q} T_{jk}^2}{n} = \dfrac{(13^2 + 16^2 + \cdots + 53^2 + 23^2)}{6} = 1801.6667$

D. Sums of Squares and Degrees of Freedom

SS_A = **(III)** − **(I)** = 1488.4167 − 1408.3333 = 80.0834

df_A = $p - 1 = 2 - 1 = 1$

SS_B = **(IV)** − **(I)** = 1629.5000 − 1408.3333 = 221.1667

df_B = $q - 1 = 4 - 1 = 3$

SS_{AB} = **(V)** + **(I)** − **(III)** − **(IV)** = 1801.6667 + 1408.3333 − 1488.4167 − 1629.5000 = 92.0833

df_{AB} = $(p - 1)(q - 1) = (1)(3) = 3$

SS_{within} = **(II)** − **(V)** = 1896.0000 − 1801.6667 = 94.3333

df_{within} = $N - pq = 48 - (2)(4) = 40$

SS_{total} = **(II)** − **(I)** = 1896.0000 − 1408.3333 = 487.6667

df_{total} = $N - 1 = 48 - 1 = 47$

E. Summary Table

Source	df	SS	MS	F
A (Rearing)	1	80.0834	80.0834	33.96**
B (Breed)	3	221.1667	73.7222	31.26**
AB (Rearing × Breed)	3	92.0833	30.6944	13.02**
Within	40	94.3333	2.3583	
Total	47	487.6667		

**$p < .01$

tities, section D displays the calculation of the sums of squares and degrees of freedom, and section E gives the summary of the analysis.

In the present example, the critical values of F for 3 and 40 degrees of freedom are 2.84 and 4.31 for tests at the .05 and .01 levels, respectively. The observed F for interaction is 13.02, a value that exceeds the critical level at $p = .01$. The F's for the main effects have been calculated for completeness of presentation. A formal summary of this analysis is presented in Table 13-9. Since there are three sources of variability

13-9

Summary of the Two-Factor Analysis of Variance (Rearing × Breed Example)

Hypotheses

Factor A	Factor B
H_0: $\alpha_1 = \alpha_2 = 0$	H_0: $\beta_1 = \beta_2 = \beta_3 = \beta_4 = 0$
H_1: Not H_0	H_1: Not H_0

AB Interaction

H_0: $\alpha\beta_{11} = \alpha\beta_{12} = \cdots = \alpha\beta_{24} = 0$

H_1: Not H_0

Assumptions and Conditions

1. The subjects are **randomly** and **independently sampled.**
2. The groups are **independent.**
3. The population variances for the groups are **homogeneous.**
4. The population distributions are **normal** in form.
5. The factors in the study are **fixed.**
6. The number of observations in each group is **equal** and **greater than 1.**

Decision Rules (.05 level; from Table E)

	Factor A ($df = 1, 40$)	Factor B ($df = 3, 40$)	AB Interaction ($df = 3, 40$)
Do not reject H_0 if:	$F_{obs} < 4.08$	$F_{obs} < 2.84$	$F_{obs} < 2.84$
Reject H_0 if:	$F_{obs} \geq 4.08$	$F_{obs} \geq 2.84$	$F_{obs} \geq 2.84$

Computation (See Table 13-8.)

$F_{obs} = 33.96$	$F_{obs} = 31.26$	$F_{obs} = 13.02$

Decision

Since the F for interaction exceeds the critical level, reject H_0. It is concluded that the effects of indulgent and disciplined rearing are not the same for each breed of dog.

to be tested, there are three sets of hypotheses, critical values, decision rules, and so on. Moreover, notice that the hypotheses are now stated in terms of **treatment effects** as a notational convenience. Observe that the symbol α_1 represents the difference in the population between the mean for a_1 and the grand mean: $\alpha_1 = \mu_{a_1} - \mu$. Under the null hypothesis, the population means for a_1 and a_2 are presumed to be equal to each other and thus to the population grand mean. Therefore, since $\mu_{a_1} = \mu$ and $\mu_{a_2} = \mu$, each treatment effect for factor A will be zero, and consequently the null hypothesis can be written H_0: $\alpha_1 = \alpha_2 = 0$ with the alternative being H_1: Not H_0. The same logic applies to the hypotheses for factor B and the AB interaction.

A plot of the data, presented in Figure 13-2, will help to interpret the significant interaction. Note that the graph is not a polygon but a **bar graph** consisting of bars for the indulged and disciplined groups within each breed. The vertical height of the bars represents the mean length of latency to eat the meat in the test situation. A tall bar indicates a long hesitation. The bar graph, rather than connected points, is used in this context because the abscissa (breed) represents discrete categories (that is, a nominal scale) and not points along an interval or higher scale.

An interaction implies that differences between levels of one factor are not the same within each level of the other factor. In terms of the example, the difference between indulged and disciplined rearing conditions is not the same within each breed of animal. Notice in the graph that the basenjis ate the food quickly, regardless of whether they were indulged or disciplined. The shetlands also showed little difference between rearing conditions but seemed to begin eating much later. In contrast, the rearing conditions did seem to affect the terriers and beagles: the indulged dogs of these two breeds took longer to go to the food than their disciplined companions— behavior opposite to that widely suspected of indulged children.

Comparisons In Chapter 10 on simple analysis of variance, procedures were described to test the difference between pairs of means within an analysis of variance. Recall that the correct procedure depended upon whether the comparison was a priori

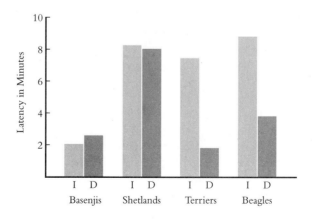

Figure 13-2 Results in which a significant interaction occurred for the rearing (indulged, disciplined) × breed experiment.

or a posteriori (that is, decided upon before or after the experiment was conducted) and the number of such comparisons to be made. The selection of appropriate procedures for making comparisons within the two-factor analysis of variance is even more complicated, and the interested student should consult more advanced texts, such as Hays and Winer et al.[10]

Size of Relationship

In Chapter 10 on the simple analysis of variance, procedures were introduced to estimate the size of a treatment effect, not just its statistical significance. This was done because a significant result of a statistical test of significance alone indicates only that some effect exists that is not likely attributable to sampling error, but it does not communicate how large that effect is. Moreover, nearly any difference between means may be statistically significant, regardless of how small, if the sample size is large enough. Consequently, some index of the size of a treatment effect is needed that will communicate the magnitude of difference represented in an effect and that is less dependent on the sample size.

There are several such indices, each with advantages and disadvantages. The one presented in this text is **omega squared,** symbolized by ω^2, which estimates the proportion of the total variance in the analysis that is associated with group differences. Readers may wish to review the logic of this index as well as its interpretation and limitations discussed on pages 260–263.

In the case of two-factor analysis of variance, a separate estimate of the proportion of total variability is derived for each possible source of variance—factor A, factor B, and the AB interaction:

$$\text{est. } \omega^2_A = \frac{SS_A - (p - 1)\, MS_{\text{within}}}{MS_{\text{within}} + SS_{\text{total}}}$$

$$\text{est. } \omega^2_B = \frac{SS_B - (q - 1)\, MS_{\text{within}}}{MS_{\text{within}} + SS_{\text{total}}}$$

$$\text{est. } \omega^2_{AB} = \frac{SS_{AB} - (p - 1)(q - 1)\, MS_{\text{within}}}{MS_{\text{within}} + SS_{\text{total}}}$$

These formulas are appropriate for two-factor analyses of variance having independent groups and fixed factors. If the F ratio for a particular treatment effect is less than 1.00, omega squared is set to zero.

To illustrate, reconsider the computational example regarding the indulged vs. disciplined rearing of four breeds of dogs presented in Table 13-8. The significance tests for factors A and B as well as the AB interaction were all significant. As discussed previously, the interpretation of a significant main effect is uncertain when the interaction is also significant, and the same caution pertains to the interpretation of estimates of the sizes of effects. So interpretive emphasis should be placed on the interaction effect. The quantities from Table 13-8 needed to estimate omega squared for the interaction are $SS_{AB} = 92.0833$, $p = 2$, $q = 4$, $MS_{\text{within}} = 2.3583$, and $SS_{\text{total}} = 487.6667$. Therefore,

---

[10] W. L. Hays, *Statistics,* 5th ed. (Fort Worth: Harcourt Brace, 1994); Winer et al., *Statistical Principles.*

$$\text{est. } \omega^2_{AB} = \frac{SS_{AB} - (p-1)(q-1) MS_{\text{within}}}{MS_{\text{within}} + SS_{\text{total}}}$$

$$= \frac{92.0833 - (2-1)(4-1) 2.3583}{2.3583 + 487.6667}$$

$$\text{est. } \omega^2_{AB} = .17$$

This means that approximately 17% of the total variability in latency to seek food in this entire study is associated with the Rearing × Breed interaction.

To determine the sizes of the main effects in this example for illustrative purposes, we require the additional quantities of $SS_A = 80.0834$ and $SS_B = 221.1667$. Then

$$\text{est. } \omega^2_A = \frac{SS_A - (p-1) MS_{\text{within}}}{MS_{\text{within}} + SS_{\text{total}}} = \frac{80.0834 - (2-1) 2.3583}{2.3583 + 487.6667} = .16$$

$$\text{est. } \omega^2_B = \frac{SS_B - (q-1) MS_{\text{within}}}{MS_{\text{within}} + SS_{\text{total}}} = \frac{221.1667 - (4-1) 2.3583}{2.3583 + 487.6667} = .44$$

Although care must be taken in interpreting these results because an interaction effect can influence the sizes of main effects, it appears that rearing condition (A), breed (B), and their interaction combine to be associated with $.17 + .16 + .44 = 77\%$ of the total variability in latency to eat a prohibited food, with most of the variance associated with breed (B).

SUMMARY

Two-factor analysis of variance is a method of testing the differences between means for groups that simultaneously represent different levels of two classification schemes, called factors. The analysis produces separate tests of the main effect of factor A, the main effect of factor B, and the interaction of factors A and B. An interaction occurs when the effect for one factor is not the same under all levels of the other factor. The assumptions and logic of two-factor analysis of variance are similar to those of simple analysis of variance.

Exercises

For Conceptual Understanding

1. Decide whether each of the following is possible or impossible:

   a. factor A significant, factor B significant, interaction significant

   b. factor A significant, factor B not significant, interaction not significant

   c. factor A not significant, factor B not significant, interaction significant

2. List the assumptions for two-factor analysis of variance and explain why they are necessary.

3. Define an interaction. If an interaction is significant, why are any significant main effects usually not interpreted?

4. The following is a fairly random set of data:

Factor B

		b_1	b_2
		3	5
		6	7
a_1		5	8
		4	2
		6	8
		9	4
a_2		4	6
		5	3

Factor A (on the left margin spanning a_1 and a_2)

Compute the analysis of variance on these data. Now add 5 to each score in both cells of the a_2 level and recompute. Explain the differences and similarities in results between the first and second computations. Now take the original data and add 10 to each score in cell ab_{11} and recompute. Compare all three analyses and explain their similarities and differences.

For Solving Problems

5. Researchers interested in different kinds of cognitive skills devised simple tests of spatial and verbal abilities. In the test of verbal abilities, subjects are asked to mentally run through the alphabet and count the number of letters, including the letter e, that contain the "ee" sound when pronounced. In the test of spatial abilities, subjects are asked to mentally run through the alphabet and count the letters that have curves when printed in capital form. Because males and females have often been found to differ in spatial versus verbal abilities, the researchers were interested in sex differences on these tests. Scores of four different groups of college students, in which a high score indicates greater speed and accuracy, are given below.[11] Assume that the two tests score subjects on the same scale.

---

[11] Based upon, but not identical to, M. Coltheart, E. Hull, and D. Slater (1975). "Sex Differences in Imagery and Reading," *Nature* 253: 438–40.

Male Subjects on the Verbal Test

4, 3, 4, 6, 3

Male Subjects on the Spatial Test

7, 6, 8, 5, 9

Female Subjects on the Verbal Test

5, 6, 8, 9, 7

Female Subjects on the Spatial Test

2, 4, 4, 7, 8

Evaluate these data with the analysis of variance. Draw up a formal summary of the analysis, then graph and interpret your results.

6. Suppose the researchers were interested in whether children showed a similar pattern of performance on the Verbal Test and Spatial Test described in Exercise 5. They gave the tests to four groups of 8-year-old children. Following the procedure used in Exercise 5, evaluate the following data and interpret the results of your analysis.

Male Subjects on the Verbal Test

1, 2, 2, 2, 3

Male Subjcts on the Spatial Test

4, 2, 3, 4, 2

Female Subjects on the Verbal Test

3, 4, 4, 3, 6

Female Subjects on the Spatial Test

3, 3, 2, 4, 2

7. Suppose an educational researcher was interested in the long-term educational histories of youth considered to be "underachievers" during high school—students who performed more poorly in school than one would expect on the basis of their mental abilities. Many counselors of underachieving students believe them to be rebelling against their parents, and advise parents of underachievers that their underachieving adolescents, typically boys, will get their acts together once they leave home. The researcher studied five groups of high school males—(1) severe underachievers, (2) moderate underachievers, (3) mild underachievers, (4) nonunderachievers who had the same grades as the underachievers (same GPA), and (5) nonunderachievers who had the same mental abilities as the underachievers (same MA). Each of the five groups was divided into high-medium, and low-ability groups on the basis of a mental test. The students were contacted 13 years after graduating from high school and asked a variety of questions, including the number of years of education after high school they had obtained. The data are presented below for the five groups. Decide whether the advice the counselors are giving is supported by the data and for whom.[12]

8. Estimate the proportion of variance associated with each source in the analysis of variance conducted in Exercise 6.

Mental Ability		Underachievers			Nonunderachievers	
		Severe	Moderate	Mild	Same GPA	Same MA
	Low	2	1	2	3	2
		0	2	1	2	3
		0	2	3	0	3
		1	0	1	2	0
	Medium	0	1	4	3	5
		1	1	2	4	3
		2	2	3	4	3
		1	3	3	2	3
	High	1	2	6	4	3
		2	1	4	4	6
		0	4	4	6	6
		1	4	3	4	4

[12] Based upon, but not identical to, R. B. McCall, C. Evahn, and L. Kratzer. *High School Underachievers: What Do They Become as Adults?* Newbury Park, CA: Sage, 1992.

For Homework

9. Given the following data, test for significant effects with the analysis of variance ($\alpha = .01$).

Factor B

	b_1	b_2
a_1	3	5
	3	4
	4	6
	2	5
a_2	8	4
	8	5
	7	7
	9	12

Factor A

10. Add 5 to every score in cell ab_{22} in Exercise 9 and recalculate the analysis. Explain any changes in sums of squares.

11. A plant nursery investigates the factors "type of sunlight" (A) and "amount of water" (B) on the heights (in feet) of young maple trees. Below are the data from this study. Present the analysis of variance summary table and interpret the results.

		Factor B	
	b_1 (driest)	b_2 (medium)	b_3 (wettest)
a_1 (shade)	1	2	5
	2	3	3
	1	5	4
	3	5	3
	3	3	5
a_2 (indirect sun)	2	4	6
	3	3	7
	1	6	6
	4	4	5
	2	5	7
a_3 (full sun)	3	6	5
	1	7	7
	3	5	5
	4	5	8
	4	4	6

Factor A

12. Estimate the proportion of total variability associated with each effect in the analysis in Exercise 11.

Nonparametric Techniques

Many hypothesis-testing techniques have been presented in the preceding chapters. To use these tests appropriately, several assumptions must be met concerning the population distributions. For example, the *t* test of the difference between two independent means requires the assumption that the two population distributions are normal and have the same variance. Since one almost never has precise knowledge about the population, one can only guess about the tenability of some of these assumptions. Furthermore, there are times when certain assumptions simply cannot be met. For example, if a very easy examination is given to a class, there may be many scores of 100% and only a few scores as low as 80%. This distribution would be markedly skewed to the left and decidedly not normal. Therefore, there is a need for statistical techniques that may be used when some of the assumptions required by the previously described procedures cannot be met.

PARAMETRIC AND NONPARAMETRIC TESTS

The statistical techniques described in the previous chapters represent tests on the values of certain parameters (for example, μ or ρ) and have made certain assumptions about other parameters (such as $\sigma_1^2 = \sigma_2^2 = \cdots = \sigma_p^2$). Consequently, they are known collectively as **parametric** statistical tests. Methods that do not test hypotheses about specific parameters and that require different (and sometimes fewer) assumptions are known as **nonparametric** tests. They are also called **distribution-free** tests, since they do not require the assumption that scores are from a normal distribution.

Although nonparametric tests are frequently used because certain assumptions cannot be made about the populations involved, a researcher may select a nonparametric technique to analyze data for other reasons as well. Sometimes the variables in question are measured with ordinal or even nominal scales (see Chapter 1). Parametric tests are not usually appropriate in such cases. For example, suppose an educator wants to correlate the scores from a new test of reading achievement with a teacher's rank ordering of pupils on reading proficiency. The teacher's rankings represent an ordinal scale, and a nonparametric index of the degree of relationship would be more appropriate than the Pearson *r*. Thus, when measurement is nominal or ordinal in character, a nonparametric test may be the only good choice.

If nonparametric techniques do not require all the assumptions that parametric techniques do, why are they not always used? In fact, parametric procedures are used more often than nonparametric ones and are traditionally preferred when everything else is equal. One reason is that parametric tests are robust with respect to violations of some of their assumptions. That is, failure to have a perfectly normal distribution is not very damaging to the accuracy of the probability values obtained with the *t* test or the analysis of variance. Even rather substantial departures from normality have a relatively minor impact on the result of the test, especially if the sample size is large. Similar statements can be made about the robustness of these techniques when group variances are not equal. Consequently, *moderate* violations of the assumptions of normality and homogeneity of variance are often not strong reasons for choosing a nonparametric over a parametric test.

In addition, parametric methods usually have a greater **power-efficiency** than do nonparametric methods. As explained in Chapter 8, the power of a statistical test is the probability that the test will reject the null hypothesis when that hypothesis is, in fact, false (that is, correctly reject H_0).[1] Power-efficiency refers to the power of a test relative to the sample size, and permits one to compare the power of two different statistical tests. If the difference between the central tendency of two groups is being considered, a t test is more likely to detect a population difference than is an appropriate nonparametric test for the given N's.

Another reason parametric methods are often preferred is that they provide information nonparametric methods do not. For example, the two-factor analysis of variance includes a test for interaction. It is more difficult to assess this type of effect with nonparametric methods.

A final consideration is more subtle. Frequently, parametric and nonparametric tests do not address precisely the same question. They are sensitive to different aspects of the data. For example, if one wants to know whether the mean of one group differs from the mean of another in the population, a parametric test makes a rather direct assessment of this question, given the assumptions. A nonparametric test designed to make a similar evaluation may actually ask whether one distribution is different from another *in any way,* and distributions may differ not only in central tendency but also in variability, skewness, and so on. Thus, although parametric and nonparametric tests are often closely related, situations can occur in which a significant difference will be obtained with one technique and not with another because parametric and nonparametric tests often evaluate slightly different aspects of the data.

To summarize, parametric tests possess the advantages of being fairly robust with respect to violations of assumptions, having more power-efficiency (everything else being equal), and sometimes providing more information about a phenomenon (such as interactions in the analysis of variance). However, when departures from normality or homogeneity of variance are severe, when the data are nominal or ordinal, or when the research question addresses issues other than the value of specific parameters, a nonparametric method may be more appropriate.

The techniques presented in this chapter are samples of some of the most common nonparametric methods. Many others exist and the interested reader is referred to textbooks that specialize on this topic.[2] Table 14-1 presents a summary of the statistical techniques, both parametric and nonparametric, presented in this text. They are divided into tests of the differences between groups and tests of association, and within each of these two categories the tests are presented according to type of sample. In the second column, the distribution, variance, and scale requirements for each test are given so that you can compare similar parametric and nonparametric tests with respect to these assumptions. The nonparametric tests listed in this table are described in detail in the remainder of this chapter.

---

[1] For a more detailed discussion of power-efficiency, see S. Siegel and N. J. Castellan Jr. (1988). *Nonparametric Statistics for the Behavioral Sciences* (2nd ed.). New York: McGraw-Hill.

[2] J. V. Bradley, *Distribution-Free Statistical Tests* (1968), Englewood Cliffs, NJ: Prentice-Hall; L. A. Marascuilo and M. McSweeney, *Nonparametric and Distribution-Free Methods for the Social Sciences* (1977), Monterey, CA: Brooks/Cole; F. Mosteller and R. Rourke, *Sturdy Statistics* (1973), Reading, MA: Addison Wesley; Siegel, *Nonparametric Statistics.*

	14-1	Summary of Statistical Tests and Their Distinguishing Characteristics

Type of Sample	Distribution, Variance, and Minimum Scale Requirements
Tests of the Differences between Groups	
Two Independent Samples	
t test	Normality, Homogeneity, Interval
Mann–Whitney *U* test	Ordinal
Two Correlated Samples	
t test	Normality, Interval
Wilcoxon test	Ordinal
Several Independent Samples	
Simple analysis of variance	Normality, Homogeneity, Interval
Two-factor analysis of variance	Normality, Homogeneity, Interval
Kruskal–Wallis test	Ordinal
Tests of Association	
One Sample	
Pearson product-moment correlation	Normality, Interval
Spearman rank-order correlation	Ordinal
Chi-square test for $r \times c$ tables	Nominal
Two Independent Samples	
Test of the difference between two Pearson correlations	Normality, Interval

TESTS ON INDEPENDENT SAMPLES

Pearson Chi-Square Test

On occasion, social and behavioral scientists collect independent and random samples of observations and want to compare them in terms of the similarity with which those observations are distributed among several discrete and mutually exclusive categories. Suppose that two or more groups of subjects are randomly selected. The groups may be boys and girls, students and nonstudents, athletes and nonathletes, and so on. In addition, suppose there is a set of several discrete categories into which any particular subject may be classified. For example, subjects may be judged to have warm, neutral, or cold personalities, or to have several levels of agreement or disagreement with a given statement or political figure. No matter what the classification system, the categories must be exhaustive and mutually exclusive: each subject must belong to one and only one category. The result is two or more groups of subjects who fall into one of two or more mutually exclusive categories. The question may be whether the groups have the same distributions of representation across the several categories. A **Pearson chi-square test,** named after Karl Pearson, the same person associated with the Pearson r, can be used to analyze data from a study with this design.

For example,[3] suppose a sample of 120 married men between 35 and 40 years of age were asked in 1960 the source of their greatest satisfaction in life—their career, their family, or some other activity. A similar sample of 110 married men were asked the same question in 1997. Notice that the two samples—1960 and 1997—are independent (that is, they are not the same or related individuals), and each person's score can be tallied in one and only one category—career, family, or other. The results are presented below:

	Career	Family	Other	Total
Men in 1960	65	42	13	120
Men in 1997	25	68	17	110
	90	110	30	230

Sometimes this format is called a **contingency table.** The research question is whether married men have changed in the source of their greatest life satisfaction. Another way to phrase the same question is to ask whether the distribution of the scores of greatest life satisfaction depends, or is *contingent,* upon the year of the survey.

In this and each of the following presentations of a statistical technique, a summary of the procedures is presented in tabular form (see Table 14-2) and their details are discussed in the text.

Hypotheses   The statistical question is whether the groups differ in the relative distribution of observations among the different categories. The null hypothesis is that in the population the two groups do not differ in their relative frequency distribution of people among the categories. The alternative hypothesis is that the groups do differ. Notice that H_1 does not specify *how* they differ but only presumes that the two relative distributions are not the same in some way.

The hypotheses apply only to a population that has the same characteristics as the sample, specifically, one that has the same marginal distributions. The **marginal dis-**

---

[3] Based on an extension and modification of results reported in S. Findlay (1984, November 5), 'Liberated' Men Grope for New Roles, *USA Today.*

tributions are given in the right and bottom margins of the above table (for example, the 120 and 110; and the 90, 110, and 30). Thus, this test of the association between year and source of life satisfaction applies only to populations in which (1) relatively equal numbers of married men exist at the two years and (2) those men generally emphasize family slightly more than career and both of these sources substantially more than others.

The marginal distributions define the nature of the population; the question is whether an association between year and source of life satisfaction exists within that population. The hypotheses begin with the statement "Given the observed marginals" to remind us of this fact.

Assumptions and Conditions   First, it must be assumed that the two samples are **independent** of each other. That usually implies that different and unrelated sets of subjects have been selected. Second, the subjects within each group must be **randomly** and **independently sampled.** Third, each observation must qualify for **one and only one category** in the classification scheme. Fourth, the **sample size must be relatively large.** The last assumption is discussed below.

Rationale and Computation   The formula used to relate a statistic to a theoretical relative frequency distribution is based upon a comparison of the frequencies actually observed and the frequencies that would be expected if the same population distribution of sources of life satisfaction existed at both years. Once again, the null hypothesis is tentatively presumed to be true—specifically, that the same population distribution exists at both years. As usual, the observed data will not conform precisely with this hypothesized state of affairs. Is the difference between the observed and expected frequencies a reasonable outcome of sampling error? If not, reject the null hypothesis.

A major requirement of this technique is to estimate the set of frequencies that would exist in the two samples of married men if the population distributions of sources of life satisfaction were the same for the two years, that is, if the null hypothesis were true. These figures can be estimated from the observed data in the following way. Consider the number of 1960 men who would have listed career as the major source if there were no differences between the 1960 and 1997 samples. From the marginals along the bottom of the table, we note that over both years $90/230 = 39.13\%$ of the married men listed career as the major source of satisfaction. Thus, if no differences existed between the years, 39.13% of the men should list career. Since the right-hand margin indicates 120 men were surveyed in 1960, 39.13% of 120, or 46.96 men should have done so in 1960. Therefore, the **expected frequency** in the top left cell is 46.96, which would have been expected, given the marginal distributions, if the distributions for the two years were identical—that is, if the null hypothesis were true.

The same process may be used to obtain the expected frequency of each cell in the table. The computation of expected frequencies is most often stated in the following terms:

To compute the **expected frequency** of any cell, multiply the marginal total for the row that contains the cell by the marginal total for the column that contains the cell, and divide this product by the total number of cases in the table.

For the 1960-career cell the expected frequency is given by

$$\frac{(120)(90)}{230} = 46.96$$

Since 90/230 is the percentage of men listing career and 120 is the total number surveyed in 1960, this formula uses the same logic as stated above.

If the expected frequencies are computed in this manner for each cell, the observed frequencies (symbolized by O) and the expected frequencies (symbolized by E) for the same data are as follows:

	Career	**Family**	**Other**
Men in 1960	$O = 65$ $E = 46.96$	$O = 42$ $E = 57.39$	$O = 13$ $E = 15.65$
Men in 1997	$O = 25$ $E = 43.04$	$O = 68$ $E = 52.61$	$O = 17$ $E = 14.35$

If the calculation of the expected frequencies is correct, the row and column sums of the expected frequencies will equal the row and column sums of the observed frequencies, although they may be off by .01 or so because of rounding. This fact can be used to check the expected frequency calculations.

We now have a set of expected frequencies, given that the null hypothesis of the equivalence of population distributions within the categories for the two groups is true and given the observed marginal values. Now the task is to determine some index of the extent to which the observed frequencies are consistent with the null hypothesis (that is, with the expected frequencies). The following expression represents such an index, and its sampling distribution approximates a theoretical distribution called **chi square** (X^2):

$$\chi^2_{obs} = \sum_{j=1}^{r} \sum_{k=1}^{c} \frac{(O_{jk} - E_{jk})^2}{E_{jk}}$$

in which

O_{jk} = the observed frequency in the cell corresponding to the intersection of the jth row and kth column

E_{jk} = the expected frequency in the cell corresponding to the intersection of the jth row and kth column

r = the number of rows

c = the number of columns

The formula directs one to take the difference between the observed and expected frequencies for each cell, square it, divide by the expected frequency, and sum these values over all cells. The resulting total is distributed as chi square with degrees of freedom

$$df = (r - 1)(c - 1)$$

in which r is the number of rows and c the number of columns in the table of frequencies.

The percentiles of the chi-square distribution for each number of degrees of freedom are known, and Table G in Appendix 2 lists values of chi square for selected significance levels at several df. In the present case, the table of data has two rows and three columns; thus, there are $(r - 1)(c - 1) = (2 - 1)(3 - 1) = 2$ degrees of freedom. The critical value for a nondirectional test at the .05 level of significance is 5.99. If the observed value of chi square, symbolized by χ^2_{obs}, exceeds the critical value, the null hypothesis, H_0, is rejected.

For the data presented, the calculation of χ^2_{obs} proceeds as follows:

$$\chi^2_{obs} = \frac{(65 - 46.96)^2}{46.96} + \frac{(42 - 57.39)^2}{57.39} + \frac{(13 - 15.65)^2}{15.65}$$

$$+ \frac{(25 - 43.04)^2}{43.04} + \frac{(68 - 52.61)^2}{52.61} + \frac{(17 - 14.35)^2}{14.35}$$

$$\chi^2_{obs} = 24.06$$

Since the observed value of chi square of 24.06 exceeds the critical value of 5.99 and, in fact, is beyond the critical value for a test at the .001 level, the null hypothesis is rejected. It can be concluded from the data that the distribution of sources of life satisfaction for married men has changed from 1960 to 1997. Apparently, family has become a more frequent source of satisfaction than career during the last 37 years. See Table 14-2 for a formal summary of this example.

When the assumptions were listed for performing such a test, it was noted that a large sample is required. The approximation to the theoretical chi-square distribution is not very good for small samples and thus the probabilities are somewhat inaccurate. Unfortunately, the closeness of the approximation is a function of many factors, and a single rule of thumb concerning sample size is not totally adequate. However, a conservative guideline would be not to have any expected frequency less than 5. If the table is only 2×2 in size, then no expected frequency should be less than 10. Notice that the requirement is based upon the *expected* frequencies, not the observed frequencies.

Special Application: Goodness of Fit   The chi–square approach to comparing observed and expected frequencies can be used to determine whether a single distribution of cases differs from an expected distribution by sampling error. In this application, *only one sample group exists,* the members of which can be placed in only one of several categories. Also, *the expected frequencies are given,* rather than being determined from the data as was the case in the contingency table described above. That is, the expected frequencies in the population are known, for example, or they can be assumed to equal a specified value depending upon the question being addressed. In general, the statistical question is whether the observed distribution of cases conforms to, or "fits," the expected distribution, and it is for this reason that such applications are called tests of **goodness of fit.** They make the same assumptions that were made for contingency tables (see Table 14–2).

For example, pollsters frequently want to demonstrate that their survey is based upon a sample that is similar, within sampling error, to the population to which they

14-2 Summary of $r \times c$ Chi-Square Test (Source of Satisfaction $\times$ Year Example)

Hypotheses

H_0: Given the observed marginals, the distributions of frequencies in the population are not different for the groups.

H_1: Given the observed marginals, these distributions are different for the groups (nondirectional).

Assumptions and Conditions

1. The groups are **independent.**
2. The subjects for each group are **randomly** and **independently sampled.**
3. Each observation must qualify for **one and only one category.**
4. The sample size must be fairly large such that **no expected frequency is less than 5 for r or c greater than 2, or less than 10 if $r = c = 2$.**

Decision Rules (from Table G)

Given: .05, a nondirectional test, and $df = (r - 1)(c - 1) = 2$

If $\chi^2_{obs} < 5.99$, do not reject H_0.

If $\chi^2_{obs} \geq 5.99$, reject H_0.

Computation (see text for data and details)

$$\chi^2_{obs} = \sum_{j=1}^{r} \sum_{k=1}^{c} \frac{(O_{jk} - E_{jk})^2}{E_{jk}}$$

in which

O_{jk} = the observed frequency in the cell corresponding to the intersection of the jth row and kth column

E_{jk} = the expected frequency in the cell corresponding to the intersection of the jth row and kth column, determined by multiplying the marginal totals of frequencies for the row and column that contain the cell and dividing by the total of the frequencies in the table.

r = the number of rows

c = the number of columns

$$\chi^2_{obs} = 24.06$$

Decision

Reject H_0.

wish to generalize their results with respect to age, gender, income, education, religious preference, ethnic, and racial composition. If they want to generalize to the entire United States population of persons 18 years or older, they may consult the latest U.S. Census to obtain the expected percentages in the population, which they will use to estimate the number of cases in each category for a sample of the size they obtained. If a telephone survey was used, for example, they may be especially interested to know whether the sample underrepresented low-income, low-education persons who do not have telephones.

Specifically, suppose a pollster conducted a survey of 1000 U.S. adults 18 years of age and older regarding their religious preference. It was especially important that the sample appropriately represent the major religious groups in the country. The religious preferences for the sample are given at the left of the table below. From the U.S. Census, the researcher found the percentages of adults in the U.S. population listing these preferences, and then multiplied those percentages by 1000 to obtain the expected frequencies for the sample of 1000. Now one has a set of observed (from the survey sample) and expected (from the U.S. Census) frequencies:

Religious Preference	Observed (Sample)	Expected (Population)
Protestant	702	670
Catholic	222	250
Jewish	34	30
Other	26	30
None	16	20

A chi square observed can be calculated using essentially the same formula as above, except one has r rows and only one column:

$$\chi^2_{obs} = \sum_{j=1}^{r} \frac{(O_j - E_j)^2}{E_j}$$

For the data presented, the observed chi square is

$$\chi^2_{obs} = \sum_{j=1}^{5} \frac{(O_j - E_j)^2}{E_j} = \frac{(702 - 670)^2}{670} + \frac{(222 - 250)^2}{250} + \frac{(34 - 30)^2}{30}$$
$$+ \frac{26 - 30)^2}{30} + \frac{(16 - 20)^2}{20}$$

$$\chi^2_{obs} = 6.53$$

This observed value of 6.53 can be compared with a critical value χ^2_{crit} having degrees of freedom equal to $(r - 1)$ *or one less than the number of categories.* In this case, the critical value from Table G in Appendix 2 for a nondirectional test at the .05 level for $df = r - 1 = 4$ is 9.49. Since the observed chi square does not exceed the critical value, the test is not significant, and the obtained sample is within sampling error of the distribution of religious preferences of the U.S. population of adults.

Sometimes the question is whether the observed distribution is different from that which would be expected by chance. Suppose a market researcher asks a sample of 120 prospective customers which one of four advertising displays for Jamaica they judged to be the best—that is, the most likely to persuade them to vacation in Jamaica. The researcher may want to know whether the distribution of responses across the four displays is different from chance; that is, do the respondents have statistically significant preferences? In this case, the expected frequencies representing "chance" responding are all equal; specifically, if no preferences exist, the 120 respondents should be distributed equally across the four categories, or $120/4 = 30$ respondents would be expected per category. The data might be:

Advertising Display	Observed (Sample)	Expected (Chance)
A	10	30
B	15	30
C	55	30
D	40	30
	120	120

Again, the same formula described above with $df = r - 1 = 4 - 1 = 3$ would be used to determine chi square observed, which is

$$\chi^2_{obs} = \sum_{j=1}^{r} \frac{(O_j - E_j)^2}{E_j} = \frac{(10 - 30)^2}{30} + \frac{(15 - 30)^2}{30}$$

$$+ \frac{(55 - 30)^2}{30} + \frac{(40 - 30)^2}{30}$$

$$\chi^2_{obs} = 45.00$$

Since the critical values for $df = 3$ are 7.82 at .05 and 16.27 at .001, respondents had significant preferences for the four advertising displays that are unlikely to be simply the consequence of sampling error.

Special Application: The Significance of a Proportion   The chi-square procedures described above can be used to test the significance of a proportion, which is a special case of the goodness-of-fit procedures when subjects from *one sample* fall into just *two categories*—agree/disagree, yes/no, male/female, etc. Usually, but not always, the expected frequencies are equal—that is, half the sample would be expected to be in each of the two categories by "chance." Of course, if it is known that some other proportion exists in the population, then that ratio could be used to determine the expected frequencies.

Suppose, for example, a researcher studying infants observes that of the 400 parents who have volunteered to bring their babies to the laboratory for several studies on play behavior, 250 or 62.5% are parents of infant boys. The researcher wonders whether parents of boys are more likely to volunteer for such studies than parents of girls or whether this observed percentage deviates from the expected by chance or sampling error alone. If the researcher assumes that boys and girls are equally represented in the local population, the expected frequencies would be 200 boys and 200 girls. In this case, the question would be whether 62.5% can be expected to deviate from 50% by sampling error alone. If the researcher knows that 104 boys are born for every 100 girls, then 104/204 = 51% of the available infants are boys, so the expected frequencies for a sample of 400 would be .51(400) = 204 boys and 196 girls. Then the question is whether 62.5% can be expected to deviate from 51% by sampling error alone. Using the latter expected frequencies, one has

Gender	Observed (Sample)	Expected (Population)
Male	250	204
Female	150	196
	400	400

The same formula given above may be used to determine chi square observed with $df = r - 1 = 1$. Specifically:

$$\chi^2_{obs} = \sum_{j=1}^{2} \frac{(O_j - E_j)^2}{E_j} = \frac{(250 - 204)^2}{204} + \frac{(150 - 196)^2}{196}$$

$$\chi^2_{obs} = 21.17$$

With $df = 1$, a nondirectional test at the .05 level requires a critical value of 3.84, so parents of boys are more likely to volunteer their infants for such studies.

Recall that each of the expected frequencies in such a table must be at least 10, and the accuracy of the probabilities increases if the expected frequencies are larger. If the expected frequencies are less than 10, another test, called **Fisher's exact test of a proportion,** is described in more advanced texts. If the expected frequencies are low but nevertheless more than 10, the accuracy of the above procedure may be improved by using **Yates' correction,** which consists of subtracting .5 from the absolute value of the difference between expected and observed frequencies before squaring those differences:

$$\chi^2_{obs} = \sum_{j=1}^{2} \frac{(|O_j - E_j| - .5)^2}{E_j}$$

For the above data, we have

$$\chi^2_{obs} = \frac{(|250 - 204| - .5)^2}{204} + \frac{(|150 - 196| - .5)^2}{196} = 20.71$$

Mann–Whitney *U* Test for the Difference between Two Independent Samples

If two samples are randomly and independently selected, if there is an underlying continuous distribution, and if there is at least an ordinal scale of measurement, then the **Mann–Whitney *U* test** may be used to evaluate the difference between the two population distributions. The *U* test is one of the most popular alternatives to the parametric *t* test for independent groups.

Hypotheses   The null hypothesis is that the populations from which the two samples have been drawn are identical. The alternative hypothesis is that these two populations are not identical. Note that this is not equivalent to testing the difference between two population means, since two distributions could be quite different in form but have identical means. Therefore, it is theoretically possible to obtain a statistically significant result with the *U* test when the means are actually identical. From a practical standpoint, the general shapes of the distributions of two groups within a single experiment are not often markedly different. When the forms of the distributions are similar, the *U* test does compare the central tendencies of the groups. Thus, if the forms of the sample distributions are similar, the results of the *U* test are often interpreted in terms of differences in central tendency; if the forms of the distributions are not similar, the results must be viewed in terms of the difference between the distributions in general.

Assumptions and Conditions   The assumptions required for the *U* test are few. **Random** and **independent sampling** with **independent groups** is required. It is also assumed that the dependent variable is **continuous.** This implies that an infinite number of values theoretically exists between any two measured values (see Chapter 1). Last, the measurement scale must be at least **ordinal** in character.

Rationale and Computation   The rationale for the test is based upon the premise that if two distributions of equal size are identical, then a listing of the observations from both groups mixed together with scores placed in rank order (that is, smallest first) should yield a sequence in which the scores from the two groups are in random order. If so, the number of scores from group *A* that precede scores from group *B* should equal, within sampling error, the number of scores from *B* that precede scores from *A*. If the scores from group *A* tend to be smaller in value than those from *B*, then more of the *A* scores will precede *B* scores when all the scores from both groups are listed in rank order.

The statistic based upon this rationale is the Mann–Whitney *U*. Suppose that the sets of scores for groups *A* and *B* are as follows:

$$A = \{5, 9, 17, 3\}$$
$$B = \{1, 8, 28, 20, 18\}$$

Arranging the scores in order beginning with the smallest score (while retaining group identity of each observation) and assigning each score a rank beginning with rank 1 for the smallest score produces the following table:

Score	1	3	5	8	9	17	18	20	28
Group	B	A	A	B	A	A	B	B	B
Rank	1	2	3	4	5	6	7	8	9

Although the value of U can be obtained by counting the number of scores of one group that precede those of another, a more convenient method requires that the total (T_j) of the ranks for one group (either group) be obtained. For the example just presented,

$$T_A = 2 + 3 + 5 + 6 = 16$$

The statistic U is then given by

$$U_{obs} = n_A n_B + \frac{n_A(n_A + 1)}{2} - T_A$$

in which n_A and n_B are the numbers of cases in groups A and B, respectively. For the present example,

$$U_{obs} = 4(5) + \frac{4(4 + 1)}{2} - 16$$

$$U_{obs} = 14$$

Size of *n* If *n* for each group is 20 or less, Table H in Appendix 2 gives the critical values for U. Notice that Table H presents tables for several different significance levels. In each case, the rows and columns represent different numbers of cases for the two groups. It makes no difference which group is used for the rows and which for the columns. At the intersection of the appropriate row and column, two values of U are given. These two values represent a type of "confidence interval," such that if the observed U falls *between* these two values, H_0 is not rejected. If the observed value is less than or equal to the lower value in the table or greater than or equal to the larger value in the table, H_0 is rejected. The critical confidence interval deals with the fact that either group may be labeled A in the formula, producing either high or low values of U_{obs} depending upon which group is chosen.

If **n for either group is greater than 20**, then Table H cannot be used. It happens that with such a large sample, the observed value of U approaches a normal distribution with

$$\text{mean} = \frac{n_A n_B}{2}$$

and

$$\text{standard deviation} = \sqrt{\frac{(n_A)(n_B)(n_A + n_B + 1)}{12}}$$

Consequently, if the size of a group is greater than 20 and the difference in sample sizes of the two groups is not too great, the significance of U_{obs} may be determined by calculating

$$z_{obs} = \frac{U_{obs} - n_A n_B / 2}{\sqrt{\dfrac{(n_A)(n_B)(n_A + n_B + 1)}{12}}},$$

which approaches the standard normal distribution. In that event, critical values of z may be obtained by consulting Table A of Appendix 2.

Ties Occasionally, the values of scores are tied.[4] For example, consider the set of scores $\{13, 15, 15, 18\}$. In this case, the score of 13 receives rank 1. The two scores of 15 are given the average of the next two ranks. The next two ranks are 2 and 3, and their average is 2.5. Consequently, each of the two scores of 15 receives a rank of 2.5. The score of 18 is then assigned a rank of 4. It is not given a rank of 3 because this rank was used in the previous averaging. If three or more scores are tied, each score receives the average rank that these scores would have received if they were distinct. The next score begins with the next unused rank. An example of ranking with several ties follows:

Score	13	13	16	19	19	22	22	22	28	28	30
Rank	1.5	1.5	3	4.5	4.5	7	7	7	9.5	9.5	11

Small-Sample Illustration One premise of Freudian theory is that a person is born with certain innate drives and needs, such as an oral need. Further, the energy system

---

[4] The method of handling tied observations suggested in this chapter is a common and convenient one, but it has certain technical liabilities. Most rank-order statistics assume an underlying continuous distribution. If this is true, ties in scores result from imprecision of measurement. That is, if more precise methods were available, no ties would exist. Consequently, in this instance the occurrence of ties is not a true reflection of what really exists. The issue of just how to handle ties is not firmly resolved, and Bradley (*Statistical Tests*) presents a good though sophisticated discussion of the alternatives. One approach advocated by Siegel and Castellan (*Nonparametric Statistics*) is to use the average-rank technique presented here and correct for ties with special formulas. However, Siegel points out, the corrections do not change the result very much even when a large proportion of the scores are tied. Bradley notes that under some conditions the average-rank approach actually biases the outcome in one direction or another rather than yielding a result that is itself a type of average or medium approximation. One alternative to this problem is to treat the tied scores as if they were not tied scores, selecting the ranks in the manner *least* favorable to rejecting the null hypothesis. Then, rerank the observations, this time treating the ties in a manner *most* favorable to rejecting the null hypothesis. As a result, one obtains two test statistics, one most and one least favorable toward rejecting the null hypothesis. If both statistics fall into the region of rejection, the null hypothesis may be unambiguously rejected. If both fall outside the rejection region, the null hypothesis may be unambiguously not rejected. However, if one value does and the other does not fall within the rejection region, no clear decision can be made. In short, *most of the methods discussed in this chapter technically assume that no ties exist.* Since ties are common in behavioral science, some procedures must be followed to handle them, and the choice of which method to follow rests on technical considerations largely beyond the scope of this text. However, the careful student will note that the method of handling ties is an issue and will be sensitive to the possible qualifications their presence may make upon the accuracy of conclusions.

of an organism is presumed to be closed, so that if a need is blocked from satisfaction the need does not dissipate but will be satisfied in other ways. For example, during the early months of life if oral needs (such as need for sucking) are not exercised, there should be more oral activity of other kinds later in life to make up for this frustration. In contrast, learning theory predicts just the opposite. A person who is not allowed to suck as an infant does not learn to suck and may suck less on other objects, for example, later in life. Which is correct, Freudian or learning theory? An experiment is run in which four infants are fed from special cups as soon as possible after birth.[5] Presumably, these infants do not satisfy their sucking instinct and should suck more later to compensate. In contrast, another six infants are fed on bottles throughout the course of infancy. These infants should receive adequate sucking experience and later should not need to suck as much. After eight months of these experiences, samples are made of the infants' behavior to determine how much of the time the infants suck their thumbs. These measures are expressed in terms of the percentage of time spent sucking during the observations.

The null hypothesis is that the population distribution for cup-fed babies is the same as for bottle-fed babies. The alternative hypothesis is that these distributions are different in some way.

The assumptions are that the subjects are **randomly** and **independently sampled,** the dependent variable is **continuous,** and the measurement scale is at least **ordinal.** The .05 level will be adopted.

In the following table, *A* indicates cup feeding and *B* bottle feeding. The scores are percentages.

Score	3	5	6	9	10	12	13	13	16	24
Group	*A*	*A*	*A*	*B*	*B*	*A*	*B*	*B*	*B*	*B*
Rank	1	2	3	4	5	6	7.5	7.5	9	10

Ranking scores is an important aspect of this and other nonparametric tests, so it is helpful to test the accuracy of the ranks assigned with two kinds of checks. First, the highest score should receive, or share with a tied score, a rank equal to that of the total number of scores (N). In the table above, the highest score, 24, has a rank of 10, which equals the total number of cases to be ranked. Second, the sum of N rankings should always be as follows:

$$\text{sum of } N \text{ ranks} = \frac{N(N + 1)}{2}$$

For $N = 10$ subjects, the sum of the ranks should equal

$$\frac{10(10 + 1)}{2} = \frac{10(11)}{2} = 55$$

---

[5] Inspired by, but not identical to, R. R. Sears and G. W. Wise (1950). "Relation of Cup Feeding in Infancy to Thumbsucking and the Oral Drive," *American Journal of Orthopsychiatry* 20: 123–38.

The sum of the ranks in the bottom row of the above table is also 55. Notice that these checks determine whether all the correct rankings have been used, but they do not test whether the ranks have been correctly assigned to individuals or groups of subjects.

U_{obs} is computed as follows:

$$U_{obs} = n_A n_B + \frac{n_A(n_A + 1)}{2} - T_A$$

$$= 4(6) + \frac{4(5)}{2} - 12$$

$$U_{obs} = 22$$

Looking at Table H of Appendix 2 for a nondirectional test at the .05 level (i.e., the second panel of the table), one finds the critical values of U for $n = (4, 6)$ are 2 and 22. The observed value of U is 22, and since this equals the critical value for U, H_0 is rejected. The distributions are probably different, and it would appear that cup-fed infants suck their thumbs less, not more, than bottle-fed infants. According to learning theory, they have not learned to need or to want to suck.

Large-Sample Illustration Suppose a similar experiment is carried out with a much larger sample, 25 cup-fed and 24 bottle-fed infants. The assumptions and hypotheses are the same as before, but since the sample sizes are large (greater than 20), the standard normal approximation will be used. For a nondirectional alternative at the .05 level, the observed z must be less than or equal to -1.96 or greater than or equal to 1.96 to reject H_0.

The data and computation are presented in Table 14-3 and a formal summary of the procedure is presented in Table 14-4. The data are arranged in increasing order of score value within each group. Then, disregarding group affiliation, the scores are rank ordered and the total of the ranks for the two groups is determined. The value of U_{obs} is computed in Table 14-3. The formula approximating the standard normal is given in Table 14-4 and the corresponding values entered. The resulting observed z is 2.07, which complies with the second decision rule, and the null hypothesis is rejected. The interpretation is that the groups probably do differ in the population, and an examination of the data indicates that learning theory is favored.

Kruskal–Wallis Test for *k* Independent Samples

The Mann–Whitney U test for two samples may be generalized to several independent samples. The most common approach to this kind of analysis is known as the **Kruskal–Wallis test.** It is analogous to the parametric simple analysis of variance but does not make all of the same assumptions. In Chapter 13, a study was described in which children were shown a movie of an adult striking a large doll. The adult was either rewarded or punished for that behavior. Suppose there had been three rather than two different films shown to the children in the experiment.[6] All three movies showed an

---

[6] Similar to A. Bandura, D. Ross, and S. A. Ross (1963). "Imitation of Film-Mediated Aggressive Models," *Journal of Abnormal and Social Psychology* 66: 3–11.

14-3 Data and Computation for Large-Sample Mann–Whitney U Test (Cup vs. Bottle Example)

Cup-Fed		Bottle-Fed	
Score	Rank	Score	Rank
8	2	3	1
10	3	22	9
11	4.5	27	10
11	4.5	29	11.5
12	6	30	13
18	7	36	14
21	8	51	23
29	11.5	51	23
45	15	51	23
46	16	59	26
49	18	65	27
49	18	74	30
49	18	76	31.5
50	20.5	76	31.5
50	20.5	81	33.5
57	25	83	35
71	28	96	40.5
73	29	98	42
81	33.5	122	44
89	36	135	45
90	37	142	46
93	38	159	47
94	39	183	48
96	40.5	190	49
109	43		
$n_A = 25$	$T_A = 521.5$	$n_B = 24$	$T_B = 703.5$

Check: $T_A + T_B = 521.5 + 703.5 \stackrel{?}{=} \dfrac{N(N + 1)}{2} = \dfrac{49(49 + 1)}{2} = 1225$

$$U_{obs} = n_A n_B + \frac{n_A(n_A + 1)}{2} - T_A$$

$$= 25(24) + \frac{25(25 + 1)}{2} - 521.5$$

$$U_{obs} = 403.5$$

14-4 Summary of Large-Sample Mann–Whitney *U* Test (Cup vs. Bottle Example)

Hypotheses

H_0: The population distributions from which the samples are drawn are identical.

H_1: These populations are different in some way (nondirectional).

Assumptions and Conditions

1. The observations are **randomly** and **independently sampled**.
2. The groups are **independent**.
3. The dependent variable is **continuous** and the measurement scale is at least **ordinal**.

Decision Rules (from Table A)

Since the sample is large ($n > 20$), the standard normal approximation will be used. For a nondirectional test at .05, the decision rules are

If $-1.96 < z_{obs} < 1.96$, do not reject H_0.

If $z_{obs} \leq -1.96$ or $z_{obs} \geq 1.96$, reject H_0.

Computation

$$U_{obs} = n_A n_B + \frac{n_A(n_A + 1)}{2} - T_A$$

in which n_A and n_B are the numbers of subjects in samples A and B, respectively, and T_A = the sum of the ranks for sample A. U_{obs} is calculated to be 403.5 in Table 14-3 for the illustrative data. The approximation to the standard normal distribution is given by

$$z_{obs} = \frac{U_{obs} - n_A n_B / 2}{\sqrt{\dfrac{(n_A)(n_B)(n_A + n_B + 1)}{12}}}$$

$$= \frac{403.5 - 24(25)/2}{\sqrt{\dfrac{(24)(25)(24 + 25 + 1)}{12}}}$$

$$z_{obs} = 2.07$$

Decision

Reject H_0.

adult striking and otherwise assaulting a large doll, Bobo. Each film depicted one of three consequences to the aggressor: the person was rewarded with praise, punished with reprimands, or nothing at all happened. The question is whether vicarious experience with different consequences would determine the number of imitative aggressive behaviors in a ten-minute test situation in which each child is left in a room with Bobo and other toys.

Hypotheses   The null hypothesis is that the three samples have identical population distributions, while the alternative hypothesis is that their population distributions are different. If the distributions have the same form, then H_1 implies that the mean scores are higher or lower in some of the groups than in others.

Assumptions and Conditions   Assume that the three **independent groups** of subjects are **randomly sampled,** the dependent variable is **continuous,** and it is measured with at least an **ordinal** scale. The conceptual dependent variable—which is continuous—is the extent to which the child imitates the adult's behavior, and it is measured by counting the number of imitative acts, which is at least an ordinal measurement. Finally, there should be **at least five observations per group** for an accurate estimation of the probability.

Rationale and Computation   The data and computation for this example are presented in Table 14-5. The method is similar to that for the Mann–Whitney test. The scores are arranged into groups in ascending order of score value. Then, without regard to group affiliation, all the observations are ranked, the smallest score value being assigned a rank of 1. If any scores are tied, each tied score receives the average of the ranks available for those scores. The ranks are totaled within each group, and this total is signified by T_j for the jth group. Once again, the accuracy of the rankings can be partly checked by making sure that the highest score receives (or shares) a rank equivalent to the total number of subjects in the analysis (N) and that the sum of the ranks in the entire analysis equals $N(N + 1)/2$.

   The general rationale is that if the groups are distributed in the population in the same way and with the same central tendency, then the average of the ranks for the several samples should be approximately equal. To the extent that the scores in one group are higher than those in the others, the several averages of ranks will be unequal. Obviously, the averages are not likely to be precisely equal, but under the null hypothesis they will differ only because of sampling error. When the differences in average ranks become so great that it is implausible to attribute them to sampling error, the tentative presumption of the null hypothesis is rejected and the existence of treatment effects in the population is suspected.

   This logic is translated into a formula for the statistic H that reflects the extent to which the sum of the ranks for the several groups differ from one another. The formula for H_{obs} is

$$H_{\text{obs}} = \left[\frac{12}{N(N + 1)}\right]\left[\sum_{j=1}^{k}\frac{T_j^2}{n_j}\right] - 3(N + 1)$$

14-5

Data and Computation for Kruskal–Wallis Test (Imitation Example)

Punished		Ignored		Rewarded	
Imitations	Rank	Imitations	Rank	Imitations	Rank
0	1	2	3	8	13.5
1	2	3	5	12	16.5
3	5	4	7.5	13	18
3	5	6	10.5	16	20.5
4	7.5	7	12	19	22
5	9	10	15	21	23
6	10.5	12	16.5	22	24
8	13.5	14	19	23	25
		16	20.5		
$n_1 = 8$	$T_1 = 53.5$	$n_2 = 9$	$T_2 = 109$	$n_3 = 8$	$T_3 = 162.5$

$$\text{Check: } \sum_{j=1}^{k} T_j = 53.5 + 109 + 162.5 \overset{?}{=} \frac{N(N+1)}{2} = \frac{25(25+1)}{2} = 325$$

$$H_{obs} = \left[\frac{12}{N(N+1)} \right] \left[\sum_{j=1}^{k} \frac{T_j^2}{n_j} \right] - 3(N+1)$$

$$= \left[\frac{12}{25(25+1)} \right] \left[\frac{(53.5)^2}{8} + \frac{(109)^2}{9} + \frac{(162.5)^2}{8} \right] - 3(25+1)$$

$$H_{obs} = 13.91 \star\star\star$$

$$df = k - 1 = 3 - 1 = 2$$

$$H_{crit} = \chi^2_{.05,df=2} = 5.99, \chi^2_{.001,df=2} = 13.82$$

in which

k = the number of groups in the analysis

n_j = the number of subjects in group j

N = the total number of subjects in the analysis

T_j^2 = the square of the total of the ranks for scores in group j

The formula directs one to sum the ranks and square this total separately for each group. Divide each squared total by the number of observations in that particular group, and sum across all groups to obtain

$$\sum_{j=1}^{k} \frac{T_j^2}{n_j}$$

Enter this and N into the above expression for H_{obs}.

The sampling distribution of the statistic H has approximately the same form as chi square with $k - 1$ degrees of freedom (k = the number of groups). This approximation is close only when there are at least five observations per group, and the accuracy of the approximation improves as N increases.

In the present case there are three groups, so the degrees of freedom for H is $k - 1 = 3 - 1 = 2$. The critical value of chi square for a nondirectional test with $df = 2$ is 5.99, as obtained from Table G in Appendix 2. Thus, if H_{obs} is greater than or equal to 5.99, the null hypothesis of equivalence of population distributions will be rejected. The computation in Table 14-5 reveals H_{obs} to be 13.91, which exceeds the critical value at .05 (and at .001 as well), so H_0 is rejected. The analysis suggests that the populations from which these three groups are drawn do differ in some way, and an examination of the data implies that a child is more likely to imitate if he or she sees the model rewarded than ignored or punished. A formal summary of this analysis is presented in Table 14-6.

TESTS ON CORRELATED SAMPLES

All tests described thus far in this chapter have been for independent samples—groups of observations on separate or unmatched subjects. Just as special procedures are required for the *t* test when the sets of observations are made on the same or matched subjects, so too a different nonparametric analysis must be made when the samples are not independent. One of the most common nonparametric tests for this situation is the **Wilcoxon test** for two correlated or matched samples.

Wilcoxon Test for Two Correlated Samples

If the same subjects are measured under two conditions (or if matched pairs of subjects provide the scores) and if the measurement is at least ordinal for both within-pair and between-pair differences, then the Wilcoxon test may be used to test the null hypothesis that the population distributions corresponding to the two sets of observations are identical.

In the sensorimotor development of organisms there seems to be a close relationship between visual experiences and the opportunity for the organism to interact physically with its environment. Some theorists have suggested that the development of normal visually guided behavior requires not only experience with visual stimuli but also feedback obtained through physical interaction with the stimulus environment. One application of this theory was made in the following experiment.[7] Eight pairs of kittens—the members of each pair coming from the same litter but each pair from a different litter—were reared in darkness until they were 10 weeks of age. They then received visual stimulation for only a short period of time each day under very

---

[7] Inspired by, but not identical to, R. Held and A. Hein (1963). "Movement-Produced Stimulation in the Development of Visually-Guided Behavior," *Journal of Comparative and Physiological Psychology* 56: 872–76.

14-6 **Summary of Kruskal–Wallis Test (Imitation Example)**

Hypotheses

H_0: The population distributions from which the groups are sampled are identical.

H_1: These distributions are different in some way.

Assumptions and Conditions

1. Subjects are **randomly** and **independently sampled** and divided into **k independent** groups with all $n_j \geq 5$.

2. The dependent variable is **continuous** and the measurement scale is at least **ordinal**.

Decision Rules (from Table G)

Given: the .05 level of significance, a nondirectional test, and that H is distributed as chi square with $k - 1 = 3 - 1 = 2$ df

If $H_{obs} < 5.99$, do not reject H_0.

If $H_{obs} \geq 5.99$, reject H_0.

Computation (see Table 14-5)

$$H_{obs} = \left[\frac{12}{N(N + 1)} \right] \left[\sum_{j=1}^{k} \frac{T_j^2}{n_j} \right] - 3(N + 1)$$

in which

k $=$ the number of groups in the analysis

n_j $=$ the number of subjects in group j

N $=$ the total number of subjects in the analysis

T_j^2 $=$ the square of the total of the ranks for scores in group j

H_{obs} $=$ 13.91

Decision

Reject H_0.

special conditions. The two kittens of each pair were placed into a vertically oriented cylindrical apparatus with stripes painted on the side. One member of each pair was allowed to move about at will, except that it wore a harness attached by means of pulleys and gears to a gondola-like carriage that held the other kitten. A post in the center of the apparatus prevented one member of the pair from seeing the other, and the circumstances were arranged so that the visual experience of the two kittens was quite similar. If the theory is correct, the active kitten should show faster development on visual–motor tasks than the kitten carried passively through the environment in the gondola. A series of tasks that measure visual–motor development was used, and each subject received a score indicating proficiency on the test battery.

Hypotheses   The null hypothesis is that the population distributions under the two conditions are identical. The alternative hypothesis is that they are not identical. Usually, if the distributions are symmetrical, the alternative hypothesis is taken to mean that the central tendency of one distribution is higher than that of the other.

Assumptions and Conditions   The assumptions required for the Wilcoxon test are less restrictive than those required for the analogous *t* test. First, the Wilcoxon test assumes **correlated groups or samples.** In the present example, pairs of measurements from related subjects—littermates—are the basic data. Because littermates—siblings—share some genes in common, they are assumed to be more similar to one another than nonlittermates, so groups composed of littermates will be correlated. In other experiments, the dependent variable might be measured twice on the same subjects—before and after an experimental treatment, for example. In addition to correlated groups, the Wilcoxon test assumes that the matched pairs or individual subjects who are measured twice are **randomly** and **independently sampled.** In the present example, each pair of kittens must be randomly and independently selected from its litter, and only one pair may come from a given litter so that each *pair* of subjects is independent of each other *pair.* Moreover, both the measurement scale and the differences between related measurements must be at least **ordinal,** so that it will be possible to rank order the differences between related pairs of scores.

Rationale and Computation   The procedure requires that one score value in each pair be subtracted from the other to obtain the difference between them, and that the algebraic sign of the difference be retained. Following this, the *absolute values* of the differences are assigned ranks starting with a rank of 1 for the smallest difference. Recall that the signs of the differences were retained but that the rankings were made without regard to sign. The ranks may now be attributed to positive and negative differences between the two groups. If there is no difference between the two groups, then the sum of the ranks associated with positive differences between the groups should be about equal to the sum of the ranks associated with negative differences between the groups. Since under this circumstance the total of all ranks would be divided relatively evenly between positive and negative differences, the smaller and the larger of these two sums would be equal within sampling error under H_0. If there is a difference between groups, then either the positive or negative sum of ranks will be quite a bit smaller than the other. As the difference between groups increases, the smaller of the two sums gets smaller and smaller. The sampling distribution of this smaller sum is known, and when the observed total becomes sufficiently small as to be unlikely to have arisen solely because of sampling error, the null hypothesis of no difference is rejected.

The eight pairs of subjects and their scores are listed in the computation section of Table 14-7. The difference between these scores (d_i) is computed. If a difference is zero (see pair D), it is dropped from the analysis, and the total N (the number of *paired* observations) is reduced by the number of such zero differences. Then the absolute

14-7

Summary of Wilcoxon Test for Correlated Groups (Sensorimotor Experience Example)

Hypotheses

H_0: The population distributions for the two correlated groups of observations are identical.

H_1: These two distributions are different in some way (nondirectional).

Assumptions and Conditions

1. The pairs of observations are **randomly** and **independently sampled,** but the two observations of a pair are made on the **same** or **matched** subjects.

2. At least **ordinal measurement** is available both between pairs and between members' scores within each pair (that is, d_i).

Decision Rules (from Table I)

Given: the .05 level, a nondirectional alternative, and $N = 7$ (see below)

If $W_{obs} > 2$, do not reject H_0.

If $W_{obs} \leq 2$, reject H_0.

Computation

Pair	Active Kitten	Passive Kitten	d_i	$|d_i|$	Rank of $|d_i|$	Signed Rank of $|d_i|$
A	14	12	2	2	5	5
B	13	10	3	3	7	7
C	11	10	1	1	2	2
D	12	12	0	0	eliminated	
E	15	13	2	2	5	5
F	11	12	−1	1	2	−2
G	13	11	2	2	5	5
H	15	14	1	1	2	2

N = number of pairs minus the number of zero differences $\quad T_+ = 26$

$N = 8 - 1 = 7$ $\qquad\qquad\qquad\qquad\qquad\qquad\qquad\quad T_- = 2$

$\qquad\qquad\qquad\qquad\qquad\qquad\qquad\qquad\qquad\qquad\qquad\quad W_{obs} = 2$

Check: $T_+ + T_- = 26 + 2 \overset{?}{=} \dfrac{N(N + 1)}{2} = \dfrac{7(7 + 1)}{2} = 28$

Decision

Reject H_0.

values of these differences ($|d_i|$) are taken and ranked, *assigning the rank of 1 to the smallest absolute difference*. If some of the absolute differences are of equal value (see pairs A, E, G), these ties are assigned the average of the ranks that they would have received if they had been distinct.

After the absolute differences have been ranked, the last column of the table merely repeats the rank that each matched pair has received but with the algebraic sign of the difference in score values attached to the rank. Now two quantities are computed: T_+ is the total of the ranks having positive signs associated with them in the last column of the table, and T_- is the sum of the ranks having negative signs. The rankings may be checked by noticing whether the largest d_i has been given (or shares) a rank equal to the total number of pairs having a nonzero difference (7 for this particular test); in addition, $T_+ + T_-$ should equal $N(N + 1)/2$.

The statistic of interest is W_{obs}, which is simply the smaller of T_+ and T_- (the signs of the ranks are now ignored). If the two groups are similar, these two totals will be about equal. As the difference between groups increases, so does the difference between T_+ and T_-, and W_{obs} (the smaller of T_+ and T_-) takes on a smaller and smaller value. When W_{obs} becomes sufficiently small relative to its sampling distribution, H_0 is rejected.

For N (the number of *paired* observations sampled *less the number of pairs having zero differences in scores*) between 5 and 50, the critical values for the sampling distribution of W as a function of N for several levels of significance and for directional and nondirectional tests are presented in Table I of Appendix 2. If W_{obs} is *less than or equal to* the tabled critical value, H_0 is rejected and a difference in the population is presumed. In the present illustration, $N = 7$, and if the test is taken to be nondirectional at the .05 level, the critical value is 2. Since W_{obs} equals this critical value, H_0 is rejected and the observed difference between the freely moving and gondola kittens is not likely to result from sampling error. Apparently, movement-produced stimulation is more enriching than stimulation that is passively experienced. A formal summary is presented in Table 14-7.

When N is greater than 50, an approximation to the standard normal may be used (actually the approximation is sufficiently good so that it can be employed with $N \geq 10$). The conversion to a standard normal deviate is given by the following:

$$z_{obs} = \frac{W_{obs} - N(N + 1)/4}{\sqrt{\dfrac{N(N + 1)(2N + 1)}{24}}}$$

RANK-ORDER CORRELATION

Spearman Rank-Order Correlation Coefficient

The Pearson product-moment correlation introduced in Chapter 6 is an appropriate index of the degree of association between two variables if the measurements are made with interval or ratio scales, not ordinal scales. But if the data are ordinal and are ranked, then applying the formula for the Pearson coefficient to the ranks provides a useful index of ordinal association called the **Spearman rank-order correlation,** symbolized by r_s.

Suppose that a group of 15 nursery-school children is observed by two judges who rank the children on their social aggressiveness. Each judge rank orders the 15 children, assigning a rank of 1 to the child presumed to have the least aggressive behavior and a rank of 15 to the child presumed to have the most aggressive behavior. The question is, to what extent do the two judges agree in their rankings? That is, how reliable are the two judges? The requirements for computing such an index are that the subjects be **randomly and independently sampled** and that the measurement be at least **ordinal.**

Suppose the data are those presented in Table 14-8. In the present case, the raw data are themselves rank orderings of the subjects by two different judges. If the data were numbers of aggressive acts, for example, *these measurements would need to be rank ordered before proceeding.* Notice that *observations are ranked only within, not across, a condition.* That is, scores for each variable (judge, in this case) are ranked separately. Ties may be handled as before, by assigning the average of the ranks that the tied observations would otherwise have received.

The method proceeds by computing the difference (d_i) between the ranks for each

14-8 **Rankings of Social Aggressiveness by Two Judges**

Pupil	Judge I	Judge II	d_i	d_i^2
A	1	3	−2	4
B	4	4	0	0
C	5	8	−3	9
D	10	5	5	25
E	8	2	6	36
F	14	15	−1	1
G	7	9	−2	4
H	2	6	−4	16
I	12	14	−2	4
J	9	7	2	4
K	15	13	2	4
L	3	1	2	4
M	13	12	1	1
N	11	10	1	1
O	6	11	−5	25

$$N = 15 \qquad \Sigma d_i = 0 \qquad \Sigma d_i^2 = 138$$

subject and then squaring each of these differences. The sum of the squared differences in ranks $\left(\sum_{i=1}^{N} d_i^2\right)$ and the number of *pairs of observations* (N) *including those with zero differences* are entered into the following formula for r_s:

$$r_S = 1 - \left[\frac{6\left(\sum_{i=1}^{N} d_i^2\right)}{N^3 - N}\right]$$

For the data presented in Table 14-8:

$$r_S = 1 - \left[\frac{6(138)}{15^3 - 15}\right]$$

$$r_S = .75$$

The formula given above for the Spearman rank-order correlation is a simplification of the formula for the Pearson product-moment correlation as applied to ranked data. It can be shown that when the Pearson formula is applied to data that have been ranked, the expression can be reduced to the formula given above for r_S. However, if ordinal data exist which are then ranked, the Pearson formula applied to ranked data (that is, r_S) will not be exactly the same value as that obtained when the formula is applied to the data before ranking takes place.

Testing the Significance of r_s

As was mentioned in the discussion of the Pearson r, it is desirable to be able to test the hypothesis that the observed value of r_s is computed on a sample drawn from a population in which the correlation is actually zero—in other words, that the observed value is merely a function of sampling error.

Using the data presented above for the similarity of the two judges' rankings, the null hypothesis would be that the observed Spearman correlation of .75 is based upon a sample from a population in which the correlation is actually zero (that is, H_0: $\rho_S = 0$). The alternative hypothesis is that a nonzero correlation actually does exist in the population (H_1: $\rho_S \neq 0$). The assumptions are that the subjects were **randomly** and **independently sampled** and that the measurement was at least **ordinal.**

Table J in Appendix 2 lists critical values for the Spearman correlation for samples of size N = 5 to N = 30 for several levels of significance and for directional and nondirectional tests.[8] If the observed correlation equals or exceeds the value in the table for the specified N, α, and type of test (directional or nondirectional), the null hypothesis is rejected. For the present case, N = 15, α = .05, and the alternative is directional since one would predict that the judges would tend to agree. Thus, the critical value of r_s is .447. Since the observed correlation of .75 is greater than the critical value, H_0 is rejected and it is concluded that the two judges do show some degree of agreement in their evaluations of the children. Table 14-9 presents a formal summary of this example.

---

[8] The values in this table are exact for $N \leq 10$ but approximate for other values of N. Note also that the entries in the table may be regarded as + (positive) or − (negative).

14-9

Summary of a Test of the Significance of the Spearman Rank-Order Correlation Coefficient (Reliability of Two Judges Example)

Hypotheses

H_0: $\rho_S \leq 0$

H_1: $\rho_S > 0$ (directional)

Assumptions and Conditions

1. The subjects are **randomly** and **independently sampled.**
2. The measurement scale is at least **ordinal.**

Decision Rules (from Table J)

Given: the .05 level, a directional alternative, and $N = 15$

If $r_S < .447$, do not reject H_0.

If $r_S \geq .447$, reject H_0.

Computation

Given: $N = 15$ and $\sum_{i=1}^{N} d_i^2 = 138$ (from Table 14-8)

$$r_S = 1 - \left[\frac{6 \sum_{i=1}^{N} d_i^2}{N^3 - N} \right]$$

in which

N = the number of pairs of scores including zero differences

d_i^2 = the squared difference in ranks for the ith pair of scores

$$r_S = 1 - \left[\frac{6(138)}{15^3 - 15} \right]$$

$$r_S = .75$$

Decision

Reject H_0.

For $N > 30$, the following expression translates r_S into an approximation to Student's t distribution with $N - 2$ degrees of freedom:

$$t = \frac{r_S \sqrt{N - 2}}{\sqrt{1 - r_S^2}}$$

SUMMARY

The statistical techniques discussed earlier in this text represent tests of the values of certain parameters, and they make assumptions about other parameters. Therefore, they are called parametric tests. Techniques that do not test hypotheses about specific parameters and that make different and sometimes fewer assumptions about parameters are known as nonparametric tests. The Pearson chi-square test can be used to assess whether the distribution of cases across the categories of one classification scheme are different depending on the category of another classification scheme. It can also determine whether a distribution conforms to a population or otherwise specified distribution and whether an observed percentage is within sampling error of a specified percentage. The Mann–Whitney U test evaluates the difference between the distributions of two independent groups, and the Kruskal–Wallis test assesses the difference between k independent samples. The Wilcoxon test compares the distributions for two correlated samples, and the Spearman rank-order correlation coefficient is a nonparametric index of correlation.

Formulas

1. Chi square for $r \times c$ tables

$$\chi^2_{obs} = \sum_{j=1}^{r} \sum_{k=1}^{c} \frac{(O_{jk} - E_{jk})^2}{E_{jk}}$$

in which

O_{jk} = the observed frequency of the jkth cell

E_{jk} = the expected frequency of the jkth cell

r = the number of rows

c = the number of columns

Refer to the chi-square distribution in Table G of Appendix 2 with

$$df = (r - 1)(c - 1)$$

2. Chi square for goodness of fit and test of a percentage

$$\chi^2_{obs} = \sum_{j=1}^{r} \frac{(O_j - E_j)^2}{E_j}$$

in which

O_j = the observed frequency of the jth row

E_j = the expected frequency of the jth row

r = the number of rows

Refer to the chi-square distribution in Table G of Appendix 2 with

$$df = r - 1$$

3. Mann–Whitney U test for two independent samples

a. For $n \leq 20$

$$U_{obs} = n_A n_B + \frac{n_A(n_A + 1)}{2} - T_A$$

in which

n_A = the number of observations in group A

n_B = the number of observations in group B

T_A = the total of the ranks for group A

Refer to the U distribution in Table H of Appendix 2.

b. For $n > 20$

$$z_{obs} = \frac{U_{obs} - n_A n_B/2}{\sqrt{\dfrac{(n_A)(n_B)(n_A + n_B + 1)}{12}}}$$

in which U_{obs}, n_A, and n_B are as defined above.

Refer to the standard normal distribution in Table A of Appendix 2.

4. **Kruskal–Wallis test for k independent samples**

$$H_{obs} = \left[\frac{12}{N(N + 1)}\right]\left[\sum_{j=1}^{k} \frac{T_j^2}{n_j}\right] - 3(N + 1)$$

in which

k = the number of groups in the analysis

n_j = the number of subjects in group j

N = the total number of subjects in the analysis

T_j^2 = the square of the total of the ranks for scores in group j

Refer H_{obs} to the chi-square distribution in Table G of Appendix 2 with

$$df = k - 1$$

5. **Wilcoxon test for two correlated samples**

a. **For $N \le 50$**

W_{obs} = the smaller of the sum of the ranks associated with positive differences in pairs of scores (T_+) and the sum of the ranks associated with negative differences (T_-)

N = the number of pairs of observations having nonzero differences

Refer to the W distribution in Table I of Appendix 2.

b. **For $N > 50$**

$$z_{obs} = \frac{W_{obs} - N(N + 1)/4}{\sqrt{\dfrac{N(N + 1)(2N + 1)}{24}}}$$

in which W_{obs} is defined as above.

Refer to the standard normal distribution in Table A of Appendix 2.

6. **Spearman rank-order correlation coefficient**

$$r_S = 1 - \left[\frac{6\left(\sum\limits_{i=1}^{N} d_i^2\right)}{N^3 - N}\right]$$

in which

N = the number of pairs of observations including zero differences

d_i = the difference in ranks for the ith pair of scores

a. **For $N \le 30$**

Refer to critical values for r_S in Table J of Appendix 2.

b. **For $N > 30$**

$$t = \frac{r_S\sqrt{N - 2}}{\sqrt{1 - r_S^2}}$$

Refer to Student's t distribution in Table B of Appendix 2 with

$$df = N - 2$$

Exercises

For Conceptual Understanding

1. Name the nonparametric test most analogous to each of the following parametric tests:

   a. independent groups *t* test

   b. correlated groups *t* test

   c. simple analysis of variance

   d. Pearson *r*

2. To what does power-efficiency refer? For a given *N*, which is more powerful, parametric tests or analogous nonparametric tests?

3. For a *t* test, the null hypothesis (nondirectional) is $\mu_1 = \mu_2$. How does the null hypothesis for an analogous nonparametric test differ?

For Solving Problems

4. A sample of children were observed from the time they were 3 until they were 12 years old and their IQs were tested periodically. It was found that 56 of the children showed increases in the general trend of their IQs over this age period, while 55 displayed essentially declining trends. The mothers of these children were seen in their homes during this period and the home visitor rated each mother on the extent to which she expected and encouraged intellectual achievement and success (data are presented below). Test the hypothesis that there is no difference in the distribution of maternal encouragement for intellectual success for those children who evidenced IQ increases and those who showed declines.[9]

		Amount of Maternal Encouragement		
		Low	**Medium**	**High**
IQ	**Up**	15	14	27
	Down	25	16	14

5. Suppose a market researcher for an automobile company suspects that there are color preference differences between male and female buyers of her company's products. Advertisements targeted to different groups should take such differences into account, if they exist. The researcher examines the sales-information files for the number of cars sold during the last year of a particular sports car that comes in three colors. Analyze the following data and state your statistical conclusions.

		Gender of Buyer	
		Male	**Female**
	White	1885	1115
Color	**Black**	2135	1265
	Red	1980	1220

6. One hypothesis about thumbsucking in infants and children is that the behavior is a learned habit. An interesting question is how children come to learn it. It happens that all normal infants have a rooting reflex. Stroking the side of the infant's mouth with a finger, for example, elicits a widening of the mouth, a turning of the head toward the stimulus, and a propensity to suck the finger. This reflex is useful in helping the infant find the mother's breast. However, until recently, sleeping infants were laid on their stomachs so that their arms are on either side of their heads. The thumb may easily touch the side of the mouth, and infants root and suck their thumbs. Perhaps infants learn the habit in this way. If infants could be prevented from rooting and sucking their thumbs when they go to sleep and wake up, presumably they would not develop as strong a habit for this behavior.[10] Suppose nylon mittens were put on the hands of nine infants just before they went

---

[9] Inspired by, but not identical to, R. B. McCall, M. I. Appelbaum, and P. S. Hogarty, Developmental Changes in Mental Performance. (1973). *Monographs of the Society for Research in Child Development* (No. 150).

[10] Inspired by, but not identical to, L. S. Benjamin (1967). "The Beginning of Thumbsucking," *Child Development* 38: 1078–88.

to sleep and then were removed when they woke up, thus preventing them from sucking their thumbs. Ten other infants were reared without the mittens. In an observation period of several hours when the infants were nine months of age, the amount of time each infant spent thumbsucking was noted. The data are presented below. Using the appropriate nonparametric technique, test the hypothesis that the wearing of mittens does not alter the amount of thumbsucking.

Mittens	No Mittens
7	10
1	5
0	6
5	12
0	6
4	11
2	9
3	4
9	8
	10

7. Given the data that follow, calculate the means and medians for the two groups and use the parametric t test for independent groups to compare the two groups. Then compute a Mann–Whitney U test on the same data. Compare the results of the two tests.

A	B
30	36
31	37
32	38
33	39
34	40
35	41
82	44

Explain any differences and attempt to draw some conclusions about when nonparametric tests might be more appropriate than parametric tests. In what way is the difference between a parametric and a nonparametric test similar to the difference between the mean and median in this example?

8. Some people, through a quirk of nature, are genetically of one sex but are born with the physical characteristics of both sexes. They are called hermaphrodites. While it is possible now to study their chromosomes and determine whether such people are genetically males or females, years ago a guess had to be made and the physical characteristics of the inappropriate sex were surgically removed. Thus, it was possible for genetic males to be reared as males or as females. So researchers, interested in the extent to which rearing environment or biology determines behavioral masculinity, compared three groups of genetic males when they were adults: (1) normal males reared as males, (2) hermaphrodite males reared as males, and (3) hermaphrodite males reared as females. If genetics was important for behavioral masculinity, the groups should be the same. But, if rearing made a difference, group 3 should be less masculine. Below are the masculinity scores for the three groups. Statistically evaluate the results.[11]

I	II	III
	Hermaphroditic Males	
Normal Males	Reared as Males	Reared as Females
41	39	25
55	44	33
37	34	28
46	49	39
50	52	21
44		35
32		

---

[11] Inspired by, but not identical to, J. Money and A. A. Ehrhardt (1973). *Man and Woman, Boy and Girl.* Baltimore MD: Johns Hopkins University Press.

9. Before instituting a new fitness program, a corporation hired a consulting psychologist to evaluate the program. It was decided to measure the effect of the program on the self-esteem of the salespeople. Their scores on a self-esteem questionnaire before and after four months of the fitness program are given below. Higher scores indicate greater self-esteem. Using the appropriate nonparametric technique, test the hypothesis that the program had no effect.

Subject	Before	After
A	23	35
B	26	28
C	33	33
D	21	27
E	30	32
F	25	22
G	33	37
H	26	28
I	29	34
J	30	34

10. For the data in Exercise 9, using r_s, determine the degree of relationship between the first and second testing of the 10 salespeople. Test the hypothesis that in the population, the relationship is zero. Does the information from this analysis affect your interpretation of the analysis in Exercise 9 in any way?

11. People who try to stop smoking must overcome a need for nicotine. Psychologist Stanley Schachter once suggested that smokers whose physiological systems have high acid levels excrete more unmetabolized nicotine and therefore have a greater need for replacement nicotine than smokers whose systems have a lower acid level. Presumably, smokers given a substance to make their systems more basic, such as bicarbonate, should be able to smoke fewer cigarettes than those given a substance that would make their systems more acidic, such as vitamin C. A group of researchers[12] decided to

test this notion. They randomly selected smokers and arbitrarily assigned them to one of three groups. The three groups were treated alike for the first three weeks of a stop-smoking program. Then one group was given vitamin C supplements, another group was furnished with bicarbonate, and a third group was given no pills. The two supplement groups did not know which substance they were receiving. The numbers of cigarettes smoked per day during the third and fifth weeks of the experiment are given below. Using nonparametric tests, assess the following questions:

a. Did the vitamin C group improve between the third and fifth week?

b. What is the correlation between the third and fifth week in the bicarbonate group? Is it significant?

c. Were the three treatment groups different in the number of cigarettes smoked at three weeks, that is, before the supplements were given? Afterward, at five weeks?

Vitamin C (Acid)		Nothing		Bicarbonate (Base)	
Third	Fifth	Third	Fifth	Third	Fifth
40	12	41	13	15	5
26	5	25	2	39	3
15	19	17	10	41	2
28	9	30	7	16	7
33	12	35	8	23	3
42	10	39	9	23	3
25	6	21	3	38	1
29	6	31	9	40	0

12. Suppose in a given high school, 27% of the students are freshmen, 23% are sophomores, 24% are juniors, and 26% are seniors. A researcher selects a sample of 260 students for a survey on

---

[12] Based on, but not identical to, research by A. J. Fix, I. Kass, J. Shipp, and J. Smith. (1979, April 14). *Science News,* 14 April 1979, p. 244.

racial attitudes, 66 of whom are freshmen, 58 are sophomores, 66 are juniors, and 70 are seniors. Does the grade distribution in the sample differ from that of the entire school? If the school has equal numbers of males and females but the sample has 140 males and 120 females, is the sample representative of the sex distribution in the school?

For Homework

13. The director of a mental health center introduced behavior modification techniques on the ward to increase desirable behaviors. A study was designed to test whether positive reinforcement for desirable behaviors, negative reinforcement for undesirable behaviors, or no change in staff behavior would lead to improvement. Each patient was administered only one regimen. Below are the numbers of persons who either improved or did not improve under these three procedures. Use the appropriate nonparametric test to determine whether the three distributions are within sampling error ($\alpha = .05$, nondirectional).

	Positive Rein-forcement	Negative Rein-forcement	No Differential Response
Improved	16	12	9
Did Not Improve	6	11	13

14. Sometimes a person's expectations about a task partly determine how well he or she performs. Below are the scores for two groups of people. Those in group A were told that the task was easy and that they would have no problem doing well; those in group B were told that the task was very difficult and that they probably would not do very well. (Higher scores indicate better performance.) Test to determine whether the two distributions of scores differ only by sampling error.

   A: 4, 6, 8, 7, 9, 8, 10, 5, 6, 9
   B: 7, 7, 5, 3, 5, 4, 2, 1, 3, 6

15. A theory of adolescence says that apathy toward the plight of others should be higher among young people because they are primarily concerned with establishing their own values, goals, and so on. Below are the scores from an apathy measure administered to four age groups. A score of 1 indicates high apathy and a score of 20 reflects low apathy (i.e., great concern for others). Use the appropriate nonparametric test to determine whether the social scientist's theory that apathy is more characteristic of the young is substantiated by the data.

Age Group	Scores
21–30	12, 8, 9, 10, 10, 11
31–40	14, 11, 7, 9, 11, 12
41–50	9, 13, 11, 12, 14, 16
51–60	12, 15, 14, 11, 10, 13

16. A special education teacher has designed a program for helping autistic children relate better to others. It has been successful with a previous group, but now the teacher wants to assess the success of the program on a new class. The number of direct eye contacts the children make with another person are recorded before and after the children attend a series of special training sessions. Below are the data for 10 children. Using the appropriate nonparametric technique, test the prediction that the number of contacts made after the program is significantly greater than the number made before.

Child	Eye Contacts Before	After
a	0	2
b	1	2
c	1	4
d	2	3
e	4	3
f	3	3
g	4	6
h	2	4
i	1	4
j	4	4

17. A television producer and a child psychologist are asked to rate from 1 (very acceptable) to 10 (not acceptable) 12 cartoons with regard to their appropriateness for young children. Below are the ratings given for each program by the two judges. Using the appropriate nonparametric test, determine the degree of relationship between the producer's and the psychologist's ratings. Is the relationship significant if one assumes that a negative relationship is impossible?

Program	Producer	Psychologist
a	3	1
b	5	4
c	9	5
d	6	8
e	4	6
f	2	6
g	1	2
h	7	9
i	7	9
j	8	6
k	3	1
l	5	2

18. In a national survey, 500 owners of each of six makes of automobile indicated whether they were satisfied with their cars. Test the hypothesis that satisfaction does not differ from owners of one make to the next.

Make	Satisfied
A	423
B	395
C	412
D	441
E	417
F	402

19. A coin is flipped 400 times and comes up heads on 55% of the flips. Test the hypothesis that the coin is fair.

Appendix 1
Math Review

A class in elementary statistics almost always includes students whose mathematical backgrounds range from a knowledge of high school algebra to facility with calculus and differential equations. It is difficult to begin a course in statistics unless some common level of mathematical experience can be assumed. This appendix reviews basic mathematical concepts and operations, including symbols, fractions, factorials, exponents, and factoring. Although some students will not need to study this section at all, others would profit from studying it fairly seriously.[1] Students should check themselves by working the exercises at the end of the section.

Most students have hand calculators that will add, subtract, multiply, divide, and take square roots, and some have access to computers that will automatically calculate all the statistics in this book. Nonetheless, a complete understanding of the material in this text requires that students be able to follow simple algebraic operations. Hand calculators and computers provide numerically accurate answers quickly, but they do not help the student understand how a formula is derived, what it means, or how to interpret the numerical result. Therefore, students are strongly encouraged to become thoroughly familiar with the material in this appendix and text and *only then* to use their calculators or computers.

SYMBOLS

Some students think that their basic problem with mathematical material is using the symbols. Usually, this problem occurs in part because the students do not take time to learn what the symbols mean. It will be helpful if readers do not proceed until they

can easily read and interpret all the symbols presented to that point.

Most symbols will be introduced as they arise in the text. However, students are assumed to be familiar with the signs for equality ($=$) and inequality ($\neq$), and with the signs for addition ($+$), subtraction ($-$), multiplication [2×3, $2 \cdot 3$, $(2)(3)$, or ab], and division ($5 \div 2$ or $5/2$).

The symbol $>$ means "is greater than." The expression $5 > 3$ is read "5 is greater than 3," and $a > b$ is read "a is greater than b." Conversely, the symbol $<$ means "is less than." The expression $3 < 5$ is read "3 is less than 5," and $b < a$ is read "b is less than a." Some students remember the difference between these two symbols by observing that the open, or larger, end of the symbol is always next to the larger quantity.

Sometimes an expression of the following type will be encountered:

$$-1.96 < t < 1.96$$

This is a short way of writing two inequalities, specifically $-1.96 < t$ and $t < 1.96$. Therefore, the statement means that the value of t is greater than -1.96 but less than 1.96. More simply, t lies between -1.96 and 1.96.

Occasionally, it is desirable to write that some quantity is "greater than or equal to" some other quantity. This fact can be stated with the symbol $\geq$, which is simply a combination of the signs $>$ and $=$. To show that z assumes values equal to 2.56 or greater, write

$$z \geq 2.56$$

Similarly, the symbol $\leq$ means "is less than or equal to." If z is less than or equal to zero, one can write

$$z \leq 0$$

In a few places, we will use the expression $|X|$, which is read "the absolute value of X." This means

---

[1] Please refer to Shiffler and Adams, *Just the Basics, Please: A Quick Review of Math for Introductory Statistics* (Belmont, CA: Duxbury Press, 1996).

that regardless of whether X is a positive or negative number, $|X|$ equals its positive, or absolute, value. Thus,

$$|-5| = 5 \quad \text{and} \quad |5| = 5$$

If $W = 4$ and $Y = 7$,

$$|W - Y| = |4 - 7| = |-3| = 3$$

SIGNED NUMBERS

Perhaps the first real problem students encounter in algebra is negative numbers and how to perform simple addition, subtraction, multiplication, and division with them. It helps to think of the number scale as running from large negative numbers at the left, through zero, to large positive numbers at the right.

-5	-4	-3	-2	-1	0	1	2	3	4	5

We use such a scale every day to measure temperature. Notice that any given number (except zero) has both quantity and sign. If you must locate 4 on the scale, you need to know whether it is a $+4$ or a -4. The value 4 is its quantity, or distance from zero, whereas the $+$ or $-$ is its sign, or direction from zero. When a number is given no sign it is assumed to be positive. Thus, 4 is understood to be $+4$. Sometimes negative numbers are written with a negative superscript before the number. Thus, -4 would be written $^-4$. This form will not be used in this text.

Addition and Subtraction Adding and subtracting are *operations,* and mathematical operations, like other kinds of operations, require a sort of movement. Adding or subtracting can be seen as moving up or down the scale a certain number of steps. Since positive numbers are to the right of zero and negative numbers are to the left of zero, adding 2 and 4 means starting at $+2$ and taking 4 steps to the right, thus ending at $+6$. Similarly, if you start at

-3 and add 2—that is, move two steps to the right—you wind up at -1. If you start at -3 and add 5, you move five steps to the right, passing through zero to $+2$.

Finally, consider $(-4) + (7)$. The parentheses are included to help you distinguish between signs that describe the location of a number and signs that instruct you to perform a given operation. This expression says to start at -4 on the scale and move 7 steps to the right: $(-4) + (7) = +3$.

To subtract, you move to the left. Therefore, $6 - 4$ directs one to start at $+6$ and move 4 steps to the left, stopping at $+2$. The problem $5 - 8$ is solved by starting at $+5$ and moving eight steps to the left, passing through zero and stopping at -3.

Signs indicating location and signs indicating operations have essentially the same meaning—plus means to the right, whereas minus means to the left. Because they are similar, they oppose each other when they occur in the same expression. As we have seen, adding a positive number is "ordinary" addition:

$$4 + (+2) = 6$$
$$3 + (+4) = 7$$
$$-5 + (+3) = -2$$

Adding a negative number is the same as subtracting that quantity:

$$4 + (-2) = 2$$
$$3 + (-4) = -1$$

Subtracting a positive number is "ordinary" subtraction:

$$4 - (+2) = 2$$
$$-5 - (+3) = -8$$

But—and here is the special case—subtracting a negative number is the same as adding that quantity:

$$4 - (-2) = 6$$
$$3 - (-4) = 7$$
$$-5 - (-3) = -2$$

Notice that the answers in the first and last sets of equations above are identical—adding a positive number is the same as subtracting a negative number. "Minus a minus is a plus."

Multiplication When a string of numbers that have different signs are multiplied, the product is positive if there is an even number of negative terms to be multiplied:

$$(-5)(-2) = 10$$
$$(-4)(1)(2)(-3) = 24$$
$$(-a)(-b)(-c)(-d) = abcd$$

The product is negative if there is an odd number of negative terms:

$$(-4)(3) = -12$$
$$(-3)(2)(5) = -30$$
$$(-a)(b)(-c)(-d) = -abcd$$

Division The same rule applies to division. If there is an even number of negative terms, the result is positive:

$$\frac{-4}{-3} = 1.33$$

$$\frac{(-a)(b)}{-c} = \frac{ab}{c} \quad \text{and} \quad \frac{(-a)(-b)}{c} = \frac{ab}{c}$$

If there is an odd number of negative terms in the division, the result is negative:

$$\frac{6}{-3} = -2$$

$$\frac{c}{(-b)(d)} = -\frac{c}{bd} \quad \text{and} \quad \frac{-c}{(b)(d)} = -\frac{c}{bd}$$

$$\text{and} \quad \frac{-c}{(-b)(-d)} = -\frac{c}{bd}$$

Students often wonder whether the statistical formulas presented in the text will work if some or all of the data (original numbers) are negative. Yes, they all work, even with negative numbers. However, the negative signs must be handled properly, as described above.

FRACTIONS

Multiplication The product of two or more fractions equals the product of the numerators divided by the product of the denominators:

$$\frac{1}{2} \cdot \frac{3}{5} = \frac{(1)(3)}{(2)(5)} = \frac{3}{10}$$

In general, multiply $\frac{a}{b}$ times $\frac{c}{d}$ as follows:

$$\frac{a}{b} \cdot \frac{c}{d} = \frac{ac}{bd}$$

Division To divide one fraction by another, invert the divisor (the fraction you want to divide by) and multiply:

$$\frac{1}{2} \div \frac{1}{3} = \frac{1}{2} \cdot \frac{3}{1} = \frac{3}{2} = 1.5$$

In general,

$$\frac{a}{b} \div \frac{c}{d} = \frac{a}{b} \cdot \frac{d}{c} = \frac{ad}{bc}$$

Reducing Fractions If large numbers are involved in fractions or if one must multiply several fractions together, one can frequently simplify the computations by reducing the fractions. To do this, divide both numerator and denominator by the same number, selecting one that goes evenly into each.

Dividing by 2,

$$\frac{2}{4} \quad \text{becomes} \quad \frac{\overset{1}{\cancel{2}}}{\underset{2}{\cancel{4}}} = \frac{1}{2}$$

Dividing by 3,

$$\frac{6}{15} \quad \text{becomes} \quad \frac{\overset{2}{\cancel{6}}}{\underset{5}{\cancel{15}}} = \frac{2}{5}$$

Sometimes you may divide several times before obtaining a numerator and denominator that cannot be divided further by the same number.

Dividing by 2,

$$\frac{12}{30} \text{ becomes } \frac{\overset{6}{\cancel{12}}}{\underset{15}{\cancel{30}}} = \frac{6}{15}$$

which, dividing by 3, becomes

$$\frac{\overset{2}{\cancel{\overset{6}{\cancel{12}}}}}{\underset{\underset{5}{\cancel{15}}}{\cancel{30}}} = \frac{2}{5}$$

Of course, if you see at the outset that 12 and 30 can both be divided by 6, the answer will be obtained sooner:

Dividing by 6,

$$\frac{12}{30} \text{ becomes } \frac{\overset{2}{\cancel{12}}}{\underset{5}{\cancel{30}}} = \frac{2}{5}$$

Essentially the same process is involved when there is a string of fractions to be multiplied. A number may be divided into both a numerator and a denominator and the process repeated until no further divisions are possible. For example:

$$\left(\frac{\overset{1}{\cancel{2}}}{\underset{1}{\cancel{3}}}\right)\left(\frac{\overset{1}{\cancel{3}}}{\underset{2}{\cancel{4}}}\right) = \frac{1}{2}$$

$$\left(\frac{\overset{1}{\cancel{3}}}{\underset{1}{\cancel{3}}}\right)\left(\frac{\overset{1}{\cancel{2}}}{\underset{1}{\cancel{3}}}\right)\left(\frac{\overset{1}{\cancel{3}}}{\underset{3}{\cancel{6}}}\right) = \frac{1}{3}$$

Occasionally, the reduction can become rather complicated:

$$\left(\frac{1}{\cancel{3}}\right)\left(\frac{\cancel{3}}{\cancel{8}}\right)\left(\frac{\overset{\cancel{3}}{\cancel{72}^{\cancel{9}}}}{\underset{\cancel{3}}{\cancel{35}}}\right)\left(\frac{7}{\cancel{3}}\right)\left(\frac{\cancel{3}}{\cancel{8}_4}\right)\left(\frac{2}{\cancel{3}}\right) = \frac{1}{4}$$

Addition and Subtraction Before fractional quantities can be added, the denominators of the two fractions must be made equal. Thus, to add $\frac{1}{2}$ and $\frac{1}{3}$ it is necessary to change both fractions to sixths. This is accomplished by multiplying the first fraction by $\frac{3}{3}$ and the second by $\frac{2}{2}$:

$$\frac{1}{2} + \frac{1}{3} = ?$$

$$\frac{1}{2}\left(\frac{3}{3}\right) + \frac{1}{3}\left(\frac{2}{2}\right) = \frac{3}{6} + \frac{2}{6} = \frac{5}{6}$$

More generally, to add $\dfrac{a}{b} + \dfrac{c}{d}$,

$$\frac{a}{b} + \frac{c}{d} = \frac{a}{b}\left(\frac{d}{d}\right) + \left(\frac{b}{b}\right)\frac{c}{d} = \frac{ad}{bd} + \frac{bc}{bd}$$

$$= \frac{ad + bc}{bd}$$

Working with Zero In some formulas requiring multiplication or division, one or more terms might be zero. It is helpful to remember that

$$\frac{0}{n} = 0$$

Thus,

$$\frac{(0)(4)(7)(2)}{(6)(14)} = 0$$

In other words, any number multiplied by 0 equals 0, and 0 divided by any nonzero number equals 0.

FACTORIALS

In some probability problems it is necessary to multiply a positive integer (a whole number) by each of the integers having a value less than that integer, ending with 1. This string of multiplications of integers is called a **factorial** and the sign "!" following the largest integer in the string is used to indicate this operation. For example, 3!, read "three factorial," would equal

$$3! = 3 \cdot 2 \cdot 1 = 6$$

In general, $n!$ means

$$n! = n(n-1)(n-2)(n-3) \cdots (1)$$

Note also that, by custom,

$$0! = 1$$

EXPONENTS

An exponent is a number written as a superscript to a base number and signifies that the base number should be multiplied by itself as many times as the exponent states. Therefore,

$$2^3 = 2 \cdot 2 \cdot 2 = 8$$

and more generally

$$n^r = \underbrace{n \cdot n \cdot n \cdots n}_{r \text{ times}}$$

Note also that

$$n^1 = n$$

and

$$n^0 = 1$$

Addition and Subtraction Generally, numbers with exponents cannot be added or subtracted without first carrying out the exponentiation of each quantity. This is true even if the numbers have the same base. For example, $2^2 + 2^3$ is handled by *first* performing the indicated exponentiation and *then* adding:

$$2^2 + 2^3 = 2 \cdot 2 + 2 \cdot 2 \cdot 2 = 4 + 8 = 12$$

Multiplication The product of two exponential quantities with the *same base number* equals that base number raised to the sum of the two exponents. For example,

$$(2^2)(2^3) = 2^{2+3} = 2^5$$

because

$$(2^2)(2^3) = (2 \cdot 2)(2 \cdot 2 \cdot 2) = 2^5$$

More generally,

$$(n^r)(n^s) = n^{r+s}$$

If the exponents *do not have the same base number* (for example, $2^4 \times 3^4$), this procedure does *not* apply and the exponentiation should be carried out first.

Division The quotient of two exponential quantities with the *same base number* is that base raised to the difference between the exponent of the numerator and that of the denominator. For example,

$$\frac{2^3}{2^2} = 2^{3-2} = 2^1 = 2$$

More generally,

$$\frac{n^r}{n^s} = n^{r-s}$$

It is helpful to remember that if s is larger than r in the expression n^r/n^s, then the result has a negative exponent. Whenever a number has a negative exponent, a reciprocal $\left(\dfrac{1}{\text{number}}\right)$ is taken before the exponentiation is carried out. For example,

$$2^{-3} = \frac{1}{2^3} = \frac{1}{2 \cdot 2 \cdot 2} = \frac{1}{8}$$

Fractions A fraction raised to a power equals the numerator raised to that power divided by the denominator raised to that power. For example,

$$\left(\frac{3}{5}\right)^2 = \frac{3^2}{5^2} = \frac{9}{25}$$

More generally,

$$\left(\frac{r}{s}\right)^n = \frac{r^n}{s^n}$$

Binomial Expansion A *binomial* is an expression involving the addition or subtraction of two quantities. Many of the algebraic manipulations presented in this text require the student to understand the expansion of a binomial, such as $(a + b)^2$:

$$(a + b)^2 = a^2 + 2ab + b^2$$

This result is obtained by taking the square of the first term in the binomial (here, a^2), plus 2 times the product of the two terms in the binomial ($2ab$), plus the square of the second term (b^2).

A more common problem is to expand $(a - b)^2$. This is accomplished in the same way but with attention to the algebraic signs:

$$(a - b)^2 = a^2 + 2a(-b) + (-b)^2 = a^2 - 2ab + b^2$$

Again, the result equals the square of the first term (a^2), plus 2 times the product of the two terms $[2(a)(-b) = -2ab]$, which will have a negative sign, plus the square of the last term $[(-b)^2 = b^2]$.

The student should remember that these procedures apply to any binomial regardless of the specific terms. For example, quite frequently it will be necessary to expand $(X_i - \overline{X})^2$. In this expression, X_i stands for any particular score in a group of scores, and $\overline{X}$ refers to the average of all the scores in the group. The result is generated just as in the examples above:

$$(X_i - \overline{X})^2 = X_i^2 - 2X_i\overline{X} + \overline{X}^2$$

Square Roots Many statistical formulas require taking a square root, the reverse operation to squaring. Since it is assumed that students have calculators for this purpose, the way to extract a square root by hand will not be reviewed here. However, notice that in addition to symbolizing a mathematical operation, the square root sign (the radical) is treated as if it were parentheses enclosing a quantity. For example,

$$\sqrt{a^2} = a \quad \text{(technically, either } +a \text{ or } -a)$$

$$3\sqrt{4} = 3\left(\sqrt{4}\right) = 3(2) = 6$$

$$2\sqrt{4 + 21} = 2\left(\sqrt{25}\right) = 2(5) = 10$$

The radical can be applied to each number in a complex term that calls for multiplication or division but not to the numbers in a term that calls for addition or subtraction. For example:

$$\sqrt{4a} = \sqrt{4}\sqrt{a} = 2\sqrt{a}$$

$$\sqrt{\frac{a + b}{c}} = \frac{\sqrt{a + b}}{\sqrt{c}}$$

But

$$\sqrt{a + b} \quad \text{does not equal} \quad \sqrt{a} + \sqrt{b}$$

and

$$\sqrt{a - b} \quad \text{does not equal} \quad \sqrt{a} - \sqrt{b}$$

FACTORING AND SIMPLIFICATION

Factoring is the process of breaking down a number into parts that, when multiplied together, equal the number. *Simplification* is the process of reducing the number of quantities in an expression. Sometimes an expression can be simplified by factoring it first.

Removing Parentheses Occasionally, algebraic expressions may be simplified by removing parentheses. When no multiplications or divisions are involved, one simply removes the parentheses and performs any required addition or subtraction.

$$4 + (2) = 4 + 2 = 6$$

$$6 + (-5) = 6 - 5 = 1$$

$$3 - (-2) + 7 = 3 + 2 + 7 = 12$$

$$a + (-b) - (-c) = a - b + c$$

Sometimes squaring or taking a square root is combined with multiplying or dividing by another quantity. In such cases, one must be very careful which operations are performed on which numbers. For example,

$$2s^2 = 2ss$$

$$(2s)^2 = (2s)(2s) = 4s^2$$

$$4(2x)^2 = 4(2x)(2x) = 4(4x^2) = 16x^2$$

Similarly,

$$\sqrt{4s^2} = \sqrt{4}\sqrt{s^2} = 2s$$

$$\sqrt{2p^2} = \sqrt{2}\sqrt{p^2} = p\sqrt{2}$$

$$\frac{1}{2}\sqrt{4x} = \frac{1}{2}\sqrt{4}\sqrt{x} = \left(\frac{1}{2}\right)(2)\sqrt{x} = \sqrt{x}$$

The sequence in which quantities are added or subtracted does not matter. For example,

$$4 + 3 = 3 + 4 = 7$$

$$4 - 3 = -3 + 4 = 1$$

$$3 + (a - 2) = (a - 2) + 3 = a + 1$$

Similarly, the sequence in which quantities are multiplied or divided does not matter:

$$4(3) = 3(4) = 12$$

$$2b = b2$$

$$2b3a = 6ab$$

$$3(4 - 5) = (-5 + 4)3 = -3$$

But when expressions become more complicated, the sequence of operations does matter. Sometimes an expression to be simplified is quite complex, involving several different terms. A term is a quantity to be added to or subtracted from another quantity. For example, each of the following expressions contains two terms:

$$4 + 3$$

$$2x + y$$

$$6\sqrt{5y} - \frac{4(x + y)^2}{3}$$

Notice that multiplication and division as well as exponentiation and taking roots can be required within a term. In a complex expression, it helps to perform the mathematical operations in a given sequence, specifically,

1. Perform any addition or subtraction *within* terms, parentheses, or radicals.
2. Calculate any exponents or roots.
3. Compute any multiplications or divisions.
4. Perform any addition or subtraction *between* terms.

For example, suppose you must simplify the expression

$$ab + 3[a - b(4 - 2)]^2$$

1. Perform any addition or subtraction within terms, parentheses, or radicals:

$$ab + 3[a - b(2)]^2$$

$$ab + 3[a - 2b]^2$$

2. Calculate any exponents or roots:

$$ab + 3[a^2 - 2(a2b) + (2b)^2]$$

$$ab + 3[a^2 - 4ab + 4b^2]$$

3. Compute any multiplication or divisions:

$$ab + 3a^2 - 3(4ab) + 3(4b^2)$$

$$ab + 3a^2 - 12ab + 12b^2$$

4. Perform any addition or subtraction between terms:

$$3a^2 - 11ab + 12b^2$$

Transposing Sometimes it is convenient to transpose a term from one side of an equation to the other. When a term is to be transposed, it is added or subtracted on both sides of the equation:

$$
\begin{array}{r}
a - b = c \\
-a = -a \\
\hline
a - b - a = c - a \\
-b = c - a
\end{array}
\quad \text{[add } -a\text{]}
$$

The operation amounts to placing the term (here, *a*) on the other side of the equation with a change of sign.

If the term to be transposed is involved in multiplication or division, the opposite operation (that is, division or multiplication) must be applied to both sides of the equations:

$$\frac{a}{b} = c$$

$$\left(\frac{a}{b}\right)\left(\frac{b}{1}\right) = c\left(\frac{b}{1}\right) \quad \text{[multiply by } b\text{]}$$

$$a = cb$$

Complex Factoring Factoring an algebraic expression involves determining the simplest set of quantities which, when multiplied together, will yield the original quantities:

$$ab + ac = a(b + c)$$

$$-ab - ac = -a(b + c)$$

Factoring is often used to simplify algebraic expressions. Consider the following example:

$$\frac{-a(b - c) - (c - ab)}{a - 1}$$

Carrying out the appropriate multiplications in the numerator, one obtains

$$\frac{-ab - a(-c) - (c - ab)}{a - 1}$$

and simplifying the signs of certain expressions, one obtains

$$\frac{-ab + ac - c + ab}{a - 1}$$

By subtracting,

$$\frac{ac - c}{a - 1}$$

factoring,

$$\frac{c(a - 1)}{a - 1}$$

and dividing, one finds the expression reduces to

$$c$$

Sometimes in the course of simplifying an expression, it helps to divide each term in the numerator by the denominator:

$$\frac{a - b}{c} = \frac{a}{c} - \frac{b}{c}$$

However, one must be cautious. The above manipulation is correct but the following manipulations are not correct:

$$\frac{c}{a - b} \quad does\ not\ equal \quad \frac{c}{a} - \frac{c}{b}$$

$$\frac{a - b}{c - d} \quad does\ not\ equal \quad \frac{a}{c} - \frac{b}{d}$$

EXERCISES

1. Perform the indicated operations:

a. $\dfrac{3}{7} + \dfrac{2}{7}$ **e.** $\dfrac{2}{5} - \dfrac{4}{5}$

b. $\dfrac{3}{6} + \dfrac{5}{6}$ **f.** $\dfrac{2}{3} - \dfrac{1}{6}$

c. $\dfrac{1}{4} + \dfrac{5}{12}$ **g.** $\dfrac{1}{4} - \dfrac{2}{9}$

d. $\dfrac{2}{3} + \dfrac{5}{7}$ **h.** $\dfrac{6}{7} - \dfrac{6}{11}$

2. Perform the indicated operations:

a. $\dfrac{1}{6} \cdot \dfrac{3}{6}$ **c.** $\dfrac{1}{2} \cdot \dfrac{9}{7} \cdot \dfrac{5}{12} \cdot \dfrac{3}{4}$

b. $\dfrac{3}{5} \cdot \dfrac{1}{4}$ **d.** $\dfrac{1}{2} \div \dfrac{1}{4}$

e. $\dfrac{3}{8} \div \dfrac{1}{4}$ **f.** $\dfrac{7}{12} \div \dfrac{2}{15}$

3. Simplify the following:

a. $7!$ **b.** $\dfrac{5!}{2!}$ **c.** $\dfrac{5!4!}{4!}$

4. Simplify the following:

a. 3^4 **g.** $2^3 \div 2^2$

b. $2^2 + 2^3$ **h.** $3^2 \div 3^3$

c. $2^1 + 3^3$ **i.** $4^2 \div 4^2$

d. $3^4 - 2^4$ **j.** $\left(\dfrac{1}{3}\right)^2$

e. $2^3 - 2^4$ **k.** $\left(\dfrac{3}{4}\right)^3 \div \left(\dfrac{1}{4}\right)$

f. $2^4 \cdot 3^2$ **l.** $\left(\dfrac{1}{3}\right)^3 \cdot \dfrac{2}{9^2}$

5. Expand and simplify the following expressions:

a. $(x + y)^2$

b. $(p - q)^2$

c. $(X + \overline{X})^2$

d. $(ab - bc)^2$

e. $\dfrac{(ab - ac) + a(b + c)}{2b}$

f. $\dfrac{a - bc}{c}$

g. $\dfrac{bc(a + c)}{abc - cb^2}$

h. $\sqrt{4X^2}$

i. $\sqrt{5X^2}$

j. $\sqrt{Y(2Y^2)}$

6. For each of the following sets, determine the sum, the square of the sum, and the sum of the squared numbers (called the sum of squares):

a. 6	**b.** 2	**c.** 3
5	5	0
2	3	−4
	4	6
		−1
		4

Appendix 2
Tables

Table A. Proportions of Area Under the Standard Normal Curve

z			z			z		
0.00	.0000	.5000	0.55	.2088	.2912	1.10	.3643	.1357
0.01	.0040	.4960	0.56	.2123	.2877	1.11	.3665	.1335
0.02	.0080	.4920	0.57	.2157	.2843	1.12	.3686	.1314
0.03	.0120	.4880	0.58	.2190	.2810	1.13	.3708	.1292
0.04	.0160	.4840	0.59	.2224	.2776	1.14	.3729	.1271
0.05	.0199	.4801	0.60	.2257	.2743	1.15	.3749	.1251
0.06	.0239	.4761	0.61	.2291	.2709	1.16	.3770	.1230
0.07	.0279	.4721	0.62	.2324	.2676	1.17	.3790	.1210
0.08	.0319	.4681	0.63	.2357	.2643	1.18	.3810	.1190
0.09	.0359	.4641	0.64	.2389	.2611	1.19	.3830	.1170
0.10	.0398	.4602	0.65	.2422	.2578	1.20	.3849	.1151
0.11	.0438	.4562	0.66	.2454	.2546	1.21	.3869	.1131
0.12	.0478	.4522	0.67	.2486	.2514	1.22	.3888	.1112
0.13	.0517	.4483	0.68	.2517	.2483	1.23	.3907	.1093
0.14	.0557	.4443	0.69	.2549	.2451	1.24	.3925	.1075
0.15	.0596	.4404	0.70	.2580	.2420	1.25	.3944	.1056
0.16	.0636	.4364	0.71	.2611	.2389	1.26	.3962	.1038
0.17	.0675	.4325	0.72	.2642	.2358	1.27	.3980	.1020
0.18	.0714	.4286	0.73	.2673	.2327	1.28	.3997	.1003
0.19	.0753	.4247	0.74	.2704	.2296	1.29	.4015	.0985
0.20	.0793	.4207	0.75	.2734	.2266	1.30	.4032	.0968
0.21	.0832	.4168	0.76	.2764	.2236	1.31	.4049	.0951
0.22	.0871	.4129	0.77	.2794	.2206	1.32	.4066	.0934
0.23	.0910	.4090	0.78	.2823	.2177	1.33	.4082	.0918
0.24	.0948	.4052	0.79	.2852	.2148	1.34	.4099	.0901
0.25	.0987	.4013	0.80	.2881	.2119	1.35	.4115	.0885
0.26	.1026	.3974	0.81	.2910	.2090	1.36	.4131	.0869
0.27	.1064	.3936	0.82	.2939	.2061	1.37	.4147	.0853
0.28	.1103	.3897	0.83	.2967	.2033	1.38	.4162	.0838
0.29	.1141	.3859	0.84	.2995	.2005	1.39	.4177	.0823
0.30	.1179	.3821	0.85	.3023	.1977	1.40	.4192	.0808
0.31	.1217	.3783	0.86	.3051	.1949	1.41	.4207	.0793
0.32	.1255	.3745	0.87	.3078	.1922	1.42	.4222	.0778
0.33	.1293	.3707	0.88	.3106	.1894	1.43	.4236	.0764
0.34	.1331	.3669	0.89	.3133	.1867	1.44	.4251	.0749
0.35	.1368	.3632	0.90	.3159	.1841	1.45	.4265	.0735
0.36	.1406	.3594	0.91	.3186	.1814	1.46	.4279	.0721
0.37	.1443	.3557	0.92	.3212	.1788	1.47	.4292	.0708
0.38	.1480	.3520	0.93	.3238	.1762	1.48	.4306	.0694
0.39	.1517	.3483	0.94	.3264	.1736	1.49	.4319	.0681
0.40	.1554	.3446	0.95	.3289	.1711	1.50	.4332	.0668
0.41	.1591	.3409	0.96	.3315	.1685	1.51	.4345	.0655
0.42	.1628	.3372	0.97	.3340	.1660	1.52	.4357	.0643
0.43	.1664	.3336	0.98	.3365	.1635	1.53	.4370	.0630
0.44	.1700	.3300	0.99	.3389	.1611	1.54	.4382	.0618
0.45	.1736	.3264	1.00	.3413	.1587	1.55	.4394	.0606
0.46	.1772	.3228	1.01	.3438	.1562	1.56	.4406	.0594
0.47	.1808	.3192	1.02	.3461	.1539	1.57	.4418	.0582
0.48	.1844	.3156	1.03	.3485	.1515	1.58	.4429	.0571
0.49	.1879	.3121	1.04	.3508	.1492	1.59	.4441	.0559
0.50	.1915	.3085	1.05	.3531	.1469	1.60	.4452	.0548
0.51	.1950	.3050	1.06	.3554	.1446	1.61	.4463	.0537
0.52	.1985	.3015	1.07	.3577	.1423	1.62	.4474	.0526
0.53	.2019	.2981	1.08	.3599	.1401	1.63	.4484	.0516
0.54	.2054	.2946	1.09	.3621	.1379	1.64	.4495	.0505

Source: P. Runyon and Audrey Haber, *Fundamentals of Behavioral Statistics*, 3rd ed., © 1976. Addison-Wesley, Reading, Massachusetts. Table A. Reprinted with permission.

Table A (continued)

z	0 z	0 z	z	0 z	0 z	z	0 z	0 z
1.65	.4505	.0495	2.22	.4868	.0132	2.79	.4974	.0026
1.66	.4515	.0485	2.23	.4871	.0129	2.80	.4974	.0026
1.67	.4525	.0475	2.24	.4875	.0125	2.81	.4975	.0025
1.68	.4535	.0465	2.25	.4878	.0122	2.82	.4976	.0024
1.69	.4545	.0455	2.26	.4881	.0119	2.83	.4977	.0023
1.70	.4554	.0446	2.27	.4884	.0116	2.84	.4977	.0023
1.71	.4564	.0436	2.28	.4887	.0113	2.85	.4978	.0022
1.72	.4573	.0427	2.29	.4890	.0110	2.86	.4979	.0021
1.73	.4582	.0418	2.30	.4893	.0107	2.87	.4979	.0021
1.74	.4591	.0409	2.31	.4896	.0104	2.88	.4980	.0020
1.75	.4599	.0401	2.32	.4898	.0102	2.89	.4981	.0019
1.76	.4608	.0392	2.33	.4901	.0099	2.90	.4981	.0019
1.77	.4616	.0384	2.34	.4904	.0096	2.91	.4982	.0018
1.78	.4625	.0375	2.35	.4906	.0094	2.92	.4982	.0018
1.79	.4633	.0367	2.36	.4909	.0091	2.93	.4983	.0017
1.80	.4641	.0359	2.37	.4911	.0089	2.94	.4984	.0016
1.81	.4649	.0351	2.38	.4913	.0087	2.95	.4984	.0016
1.82	.4656	.0344	2.39	.4916	.0084	2.96	.4985	.0015
1.83	.4664	.0336	2.40	.4918	.0082	2.97	.4985	.0015
1.84	.4671	.0329	2.41	.4920	.0080	2.98	.4986	.0014
1.85	.4678	.0322	2.42	.4922	.0078	2.99	.4986	.0014
1.86	.4686	.0314	2.43	.4925	.0075	3.00	.4987	.0013
1.87	.4693	.0307	2.44	.4927	.0073	3.01	.4987	.0013
1.88	.4699	.0301	2.45	.4929	.0071	3.02	.4987	.0013
1.89	.4706	.0294	2.46	.4931	.0069	3.03	.4988	.0012
1.90	.4713	.0287	2.47	.4932	.0068	3.04	.4988	.0012
1.91	.4719	.0281	2.48	.4934	.0066	3.05	.4989	.0011
1.92	.4726	.0274	2.49	.4936	.0064	3.06	.4989	.0011
1.93	.4732	.0268	2.50	.4938	.0062	3.07	.4989	.0011
1.94	.4738	.0262	2.51	.4940	.0060	3.08	.4990	.0010
1.95	.4744	.0256	2.52	.4941	.0059	3.09	.4990	.0010
1.96	.4750	.0250	2.53	.4943	.0057	3.10	.4990	.0010
1.97	.4756	.0244	2.54	.4945	.0055	3.11	.4991	.0009
1.98	.4761	.0239	2.55	.4946	.0054	3.12	.4991	.0009
1.99	.4767	.0233	2.56	.4948	.0052	3.13	.4991	.0009
2.00	.4772	.0228	2.57	.4949	.0051	3.14	.4992	.0008
2.01	.4778	.0222	2.58	.4951	.0049	3.15	.4992	.0008
2.02	.4783	.0217	2.59	.4952	.0048	3.16	.4992	.0008
2.03	.4788	.0212	2.60	.4953	.0047	3.17	.4992	.0008
2.04	.4793	.0207	2.61	.4955	.0045	3.18	.4993	.0007
2.05	.4798	.0202	2.62	.4956	.0044	3.19	.4993	.0007
2.06	.4803	.0197	2.63	.4957	.0043	3.20	.4993	.0007
2.07	.4808	.0192	2.64	.4959	.0041	3.21	.4993	.0007
2.08	.4812	.0188	2.65	.4960	.0040	3.22	.4994	.0006
2.09	.4817	.0183	2.66	.4961	.0039	3.23	.4994	.0006
2.10	.4821	.0179	2.67	.4962	.0038	3.24	.4994	.0006
2.11	.4826	.0174	2.68	.4963	.0037	3.25	.4994	.0006
2.12	.4830	.0170	2.69	.4964	.0036	3.30	.4995	.0005
2.13	.4834	.0166	2.70	.4965	.0035	3.35	.4996	.0004
2.14	.4838	.0162	2.71	.4966	.0034	3.40	.4997	.0003
2.15	.4842	.0158	2.72	.4967	.0033	3.45	.4997	.0003
2.16	.4846	.0154	2.73	.4968	.0032	3.50	.4998	.0002
2.17	.4850	.0150	2.74	.4969	.0031	3.60	.4998	.0002
2.18	.4854	.0146	2.75	.4970	.0030	3.70	.4999	.0001
2.19	.4857	.0143	2.76	.4971	.0029	3.80	.4999	.0001
2.20	.4861	.0139	2.77	.4972	.0028	3.90	.49995	.00005
2.21	.4864	.0136	2.78	.4973	.0027	4.00	.49997	.00003

Table B. Critical Values of *t*

df	Level of significance for a directional (one-tailed) test					
	.10	.05	.025	.01	.005	.0005
	Level of significance for a nondirectional (two-tailed) test					
	.20	.10	.05	.02	.01	.001
1	3.078	6.314	12.706	31.821	63.657	636.619
2	1.886	2.920	4.303	6.965	9.925	31.598
3	1.638	2.353	3.182	4.541	5.841	12.941
4	1.533	2.132	2.776	3.747	4.604	8.610
5	1.476	2.015	2.571	3.365	4.032	6.859
6	1.440	1.943	2.447	3.143	3.707	5.959
7	1.415	1.895	2.365	2.998	3.499	5.405
8	1.397	1.860	2.306	2.896	3.355	5.041
9	1.383	1.833	2.262	2.821	3.250	4.781
10	1.372	1.812	2.228	2.764	3.169	4.587
11	1.363	1.796	2.201	2.718	3.106	4.437
12	1.356	1.782	2.179	2.681	3.055	4.318
13	1.350	1.771	2.160	2.650	3.012	4.221
14	1.345	1.761	2.145	2.624	2.977	4.140
15	1.341	1.753	2.131	2.602	2.947	4.073
16	1.337	1.746	2.120	2.583	2.921	4.015
17	1.333	1.740	2.110	2.567	2.898	3.965
18	1.330	1.734	2.101	2.552	2.878	3.922
19	1.328	1.729	2.093	2.539	2.861	3.883
20	1.325	1.725	2.086	2.528	2.845	3.850
21	1.323	1.721	2.080	2.518	2.831	3.819
22	1.321	1.717	2.074	2.508	2.819	3.792
23	1.319	1.714	2.069	2.500	2.807	3.767
24	1.318	1.711	2.064	2.492	2.797	3.745
25	1.316	1.708	2.060	2.485	2.787	3.725
26	1.315	1.706	2.056	2.479	2.779	3.707
27	1.314	1.703	2.052	2.473	2.771	3.690
28	1.313	1.701	2.048	2.467	2.763	3.674
29	1.311	1.699	2.045	2.462	2.756	3.659
30	1.310	1.697	2.042	2.457	2.750	3.646
40	1.303	1.684	2.021	2.423	2.704	3.551
60	1.296	1.671	2.000	2.390	2.660	3.460
120	1.289	1.658	1.980	2.358	2.617	3.373
∞	1.282	1.645	1.960	2.326	2.576	3.291

The value listed in the table is the critical value of *t* for the number of degrees of freedom listed in the left column for a directional (one-tailed) or nondirectional (two-tailed) test at the significance level indicated at the top of each column. If the observed *t* is *greater than or equal to* the tabled value, reject H_0. Since the *t* distribution is symmetrical about $t = 0$, these critical values represent both + and − values for nondirectional tests.

Source: Table B is taken from Table III of Fisher and Yates, *Statistical Tables for Biological, Agricultural and Medical Research*, published by Longman Group Ltd., London (previously published by Oliver and Boyd, Ltd., Edinburgh), and by permission of the authors and publishers.

Table C. Critical Values of the Pearson Product-Moment Correlation Coefficient, *r*

	Level of significance for a directional (one-tailed) test				
	.05	.025	.01	.005	.0005
	Level of significance for a nondirectional (two-tailed) test				
df = N−2	.10	.05	.02	.01	.001
1	.9877	.9969	.9995	.9999	1.0000
2	.9000	.9500	.9800	.9900	.9990
3	.8054	.8783	.9343	.9587	.9912
4	.7293	.8114	.8822	.9172	.9741
5	.6694	.7545	.8329	.8745	.9507
6	.6215	.7067	.7887	.8343	.9249
7	.5822	.6664	.7498	.7977	.8982
8	.5494	.6319	.7155	.7646	.8721
9	.5214	.6021	.6851	.7348	.8471
10	.4973	.5760	.6581	.7079	.8233
11	.4762	.5529	.6339	.6835	.8010
12	.4575	.5324	.6120	.6614	.7800
13	.4409	.5139	.5923	.6411	.7603
14	.4259	.4973	.5742	.6226	.7420
15	.4124	.4821	.5577	.6055	.7246
16	.4000	.4683	.5425	.5897	.7084
17	.3887	.4555	.5285	.5751	.6932
18	.3783	.4438	.5155	.5614	.6787
19	.3687	.4329	.5034	.5487	.6652
20	.3598	.4227	.4921	.5368	.6524
25	.3233	.3809	.4451	.4869	.5974
30	.2960	.3494	.4093	.4487	.5541
35	.2746	.3246	.3810	.4182	.5189
40	.2573	.3044	.3578	.3932	.4896
45	.2428	.2875	.3384	.3721	.4648
50	.2306	.2732	.3218	.3541	.4433
60	.2108	.2500	.2948	.3248	.4078
70	.1954	.2319	.2737	.3017	.3799
80	.1829	.2172	.2565	.2830	.3568
90	.1726	.2050	.2422	.2673	.3375
100	.1638	.1946	.2301	.2540	.3211

If the observed value of *r* is *greater than or equal to* the tabled value for the appropriate level of significance (columns) and degrees of freedom (rows), reject H_0. The degrees of freedom are the number of pairs of scores minus two, or $N - 2$. The critical values in the table are both + and − for nondirectional (two-tailed) tests.

Source: Table C is taken from Table VII of Fisher and Yates, *Statistical Tables for Biological, Agricultural, and Medical Research*, published by Longman Group Ltd., London (previously published by Oliver and Boyd, Ltd., Edinburgh), and by permission of the authors and publishers.

Table D.   Transformation of *r* to z_r

r	z_r	r	z_r	r	z_r	r	z_r	r	z_r
.000	.000	.200	.203	.400	.424	.600	.693	.800	1.099
.005	.005	.205	.208	.405	.430	.605	.701	.805	1.113
.010	.010	.210	.213	.410	.436	.610	.709	.810	1.127
.015	.015	.215	.218	.415	.442	.615	.717	.815	1.142
.020	.020	.220	.224	.420	.448	.620	.725	.820	1.157
.025	.025	.225	.229	.425	.454	.625	.733	.825	1.172
.030	.030	.230	.234	.430	.460	.630	.741	.830	1.188
.035	.035	.235	.239	.435	.466	.635	.750	.835	1.204
.040	.040	.240	.245	.440	.472	.640	.758	.840	1.221
.045	.045	.245	.250	.445	.478	.645	.767	.845	1.238
.050	.050	.250	.255	.450	.485	.650	.775	.850	1.256
.055	.055	.255	.261	.455	.491	.655	.784	.855	1.274
.060	.060	.260	.266	.460	.497	.660	.793	.860	1.293
.065	.065	.265	.271	.465	.504	.665	.802	.865	1.313
.070	.070	.270	.277	.470	.510	.670	.811	.870	1.333
.075	.075	.275	.282	.475	.517	.675	.820	.875	1.354
.080	.080	.280	.288	.480	.523	.680	.829	.880	1.376
.085	.085	.285	.293	.485	.530	.685	.838	.885	1.398
.090	.090	.290	.299	.490	.536	.690	.848	.890	1.422
.095	.095	.295	.304	.495	.543	.695	.858	.895	1.447
.100	.100	.300	.310	.500	.549	.700	.867	.900	1.472
.105	.105	.305	.315	.505	.556	.705	.877	.905	1.499
.110	.110	.310	.321	.510	.563	.710	.887	.910	1.528
.115	.116	.315	.326	.515	.570	.715	.897	.915	1.557
.120	.121	.320	.332	.520	.576	.720	.908	.920	1.589
.125	.126	.325	.337	.525	.583	.725	.918	.925	1.623
.130	.131	.330	.343	.530	.590	.730	.929	.930	1.658
.135	.136	.335	.348	.535	.597	.735	.940	.935	1.697
.140	.141	.340	.354	.540	.604	.740	.950	.940	1.738
.145	.146	.345	.360	.545	.611	.745	.962	.945	1.783
.150	.151	.350	.365	.550	.618	.750	.973	.950	1.832
.155	.156	.355	.371	.555	.626	.755	.984	.955	1.886
.160	.161	.360	.377	.560	.633	.760	.996	.960	1.946
.165	.167	.365	.383	.565	.640	.765	1.008	.965	2.014
.170	.172	.370	.388	.570	.648	.770	1.020	.970	2.092
.175	.177	.375	.394	.575	.655	.775	1.033	.975	2.185
.180	.182	.380	.400	.580	.662	.780	1.045	.980	2.298
.185	.187	.385	.406	.585	.670	.785	1.058	.985	2.443
.190	.192	.390	.412	.590	.678	.790	1.071	.990	2.647
.195	.198	.395	.418	.595	.685	.795	1.085	.995	2.994

Table E. Critical Values of F (.05 Level in Roman Type, .01 Level in Boldface)

Degrees of freedom for the numerator

df₂ \ df₁	1	2	3	4	5	6	7	8	9	10	11	12	14	16	20	24	30	40	50	75	100	200	500	∞
1	161 **4,052**	200 **4,999**	216 **5,403**	225 **5,625**	230 **5,764**	234 **5,859**	237 **5,928**	239 **5,981**	241 **6,022**	242 **6,056**	243 **6,082**	244 **6,106**	245 **6,142**	246 **6,169**	248 **6,208**	249 **6,234**	250 **6,261**	251 **6,286**	252 **6,302**	253 **6,323**	253 **6,334**	254 **6,352**	254 **6,361**	254 **6,366**
2	18.51 **98.49**	19.00 **99.00**	19.16 **99.17**	19.25 **99.25**	19.30 **99.30**	19.33 **99.33**	19.36 **99.36**	19.37 **99.37**	19.38 **99.39**	19.39 **99.40**	19.40 **99.41**	19.41 **99.42**	19.42 **99.43**	19.43 **99.44**	19.44 **99.45**	19.45 **99.46**	19.46 **99.47**	19.47 **99.48**	19.47 **99.48**	19.48 **99.49**	19.49 **99.49**	19.49 **99.49**	19.50 **99.50**	19.50 **99.50**
3	10.13 **34.12**	9.55 **30.82**	9.28 **29.46**	9.12 **28.71**	9.01 **28.24**	8.94 **27.91**	8.88 **27.67**	8.84 **27.49**	8.81 **27.34**	8.78 **27.23**	8.76 **27.13**	8.74 **27.05**	8.71 **26.92**	8.69 **26.83**	8.66 **26.69**	8.64 **26.60**	8.62 **26.50**	8.60 **26.41**	8.58 **26.35**	8.57 **26.27**	8.56 **26.23**	8.54 **26.18**	8.54 **26.14**	8.53 **26.12**
4	7.71 **21.20**	6.94 **18.00**	6.59 **16.69**	6.39 **15.98**	6.26 **15.52**	6.16 **15.21**	6.09 **14.98**	6.04 **14.80**	6.00 **14.66**	5.96 **14.54**	5.93 **14.45**	5.91 **14.37**	5.87 **14.24**	5.84 **14.15**	5.80 **14.02**	5.77 **13.93**	5.74 **13.83**	5.71 **13.74**	5.70 **13.69**	5.68 **13.61**	5.66 **13.57**	5.65 **13.52**	5.64 **13.48**	5.63 **13.46**
5	6.61 **16.26**	5.79 **13.27**	5.41 **12.06**	5.19 **11.39**	5.05 **10.97**	4.95 **10.67**	4.88 **10.45**	4.82 **10.29**	4.78 **10.15**	4.74 **10.05**	4.70 **9.96**	4.68 **9.89**	4.64 **9.77**	4.60 **9.68**	4.56 **9.55**	4.53 **9.47**	4.50 **9.38**	4.46 **9.29**	4.44 **9.24**	4.42 **9.17**	4.40 **9.13**	4.38 **9.07**	4.37 **9.04**	4.36 **9.02**
6	5.99 **13.74**	5.14 **10.92**	4.76 **9.78**	4.53 **9.15**	4.39 **8.75**	4.28 **8.47**	4.21 **8.26**	4.15 **8.10**	4.10 **7.98**	4.06 **7.87**	4.03 **7.79**	4.00 **7.72**	3.96 **7.60**	3.92 **7.52**	3.87 **7.39**	3.84 **7.31**	3.81 **7.23**	3.77 **7.14**	3.75 **7.09**	3.72 **7.02**	3.71 **6.99**	3.69 **6.94**	3.68 **6.90**	3.67 **6.88**
7	5.59 **12.25**	4.74 **9.55**	4.35 **8.45**	4.12 **7.85**	3.97 **7.46**	3.87 **7.19**	3.79 **7.00**	3.73 **6.84**	3.68 **6.71**	3.63 **6.62**	3.60 **6.54**	3.57 **6.47**	3.52 **6.35**	3.49 **6.27**	3.44 **6.15**	3.41 **6.07**	3.38 **5.98**	3.34 **5.90**	3.32 **5.85**	3.29 **5.78**	3.28 **5.75**	3.25 **5.70**	3.24 **5.67**	3.23 **5.65**
8	5.32 **11.26**	4.46 **8.65**	4.07 **7.59**	3.84 **7.01**	3.69 **6.63**	3.58 **6.37**	3.50 **6.19**	3.44 **6.03**	3.39 **5.91**	3.34 **5.82**	3.31 **5.74**	3.28 **5.67**	3.23 **5.56**	3.20 **5.48**	3.15 **5.36**	3.12 **5.28**	3.08 **5.20**	3.05 **5.11**	3.03 **5.06**	3.00 **5.00**	2.98 **4.96**	2.96 **4.91**	2.94 **4.88**	2.93 **4.86**
9	5.12 **10.56**	4.26 **8.02**	3.86 **6.99**	3.63 **6.42**	3.48 **6.06**	3.37 **5.80**	3.29 **5.62**	3.23 **5.47**	3.18 **5.35**	3.13 **5.26**	3.10 **5.18**	3.07 **5.11**	3.02 **5.00**	2.98 **4.92**	2.93 **4.80**	2.90 **4.73**	2.86 **4.64**	2.82 **4.56**	2.80 **4.51**	2.77 **4.45**	2.76 **4.41**	2.73 **4.36**	2.72 **4.33**	2.71 **4.31**
10	4.96 **10.04**	4.10 **7.56**	3.71 **6.55**	3.48 **5.99**	3.33 **5.64**	3.22 **5.39**	3.14 **5.21**	3.07 **5.06**	3.02 **4.95**	2.97 **4.85**	2.94 **4.78**	2.91 **4.71**	2.86 **4.60**	2.82 **4.52**	2.77 **4.41**	2.74 **4.33**	2.70 **4.25**	2.67 **4.17**	2.64 **4.12**	2.61 **4.05**	2.59 **4.01**	2.56 **3.96**	2.55 **3.93**	2.54 **3.91**
11	4.84 **9.65**	3.98 **7.20**	3.59 **6.22**	3.36 **5.67**	3.20 **5.32**	3.09 **5.07**	3.01 **4.88**	2.95 **4.74**	2.90 **4.63**	2.86 **4.54**	2.82 **4.46**	2.79 **4.40**	2.74 **4.29**	2.70 **4.21**	2.65 **4.10**	2.61 **4.02**	2.57 **3.94**	2.53 **3.86**	2.50 **3.80**	2.47 **3.74**	2.45 **3.70**	2.42 **3.66**	2.41 **3.62**	2.40 **3.60**
12	4.75 **9.33**	3.88 **6.93**	3.49 **5.95**	3.26 **5.41**	3.11 **5.06**	3.00 **4.82**	2.92 **4.65**	2.85 **4.50**	2.80 **4.39**	2.76 **4.30**	2.72 **4.22**	2.69 **4.16**	2.64 **4.05**	2.60 **3.98**	2.54 **3.86**	2.50 **3.78**	2.46 **3.70**	2.42 **3.61**	2.40 **3.56**	2.36 **3.49**	2.35 **3.46**	2.32 **3.41**	2.31 **3.38**	2.30 **3.36**
13	4.67 **9.07**	3.80 **6.70**	3.41 **5.74**	3.18 **5.20**	3.02 **4.86**	2.92 **4.62**	2.84 **4.44**	2.77 **4.30**	2.72 **4.19**	2.67 **4.10**	2.63 **4.02**	2.60 **3.96**	2.55 **3.85**	2.51 **3.78**	2.46 **3.67**	2.42 **3.59**	2.38 **3.51**	2.34 **3.42**	2.32 **3.37**	2.28 **3.30**	2.26 **3.27**	2.24 **3.21**	2.22 **3.18**	2.21 **3.16**

Degrees of freedom for the denominator

The values in the table are the critical values of F for the degrees of freedom listed over the columns (the degrees of freedom for the numerator of the F ratio) and the degrees of freedom listed for the rows (the degrees of freedom for the denominator of the F ratio). The critical value for the .05 level of significance is presented first (roman type) followed by the critical value at the .01 level (boldface). If the observed value is *greater than or equal to* the tabled value, reject H_0. F values are always positive.

Source: Reprinted by permission from *Statistical Methods* by George B. Snedecor and William G. Cochran, Sixth Edition © 1967 by The Iowa State University Press, Ames, Iowa 50010.

Table E   (continued)

Degrees of freedom for the numerator

(den.)	1	2	3	4	5	6	7	8	9	10	11	12	14	16	20	24	30	40	50	75	100	200	500	∞
14	4.60 8.86	3.74 6.51	3.34 5.56	3.11 5.03	2.96 4.69	2.85 4.46	2.77 4.28	2.70 4.14	2.65 4.03	2.60 3.94	2.56 3.86	2.53 3.80	2.48 3.70	2.44 3.62	2.39 3.51	2.35 3.43	2.31 3.34	2.27 3.26	2.24 3.21	2.21 3.14	2.19 3.11	2.16 3.06	2.14 3.02	2.13 3.00
15	4.54 8.68	3.68 6.36	3.29 5.42	3.06 4.89	2.90 4.56	2.79 4.32	2.70 4.14	2.64 4.00	2.59 3.89	2.55 3.80	2.51 3.73	2.48 3.67	2.43 3.56	2.39 3.48	2.33 3.36	2.29 3.29	2.25 3.20	2.21 3.12	2.18 3.07	2.15 3.00	2.12 2.97	2.10 2.92	2.08 2.89	2.07 2.87
16	4.49 8.53	3.63 6.23	3.24 5.29	3.01 4.77	2.85 4.44	2.74 4.20	2.66 4.03	2.59 3.89	2.54 3.78	2.49 3.69	2.45 3.61	2.42 3.55	2.37 3.45	2.33 3.37	2.28 3.25	2.24 3.18	2.20 3.10	2.16 3.01	2.13 2.96	2.09 2.89	2.07 2.86	2.04 2.80	2.02 2.77	2.01 2.75
17	4.45 8.40	3.59 6.11	3.20 5.18	2.96 4.67	2.81 4.34	2.70 4.10	2.62 3.93	2.55 3.79	2.50 3.68	2.45 3.59	2.41 3.52	2.38 3.45	2.33 3.35	2.29 3.27	2.23 3.16	2.19 3.08	2.15 3.00	2.11 2.92	2.08 2.86	2.04 2.79	2.02 2.76	1.99 2.70	1.97 2.67	1.96 2.65
18	4.41 8.28	3.55 6.01	3.16 5.09	2.93 4.58	2.77 4.25	2.66 4.01	2.58 3.85	2.51 3.71	2.46 3.60	2.41 3.51	2.37 3.44	2.34 3.37	2.29 3.27	2.25 3.19	2.19 3.07	2.15 3.00	2.11 2.91	2.07 2.83	2.04 2.78	2.00 2.71	1.98 2.68	1.95 2.62	1.93 2.59	1.92 2.57
19	4.38 8.18	3.52 5.93	3.13 5.01	2.90 4.50	2.74 4.17	2.63 3.94	2.55 3.77	2.48 3.63	2.43 3.52	2.38 3.43	2.34 3.36	2.31 3.30	2.26 3.19	2.21 3.12	2.15 3.00	2.11 2.92	2.07 2.84	2.02 2.76	2.00 2.70	1.96 2.63	1.94 2.60	1.91 2.54	1.90 2.51	1.88 2.49
20	4.35 8.10	3.49 5.85	3.10 4.94	2.87 4.43	2.71 4.10	2.60 3.87	2.52 3.71	2.45 3.56	2.40 3.45	2.35 3.37	2.31 3.30	2.28 3.23	2.23 3.13	2.18 3.05	2.12 2.94	2.08 2.86	2.04 2.77	1.99 2.69	1.96 2.63	1.92 2.56	1.90 2.53	1.87 2.47	1.85 2.44	1.84 2.42
21	4.32 8.02	3.47 5.78	3.07 4.87	2.84 4.37	2.68 4.04	2.57 3.81	2.49 3.65	2.42 3.51	2.37 3.40	2.32 3.31	2.28 3.24	2.25 3.17	2.20 3.07	2.15 2.99	2.09 2.88	2.05 2.80	2.00 2.72	1.96 2.63	1.93 2.58	1.89 2.51	1.87 2.47	1.84 2.42	1.82 2.38	1.81 2.36
22	4.30 7.94	3.44 5.72	3.05 4.82	2.82 4.31	2.66 3.99	2.55 3.76	2.47 3.59	2.40 3.45	2.35 3.35	2.30 3.26	2.26 3.18	2.23 3.12	2.18 3.02	2.13 2.94	2.07 2.83	2.03 2.75	1.98 2.67	1.93 2.58	1.91 2.53	1.87 2.46	1.84 2.42	1.81 2.37	1.80 2.33	1.78 2.31
23	4.28 7.88	3.42 5.66	3.03 4.76	2.80 4.26	2.64 3.94	2.53 3.71	2.45 3.54	2.38 3.41	2.32 3.30	2.28 3.21	2.24 3.14	2.20 3.07	2.14 2.97	2.10 2.89	2.04 2.78	2.00 2.70	1.96 2.62	1.91 2.53	1.88 2.48	1.84 2.41	1.82 2.37	1.79 2.32	1.77 2.28	1.76 2.26
24	4.26 7.82	3.40 5.61	3.01 4.72	2.78 4.22	2.62 3.90	2.51 3.67	2.43 3.50	2.36 3.36	2.30 3.25	2.26 3.17	2.22 3.09	2.18 3.03	2.13 2.93	2.09 2.85	2.02 2.74	1.98 2.66	1.94 2.58	1.89 2.49	1.86 2.44	1.82 2.36	1.80 2.33	1.76 2.27	1.74 2.23	1.73 2.21
25	4.24 7.77	3.38 5.57	2.99 4.68	2.76 4.18	2.60 3.86	2.49 3.63	2.41 3.46	2.34 3.32	2.28 3.21	2.24 3.13	2.20 3.05	2.16 2.99	2.11 2.89	2.06 2.81	2.00 2.70	1.96 2.62	1.92 2.54	1.87 2.45	1.84 2.40	1.80 2.32	1.77 2.29	1.74 2.23	1.72 2.19	1.71 2.17
26	4.22 7.72	3.37 5.53	2.98 4.64	2.74 4.14	2.59 3.82	2.47 3.59	2.39 3.42	2.32 3.29	2.27 3.17	2.22 3.09	2.18 3.02	2.15 2.96	2.10 2.86	2.05 2.77	1.99 2.66	1.95 2.58	1.90 2.50	1.85 2.41	1.82 2.36	1.78 2.28	1.76 2.25	1.72 2.19	1.70 2.15	1.69 2.13

Degrees of freedom for the denominator

The function, $F = e$ with exponent $2z$, is computed in part from Fisher's table VI. Additional entries are by interpolation, mostly graphical.

Table E (continued)

Degrees of freedom for the numerator

df	1	2	3	4	5	6	7	8	9	10	11	12	14	16	20	24	30	40	50	75	100	200	500	∞
27	4.21 / 7.68	3.35 / 5.49	2.96 / 4.60	2.73 / 4.11	2.57 / 3.79	2.46 / 3.56	2.37 / 3.39	2.30 / 3.26	2.25 / 3.14	2.20 / 3.06	2.16 / 2.98	2.13 / 2.93	2.08 / 2.83	2.03 / 2.74	1.97 / 2.63	1.93 / 2.55	1.88 / 2.47	1.84 / 2.38	1.80 / 2.33	1.76 / 2.25	1.74 / 2.21	1.71 / 2.16	1.68 / 2.12	1.67 / 2.10
28	4.20 / 7.64	3.34 / 5.45	2.95 / 4.57	2.71 / 4.07	2.56 / 3.76	2.44 / 3.53	2.36 / 3.36	2.29 / 3.23	2.24 / 3.11	2.19 / 3.03	2.15 / 2.95	2.12 / 2.90	2.06 / 2.80	2.02 / 2.71	1.96 / 2.60	1.91 / 2.52	1.87 / 2.44	1.81 / 2.35	1.78 / 2.30	1.75 / 2.22	1.72 / 2.18	1.69 / 2.13	1.67 / 2.09	1.65 / 2.06
29	4.18 / 7.60	3.33 / 5.42	2.93 / 4.54	2.70 / 4.04	2.54 / 3.73	2.43 / 3.50	2.35 / 3.33	2.28 / 3.20	2.22 / 3.08	2.18 / 3.00	2.14 / 2.92	2.10 / 2.87	2.05 / 2.77	2.00 / 2.68	1.94 / 2.57	1.90 / 2.49	1.85 / 2.41	1.80 / 2.32	1.77 / 2.27	1.73 / 2.19	1.71 / 2.15	1.68 / 2.10	1.65 / 2.06	1.64 / 2.03
30	4.17 / 7.56	3.32 / 5.39	2.92 / 4.51	2.69 / 4.02	2.53 / 3.70	2.42 / 3.47	2.34 / 3.30	2.27 / 3.17	2.21 / 3.06	2.16 / 2.98	2.12 / 2.90	2.09 / 2.84	2.04 / 2.74	1.99 / 2.66	1.93 / 2.55	1.89 / 2.47	1.84 / 2.38	1.79 / 2.29	1.76 / 2.24	1.72 / 2.16	1.69 / 2.13	1.66 / 2.07	1.64 / 2.03	1.62 / 2.01
32	4.15 / 7.50	3.30 / 5.34	2.90 / 4.46	2.67 / 3.97	2.51 / 3.66	2.40 / 3.42	2.32 / 3.25	2.25 / 3.12	2.19 / 3.01	2.14 / 2.94	2.10 / 2.86	2.07 / 2.80	2.02 / 2.70	1.97 / 2.62	1.91 / 2.51	1.86 / 2.42	1.82 / 2.34	1.76 / 2.25	1.74 / 2.20	1.69 / 2.12	1.67 / 2.08	1.64 / 2.02	1.61 / 1.98	1.59 / 1.96
34	4.13 / 7.44	3.28 / 5.29	2.88 / 4.42	2.65 / 3.93	2.49 / 3.61	2.38 / 3.38	2.30 / 3.21	2.23 / 3.08	2.17 / 2.97	2.12 / 2.89	2.08 / 2.82	2.05 / 2.76	2.00 / 2.66	1.95 / 2.58	1.89 / 2.47	1.84 / 2.38	1.80 / 2.30	1.74 / 2.21	1.71 / 2.15	1.67 / 2.08	1.64 / 2.04	1.61 / 1.98	1.59 / 1.94	1.57 / 1.91
36	4.11 / 7.39	3.26 / 5.25	2.86 / 4.38	2.63 / 3.89	2.48 / 3.58	2.36 / 3.35	2.28 / 3.18	2.21 / 3.04	2.15 / 2.94	2.10 / 2.86	2.06 / 2.78	2.03 / 2.72	1.98 / 2.62	1.93 / 2.54	1.87 / 2.43	1.82 / 2.35	1.78 / 2.26	1.72 / 2.17	1.69 / 2.12	1.65 / 2.04	1.62 / 2.00	1.59 / 1.94	1.56 / 1.90	1.55 / 1.87
38	4.10 / 7.35	3.25 / 5.21	2.85 / 4.34	2.62 / 3.86	2.46 / 3.54	2.35 / 3.32	2.26 / 3.15	2.19 / 3.02	2.14 / 2.91	2.09 / 2.82	2.05 / 2.75	2.02 / 2.69	1.96 / 2.59	1.92 / 2.51	1.85 / 2.40	1.80 / 2.32	1.76 / 2.22	1.71 / 2.14	1.67 / 2.08	1.63 / 2.00	1.60 / 1.97	1.57 / 1.90	1.54 / 1.86	1.53 / 1.84
40	4.08 / 7.31	3.23 / 5.18	2.84 / 4.31	2.61 / 3.83	2.45 / 3.51	2.34 / 3.29	2.25 / 3.12	2.18 / 2.99	2.12 / 2.88	2.07 / 2.80	2.04 / 2.73	2.00 / 2.66	1.95 / 2.56	1.90 / 2.49	1.84 / 2.37	1.79 / 2.29	1.74 / 2.20	1.69 / 2.11	1.66 / 2.05	1.61 / 1.97	1.59 / 1.94	1.55 / 1.88	1.53 / 1.84	1.51 / 1.81
42	4.07 / 7.27	3.22 / 5.15	2.83 / 4.29	2.59 / 3.80	2.44 / 3.49	2.32 / 3.26	2.24 / 3.10	2.17 / 2.96	2.11 / 2.86	2.06 / 2.77	2.02 / 2.70	1.99 / 2.64	1.94 / 2.54	1.89 / 2.46	1.82 / 2.35	1.78 / 2.26	1.73 / 2.17	1.68 / 2.08	1.64 / 2.02	1.60 / 1.94	1.57 / 1.91	1.54 / 1.85	1.51 / 1.80	1.49 / 1.78
44	4.06 / 7.24	3.21 / 5.12	2.82 / 4.26	2.58 / 3.78	2.43 / 3.46	2.31 / 3.24	2.23 / 3.07	2.16 / 2.94	2.10 / 2.84	2.05 / 2.75	2.01 / 2.68	1.98 / 2.62	1.92 / 2.52	1.88 / 2.44	1.81 / 2.32	1.76 / 2.24	1.72 / 2.15	1.66 / 2.06	1.63 / 2.00	1.58 / 1.92	1.56 / 1.88	1.52 / 1.82	1.50 / 1.78	1.48 / 1.75
46	4.05 / 7.21	3.20 / 5.10	2.81 / 4.24	2.57 / 3.76	2.42 / 3.44	2.30 / 3.22	2.22 / 3.05	2.14 / 2.92	2.09 / 2.82	2.04 / 2.73	2.00 / 2.66	1.97 / 2.60	1.91 / 2.50	1.87 / 2.42	1.80 / 2.30	1.75 / 2.22	1.71 / 2.13	1.65 / 2.04	1.62 / 1.98	1.57 / 1.90	1.54 / 1.86	1.51 / 1.80	1.48 / 1.76	1.46 / 1.72
48	4.04 / 7.19	3.19 / 5.08	2.80 / 4.22	2.56 / 3.74	2.41 / 3.42	2.30 / 3.20	2.21 / 3.04	2.14 / 2.90	2.08 / 2.80	2.03 / 2.71	1.99 / 2.64	1.96 / 2.58	1.90 / 2.48	1.86 / 2.40	1.79 / 2.28	1.74 / 2.20	1.70 / 2.11	1.64 / 2.02	1.61 / 1.96	1.56 / 1.88	1.53 / 1.84	1.50 / 1.78	1.47 / 1.73	1.45 / 1.70

Degrees of freedom for the denominator

Table E (continued)

Degrees of freedom for the numerator

df (denom.)	1	2	3	4	5	6	7	8	9	10	11	12	14	16	20	24	30	40	50	75	100	200	500	∞
50	4.03 7.17	3.18 5.06	2.79 4.20	2.56 3.72	2.40 3.41	2.29 3.18	2.20 3.02	2.13 2.88	2.07 2.78	2.02 2.70	1.98 2.62	1.95 2.56	1.90 2.46	1.85 2.39	1.78 2.26	1.74 2.18	1.69 2.10	1.63 2.00	1.60 1.94	1.55 1.86	1.52 1.82	1.48 1.76	1.46 1.71	1.44 1.68
55	4.02 7.12	3.17 5.01	2.78 4.16	2.54 3.68	2.38 3.37	2.27 3.15	2.18 2.98	2.11 2.85	2.05 2.75	2.00 2.66	1.97 2.59	1.93 2.53	1.88 2.43	1.83 2.35	1.76 2.23	1.72 2.15	1.67 2.06	1.61 1.96	1.58 1.90	1.52 1.82	1.50 1.78	1.46 1.71	1.43 1.66	1.41 1.64
60	4.00 7.08	3.15 4.98	2.76 4.13	2.52 3.65	2.37 3.34	2.25 3.12	2.17 2.95	2.10 2.82	2.04 2.72	1.99 2.63	1.95 2.56	1.92 2.50	1.86 2.40	1.81 2.32	1.75 2.20	1.70 2.12	1.65 2.03	1.59 1.93	1.56 1.87	1.50 1.79	1.48 1.74	1.44 1.68	1.41 1.63	1.39 1.60
65	3.99 7.04	3.14 4.95	2.75 4.10	2.51 3.62	2.36 3.31	2.24 3.09	2.15 2.93	2.08 2.79	2.02 2.70	1.98 2.61	1.94 2.54	1.90 2.47	1.85 2.37	1.80 2.30	1.73 2.18	1.68 2.09	1.63 2.00	1.57 1.90	1.54 1.84	1.49 1.76	1.46 1.71	1.42 1.64	1.39 1.60	1.37 1.56
70	3.98 7.01	3.13 4.92	2.74 4.08	2.50 3.60	2.35 3.29	2.23 3.07	2.14 2.91	2.07 2.77	2.01 2.67	1.97 2.59	1.93 2.51	1.89 2.45	1.84 2.35	1.79 2.28	1.72 2.15	1.67 2.07	1.62 1.98	1.56 1.88	1.53 1.82	1.47 1.74	1.45 1.69	1.40 1.62	1.37 1.56	1.35 1.53
80	3.96 6.96	3.11 4.88	2.72 4.04	2.48 3.56	2.33 3.25	2.21 3.04	2.12 2.87	2.05 2.74	1.99 2.64	1.95 2.55	1.91 2.48	1.88 2.41	1.82 2.32	1.77 2.24	1.70 2.11	1.65 2.03	1.60 1.94	1.54 1.84	1.51 1.78	1.45 1.70	1.42 1.65	1.38 1.57	1.35 1.52	1.32 1.49
100	3.94 6.90	3.09 4.82	2.70 3.98	2.46 3.51	2.30 3.20	2.19 2.99	2.10 2.82	2.03 2.69	1.97 2.59	1.92 2.51	1.88 2.43	1.85 2.36	1.79 2.26	1.75 2.19	1.68 2.06	1.63 1.98	1.57 1.89	1.51 1.79	1.48 1.73	1.42 1.64	1.39 1.59	1.34 1.51	1.30 1.46	1.28 1.43
125	3.92 6.84	3.07 4.78	2.68 3.94	2.44 3.47	2.29 3.17	2.17 2.95	2.08 2.79	2.01 2.65	1.95 2.56	1.90 2.47	1.86 2.40	1.83 2.33	1.77 2.23	1.72 2.15	1.65 2.03	1.60 1.94	1.55 1.85	1.49 1.75	1.45 1.68	1.39 1.59	1.36 1.54	1.31 1.46	1.27 1.40	1.25 1.37
150	3.91 6.81	3.06 4.75	2.67 3.91	2.43 3.44	2.27 3.14	2.16 2.92	2.07 2.76	2.00 2.62	1.94 2.53	1.89 2.44	1.85 2.37	1.82 2.30	1.76 2.20	1.71 2.12	1.64 2.00	1.59 1.91	1.54 1.83	1.47 1.72	1.44 1.66	1.37 1.56	1.34 1.51	1.29 1.43	1.25 1.37	1.22 1.33
200	3.89 6.76	3.04 4.71	2.65 3.88	2.41 3.41	2.26 3.11	2.14 2.90	2.05 2.73	1.98 2.60	1.92 2.50	1.87 2.41	1.83 2.34	1.80 2.28	1.74 2.17	1.69 2.09	1.62 1.97	1.57 1.88	1.52 1.79	1.45 1.69	1.42 1.62	1.35 1.53	1.32 1.48	1.26 1.39	1.22 1.33	1.19 1.28
400	3.86 6.70	3.02 4.66	2.62 3.83	2.39 3.36	2.23 3.06	2.12 2.85	2.03 2.69	1.96 2.55	1.90 2.46	1.85 2.37	1.81 2.29	1.78 2.23	1.72 2.12	1.67 2.04	1.60 1.92	1.54 1.84	1.49 1.74	1.42 1.64	1.38 1.57	1.32 1.47	1.28 1.42	1.22 1.32	1.16 1.24	1.13 1.19
1000	3.85 6.66	3.00 4.62	2.61 3.80	2.38 3.34	2.22 3.04	2.10 2.82	2.02 2.66	1.95 2.53	1.89 2.43	1.84 2.34	1.80 2.26	1.76 2.20	1.70 2.09	1.65 2.01	1.58 1.89	1.53 1.81	1.47 1.71	1.41 1.61	1.36 1.54	1.30 1.44	1.26 1.38	1.19 1.28	1.13 1.19	1.08 1.11
∞	3.84 6.64	2.99 4.60	2.60 3.78	2.37 3.32	2.21 3.02	2.09 2.80	2.01 2.64	1.94 2.51	1.88 2.41	1.83 2.32	1.79 2.24	1.75 2.18	1.69 2.07	1.64 1.99	1.57 1.87	1.52 1.79	1.46 1.69	1.40 1.59	1.35 1.52	1.28 1.41	1.24 1.36	1.17 1.25	1.11 1.15	1.00 1.00

Degrees of freedom for the denominator

Table F.   Critical Values of the Studentized Range Statistic, *q*

df_{within}	$1-\alpha$														
		r = number of groups in design													
		2	3	4	5	6	7	8	9	10	11	12	13	14	15
1	.95	18.0	27.0	32.8	37.1	40.4	43.1	45.4	47.4	49.1	50.6	52.0	53.2	54.3	55.4
	.99	90.0	135	164	186	202	216	227	237	246	253	260	266	272	277
2	.95	6.09	8.3	9.8	10.9	11.7	12.4	13.0	13.5	14.0	14.4	14.7	15.1	15.4	15.7
	.99	14.0	19.0	22.3	24.7	26.6	28.2	29.5	30.7	31.7	32.6	33.4	34.1	34.8	35.4
3	.95	4.50	5.91	6.82	7.50	8.04	8.48	8.85	9.18	9.46	9.72	9.95	10.2	10.4	10.5
	.99	8.26	10.6	12.2	13.3	14.2	15.0	15.6	16.2	16.7	17.1	17.5	17.9	18.2	18.5
4	.95	3.93	5.04	5.76	6.29	6.71	7.05	7.35	7.60	7.83	8.03	8.21	8.37	8.52	8.66
	.99	6.51	8.12	9.17	9.96	10.6	11.1	11.5	11.9	12.3	12.6	12.8	13.1	13.3	13.5
5	.95	3.64	4.60	5.22	5.67	6.03	6.33	6.58	6.80	6.99	7.17	7.32	7.47	7.60	7.72
	.99	5.70	6.97	7.80	8.42	8.91	9.32	9.67	9.97	10.2	10.5	10.7	10.9	11.1	11.2
6	.95	3.46	4.34	4.90	5.31	5.63	5.89	6.12	6.32	6.49	6.65	6.79	6.92	7.03	7.14
	.99	5.24	6.33	7.03	7.56	7.97	8.32	8.61	8.87	9.10	9.30	9.49	9.65	9.81	9.95
7	.95	3.34	4.16	4.69	5.06	5.36	5.61	5.82	6.00	6.16	6.30	6.43	6.55	6.66	6.76
	.99	4.95	5.92	6.54	7.01	7.37	7.68	7.94	8.17	8.37	8.55	8.71	8.86	9.00	9.12
8	.95	3.26	4.04	4.53	4.89	5.17	5.40	5.60	5.77	5.92	6.05	6.18	6.29	6.39	6.48
	.99	4.74	5.63	6.20	6.63	6.96	7.24	7.47	7.68	7.87	8.03	8.18	8.31	8.44	8.55
9	.95	3.20	3.95	4.42	4.76	5.02	5.24	5.43	5.60	5.74	5.87	5.98	6.09	6.19	6.28
	.99	4.60	5.43	5.96	6.35	6.66	6.91	7.13	7.32	7.49	7.65	7.78	7.91	8.03	8.13
10	.95	3.15	3.88	4.33	4.65	4.91	5.12	5.30	5.46	5.60	5.72	5.83	5.93	6.03	6.11
	.99	4.48	5.27	5.77	6.14	6.43	6.67	6.87	7.05	7.21	7.36	7.48	7.60	7.71	7.81
11	.95	3.11	3.82	4.26	4.57	4.82	5.03	5.20	5.35	5.49	5.61	5.71	5.81	5.90	5.99
	.99	4.39	5.14	5.62	5.97	6.25	6.48	6.67	6.84	6.99	7.13	7.26	7.36	7.46	7.56
12	.95	3.08	3.77	4.20	4.51	4.75	4.95	5.12	5.27	5.40	5.51	5.62	5.71	5.80	5.88
	.99	4.32	5.04	5.50	5.84	6.10	6.32	6.51	6.67	6.81	6.94	7.06	7.17	7.26	7.36

Table F (continued)

df_{within}	$1-\alpha$	\multicolumn{14}{c}{r = number of groups in design}													
		2	3	4	5	6	7	8	9	10	11	12	13	14	15
13	.95	3.06	3.73	4.15	4.45	4.69	4.88	5.05	5.19	5.32	5.43	5.53	5.63	5.71	5.79
	.99	4.26	4.96	5.40	5.73	5.98	6.19	6.37	6.53	6.67	6.79	6.90	7.01	7.10	7.19
14	.95	3.03	3.70	4.11	4.41	4.64	4.83	4.99	5.13	5.25	5.36	5.46	5.55	5.64	5.72
	.99	4.21	4.89	5.32	5.63	5.88	6.08	6.26	6.41	6.54	6.66	6.77	6.87	6.96	7.05
16	.95	3.00	3.65	4.05	4.33	4.56	4.74	4.90	5.03	5.15	5.26	5.35	5.44	5.52	5.59
	.99	4.13	4.78	5.19	5.49	5.72	5.92	6.08	6.22	6.35	6.46	6.56	6.66	6.74	6.82
18	.95	2.97	3.61	4.00	4.28	4.49	4.67	4.82	4.96	5.07	5.17	5.27	5.35	5.43	5.50
	.99	4.07	4.70	5.09	5.38	5.60	5.79	5.94	6.08	6.20	6.31	6.41	6.50	6.58	6.65
20	.95	2.95	3.58	3.96	4.23	4.45	4.62	4.77	4.90	5.01	5.11	5.20	5.28	5.36	5.43
	.99	4.02	4.64	5.02	5.29	5.51	5.69	5.84	5.97	6.09	6.19	6.29	6.37	6.45	6.52
24	.95	2.92	3.53	3.90	4.17	4.37	4.54	4.68	4.81	4.92	5.01	5.10	5.18	5.25	5.32
	.99	3.96	4.54	4.91	5.17	5.37	5.54	5.69	5.81	5.92	6.02	6.11	6.19	6.26	6.33
30	.95	2.89	3.49	3.84	4.10	4.30	4.46	4.60	4.72	4.83	4.92	5.00	5.08	5.15	5.21
	.99	3.89	4.45	4.80	5.05	5.24	5.40	5.54	5.56	5.76	5.85	5.93	6.01	6.08	6.14
40	.95	2.86	3.44	3.79	4.04	4.23	4.39	4.52	4.63	4.74	4.82	4.91	4.98	5.05	5.11
	.99	3.82	4.37	4.70	4.93	5.11	5.27	5.39	5.50	5.60	5.69	5.77	5.84	5.90	5.96
60	.95	2.83	3.40	3.74	3.98	4.16	4.31	4.44	4.55	4.65	4.73	4.81	4.88	4.94	5.00
	.99	3.76	4.28	4.60	4.82	4.99	5.13	5.25	5.36	5.45	5.53	5.60	5.67	5.73	5.79
120	.95	2.80	3.36	3.69	3.92	4.10	4.24	4.36	4.48	4.56	4.64	4.72	4.78	4.84	4.90
	.99	3.70	4.20	4.50	4.71	4.87	5.01	5.12	5.21	5.30	5.38	5.44	5.51	5.56	5.61
∞	.95	2.77	3.31	3.63	3.86	4.03	4.17	4.29	4.39	4.47	4.55	4.62	4.68	4.74	4.80
	.99	3.64	4.12	4.40	4.60	4.76	4.88	4.99	5.08	5.16	5.23	5.29	5.35	5.40	5.45

Source: This table is abridged from Table II.2 in *The Probability Integrals of the Range and of the Studentized Range*, prepared by H. Leon Harter, Donald S. Clemm, and Eugene H. Guthrie. These tables are published in WADC Tech. Rep. 58-484, vol. 2, 1959, Wright Air Development Center, and are reproduced with the kind permission of the authors.

Table G.   Critical Values of Chi Square, χ^2

df	Level of significance for a directional test					
	.10	.05	.025	.01	.005	.0005
	Level of significance for a nondirectional test					
	.20	.10	.05	.02	.01	.001
1	1.64	2.71	3.84	5.41	6.64	10.83
2	3.22	4.60	5.99	7.82	9.21	13.82
3	4.64	6.25	7.82	9.84	11.34	16.27
4	5.99	7.78	9.49	11.67	13.28	18.46
5	7.29	9.24	11.07	13.39	15.09	20.52
6	8.56	10.64	12.59	15.03	16.81	22.46
7	9.80	12.02	14.07	16.62	18.48	24.32
8	11.03	13.36	15.51	18.17	20.09	26.12
9	12.24	14.68	16.92	19.68	21.67	27.88
10	13.44	15.99	18.31	21.16	23.21	29.59
11	14.63	17.28	19.68	22.62	24.72	31.26
12	15.81	18.55	21.03	24.05	26.22	32.91
13	16.98	19.81	22.36	25.47	27.69	34.53
14	18.15	21.06	23.68	26.87	29.14	36.12
15	19.31	22.31	25.00	28.26	30.58	37.70
16	20.46	23.54	26.30	29.63	32.00	39.29
17	21.62	24.77	27.59	31.00	33.41	40.75
18	22.76	25.99	28.87	32.35	34.80	42.31
19	23.90	27.20	30.14	33.69	36.19	43.82
20	25.04	28.41	31.41	35.02	37.57	45.32
21	26.17	29.62	32.67	36.34	38.93	46.80
22	27.30	30.81	33.92	37.66	40.29	48.27
23	28.43	32.01	35.17	38.97	41.64	49.73
24	29.55	33.20	36.42	40.27	42.98	51.18
25	30.68	34.38	37.65	41.57	44.31	52.62
26	31.80	35.56	38.88	42.86	45.64	54.05
27	32.91	36.74	40.11	44.14	46.96	55.48
28	34.03	37.92	41.34	45.42	48.28	56.89
29	35.14	39.09	42.69	46.69	49.59	58.30
30	36.25	40.26	43.77	47.96	50.89	59.70
32	38.47	42.59	46.19	50.49	53.49	62.49
34	40.68	44.90	48.60	53.00	56.06	65.25
36	42.88	47.21	51.00	55.49	58.62	67.99
38	45.08	49.51	53.38	57.97	61.16	70.70
40	47.27	51.81	55.76	60.44	63.69	73.40
44	51.64	56.37	60.48	65.34	68.71	78.75
48	55.99	60.91	65.17	70.20	73.68	84.04
52	60.33	65.42	69.83	75.02	78.62	89.27
56	64.66	69.92	74.47	79.82	83.51	94.46
60	68.97	74.40	79.08	84.58	88.38	99.61

The table lists the critical values of chi square for the degrees of freedom shown at the left for tests corresponding to those significance levels which head each column. If the observed value of χ^2_{obs} is *greater than or equal to* the tabled value, reject H_0. All chi squares are positive.

Source: Table G is taken from Table IV of Fisher and Yates, *Statistical Tables for Biological, Agricultural and Medical Research*, published by Longman Group Ltd., London (previously published by Oliver and Boyd, Ltd., Edinburgh), and by permission of the authors and publishers.

Table H. Critical Values of the Mann–Whitney U for a Directional Test at .05 or a Nondirectional Test at .10

n_B \\ n_A	1	2	3	4	5	6	7	8	9	10	11	12	13	14	15	16	17	18	19	20
1	--	--	--	--	--	--	--	--	--	--	--	--	--	--	--	--	--	--	0/19	0/20
2	--	--	--	--	0/10	0/12	0/14	1/15	1/17	1/19	1/21	2/22	2/24	2/26	3/27	3/29	3/31	4/32	4/34	4/36
3	--	--	0/9	0/12	1/14	2/16	2/19	3/21	3/24	4/26	5/28	5/31	6/33	7/35	7/38	8/40	9/42	9/45	10/47	11/49
4	--	--	0/12	1/15	2/18	3/21	4/24	5/27	6/30	7/33	8/36	9/39	10/42	11/45	12/48	14/50	15/53	16/56	17/59	18/62
5	--	0/10	1/14	2/18	4/21	5/25	6/29	8/32	9/36	11/39	12/43	13/47	15/50	16/54	18/57	19/61	20/65	22/68	23/72	25/75
6	--	0/12	2/16	3/21	5/25	7/29	8/34	10/38	12/42	14/46	16/50	17/55	19/59	21/63	23/67	25/71	26/76	28/80	30/84	32/88
7	--	0/14	2/19	4/24	6/29	8/34	11/38	13/43	15/48	17/53	19/58	21/63	24/67	26/72	28/77	30/82	33/86	35/91	37/96	39/101
8	--	1/15	3/21	5/27	8/32	10/38	13/43	15/49	18/54	20/60	23/65	26/70	28/76	31/81	33/87	36/92	39/97	41/103	44/108	47/113
9	--	1/17	3/24	6/30	9/36	12/42	15/48	18/54	21/60	24/66	27/72	30/78	33/84	36/90	39/96	42/102	45/108	48/114	51/120	54/126
10	--	1/19	4/26	7/33	11/39	14/46	17/53	20/60	24/66	27/73	31/79	34/86	37/93	41/99	44/106	48/112	51/119	55/125	58/132	62/138
11	--	1/21	5/28	8/36	12/43	16/50	19/58	23/65	27/72	31/79	34/87	38/94	42/101	46/108	50/115	54/122	57/130	61/137	65/144	69/151
12	--	2/22	5/31	9/39	13/47	17/55	21/63	26/70	30/78	34/86	38/94	42/102	47/109	51/117	55/125	60/132	64/140	68/148	72/156	77/163
13	--	2/24	6/33	10/42	15/50	19/59	24/67	28/76	33/84	37/93	42/101	47/109	51/118	56/126	61/134	65/143	70/151	75/159	80/167	84/176
14	--	2/26	7/35	11/45	16/54	21/63	26/72	31/81	36/90	41/99	46/108	51/117	56/126	61/135	66/144	71/153	77/161	82/170	87/179	92/188
15	--	3/27	7/38	12/48	18/57	23/67	28/77	33/87	39/96	44/106	50/115	55/125	61/134	66/144	72/153	77/163	83/172	88/182	94/191	100/200
16	--	3/29	8/40	14/50	19/61	25/71	30/82	36/92	42/102	48/112	54/122	60/132	65/143	71/153	77/163	83/173	89/183	95/193	101/203	107/213
17	--	3/31	9/42	15/53	20/65	26/76	33/86	39/97	45/108	51/119	57/130	64/140	70/151	77/161	83/172	89/183	96/193	102/204	109/214	115/225
18	--	4/32	9/45	16/56	22/68	28/80	35/91	41/103	48/114	55/123	61/137	68/148	75/159	82/170	88/182	95/193	102/204	109/215	116/226	123/237
19	0/19	4/34	10/47	17/59	23/72	30/84	37/96	44/108	51/120	58/132	65/144	72/156	80/167	87/179	94/191	101/203	109/214	116/226	123/238	130/250
20	0/20	4/36	11/49	18/62	25/75	32/88	39/101	47/113	54/126	62/138	69/151	77/163	84/176	92/188	100/200	107/213	115/225	123/237	130/250	138/262

(Dashes in the body of the table indicate that no decision is possible at the stated level of significance.)

If the observed value of U falls between the two values presented in the table for n_A and n_B, do not reject H_0. Otherwise, reject H_0.

Table H. Critical Values of the Mann–Whitney *U* for a Directional Test at .025 or a Nondirectional Test at .05

n_B \ n_A	1	2	3	4	5	6	7	8	9	10	11	12	13	14	15	16	17	18	19	20
1	--	--	--	--	--	--	--	--	--	--	--	--	--	--	--	--	--	--	--	--
2	--	--	--	--	--	--	--	0/16	0/18	0/20	0/22	1/23	1/25	1/27	1/29	1/31	2/32	2/34	2/36	2/38
3	--	--	--	--	0/15	1/17	1/20	2/22	2/25	3/27	3/30	4/32	4/35	5/37	5/40	6/42	6/45	7/47	7/50	8/52
4	--	--	--	0/16	1/19	2/22	3/25	4/28	4/32	5/35	6/38	7/41	8/44	9/47	10/50	11/53	11/57	12/60	13/63	13/67
5	--	--	0/15	1/19	2/23	3/27	5/30	6/34	7/38	8/42	9/46	11/49	12/53	13/57	14/61	15/65	17/68	18/72	19/76	20/80
6	--	--	1/17	2/22	3/27	5/31	6/36	8/40	10/44	11/49	13/53	14/58	16/62	17/67	19/71	21/75	22/80	24/84	25/89	27/93
7	--	--	1/20	3/25	5/30	6/36	8/41	10/46	12/51	14/56	16/61	18/66	20/71	22/76	24/81	26/86	28/91	30/96	32/101	34/106
8	--	0/16	2/22	4/28	6/34	8/40	10/46	13/51	15/57	17/63	19/69	22/74	24/80	26/86	29/91	31/97	34/102	36/108	38/111	41/119
9	--	0/18	2/25	4/32	7/38	10/44	12/51	15/57	17/64	20/70	23/76	26/82	28/89	31/95	34/101	37/107	39/114	42/120	45/126	48/132
10	--	0/20	3/27	5/35	8/42	11/49	14/56	17/63	20/70	23/77	26/84	29/91	33/97	36/104	39/111	42/118	45/125	48/132	52/138	55/145
11	--	0/22	3/30	6/38	9/46	13/53	16/61	19/69	23/76	26/84	30/91	33/99	37/106	40/114	44/121	47/129	51/136	55/143	58/151	62/158
12	--	1/23	4/32	7/41	11/49	14/58	18/66	22/74	26/82	29/91	33/99	37/107	41/115	45/123	49/131	53/139	57/147	61/155	65/163	69/171
13	--	1/25	4/35	8/44	12/53	16/62	20/71	24/80	28/89	33/97	37/106	41/115	45/124	50/132	54/141	59/149	63/158	67/167	72/175	76/184
14	--	1/27	5/37	9/47	13/51	17/67	22/76	26/86	31/95	36/104	40/114	45/123	50/132	55/141	59/151	64/160	67/171	74/178	78/188	83/197
15	--	1/29	5/40	10/50	14/61	19/71	24/81	29/91	34/101	39/111	44/121	49/131	54/141	59/151	64/161	70/170	75/180	80/190	85/200	90/210
16	--	1/31	6/42	11/53	15/65	21/75	26/86	31/97	37/107	42/118	47/129	53/139	59/149	64/160	70/170	75/181	81/191	86/202	92/212	98/222
17	--	2/32	6/45	11/57	17/68	22/80	28/91	34/102	39/114	45/125	51/136	57/147	63/158	67/171	75/180	81/191	87/202	93/213	99/224	105/235
18	--	2/34	7/47	12/60	18/72	24/84	30/96	36/108	42/120	48/132	55/143	61/155	67/167	74/178	80/190	86/202	93/213	99/225	106/236	112/248
19	--	2/36	7/50	13/63	19/76	25/89	32/101	38/114	45/126	52/138	58/151	65/163	72/175	78/188	85/200	92/212	99/224	106/236	113/248	119/261
20	--	2/38	8/52	13/67	20/80	27/93	34/106	41/119	48/132	55/145	62/158	69/171	76/184	83/197	90/210	98/222	105/235	112/248	119/261	127/273

(Dashes in the body of the table indicate that no decision is possible at the stated level of significance.)

Source: From Mann, H. B., and Whitney, D. R., "On a Test of Whether One of Two Random Variables Is Stochastically Larger Than the Other," *Annals of Mathematical Statistics* 18 (1947): 50–60, and Auble, D., "Extended Tables for the Mann-Whitney Statistic," *Bulletin of the Institute of Educational Research at Indiana University*, vol. 1, no. 2 (1953), as used in Runyon and Haber, *Fundamentals of Behavioral Statistics*, 3rd ed., Addison-Wesley, Reading, Mass., 1976. Reprinted by permission.

Table H. Critical Values of the Mann–Whitney *U* for a Directional Test at .01 or a Nondirectional Test at .02

n_B \ n_A	1	2	3	4	5	6	7	8	9	10	11	12	13	14	15	16	17	18	19	20
1	--	--	--	--	--	--	--	--	--	--	--	--	--	--	--	--	--	--	--	--
2	--	--	--	--	--	--	--	--	--	--	--	--	0/26	0/28	0/30	0/32	0/34	0/36	1/37	1/39
3	--	--	--	--	--	--	0/21	0/24	1/26	1/29	1/32	2/34	2/37	2/40	3/42	3/45	4/47	4/50	4/52	5/55
4	--	--	--	--	0/20	1/23	1/27	2/30	3/33	3/37	4/40	5/43	5/47	6/50	7/53	7/57	8/60	9/63	9/67	10/70
5	--	--	--	0/20	1/24	2/28	3/32	4/36	5/40	6/44	7/48	8/52	9/56	10/60	11/64	12/68	13/72	14/76	15/80	16/84
6	--	--	--	1/23	2/28	3/33	4/38	6/42	7/47	8/52	9/57	11/61	12/66	13/71	15/75	16/80	18/84	19/89	20/94	22/98
7	--	--	0/21	1/27	3/32	4/38	6/43	7/49	9/54	11/59	12/65	14/70	16/75	17/81	19/86	21/91	23/96	24/102	26/107	28/112
8	--	--	0/24	2/30	4/36	6/42	7/49	9/55	11/61	13/67	15/73	17/79	20/84	22/90	24/96	26/102	28/108	30/114	32/120	34/126
9	--	--	1/26	3/33	5/40	7/47	9/54	11/61	14/67	16/74	18/81	21/87	23/94	26/100	28/107	31/113	33/120	36/126	38/133	40/140
10	--	--	1/29	3/37	6/44	8/52	11/59	13/67	16/74	19/81	22/88	24/96	27/103	30/110	33/117	36/124	38/132	41/139	44/146	47/153
11	--	--	1/32	4/40	7/48	9/57	12/65	15/73	18/81	22/88	25/96	28/104	31/112	34/120	37/128	41/135	44/143	47/151	50/159	53/167
12	--	--	2/34	5/43	8/52	11/61	14/70	17/79	21/87	24/96	28/104	31/113	35/121	38/130	42/138	46/146	49/155	53/163	56/172	60/180
13	--	0/26	2/37	5/47	9/56	12/66	16/76	20/84	23/94	27/103	31/112	35/121	39/130	43/139	47/148	51/157	55/166	59/175	63/184	67/193
14	--	0/28	2/40	6/50	10/60	13/71	17/81	22/90	26/100	30/110	34/120	38/130	43/139	47/149	51/159	56/168	60/178	65/187	69/197	73/207
15	--	0/30	3/42	7/53	11/64	15/75	19/86	24/96	28/107	33/117	37/128	42/138	47/148	51/159	56/169	61/179	66/189	70/200	75/210	80/220
16	--	0/32	3/45	7/57	12/68	16/80	21/91	26/102	31/113	36/124	41/135	46/146	51/157	56/168	61/179	66/190	71/201	76/212	82/222	87/233
17	--	0/34	4/47	8/60	13/72	18/84	23/96	28/108	33/120	38/132	44/143	49/155	55/166	60/178	66/189	71/201	77/212	82/224	88/234	93/247
18	--	0/36	4/50	9/63	14/76	19/89	24/102	30/114	36/126	41/139	47/151	53/163	59/175	65/187	70/200	76/212	82/224	88/236	94/248	100/260
19	--	1/37	4/53	9/67	15/80	20/94	26/107	32/120	38/133	44/146	50/159	56/172	63/184	69/197	75/210	82/222	88/235	94/248	101/260	107/273
20	--	1/39	5/55	10/70	16/84	22/98	28/112	34/126	40/140	47/153	53/167	60/180	67/193	73/207	80/220	87/233	93/247	100/260	107/273	114/286

(Dashes in the body of the table indicate that no decision is possible at the stated level of significance.)

Table H. Critical Values of the Mann–Whitney U for a Directional Test at .005 or a Nondirectional Test at .01

n_B \ n_A	1	2	3	4	5	6	7	8	9	10	11	12	13	14	15	16	17	18	19	20
1	--	--	--	--	--	--	--	--	--	--	--	--	--	--	--	--	--	--	--	--
2	--	--	--	--	--	--	--	--	--	--	--	--	--	--	--	--	--	--	0 / 38	0 / 40
3	--	--	--	--	--	--	--	--	0 / 27	0 / 30	0 / 33	1 / 35	1 / 38	1 / 41	2 / 43	2 / 46	2 / 49	2 / 52	3 / 54	3 / 57
4	--	--	--	--	--	0 / 24	0 / 28	1 / 31	1 / 35	2 / 38	2 / 42	3 / 45	3 / 49	4 / 52	5 / 55	5 / 59	6 / 62	6 / 66	7 / 69	8 / 72
5	--	--	--	--	0 / 25	1 / 29	1 / 34	2 / 38	3 / 42	4 / 46	5 / 50	6 / 54	7 / 58	7 / 63	8 / 67	9 / 71	10 / 75	11 / 79	12 / 83	13 / 87
6	--	--	--	0 / 24	1 / 29	2 / 34	3 / 39	4 / 44	5 / 49	6 / 54	7 / 59	9 / 63	10 / 68	11 / 73	12 / 78	13 / 83	15 / 87	16 / 92	17 / 97	18 / 102
7	--	--	--	0 / 28	1 / 34	3 / 39	4 / 45	6 / 50	7 / 56	9 / 61	10 / 67	12 / 72	13 / 78	15 / 83	16 / 89	18 / 94	19 / 100	21 / 105	22 / 111	24 / 116
8	--	--	--	1 / 31	2 / 38	4 / 44	6 / 50	7 / 57	9 / 63	11 / 69	13 / 75	15 / 81	17 / 87	18 / 94	20 / 100	22 / 106	24 / 112	26 / 118	28 / 124	30 / 130
9	--	--	0 / 27	1 / 35	3 / 42	5 / 49	7 / 56	9 / 63	11 / 70	13 / 77	16 / 83	18 / 90	20 / 97	22 / 104	24 / 111	27 / 117	29 / 124	31 / 131	33 / 138	36 / 144
10	--	--	0 / 30	2 / 38	4 / 46	6 / 54	9 / 61	11 / 69	13 / 77	16 / 84	18 / 92	21 / 99	24 / 106	26 / 114	29 / 121	31 / 129	34 / 136	37 / 143	39 / 151	42 / 158
11	--	--	0 / 33	2 / 42	5 / 50	7 / 59	10 / 67	13 / 75	16 / 83	18 / 92	21 / 100	24 / 108	27 / 116	30 / 124	33 / 132	36 / 140	39 / 148	42 / 156	45 / 164	48 / 172
12	--	--	1 / 35	3 / 45	6 / 54	9 / 63	12 / 72	15 / 81	18 / 90	21 / 99	24 / 108	27 / 117	31 / 125	34 / 134	37 / 143	41 / 151	44 / 160	47 / 169	51 / 177	54 / 186
13	--	--	1 / 38	3 / 49	7 / 58	10 / 68	13 / 78	17 / 87	20 / 97	24 / 106	27 / 116	31 / 125	34 / 125	38 / 144	42 / 153	45 / 163	49 / 172	53 / 181	56 / 191	60 / 200
14	--	--	1 / 41	4 / 52	7 / 63	11 / 73	15 / 83	18 / 94	22 / 104	26 / 114	30 / 124	34 / 134	38 / 144	42 / 154	46 / 164	50 / 174	54 / 184	58 / 194	63 / 203	67 / 213
15	--	--	2 / 43	5 / 55	8 / 67	12 / 78	16 / 89	20 / 100	24 / 111	29 / 121	33 / 132	37 / 143	42 / 153	46 / 164	51 / 174	55 / 185	60 / 195	64 / 206	69 / 216	73 / 227
16	--	--	2 / 46	5 / 59	9 / 71	13 / 83	18 / 94	22 / 106	27 / 117	31 / 129	36 / 140	41 / 151	45 / 163	50 / 174	55 / 185	60 / 196	65 / 207	70 / 218	74 / 230	79 / 241
17	--	--	2 / 49	6 / 62	10 / 75	15 / 87	19 / 100	24 / 112	29 / 124	34 / 148	39 / 148	44 / 160	49 / 172	54 / 184	60 / 195	65 / 207	70 / 219	75 / 231	81 / 242	86 / 254
18	--	--	2 / 52	6 / 66	11 / 79	16 / 92	21 / 105	26 / 118	31 / 131	37 / 143	42 / 156	47 / 169	53 / 181	58 / 194	64 / 206	70 / 218	75 / 231	81 / 243	87 / 255	92 / 268
19	--	0 / 38	3 / 54	7 / 69	12 / 83	17 / 97	22 / 111	28 / 124	33 / 138	39 / 151	45 / 164	51 / 177	56 / 191	63 / 203	69 / 216	74 / 230	81 / 242	87 / 255	93 / 268	99 / 281
20	--	0 / 40	3 / 57	8 / 72	13 / 87	18 / 102	24 / 116	30 / 130	36 / 144	42 / 158	48 / 172	54 / 186	60 / 200	67 / 213	73 / 227	79 / 241	86 / 254	92 / 268	99 / 281	105 / 295

(Dashes in the body of the table indicate that no decision is possible at the stated level of significance.)

Table I.   Critical Values of *W* for the Wilcoxon Test

	Level of significance for a directional test					Level of significance for a directional test			
	.05	.025	.01	.005		.05	.025	.01	.005
	Level of significance for a nondirectional test					Level of significance for a nondirectional test			
N	.10	.05	.02	.01	*N*	.10	.05	.02	.01
5	0	--	--	--	28	130	116	101	91
6	2	0	--	--	29	140	126	110	100
7	3	2	0	--	30	151	137	120	109
8	5	3	1	0	31	163	147	130	118
9	8	5	3	1	32	175	159	140	128
10	10	8	5	3	33	187	170	151	138
11	13	10	7	5	34	200	182	162	148
12	17	13	9	7	35	213	195	173	159
13	21	17	12	9	36	227	208	185	171
14	25	21	15	12	37	241	221	198	182
15	30	25	19	15	38	256	235	211	194
16	35	29	23	19	39	271	249	224	207
17	41	34	27	23	40	286	264	238	220
18	47	40	32	27	41	302	279	252	233
19	53	46	37	32	42	319	294	266	247
20	60	52	43	37	43	336	310	281	261
21	67	58	49	42	44	353	327	296	276
22	75	65	55	48	45	371	343	312	291
23	83	73	62	54	46	389	361	328	307
24	91	81	69	61	47	407	378	345	322
25	100	89	76	68	48	426	396	362	339
26	110	98	84	75	49	446	415	379	355
27	119	107	92	83	50	466	434	397	373

For a given N (the number of pairs of scores minus the pairs having zero differences), if the observed value is *less than or equal to* the value in the table for the appropriate level of significance, reject H_0.

Source: From F. Wilcoxon, S. Katte, and R. A. Wilcox, *Critical Values and Probability Levels for the Wilcoxon Rank Sum Test and the Wilcoxon Signed Rank Test*, New York, American Cyanamid Co., 1963, and F. Wilcoxon and R. A. Wilcox, *Some Rapid Approximate Statistical Procedures*, New York, Lederle Laboratories, 1964, as used in Runyon and Haber, *Fundamentals of Behavioral Statistics*, 3rd ed., Addison-Wesley, Reading, Mass., 1976. Reprinted by permission of the American Cyanamid Company.

Table J. Critical Values for the Spearman, Rank-Order Correlation Coefficient, r_s

N	Significance level for a directional test at			
	.05	.025	.005	.001
	Significance level for a nondirectional test at			
	.10	.05	.01	.002
5	.900	1.000		
6	.829	.886	1.000	
7	.715	.786	.929	1.000
8	.620	.715	.881	.953
9	.600	.700	.834	.917
10	.564	.649	.794	.879
11	.537	.619	.764	.855
12	.504	.588	.735	.826
13	.484	.561	.704	.797
14	.464	.539	.680	.772
15	.447	.522	.658	.750
16	.430	.503	.636	.730
17	.415	.488	.618	.711
18	.402	.474	.600	.693
19	.392	.460	.585	.676
20	.381	.447	.570	.661
21	.371	.437	.556	.647
22	.361	.426	.544	.633
23	.353	.417	.532	.620
24	.345	.407	.521	.608
25	.337	.399	.511	.597
26	.331	.391	.501	.587
27	.325	.383	.493	.577
28	.319	.376	.484	.567
29	.312	.369	.475	.558
30	.307	.363	.467	.549

If the observed value of r_s is *greater than or equal to* the tabled value for the appropriate level of significance, reject H_0. Note that the left-hand column is the number of pairs of scores, not the number of degrees of freedom. The critical values listed are both + and − for nondirectional tests.

Source: Glasser, G. J., and Winter, R. F., "Critical Values of the Coefficient of Rank Correlation for Testing the Hypothesis of Independence," *Biometrika* 48 (1961): 444.

Table K.   Random Numbers

```
22 17 68 65 84    68 95 23 92 35    87 02 22 57 51    61 09 43 95 06    58 24 82 03 47
19 36 27 59 46    13 79 93 37 55    39 77 32 77 09    85 52 05 30 62    47 83 51 62 74
16 77 23 02 77    09 61 87 25 21    28 06 24 25 93    16 71 13 59 78    23 05 47 47 25
78 43 76 71 61    20 44 90 32 64    97 67 63 99 61    46 38 03 93 22    69 81 21 99 21
03 28 28 26 08    73 37 32 04 05    69 30 16 09 05    88 69 58 28 99    35 07 44 75 47

93 22 53 64 39    07 10 63 76 35    87 03 04 79 88    08 13 13 85 51    55 34 57 72 69
78 76 58 54 74    92 38 70 96 92    52 06 79 79 45    82 63 18 27 44    69 66 92 19 09
23 68 35 26 00    99 53 93 61 28    52 70 05 48 34    56 65 05 61 86    90 92 10 70 80
15 39 25 70 99    93 86 52 77 65    15 33 59 05 28    22 87 26 07 47    86 96 98 29 06
58 71 96 30 24    18 46 23 34 27    85 13 99 24 44    49 18 09 79 49    74 16 32 23 02

57 35 27 33 72    24 53 63 94 09    41 10 76 47 91    44 04 95 49 66    39 60 04 59 81
48 50 86 54 48    22 06 34 72 52    82 21 15 65 20    33 29 94 71 11    15 91 29 12 03
61 96 48 95 03    07 16 39 33 66    98 56 10 56 79    77 21 30 27 12    90 49 22 23 62
36 93 89 41 26    29 70 83 63 51    99 74 20 52 36    87 09 41 15 09    98 60 16 03 03
18 87 00 42 31    57 90 12 02 07    23 47 37 17 31    54 08 01 88 63    39 41 88 92 10

88 56 53 27 59    33 35 72 67 47    77 34 55 45 70    08 18 27 38 90    16 95 86 70 75
09 72 95 84 29    49 41 31 06 70    42 38 06 45 18    64 84 73 31 65    52 53 37 97 15
12 96 88 17 31    65 19 69 02 83    60 75 86 90 68    24 64 19 35 51    56 61 87 39 12
85 94 57 24 16    92 09 84 38 76    22 00 27 69 85    29 81 94 78 70    21 94 47 90 12
38 64 43 59 98    98 77 87 68 07    91 51 67 62 44    40 98 05 93 78    23 32 65 41 18

53 44 09 42 72    00 41 86 79 79    68 47 22 00 20    35 55 31 51 51    00 83 63 22 55
40 76 66 26 84    57 99 99 90 37    36 63 32 08 58    37 40 13 68 97    87 64 81 07 83
02 17 79 18 05    12 59 52 57 02    22 07 90 47 03    28 14 11 30 79    20 69 22 40 98
95 17 82 06 53    31 51 10 96 46    92 06 88 07 77    56 11 50 81 69    40 23 72 51 39
35 76 22 42 92    96 11 83 44 80    34 68 35 48 77    33 42 40 90 60    73 96 53 97 86

26 29 13 56 41    85 47 04 66 08    34 72 57 59 13    82 43 80 46 15    38 26 61 70 04
77 80 20 75 82    72 82 32 99 90    63 95 73 76 63    89 73 44 99 05    48 67 26 43 18
46 40 66 44 52    91 36 74 43 53    30 82 13 54 00    78 45 63 98 35    55 03 36 67 68
37 56 08 18 09    77 53 84 46 47    31 91 18 95 58    24 16 74 11 53    44 10 13 85 57
61 65 61 68 66    37 27 47 39 19    84 83 70 07 48    53 21 40 06 71    95 06 79 88 54

93 43 69 64 07    34 18 04 52 35    56 27 09 24 86    61 85 53 83 45    19 90 70 99 00
21 96 60 12 99    11 20 99 45 18    48 13 93 55 34    18 37 79 49 90    65 97 38 20 46
95 20 47 97 97    27 37 83 28 71    00 06 41 41 74    45 89 09 39 84    51 67 11 52 49
97 86 21 78 73    10 65 81 92 59    58 76 17 14 97    04 76 62 16 17    17 95 70 45 80
69 92 06 34 13    59 71 74 17 32    27 55 10 24 19    23 71 82 13 74    63 52 52 01 41

04 31 17 21 56    33 73 99 19 87    26 72 39 27 67    53 77 57 68 93    60 61 97 22 61
61 06 98 03 91    87 14 77 43 96    43 00 65 98 50    45 60 33 01 07    98 99 46 50 47
85 93 85 86 88    72 87 08 62 40    16 06 10 89 20    23 21 34 74 97    76 38 03 29 63
21 74 32 47 45    73 96 07 94 52    09 65 90 77 47    25 76 16 19 33    53 05 70 53 30
15 69 53 82 80    79 96 23 53 10    65 39 07 16 29    45 33 02 43 70    02 87 40 41 45

02 89 08 04 49    20 21 14 68 86    87 63 93 95 17    11 29 01 95 80    35 14 97 35 33
87 18 15 89 79    85 43 01 72 73    08 61 74 51 69    89 74 39 82 15    94 51 33 41 67
98 83 71 94 22    59 97 50 99 52    08 52 85 08 40    87 80 61 65 31    91 51 80 32 44
10 08 58 21 66    72 68 49 29 31    89 85 84 46 06    59 73 19 85 23    65 09 29 75 63
47 90 56 10 08    88 02 84 27 83    42 29 72 23 19    66 56 45 65 79    20 71 53 20 25

22 85 61 68 90    49 64 92 85 44    16 40 12 89 88    50 14 49 81 06    01 82 77 45 12
67 80 43 79 33    12 83 11 41 16    25 58 19 68 70    77 02 54 00 52    53 43 37 15 26
27 62 50 96 72    79 44 61 40 15    14 53 40 65 39    27 31 58 50 28    11 39 03 34 25
33 78 80 87 15    38 30 06 38 21    14 47 47 07 26    54 96 87 53 32    40 36 40 96 76
13 13 92 66 99    47 24 49 57 74    32 25 43 62 17    10 97 11 69 84    99 63 22 32 98
```

Source: Table K is taken from Table XXXIII of Fisher and Yates, *Statistical Tables for Biological, Agricultural and Medical Research*, published by Longman Group Ltd., London (previously published by Oliver and Boyd, Ltd., Edinburgh), and by permission of the authors and publishers.

Table K (continued)

```
10 27 53 96 23    71 50 54 36 23    54 31 04 82 98    04 14 12 15 09    26 78 25 47 47
28 41 50 61 88    64 85 27 20 18    83 36 36 05 56    39 71 65 09 62    94 76 62 11 89
34 21 42 57 02    59 19 18 97 48    80 30 03 30 98    05 24 67 70 07    84 97 50 87 46
61 81 77 23 23    82 82 11 54 08    53 28 70 58 96    44 07 39 55 43    42 34 43 39 28
61 15 18 13 54    16 86 20 26 88    90 74 80 55 09    14 53 90 51 17    52 01 63 01 59

91 76 21 64 64    44 91 13 32 97    75 31 62 66 54    84 80 32 75 77    56 08 25 70 29
00 97 79 08 06    37 30 28 59 85    53 56 68 53 40    01 74 39 59 73    30 19 99 85 48
36 46 18 34 94    75 20 80 27 77    78 91 69 16 00    08 43 18 73 68    67 69 61 34 25
88 98 99 60 50    65 95 79 42 94    93 62 40 89 96    43 56 47 71 66    46 76 29 67 02
04 37 59 87 21    05 02 03 24 17    47 97 81 56 51    92 34 86 01 82    55 51 33 12 91

63 62 06 34 41    94 21 78 55 09    72 76 45 16 94    29 95 81 83 83    79 88 01 97 30
78 47 23 53 90    34 41 92 45 71    09 23 70 70 07    12 38 92 79 43    14 85 11 47 23
87 68 62 15 43    53 14 36 59 25    54 47 33 70 15    59 24 48 40 35    50 03 42 99 36
47 60 92 10 77    88 59 53 11 52    66 25 69 07 04    48 68 64 71 06    61 65 70 22 12
56 88 87 59 41    65 28 04 67 53    95 79 88 37 31    50 41 06 94 76    81 83 17 16 33

02 57 45 86 67    73 43 07 34 48    44 26 87 93 29    77 09 61 67 84    06 69 44 77 75
31 54 14 13 17    48 62 11 90 60    68 12 93 64 28    46 24 79 16 76    14 60 25 51 01
28 50 16 43 36    28 97 85 58 99    67 22 52 76 23    24 70 36 54 54    59 28 61 71 96
63 29 62 66 50    02 63 45 52 38    67 63 47 54 75    83 24 78 43 20    92 63 13 47 48
45 65 58 26 51    76 96 59 38 72    86 57 45 71 46    44 67 76 14 55    44 88 01 62 12

39 65 36 63 70    77 45 85 50 51    74 13 39 35 22    30 53 36 02 95    49 34 88 73 61
73 71 98 16 04    29 18 94 51 23    76 51 94 84 86    79 93 96 38 63    08 58 25 58 94
72 20 56 20 11    72 65 71 08 86    79 57 95 13 91    97 48 72 66 48    09 71 17 24 89
75 17 26 99 76    89 37 20 70 01    77 31 61 95 46    26 97 05 73 51    53 33 18 72 87
37 48 60 82 29    81 30 15 39 14    48 38 75 93 29    06 87 37 78 48    45 56 00 84 47

68 08 02 80 72    83 71 46 30 49    89 17 95 88 29    02 39 56 03 46    97 74 06 56 17
14 23 98 61 67    70 52 85 01 50    01 84 02 78 43    10 62 98 19 41    18 83 99 47 99
49 08 96 21 44    25 27 99 41 28    07 41 08 34 66    19 42 74 39 91    41 96 53 78 72
78 37 06 08 43    63 61 62 42 29    39 68 95 10 96    09 24 23 00 62    56 12 80 73 16
37 21 34 17 68    68 96 83 23 56    32 84 60 15 31    44 73 67 34 77    91 15 79 74 58

14 29 09 34 04    87 83 07 55 07    76 58 30 83 64    87 29 25 58 84    86 50 60 00 25
58 43 28 06 36    49 52 83 51 14    47 56 91 29 34    05 87 31 06 95    12 45 57 09 09
10 43 67 29 70    80 62 80 03 42    10 80 21 38 84    90 56 35 03 09    43 12 74 49 14
44 38 88 39 54    86 97 37 44 22    00 95 01 31 76    17 16 29 56 63    38 78 94 49 81
90 69 59 19 51    85 39 52 85 13    07 28 37 07 61    11 16 36 27 03    78 86 72 04 95

41 47 10 25 62    97 05 31 03 61    20 26 36 31 62    68 69 86 95 44    84 95 48 46 45
91 94 14 63 19    75 89 11 47 11    31 56 34 19 09    79 57 92 36 59    14 93 87 81 40
80 06 54 18 66    09 18 94 06 19    98 40 07 17 81    22 45 44 84 11    24 62 20 42 31
67 72 77 63 48    84 08 31 55 58    24 33 45 77 58    80 45 67 93 82    75 70 16 08 24
59 40 24 13 27    79 26 88 86 30    01 31 60 10 39    53 58 47 70 93    85 81 56 39 38

05 90 35 89 95    01 61 16 96 94    50 78 13 69 36    37 68 53 37 31    71 26 35 03 71
44 43 80 69 98    46 68 05 14 82    90 78 50 05 62    77 79 13 57 44    59 60 10 39 66
61 81 31 96 82    00 57 25 60 59    46 72 60 18 77    55 66 12 62 11    08 99 55 64 57
42 88 07 10 05    24 98 65 63 21    47 21 61 88 32    27 80 30 21 60    10 92 35 36 12
77 94 30 05 39    28 10 99 00 27    12 73 73 99 12    49 99 57 94 82    96 88 57 17 91

78 83 19 76 16    94 11 68 84 26    23 54 20 86 85    23 86 66 99 07    36 37 34 92 09
87 76 59 61 81    43 63 64 61 61    65 76 36 95 90    18 48 27 45 68    27 23 65 30 72
91 43 05 96 47    55 78 99 95 24    37 55 85 78 78    01 48 41 19 10    35 19 54 07 73
84 97 77 72 73    09 62 06 65 72    87 12 49 03 60    41 15 20 76 27    50 47 02 29 16
87 41 60 76 83    44 88 96 07 80    83 05 83 38 96    73 70 66 81 90    30 56 10 48 59
```

Appendix 3
Symbols

The following glossary is arranged in approximately alphabetical order, with English and Greek letters intermixed. Nonletter symbols have been placed at the end.

Symbol	Meaning	Symbol	Meaning
a	Regression constant representing the y-intercept.	E_{jk}	Expected frequency for the jkth cell.
		E_j	Expected frequency for the jth row.
α	Greek alpha, the significance level in hypothesis testing, which is also the probability of a type I error.	e	Base of Napierian logarithms, $e = 2.7183.\ldots$
		F	Depth of the fourths.
b	Regression constant indicating the slope of the regression line. It is also written with multiple subscripts ($b_{z_y z_x}$) indicating the value is for standardized variables z_y and z_x.	F	Test statistic, usually the ratio of two independent variance estimates. It is sometimes subscripted to indicate whether the value is an observed (F_{obs}) or a critical one (F_{crit}).
β	Unsubscripted Greek beta represents the probability of a type II error.	F_U	The upper fourth.
		F_L	The lower fourth.
c	Often used to denote any nonzero constant.	f	Frequency.
$°C$	Degrees Celsius.	$°F$	Degrees Fahrenheit.
$_nC_r$	Number of combinations (order irrelevant) of n things taken r at a time.	H	Test statistic for the Kruskal–Wallis test.
		H_0	Null hypothesis.
χ^2	Greek chi squared, a test statistic used in several nonparametric tests. It is often subscripted to indicate whether the value is an observed (χ^2_{obs}) or a critical one (χ^2_{crit}).	H_1	Alternative hypothesis.
		i	Size of the score value measurement unit (whole numbers, $i = 1$; tenths, $i = .1$). Also used as a subscript indicating the ith subject or score.
$Cum\ f$	Cumulative frequency.	k	Often used to denote any nonzero constant and also the number of groups or independent samples (Kruskal–Wallis test).
$Cum\ Rel\ f$	Cumulative relative frequency.		
D_i	Difference between pairs of scores (for example, in the t test for the difference between means for correlated groups).		
		L	The lower real limit of the score value containing the required score value.
d_i	Difference between scores (Wilcoxon test) or between ranks of paired scores (for example, in Spearman's rank-order correlation).	LEx	Lower extreme score.
		$\log_e$	Logarithm to the base e, also called the natural or Napierian logarithm. Sometimes written ln in other contexts.
df	Degrees of freedom.		
E	Depth of the extreme scores, always equal to 1.	$\log_{10}$	Logarithm to the base 10. Most frequently written without the subscript 10.

Symbol	Meaning	Symbol	Meaning
μ	Greek mu, the population mean. It is also written with subscripts (for example, $\mu_{\bar{x}}$, indicating the population mean of the distribution of sample means).	π	Greek pi, a constant equal to 3.1415. . . .
M	Alternative symbol for the mean, called $\bar{X}$ in this text.	q	Studentized Range Statistic. It is sometimes written with subscripts indicating the value is an observed (q_{obs}) or a critical one (q_{crit}). Also refers to the probability of a failure in binomial probability.
M	The depth of the median.		
M_d	Median.		
M_o	Mode.	r	The sample Pearson product-moment correlation coefficient.
MS	Mean square or variance estimate in the analysis of variance. It is often written with subscripts indicating the mean square for a particular source of variability (for example, MS_{between}, MS_{AB}).	r^2	Estimated proportion of variance in Y associated with X (the square of the correlation coefficient).
		ρ	Greek rho, the population product-moment correlation coefficient.
N	Total number of subjects, observations, or paired observations (depending on the statistical context).	r_S	The sample Spearman rank-order correlation coefficient.
n	Number of subjects or observations within a specific subgroup of a larger sample. Sometimes written n_j to indicate the number of subjects in the jth group.	ρ_S	The population Spearman rank-order correlation coefficient.
		$Rel\, f$	Relative frequency.
		S	Universal set or sample space.
n_b	Number of scores falling below the lower limit of the interval containing the desired value.	s	The sample standard deviation. It is also written with subscripts that indicate the variable or statistic to which the standard deviation applies (for example, s_x, the standard deviation of the X's; $s_{\bar{x}}$, the standard error of the mean; $s_{\bar{x}_1 - \bar{x}_2}$, the standard error of the difference between two means).
n_w	Number of scores within the interval containing the desired value.		
O_{jk}	Observed frequency for the jkth cell.		
O_j	Observed frequency for the jth row.		
est. ω^2	Greek omega squared, an estimate of the proportion of total variability in an analysis associated with certain groups or effects.	s^2	The sample variance. It is also written with subscripts (see s).
		σ	Lower-case Greek sigma, the population standard deviation. It is also written with subscripts (see s).
$P(A)$	Probability of event A.		
$P(B \mid A)$	Conditional probability of the event B given that event A has already occurred.	σ^2	Lower-case Greek sigma squared, the population variance. It is also written with subscripts (see s).
p	Probability that the observed data could be obtained if the null hypothesis is true. Also refers to the probability of a success in binomial probability.	$s_{y \cdot x}$	The sample standard error of estimate in regression, read "s sub y dot x."
		$\sigma_{y \cdot x}$	The population standard error of estimate, read "sigma sub y dot x."
P_n	The nth percentile point.		
$_nP_r$	Number of permutations (order considered) of n things taken r at a time.	SD	Alternative symbol for the sample standard deviation, called s in this text.

Symbol	Meaning	Symbol	Meaning
SE	Alternative symbol for the standard error (for example, for the mean, called $s_{\bar{x}}$ in this text).	W	Test statistic for the Wilcoxon test. It is also written with subscripts indicating the value is an observed (W_{obs}) or a critical one (W_{crit}).
SS	Sum of squares. It is often written with subscripts indicating the particular source of variability (for example, $SS_{between}$, SS_{AB}).	z	Standard score or standard normal deviate. It is also written with subscripts indicating the value is an observed (z_{obs}) or a critical one (z_{crit}).
$\displaystyle\sum_{i=1}^{N} X_i$	Capital Greek sigma means to sum the X_i for $i = 1$ to N. It is also written without limits on the summation sign and subscripts, ΣX.	z_r	Transformed value of the correlation coefficient, r.
T	Total, sometimes written with subscripts indicating which scores are summed.	A'	The set complement to A (i.e., all elements not in A).
t	Student's t test statistic. It is also written with subscripts indicating the value is an observed (t_{obs}) or a critical one (t_{crit}), or indicating the value of t at $p = .05(t_{.05})$.	$\subseteq$	The sign $\subseteq$ in $A \subseteq B$ indicates that A is a subset of B.
U	Test statistic for the Mann–Whitney U test. It is also written with subscripts indicating the value is an observed (U_{obs}) or a critical one (U_{crit}).	$\cup$	The sign $\cup$ in $A \cup B$ indicates union, the set of elements that are in A, in B, or in both A and B.
UEx	Upper extreme score.	$\cap$	The sign $\cap$ in $A \cap B$ indicates intersection, the set of elements that are in both A and in B.
X_i	The ith score of variable X. Other letters (for example, Y_i, W_i) are also used to denote variables.	$=$	Is equal to.
		$\neq$	Is not equal to.
$\bar{X}$	The sample mean. Any variable with a bar over it signifies the mean of that variable (for example, $\bar{Y}$). It is also written with subscripts indicating the levels and factors involved (for example, $\bar{X}_{j.}$, $\bar{X}_{.k}$, $\bar{X}$).	$<$	Is less than.
		$\leq$	Is less than or equal to.
		$>$	Is greater than.
		$\geq$	Is greater than or equal to.
		$\pm$	Plus or minus.
		$\sqrt{}$	Square root of.
$\hat{Y}_i$	Value of Y predicted on the basis of the regression line for subject i, read "Y hat."	$\lvert c \rvert$	The absolute value of c.
		∞	Infinity.
		$\varnothing$	Null, or empty, set.
		*, **, ***	The observed value of the test statistic is significant at the .05(*), the .01(**), or the .001(***) level.

Appendix 4
Terms

a posteriori   After the fact (see *comparison*).

a priori   Before the fact (see *comparison*).

abscissa   The horizontal axis.

absolute zero point   A value that indicates nothing at all exists of the attribute being measured.

alternative hypothesis   The hypothesis that is not tentatively held to be true, symbolized by H_1.

analysis of variance   A statistical procedure that assesses the likelihood that the means of groups are equal to a common population mean by comparing an estimate of the population variance determined between groups with an estimate of the same population variance determined within groups.

assumptions   Statements presumed to be true for the duration of the statistical analysis.

asymptotic   A characteristic of the graph of a theoretical distribution in which the tails of the distribution get closer to the X-axis but never touch it as they get further from the mean.

axes   The vertical and horizontal dimensions intersecting at the origin of a plot (for example, the abscissa and ordinate).

bar graph   A graph consisting of bars representing qualitatively different groups, typically on a nominal scale, rising from the abscissa to a height representing a value on the ordinate.

batch   A distribution of scores.

between-groups variance estimate   An estimate of the population variance based upon the deviation of group means about the grand mean.

bimodal   The characteristic of a distribution having two modes or peaks.

binomial probability   The probability of exactly r successes in n independent trials that have only two possible outcomes (arbitrarily called "success" and "failure") with the probability p of a success and the probability q of a failure ($q = 1 - p$).

boxplot   A graphical display of several resistant indicators of a batch.

causal relationship   A relationship in which the independent variable in some way actually influences or determines the value of the dependent variable.

cell   Any single group in an analysis of variance design.

Central Limit Theorem   The proposition that the sampling distribution of the mean will approach a normal form as the size of the samples of raw scores increases.

central tendency   A point on a scale corresponding to a typical, representative, or central score in a distribution.

chi square   A theoretical sampling distribution, symbolized by χ^2.

chi-square test   Test of the null hypothesis that the distributions of frequencies in a two-factor classification are independent.

class interval   A segment of the measurement scale that contains several possible score values.

combination   Any set or subset of objects or events without regard to their internal order, symbolized by $_nC_r$, which is the number of combinations of r objects that can be selected from n objects.

comparison   The difference between two or more means assessed in the context of a more general analysis (also called a *contrast*):

  a posteriori   A comparison that the researcher decides to carry out after the general analysis is conducted and partly as a consequence of seeing the results of that analysis.

  a priori   A comparison that a researcher decides to conduct before performing the experiment.

complement   The set of all elements in S which are not in Set A are in its complement, symbolized by A'.

conditional probability   The probability that an event B will occur given that some other event A has already occurred, symbolized by $P(B|A)$ and read "the probability of B given A."

confidence interval   The interval of values that is likely to contain a population parameter with a given probability (called a level of confidence).

confidence limits   The score values defining the range of a confidence interval.

confounding   The situation in which an extraneous variable influences the dependent variable.

consistent estimator   An estimator that tends to get closer and closer to the value of the population parameter as the size of the sample increases.

constant   A quantity that does not change its value within a given context, symbolized with a lower-case italicized Roman letter (for example, c or k).

contingency table   A table of independent frequencies classified simultaneously according to two categorization schemes.

continuous variable   A variable that theoretically can assume an infinite number of values between any two points (regardless of how it is actually measured.)

contrast   See *comparison*.

control group   A group in an experiment that is treated in the same way as the experimental group but without experiencing the treatment of interest.

correlated groups or samples   Samples in which the same subjects or independent subjects matched on a relevant characteristic compose the groups.

correlation coefficient (Pearson product-moment)   An index ranging between -1.0 and $+1.0$ that reflects the direction and degree of linear relationship between two variables, symbolized by r.

correlation coefficient (Spearman rank-order)   The Pearson product-moment correlation coefficient determined on ranked data, symbolized by r_S.

critical level   See *significance level*.

critical ratio   Formulas that yield values that define a critical region.

critical region   Those values of an observed statistic that will result in rejection of the null hypothesis.

critical value   A value of a statistic that defines whether an observed value of that statistic will or will not result in rejection of the null hypothesis.

cross product   The product of two different variables (for example, XY).

cumulative frequency distribution   A distribution in which the entry for any score value or class interval is the sum of the frequencies at that value or that interval plus the frequencies of all lower score values or intervals.

cumulative relative frequency distribution   A distribution in which the entry for any score value or class interval expresses that value's or that interval's cumulative frequency as a portion of the total number of cases.

curvilinear relationship   An association between two variables that may be accurately represented on a graph by a curved line.

decision errors   Errors made in the statistical process of deciding whether to reject the null hypothesis.

decision rules   Statements, phrased in terms of statistics to be calculated, that stipulate precisely when the null hypothesis will be rejected and when it will not.

degrees of freedom   The number of components in the calculation of a statistic that are free to vary, symbolized by df.

density function   The equation for a theoretical relative frequency distribution (also called a *probability function*).

dependent events   Two events such that the occurrence of one event alters the probability of the occurrence of the other.

dependent variable   The variable whose values are thought to depend on another variable.

depth   The cumulative frequency of a line in a stem-and-leaf display taken either from the top or bottom of the group, whichever is smallest.

depth of the fourths   The depth of the cases approximately defining the upper and lower one-fourths of a batch.

descriptive statistics   Procedures for organizing, summarizing, and describing quantitative information or data.

direct relationship   A positive relationship.

directional test   A statistical test in which the alternative hypothesis specifies the direction of the departure from what is expected under the null hypothesis. (This is sometimes called a *one-tailed test*.)

discrete variable   A variable that theoretically can assume only a countable number of values between any two points.

disjoint sets   Two sets that share no common elements (that is, whose intersection is the null set).

distribution   See types by name (for example, *frequency, relative frequency, cumulative; t, F; population, sample, sampling, theoretical sampling*).

distribution-free tests   See *nonparametric tests*.

double blind   A research design in which neither subjects nor observers know the research conditions.

effect   A difference among population means, see *treatment effect.*

efficiency   See *relatively efficient estimator, power-efficiency.*

element   A member of a set.

elementary event   Any of the simplest elements or outcomes of the sample space of an idealized experiment.

empirical   Signifies "experienced" or "observed," and is sometimes used to describe distributions of scores that are actually observed rather than those that are theoretical.

empty set   A set containing no elements, symbolized by $\varnothing$ (also called the *null set*).

endogenous parameter   See *subject parameter.*

equal intervals   A property of a scale such that the magnitude of the attribute represented by a unit on the scale is the same regardless of where on the scale the unit falls.

equality of sets   Two sets, A and B, are equal if every element in A is also an element of B and if every element in B is also an element of A.

estimation   The procedures used to estimate population parameters by using sample statistics. See also characteristics of estimators (for example, *consistency, relative efficiency, sufficiency, unbiasedness*) and types of estimation (for example, *interval, point*).

eta   A correlation coefficient that reflects the degree of nonlinear association between two variables.

event   A subset of elementary events or outcomes in the sample space of an idealized experiment.

exogenous parameter   See *situational parameter.*

expectation theory   Mathematical procedures used to determine the expected value of a statistic under certain conditions of unlimited sampling.

expected frequency   The frequency for a cell in a contingency table determined by multiplying the marginal total for the row by the marginal total for the column containing the cell and dividing by the grand total, symbolized by E_{jk}.

experimental group   A group in an experiment that receives the treatment of interest.

experimental research   Research intended to determine if an association between two variables is causal.

exploratory data analysis   An approach to analyzing data that includes resistant indicators.

extraneous variable   A variable other than the independent variable that may influence the dependent variable.

extreme scores   The highest and lowest cases in a batch excluding any outliers.

F   A theoretical relative frequency distribution of the ratio of two independent sample variances.

factor   A classification scheme defining an independent variable in the analysis of variance. See also *fixed, random.*

five-number summary   A tabular display of several resistant indicators of a batch.

fixed factor (or set of groups)   A set of conditions or groups defining a factor in which the nature of the conditions or groups is established and "fixed" by the researcher.

fixed model   An analysis of variance having two or more factors in which all factors are fixed.

fourth-spread   The difference in score values between the upper and lower fourths of a batch; approximately the range of scores of the middle half of the batch.

fourths   The values of the cases approximately defining the upper and lower one-fourths of a batch.

frequency distribution   A tally of the number of times each score value or interval of score values occurs in a group of scores.

frequency histogram   A graph of a frequency distribution presented in terms of bars spanning a score value or class interval and extending a distance corresponding to the frequency count for that interval. The term may also apply to relative and cumulative distributions.

frequency polygon   A graphic presentation of a frequency distribution in which points corresponding to the frequencies of each score value or class interval are connected by straight lines. The term may also apply to relative and cumulative distributions.

gambler's fallacy   The failure to appreciate the independence of some sequential events (for example, "because I have lost at dice so many times in a row, it must mean the probability is greater that I will win the next time").

goodness of fit A statistical procedure that tests the likelihood that a distribution of observed frequencies "fits" a specified distribution.

grouped data Data that are presented as frequencies corresponding to class intervals as opposed to individual score values.

homogeneity of variance The assumption that the variances of populations are equal.

hypotheses A set of two or more contradictory and often exhaustive possibilities, only one of which can actually be the case. See also *alternative, null.*

hypothesis testing The statistical procedures for testing hypotheses.

idealized experiment An imagined experiment in which a given phenomenon is conceived to be repeated an indefinite number of times under ideal conditions such that every outcome occurs a representative number of times.

independence, in sampling A characteristic of sampling in which the inclusion in the sample of any element in the population does not alter the likelihood of including any other element from the population in the sample.

independent events Two events such that the occurrence of one event does not alter the probability of the occurrence of the other event.

independent groups or samples Samples in which the subjects in the different groups are different individuals and not deliberately matched on any relevant characteristic.

independent variable The variable thought to influence another variable.

individual differences The fact that individuals are not the same and typically score differently on the same measurement.

inferential statistics Methods for making inferences about a larger group of individuals on the basis of data gathered on a smaller group.

interaction A treatment effect such that the nature of the effect for one factor is not the same within all levels of another factor.

interquartile range The score values that separate the middle 50% from the remainder of the distribution.

intersection The intersection of sets *A* and *B* contains all the elements that are in both *A* and *B* but not in *A* or *B* alone, symbolized $A \cap B$.

interval estimation The process in which a range of values is stipulated to contain a population parameter with a specified probability.

interval scale Any scale of measurement possessing the properties of magnitude and equal intervals but not an absolute zero point.

inverse relationship A negative relationship.

Kruskal–Wallis test A nonparametric test for *k* independent samples having a continuous dependent variable measured with at least an ordinal scale.

kurtosis The curvedness or peakedness of a graph of a frequency distribution; also the third moment about the mean.

leaf The smallest digit of numbers in a group; used to define the width of lines in a stem-and-leaf display.

least squares criterion A criterion for determining the regression line that requires the sum of the squared deviations between points and the regression line to be a minimum.

leptokurtic A characteristic of thinness or peakedness in a distribution.

level A condition or group that represents one qualitative or quantitative classification within a factor.

limits of the summation The notation above and below the summation sign that direct the summation across a set of values for the subscript indicated. For example, $\sum_{i=1}^{N} X_i$ directs one to sum the X_i from $i = 1$ to $i = N$.

line A line or class interval in a stem-and-leaf display.

line width The size of the class interval; that is, the number of possible leaves per line (stem), in stem-and-leaf displays.

linear relationship An association between two variables that may be accurately represented on a graph by a straight line.

lower quartile point The score value that separates the bottom 25% from the remainder of the distribution.

lower real limit For a number, that point half a measurement unit below the number; for a class interval, the lower real limit of the lowest score value contained in that interval.

magnitude A property of a scale such that one instance of the attribute being measured can be judged to be greater than, less than, or equal to another instance of the attribute.

main effect A difference among population means for levels of a factor ignoring other factors in an analysis of variance.

Mann–Whitney *U* test A nonparametric statistical test for the difference between two independent samples having a continuous dependent variable measured with at least an ordinal scale.

marginals The values in the margins of classification tables representing totals summed over rows or columns.

mean A measure of central tendency, commonly known as the arithmetic average, determined by summing all the scores in a distribution and dividing by the number of such scores, symbolized by $\overline{X}$ for the variable X.

mean square A sum of squares divided by its degrees of freedom; a variance estimate; symbolized by MS.

measurement The orderly assignment of a numerical value to a characteristic.

measurement error Error in the measurement of a characteristic.

median The point that divides a distribution into two parts such that an equivalent number of scores fall above and below that point, symbolized by M_d.

midpoint of a class interval The point halfway between the interval's real limits, determined by adding one-half the size of the interval to its lower real limit.

mixed model An analysis of variance of two or more factors in which at least one factor is fixed and at least one factor is random.

mode The most frequently occurring score in a distribution, symbolized by M_o.

multimodal The characteristic of having two or more modes or peaks in a distribution.

multiple correlation A correlation coefficient that reflects the degree of relationship between a weighted set of predictor variables and one criterion variable.

mutually exclusive events Two events in the same idealized experiment that share no elementary event (for example, their sets are *disjoint*).

negative relationship A relationship between two variables in which the slope of the regression line is negative (that is, it runs from top left to bottom right on the graph); high values on one variable are associated with low values on the other variable.

nominal scale A scale that results from the classification of items into discrete (that is, mutually exclusive) groups that do not bear any magnitude relationship to one another.

noncausal relationship A relationship in which the independent variable does not actually produce, cause, or influence the dependent variable, although the values of the independent and dependent variables are in some way related.

nondirectional test A statistical test in which the alternative hypothesis does not specify the direction of the departure from what is expected under the null hypothesis (also called *two-tailed test*).

nonlinear relationship An association between two variables that may not be accurately represented on a graph by a straight line.

nonparametric tests Statistical techniques that do not represent tests of the values of certain parameters and often make different and fewer assumptions than parametric tests (also called *distribution-free tests*).

normal distribution A theoretical distribution defined by a mathematical formula and which typically has a bell-shaped form when graphed.

null hypothesis The hypothesis that is tentatively held to be true, symbolized by H_0.

null set A set containing no elements, symbolized by $\varnothing$ (also called the *empty set*).

observational research Research the purpose of which is to find out whether a relationship—causal or noncausal—exists between the variables of interest.

observer bias The influence on the dependent variable of the observer instead of, or in addition to, the independent variable.

omega squared A statistic, symbolized by Greek ω^2, that estimates the proportion of the total variability in an analysis that is associated with groups or a treatment effect.

one–tailed test See *directional test*.

operationalize To define in terms of specific acts or operations.

ordinal scale Any scale of measurement possessing only magnitude and not the properties of equal intervals or an absolute zero point.

ordinate The vertical axis.

origin The point (0, 0) on a scale or graph.

outcome See *elementary event*.

outliers Cases that fall beyond the fourths ± 1.5 (fourth-spread) in a batch.

parameter A quantitative characteristic of a population, symbolized with a Greek letter. Also, a characteristic that influences a dependent variable. See also *situational, subject, endogenous, exogenous.*

parametric tests Statistical techniques that represent tests of the values of certain parameters and which usually make certain assumptions about other parameters.

partial correlation A correlation coefficient that reflects the degree of relationship between two variables after each variable's relationship with a third variable has been removed.

Pearson chi-square test See *chi-square test.*

Pearson product–moment correlation coefficient See *correlation coefficient.*

percentile point The score value below which the proportion *P* of the cases in a distribution fall.

percentile rank The percentile corresponding to a given percentile point.

permutation An ordered sequence of objects or events from a set, symbolized $_nP_r$, which is the number of permutations of *n* things taken *r* at a time.

pilot experiment A trial or practice experiment in which a relatively few subjects are observed under each of the experimental conditions as a means of assessing the feasibility of the research procedures.

pilot subjects Subjects in a pilot experiment.

placebo An object or treatment that the subject believes to be effective in producing a particular consequence but which actually does not produce that consequence.

platykurtic The characteristic of flatness in describing a distribution.

point estimation The process by which a single value is used to estimate a population parameter.

population A collection of subjects, events, or scores that have some common characteristic of interest.

population distribution A collection of scores from a population.

positive relationship A relationship between two variables in which the line describing the relationship has a positive slope (that is, it runs from bottom left to top right on a graph); high values on one variable are associated with high values on the other variable.

power The probability that a statistical test will lead to a correct decision to reject the null hypothesis when it is indeed false.

power–efficiency An index that reflects the probability that a statistical test will correctly reject the null hypothesis relative to the size of the sample involved.

probability The likelihood of one event relative to all possible events; the quantification of uncertainty (see *binomial, conditional, simple classical*).

probability function The equation for a theoretical relative frequency distribution (also called a *density function*).

proportional stratified random sample A sample produced by a process in which major subgroups within the population are proportionately represented in the sample and obtained by simple random sampling within those subgroups.

random factor (or set of groups) A set of conditions or groups defining a factor that are selected at random from a much larger potential set.

random sample See *simple random sample.*

range A measure of variability determined by taking the largest score minus the smallest score in a distribution.

ratio scale A scale of measurement possessing magnitude, equal intervals, and an absolute zero point.

real limits Those points falling one-half measurement unit above and one-half measurement unit below a particular number.

region of rejection See *critical region.*

regression Techniques involving the prediction of one variable from another.

regression constants The slope and *y*-intercept of a regression line.

regression line The line describing a linear relationship between two variables.

regression toward the mean The tendency for individuals who are extreme on the predictor variable to be less extreme on the trait being predicted.

relationship, causal See *causal relationship.*

relationship, noncausal See *noncausal relationship.*

relatively efficient estimator   An estimator whose sampling distribution has a smaller standard error than another estimator for samples of a particular size.

relative frequency distribution   A distribution of the proportions of the total number of cases observed at each score value or interval of score values.

reliability   The relative extent to which the measurement procedures assign the same value to a characteristic of an individual each time that it is measured under essentially the same circumstances.

repeated measures   A condition in which the same subjects are measured in the same way on more than one occasion or under more than one circumstance.

representative sample   A sample in which all major groups in the population are accurately and proportionately represented.

research design   The methods scientists use to make observations that will produce empirical information, usually about the relationship between two or more variables.

resistant indicator   An indicator of central tendency or variability, for example, that changes relatively little in value if a small portion of the data is replaced with new numbers that may be very different than the original ones.

sample   A subgroup of a population (see also *proportional stratified, simple random*).

sample distribution   A collection of raw scores obtained from a sample or subgroup of a population.

sample space   A collection of all the elements or outcomes of an idealized experiment, symbolized by S.

sample statistics   Quantities that characterize samples of raw scores.

sampling distribution   The distribution of a statistic determined on separate independent samples of size N drawn from a given population.

sampling error   The error that results when a sample statistic is used to estimate a population parameter.

sampling statistics   Quantities that characterize sampling distributions of statistics.

sampling with replacement   Sampling elements from a sample space such that each element is returned to the sample space after selection.

sampling without replacement   Sampling elements from a sample space such that elements are not returned to the sample space after selection.

scale of measurement   The numerical system by which measurements are made.

scatterplot   A graph of a collection of pairs of scores.

set   A well-defined collection of things, typically objects or events.

significance level   The probability value that forms the boundary between rejecting and not rejecting the null hypothesis, symbolized by α (also called *critical level*).

simple classical probability   In a sample space containing equally likely elementary events, the probability of a given event A is defined to be the number of outcomes in A divided by the total number of outcomes in the sample space, symbolized by $P(A) = \#(A)/\#(S)$.

simple random sample   A sample resulting from a selection process in which all elements of the population have an equal probability of being selected for the sample.

single blind   A research design in which either the subjects or the observers, but not both, are not aware of the research conditions.

situational parameter   A parameter that exists in the environmental situation external to the subject (also called *exogenous*).

size of a class interval   Obtained by subtracting the lower real limit of the class interval from its upper real limit, symbolized by i (when not used as a subscript).

size of an effect or relationship   Procedures that estimate how much of an effect or relationship exists, often in terms of the proportion of the total variability in the analysis associated with that effect or relationship.

skewness   An asymmetry in a distribution in which the scores are bunched on one side of the central tendency and trail out on the other side. If the trail is to the right or positive end of the scale, the distribution is said to be *positively skewed* or *skewed to the right*. If the distribution trails off to the left or negative side of the scale, it is said to be *negatively skewed* or *skewed to the left*.

slope   The tilt of a line defined to be the vertical distance divided by the horizontal distance between any two points on the line, symbolized by b.

Spearman rank-order correlation coefficient   A nonparametric measure of association between two variables, each of which is measured with at least an ordinal scale (also the Pearson product-moment correlation applied to ranked data), symbolized by r_s.

standard deviation An index of variability defined to be the positive square root of the variance, symbolized by s.

standard deviation unit A unit of measurement equal to the size of the standard deviation for a given distribution.

standard error The standard deviation of a sampling distribution of a statistic. For example, the standard deviation of the sampling distribution of the mean is the standard error of the mean.

standard error of estimate An index of the degree of variability of the data points about a regression line determined to be the square root of the sum of squared deviations of the points about the line divided by $N - 2$, symbolized by $s_{y \cdot x}$.

standard error of the difference between means The standard deviation of the sampling distribution of the differences between pairs of sample means, symbolized by $s_{\bar{x}_1 - \bar{x}_2}$.

standard error of the mean The error or variability in the sampling distribution of the mean, usually measured by its standard deviation, symbolized by $s_{\bar{x}}$.

standard normal deviate Score (z) from the standard normal distribution, which indicates the number of standard deviations that separates a score from the mean.

standard normal distribution A theoretical normal distribution with mean of zero and standard deviation of one for which percentiles are tabled, symbolized by z.

standard score A score typically obtained by subtracting the mean of a distribution from each score and dividing by the standard deviation.

stated limits The highest and lowest score values contained in a class interval.

statistic A quantitative characteristic of a sample, symbolized by a Roman letter.

statistics The study of methods for describing and interpreting quantitative information, including techniques of organizing and summarizing data and techniques for making generalizations and inferences from data.

stem The larger digits of numbers in a group; used to define lines in a stem-and-leaf display.

stem-and-leaf display A method of displaying scores in a distribution or batch that combines elements of a frequency distribution, histogram, and cumulative frequency distribution and that retains more detailed information about the original data.

Studentized range statistic A test statistic used when all possible a posteriori pairwise comparisons are to be made after an analysis of variance, symbolized by q.

Student's *t* distribution See *t* distribution.

subject bias The influence exerted on the dependent variable by the subjects instead of (or in addition to) that exerted by the independent variable (such as the bias shown when subjects know—or think they know—the intent of a study, and that knowledge influences their behavior).

subject parameter A parameter that resides within the subject (also called *endogenous*).

subset A portion of a set such that if every element in A is also an element of B, then A is a subset of B, symbolized by $A \subseteq B$.

sufficient estimator An estimator whose accuracy cannot be improved by using any other aspects of the sample data that are not already involved in its definition.

sum of squares The sum of squared deviations of values about their mean, symbolized by SS.

_t_ distribution A theoretical relative frequency distribution in which the standard error of the mean is estimated from sample values (also called *Student's t distribution*).

theoretical distribution A hypothetical or mathematical distribution based on a population of an infinite number of cases.

theoretical sampling distribution A theoretical distribution of a statistic, the characteristics of which are determined mathematically rather than by actually making observations.

treatment effect A nonzero difference between the population mean for a group and the average of the population means for all groups, estimated by the difference between a sample group mean and the grand mean.

two-factor design A research design in which all subjects are classified simultaneously according to two classification schemes, or factors.

two-tailed test See *nondirectional test*.

type I error A decision error in which the statistical decision is to reject the null hypothesis when it is actually true.

type II error A decision error in which the decision is not to reject the null hypothesis when it is actually false.

unbiased estimator An estimator of a population parameter whose average over all possible random samples of a given size equals the value of the parameter.

union The union of sets A and B is the set of all elements that are in A, in B, or in both A and B; symbolized $A \cup B$.

unit The smallest degree of measurement in a given situation.

unreliability The fact that the same measurement on the same person on two different occasions may not produce the same value.

universal set Includes all things to be considered in any one discussion, symbolized by S (also called a *sample space* in probability).

upper quartile point The score value that separates the top 25% from the remainder of the distribution.

upper real limit For a number, that point half a measurement unit above the number; for a class interval, the upper real limit of the highest score value contained in that interval.

validity The extent to which the measurement procedures accurately reflect the variable being measured.

variable The general characteristic being measured on a set of objects, people, or events, the members of which may take on different numerical values (see *dependent, extraneous, independent*).

variability The extent to which scores in a distribution deviate from central tendency or from one another.

variance An index that reflects the degree of variability in a group of scores, determined by squaring the deviations of each score about the mean and dividing by one less than the number of cases in the distribution, symbolized by s^2. See also *between-groups, within-groups*.

Wilcoxon test A nonparametric test for two correlated samples having a dependent variable measured with at least an ordinal scale.

within-groups variance estimate An estimate of a population variance based upon the deviation of scores within each group about the mean of that group.

y-intercept The point at which a line intersects the y axis, symbolized by a.

Y-predicted The variable that is predicted by a regression equation (that is, the regression line), symbolized by $\hat{Y}_i$.

Appendix 5
Answers

CHAPTER 1   p. 3

1a. Nominal; **1b.** ratio; **1c.** ordinal; **1d.** ratio; **1e.** ordinal; **1f.** interval; **1g.** nominal. **2.** The difference centers on the fact that the Celsius scale is interval, whereas the Kelvin scale is ratio because it has a zero point. **3.** Student 6 has a rank of 2 and student 5 a rank of 3. Their ranks as well as their scores (24 and 25, respectively) differ by one point. In contrast, student 4 also differs from student 5 by a rank of one, but has 16 more score points. Thus, when scores are transformed to ranks, the equal-interval property of the measurement scale is lost. **4a.** Continuous; **4b.** discrete; **4c.** continuous; **4d.** continuous; **4e.** discrete; **4f.** continuous; **4g.** continuous; **4h.** discrete. **5.** The three sources of variability are individual differences, measurement error, and unreliability. Statistics are used to quantify variability and uncertainty, helping scientists to draw conclusions when the answer is not certain (see p. 6). **6.** Inferential statistics permits scientists to draw conclusions about what exists in general (that is, in the population) on the basis of a relatively small sample (see pp. 6–9). **7a.** 21.15 to 21.25; **7b.** −2.225 to −2.215; **7c.** 99.985 to 99.995; **7d.** 20.5 to 21.5; **7e.** 1.45 to 1.55; **7f.** .405 to .415; **7g.** 21.195 to 21.205; **7h.** 14.0005 to 14.0015; **7i.** −1.5 to −.5; **7j.** 15.75 to 15.85; **7k.** 6.00005 to 6.00015; **7l.** −93.425 to −93.415; **7m.** 27.695 to 27.705. **8a.** 1.6; **8b.** 3.2; **8c.** 6.0; **8d.** 9.4; **8e.** 10.0; **8f.** 6.3; **8g.** 11.5; **8h.** 11.4; **8i.** 11.4. **9a.** 23; **9b.** 17; **9c.** 13; **9d.** 32; **9e.** 55; **9f.** 148; **9g.** 137; **9h.** 529; **9i.** 23; **9j.** 228; **9k.** 1024; **9l.** 55. **10a.** 69; **10b.** 444; **10c.** 165; **10d.** 15; **10e.** 25. **11a.** $c(k + 1)$; **11b.** 1; **11c.**

$$1 + \frac{\Sigma Z}{Nk(\Sigma Z + 1)}$$

CHAPTER 2   p. 27

1. See pp. 28–33. **2.** The lines of a frequency polygon must touch the X-axis to make it a closed figure, which is the definition of a polygon. Also, they indicate no frequencies at these score values. **3.** Generally, a histogram is used when the scale of measurement is nominal (that is, categorical), ordinal, or discrete. Polygons should be used when the measurement scale is interval, ratio, or continuous. See pp. 40–43. **4.** No space is put between the bars, because the bars are intended to communicate the frequency for each score value or class interval. A space would indicate no frequencies for some score values. In the case of a nominal scale, the score values are points, not intervals, so spaces may be put between the bars. **5.** Since a nominal scale does not possess the characteristic of magnitude, it usually makes no sense to accumulate across categories that have no magnitude relation to one another. **6.** Usually, the greater the skewness, the greater the variability, but it is not necessarily the case that the greater the variability the greater the skewness. Platykurtic distributions tend to have more variability than leptokurtic, but again the reverse is not necessarily true. Draw an example in which the variability is great but the distribution is not platykurtic. **7.** Stem-and-leaf displays have the advantages that they combine a frequency distribution and histogram in one display, and they retain all of the information in the raw data. They have the disadvantage that they can use line widths (class interval sizes) of only 2, 5, or 10 times a power of 10.

8.

Class Interval	Real Limits	Interval Size	Midpoint	f	Rel f	Cum f	Cum Rel f
95–99	94.5–99.5	5	97.0	3	.06	50	1.00
90–94	89.5–94.5	5	92.0	4	.08	47	.94
85–89	84.5–89.5	5	87.0	12	.24	43	.86
80–84	79.5–84.5	5	82.0	16	.32	31	.62
75–79	74.5–79.5	5	77.0	6	.12	15	.30
70–74	69.5–74.5	5	72.0	3	.06	9	.18
65–69	64.5–69.5	5	67.0	2	.04	6	.12
60–64	59.5–64.5	5	62.0	2	.04	4	.08
55–59	54.5–59.5	5	57.0	1	.02	2	.04
50–54	49.5–54.5	5	52.0	0	.00	1	.02
45–49	44.5–49.5	5	47.0	1	.02	1	.02

$N = 50$

Eleven intervals of size 5 were selected, but 9 to 13 intervals might have been picked as long as they covered the range of scores, and the lower stated limit of the first interval was evenly divisible by the interval size. See Tables 2–7 and 2–8, pp. 35 and 36. **9.** See examples of graphs in text, paying close attention to the points mentioned in the text.

10.

Class Interval	Real Limits	Interval Size	Midpoint	f	Rel f	Cum f	Cum Rel f
45–49	44.5–49.5	5	47	3	.12	25	1.00
40–44	39.5–44.5	5	42	7	.28	22	.88
35–39	34.5–39.5	5	37	8	.32	15	.60
30–34	29.5–34.5	5	32	4	.16	7	.28
25–29	24.5–29.5	5	27	3	.12	3	.12

$N = 25$

11.

Class Interval	Real Limits	Interval Size	Midpoint	f	Rel f	Cum f	Cum Rel f
4.2–4.7	4.15–4.75	.6	4.45	1	.04	28	1.00
3.6–4.1	3.55–4.15	.6	3.85	1	.04	27	.96
3.0–3.5	2.95–3.55	.6	3.25	3	.11	26	.93
2.4–2.9	2.35–2.95	.6	2.65	5	.18	23	.82
1.8–2.3	1.75–2.35	.6	2.05	8	.29	18	.64
1.2–1.7	1.15–1.75	.6	1.45	7	.25	10	.36
.6–1.1	.55–1.15	.6	.85	3	.11	3	.11

$N = 28$

Because of rounding, the *Rel f* column of Exercise 11 does not sum to 1.00. **12.** The distribution in Exercise 8 is negatively skewed; the distribution in Exercise 11 is positively skewed.

13.

Depth (N = 50)	**Statistics Exam Scores** Unit = 1 Point	
	Stem	**Leaves**
3	9•	5 5 9
7	9⋆	0 2 3 4
19	8•	5 5 6 6 7 8 8 8 9 9 9 9
(16)	8⋆	0 0 1 1 2 2 2 2 2 2 2 2 4 4 4 4
15	7•	7 7 7 9 9 9
9	7⋆	2 3 3
6	6•	7 7
4	6⋆	1 4
2	5•	6
1	5⋆	
1	4•	7

Depth (N = 25)	**Aggressive Acts** Unit = 1 Act	
	Stem	**Leaves**
3	40•	7 7 8
10	40⋆	0 0 1 2 3 4 4
(8)	30•	5 6 7 7 7 8 8 9
7	30⋆	2 3 3 4
3	20•	6 7 8

14. Depth

(N = 28)	Stem	Leaves
1	4f	4
1	t	
1	4*	
2	3•	8
2	s	
2	f	
2	t	
5	3*	0 1 1
6	2•	8
7	s	7
10	f	4 5 5
14	t	2 2 3 3
14	2*	0 0 1
11	1•	9
10	s	6 7
8	f	5 5
6	t	2 3 3
3	1*	1
2	0•	8 9

CHAPTER 3   p. 58

1. The mean is appropriate as a measure of central tendency because the sum of the deviations of the scores about the mean is zero and the sum of the square of the deviations of the scores about the mean is less than about any other value (that is, the least squares sense). **2.** The variance is an appropriate measure of variability because it reflects the sum of squared deviations of scores about their most central value (that is, the mean), and it is proportional to the sum of squared differences between each score and every other score. **3a.** Median because the distribution is skewed to the right; **3b.** mode because the distribution is likely to be bimodal as a result of having both boys and girls in it; **3c.** mean, median, and mode are likely to be approximately the same. **4.** See graphs at end of answers. **5.** For example: (4, 5, 6), (3, 5, 7), and (2, 5, 8). **6.** The range is based upon only the two most extreme scores, not all the scores. **7.** See pp. 71–77 for a discussion of this issue. **8.** Resistant measures are less influenced by changes in small portions of the data, especially atypical deviant scores. **9.** The denominator $N - 1$ is used so that s^2 is an unbiased estimator of σ^2. **10.** Means = 5.00, 3.50, 5.00, 4.375; medians = 4.00, 3.50 4.00, 4.00; modes = 2, 1, 2, 4. **11.** The sum of squared deviations about the mean is 82; about the median is 91. **12a.** 6; **12b.** 4.5; **12c.** 3.0; **12d.** 6.00; **12e.** 3.00; **12f.** 3.3; **12g.** 3.3; **12h.** 3.3. **13a.** $s^2 = 14.00$, $s = 3.74$; **13b.** $s^2 = 5.71$, $s = 2.39$; **13c.** $s^2 = 10.25$, $s = 3.20$; **13d.** $s^2 = 3.70$, $s = 1.92$; **14a.** $\overline{X} = 4.00$, $s^2 = 2.67$, $s = 1.63$; **14b.** $\overline{X} = 4.00$, $s^2 = 18.67$, $s = 4.32$; **14c.** $\overline{X} = 4.00$, $s^2 = 46.89$, $s = 6.85$. **15.** The distribution and statistics are:

X	f	All subjects	Those who drink something
6	1	$\overline{X} = 2.23$	3.35
5	3	$M_d = 2.50$	3
4	5	$M_0 = 0$	3
3	6	Range = 6	5
2	3	$s^2 = 3.77$	1.82
1	2	$s = 1.94$	1.35
0	10		
	$N = 30$		

The modal behavior is not to drink at all—10/30 = 33% of students did not drink anything. This means that the statistics calculated on the entire sample are misleading, at least with respect to that portion of the sample who did drink. For example, the average number of drinks for all students was 2.23, but among those who drank anything, it was 3.35 with a mode of 3. Therefore, college seniors who drink at all do not tend to have only one or two drinks, but 3 to 4.

16.

Golf Scores
(Unit = 1 Stroke)

Depth (N = 40)	Stem	Leaves
1	310•	9
1	s	
2	f	5
2	t	
2	310*	
2	300•	
4	s	6 6
6	f	5 5
9	t	2 2 3
12	300*	0 0 1
15	290•	8 9 9
18	s	6 6 7
(3)	f	4 4 5
19	t	2 3 3 3
15	290*	0 0 1 1
11	280•	8 8 9
8	s	6 7 7
5	f	4 5 5
2	t	2
1	280*	0

Golf Strokes

$\overline{X}$
M_d
M_o
(4a)

$M_o \quad \overline{X} \quad M_o$
M_d
(4b)

$M_d \ \overline{X}$
(4c)

$\overline{X} \ M_d$
(4d)

CHAPTER 4   p. 91

1a. 53, 25, 5; **1b.** 45, 25, 5; **1c.** 500, 2500, 50; **1d.** 12.5, 1.5625, 1.25; **1e.** 5, 6.25, 2.5. **2a.** 1, .11, .33; **2b.** 14, 16, 4; **2c.** 25, 100, 10. **3.** Percentiles reflect only ordinal position (that is, they specify the proportion of the group falling below a given score) and do not indicate how far the other scores are from a given percentile value. Standard scores take into account the variability of the distribution. **4.** See pp. 104–107. **5.** Relative frequency, or the proportion of cases falling between two specified values on the abscissa. **6.** Section (d), because a score of 90 in this distribution yields the highest $z = 2.5$. **7a.** 71.83; **7b.** 55.25; **7c.** 55.50; **7d.** 76.50; **7e.** 80.50; **7f.** 77.75; **7g.** 91.50; **7h.** 72.50. **8a.** $P_{.1125}$; **8b.** $P_{.2625}$; **8c.** $P_{.4625}$; **8d.** $P_{.5125}$; **8e.** $P_{.55}$; **8f.** $P_{.65}$; **8g.** $P_{.8125}$; **8h.** $P_{.9875}$. **9.** $\overline{X} = 5.00$; $s = 5$; $z_i = -1.0, -.6, -.2, .2, 1.6$; $\overline{z} = 0$; $s_z = 1$; yes, because a distribution of z scores has a mean of zero and a standard deviation of 1. **10.** .50; .50; .1587; .8413. **11a.** .6826; **11b.** .2417; **11c.** .7734; **11d.** .0500; **11e.** .0100. **12.** ± 3 standard deviations; 37 and 49; ± 1 standard deviation or between score values 41 and 45. **13.** $P_{.0228}$; $P_{.3707}$; $P_{.9452}$. **14.** 27.2; 15.15. **15.** .3721; .3345; .1974. **16a.** .50; **16b.** .84; **16c.** .98; **16d.** .62%. **17.** Standard scores for the six competitors are:

Competitor	Swim	Bicycle	Run	Total z
A	−1.12	.78	.85	.51
B	−.22	.23	−.02	−.01
C	−.40	.72	−1.32	−1.00
D	−.01	−.06	−.86	−.93
E	−.12	.26	1.34	1.48
F	1.88	−1.93	.01	−.04

When times are simply added, the order of finish is F, D, C, B, A, and E. But standard scores for each event consider a person's time with respect to the mean and variability of other competitors in that event and weight each event equally. If these are calculated and added as above, the order of "finish" is C, D, F, B, A, and E. Competitor F does not win the standardized race because the poor showing in the swim is not as penalizing when actual times are used as much as it is when standardized scores are used. This is because the swim is the shortest segment, and a poor performance costs a competitor only about 2 minutes actual time but a very large negative z score.

CHAPTER 5   p. 123

1. Money earned = .05(sales) + 800. **2.** $F° = 1.8C° + 32$. **3.** See pp. 135–136. **4.** Both are measures of variability, but one reflects variability about a mean and the other variability about a regression line. See p. 142. **5a.** Impossible, the line with a positive b and positive a would not intercept X-axis at 5; **5b.** possible; **5c.** impossible, $s_{y \cdot x}$ must be less than or equal to s_y. **6.** Y on X: $(\mathbf{I_{XY}}) = 156$, $(\mathbf{II_X}) = 300$, $(\mathbf{III_Y}) = 176$, $a = 1.07$, $b = .52$, $\hat{Y} = .52X + 1.07$; W on X: $(\mathbf{I_{XW}}) = -228$, $(\mathbf{II_X}) = 300$, $(\mathbf{III_W}) = 212$, $a = 8.47$, $b = -.76$, $\hat{W} = -.76X + 8.47$; W on Y: $(\mathbf{I_{YW}}) = -100$, $(\mathbf{II_Y}) = 176$, $(\mathbf{III_W}) = 212$, $a = 6.75$ (or 6.76 if intermediate values are rounded), $b = -.57$, $\hat{W} = -.57Y + 6.75$. **7a.** 4.19; **7b.** 6.19; **7c.** 5.04; **7d.** $X = 12$ beyond range of original scores. **8.** $s_{y \cdot x} = 1.99$ (or 1.95 with rounding), $s_{w \cdot x} = 1.27$ (or 1.24), $s_{w \cdot y} = 2.54$. **9.** $\hat{Y} = 1.5X + 95.005$; $s_{y \cdot x} = 7.31$; for $X = 7$, $\hat{Y} = 105.505$; 2 out of 3 would be within 105.505 ± 7.31 or between 98.195 and 112.815.

CHAPTER 6   p. 151

1. $\Sigma(\hat{Y}_i - \overline{Y})^2$. The total sum of squares of points about the mean can be partitioned into the following components:

$$\Sigma(Y_i - \overline{Y})^2 = \Sigma(\hat{Y}_i - \overline{Y})^2 + \Sigma(Y_i - \hat{Y}_i)^2$$

The $\Sigma(Y_i - \hat{Y}_i)^2$ represents the squared deviations of the points about the regression line, that is, the error remaining after predicting Y from X. The $\Sigma(\hat{Y}_i - \overline{Y})^2$ is the portion of the total that is not error, that is, the variability in Y_i associated with X. **2.** Since $r = 1 - s^2_{y \cdot x}/s_y^2$, the correlation becomes larger as $s^2_{y \cdot x}/s_y^2$ becomes smaller. **3.** .49. **4.** The direction of prediction determines the regression constants, which are specific to the predicted variable, but the value of r is not influenced by the direction of prediction. **5.** Since $r = b_{yx}(s_x/s_y)$, when scores are standardized, $s_x = s_y = 1$, making $s_x/s_y = 1$. The result is $r = b_{z_x z_y}$. **6.** See p. 166. **7.** Extreme points influence the regression constants and r more than points near $(\overline{X}, \overline{Y})$ because their deviations from $(\overline{X}, \overline{Y})$, being large, contribute disproportionately to the numerators of b and r when they are squared. **8a.** Possible; **8b.** impossible, r and b must be the same sign; **8c.** impossible, $s_{y \cdot x}$ must be $\leq s_y$; **8d.** impossible, since $r = b(s_x/s_y)$, r should be .80 rather than .15 given the other information; **8e.** impossible, if $r = 1.00$, $s_{y \cdot x}$ must be 0; **8f.** possible (in fact, always true). **9.** For A and B: $(\mathbf{I_{AB}}) = -421$, $(\mathbf{II_A}) = 698$, $(\mathbf{III_B}) = 458$, $r = -.74$; for A and C: $(\mathbf{I_{AC}}) = 87$, $(\mathbf{II_A}) = 698$, $(\mathbf{III_C}) = 684$, $r = .13$; for B and C: $(\mathbf{I_{BC}}) = -147$, $(\mathbf{II_B}) = 458$, $(\mathbf{III_C}) = 684$, $r = -.26$; r does not change with changes in the origins or units of the scales of measurement. **10.** $(\mathbf{I_{XY}}) = 49$, $(\mathbf{II_X}) = 124$, $(\mathbf{III_Y}) = 140$, $r = .37$; After adding score (12, 8): $(\mathbf{I_{XY}}) = 391$, $(\mathbf{II_X}) = 783$, $(\mathbf{III_Y}) = 335$, $r = .76$; a score extremely deviant from the point $(\overline{X}, \overline{Y})$ will likely alter r considerably; adding score $(-12, 8)$ will produce $r = -.61$. **11.** $(\mathbf{I_{XY}}) = -1637$, $(\mathbf{II_X}) = 443$, $(\mathbf{III_Y}) = 13{,}115$; $\overline{X} = 3.08$, $s_x^2 = 3.36$, $s_x = 1.83$; $\overline{Y} = 100.42$, $s_y^2 = 99.36$, $s_y = 9.97$; $b = -3.70$, $a = 111.82$, $\hat{Y} = -3.70X + 111.82$; $s_{y \cdot x} = 7.67$; $r = -.68$; For $X = 5$, $\hat{Y} = 93.32$; For $X = 2$, $\hat{Y} = 104.42$.

CHAPTER 7   p. 181

1. A class held at 8 A.M. would likely appeal to only some of the students. Therefore, one might wonder whether the sample of students in an 8 A.M. class would be typical of all college students and whether it would be composed of a different type of student than a 2 P.M. class. Also, students in an 8 A.M. class might be less alert than those in a 2 P.M. class. **1b.** Are those students who volunteer for a jury experiment, as opposed to some other type of experiment, typical of all college students? Are college students typical of all young adults? Are jury members representative of young adults or of all adults? Are those adults who actually serve on juries representative of all adults in the pool of possible jury members? **1c.** Are people who live in a particular major city and who also have telephones typical of people who live in other cities? Of people who live in smaller towns? Of people who do not have telephones? **2.** The standard error of the mean is a measure of the sampling error of the sample mean as an estimator of the population mean. **3.** If the population distribution of raw scores is normal or if the N is large, the sampling distribution of the mean will be normal in form or will at least approach normality. **4.** Two variables are independent if they are unrelated in such a way that the value of one does not influence (or relate to) the value of the other. $\overline{X}$ and s_x^2 are independent if the population distribution of X's is normal (actually, symmetrical) and if observations are randomly and independently sampled. **5.** The theoretical relative frequency of an event in an idealized experiment is the probability of that event. **6.** See pp. 197–199. **7a.** 4; **7b.** 2; **7c.** 10; **7d.** 5; **7e.** 5; **7f.** .21; **7g.** 50. **8a.** .2420; **8b.** .0668; **8c.** .9332, .0668; **8d.** .4592. **9a.** .50; **9b.** .50; **9c.** .6826; **9d.** .1056; **9e.** .6247. **10a.** .1357; **10b.** .0808; **10c.** .9599, .0401; **10d.** .3326. **11a.** .0968; **11b.** .4744; **11c.** .2578; **11d.** .6454; **11e.** .2531. **12a.** 95%: 52.36 to 87.64; 99%: 46.825 to 93.175; **12b.** 95%: 86.48 to 133.52; 99%: 79.10 to 140.90; **12c.** 95%: 127.08 to 232.92; 99%: 110.475

to 249.525. **13.** The 90% confidence limits are -1.645 and $+1.645$ standard errors of the predicted value: $71 - 1.645(1.33)$ and $71 + 1.645(1.33)$ or between 68.81 and 73.19 inches.

CHAPTER 8 p.207

1. Assumptions are held to be true throughout the hypothesis-testing procedure, whereas hypotheses are a set of mutually exclusive and often exhaustive alternatives, one of which is being tested and tentatively held to be true until the evidence suggests otherwise. **2.** By estimating the probability that a particular outcome would occur by chance alone—that is, simply as a result of the variability inherent in sampling—the statistical test can help decide the tenability of explaining the result as due to chance. However, by its basic logic, the statistical test cannot prove that the result was not due to chance, only that, in the case of a significant test, the observed outcome would very seldom occur by chance alone. **3.** H_0, not H_1, is being tested. Therefore, H_0 can be rejected, but H_1 cannot be accepted. **4.** Type I error occurs when H_0 is erroneously rejected; Type II error occurs when H_0 is erroneously not rejected. **5.** Power is the probability that the test will correctly reject H_0. **6a.** Power would increase; **6b.** β would decrease; **6c.** α is decided by the researcher; it is not affected by N. **7.** β would increase and power would decrease. **8.** The consequences of the two types of error should determine the level of α. See p. 220. **9.** If there is considerable evidence or a decisive theory that indicates that the result of an experiment will be in a specific direction (for example, that mean A will be larger than mean B), then a directional test can be performed. Otherwise, use a nondirectional test. **10.** Use z if the population standard error is available; use t if it must be estimated. **11a.** *Assumptions.* The members of the sample are randomly and independently selected and the population involved is normal with $\mu = 81$ and $\sigma_x = 10$. *Hypotheses.* H_0: $\overline{X}$ is computed on a sample drawn from a population with $\mu = 81$. H_1: $\overline{X}$ is computed on a sample drawn from a population with $\mu \neq 81$. *Formula.*

$$z = \frac{\overline{X} - \mu}{\sigma_{\bar{x}}} = \frac{\overline{X} - \mu}{\sigma_x / \sqrt{N}}$$

Significance level. Assume $\alpha = .05$. *Critical values.* -1.96 and $+1.96$ for a two-tailed test. *Decision rules.* If z_{obs} is between -1.96 and $+1.96$, do not reject H_0; if z_{obs} is less than -1.96 or greater than $+1.96$, reject H_0. *Computation.* $z_{obs} = 2.00$. *Decision.* Reject H_0. **11b.** One would then use the formula for t with $df = 24$ rather than the z distribution. The critical values of t would be ± 2.064 for a two-tailed test at $\alpha = .05$, and the decision rules would be: If t_{obs} is between -2.064 and $+2.064$, do not reject H_0. If t_{obs} is less than -2.064 or greater than $+2.064$, reject H_0. Because s_x is also 10, the observed t would be calculated by

$$t_{obs} = \frac{\overline{X} - \mu}{s_x / \sqrt{N}} = \frac{85 - 81}{10 / \sqrt{25}} = 2.00$$

This would result in the decision not to reject H_0. **11c.** *Assumptions.* The members of the sample are randomly and independently selected and the population of nonpierced males is normal with a mean of 62 inches. *Hypotheses.* H_0: $\overline{X}$ is computed on a sample from a population with $\mu \leq 62$. H_1: $\overline{X}$ is computed on a sample from a population with $\mu > 62$. (Notice that this is a directional alternative. Why?) *Formula.*

$$t = \frac{\overline{X} - \mu}{s_{\bar{x}}} = \frac{\overline{X} - \mu}{s_x / \sqrt{N}}$$

Significance level. Assume $\alpha = .05$. *Critical value.* From Table B with $df = N - 1 = 35$, one-tailed, $\alpha = .05$: $t_{crit} = +1.691$ (approximately, obtained by interpolation). *Decision rules.* If t_{obs} is less than 1.691, do not reject H_0. If t_{obs} is greater than or equal to 1.691, reject H_0. *Computation.*

$$t_{obs} = \frac{\overline{X} - \mu}{s_x/\sqrt{N}} = \frac{64.5 - 62}{7/\sqrt{36}} = 2.14$$

Decision. Reject H_0, since the observed mean is greater than would be expected to occur from chance effects of sampling from a population with a mean of 62. This result does not constitute proof that the piercing and molding causes increased height. Since the design of the study was observational rather than experimental, alternative explanations, such as diet or genetics, cannot be ruled out. **11d.** *Assumptions.* The sample of students is randomly and independently selected and the mean of the current student population is 113. *Hypotheses.* H_0: $\overline{X}$ is computed on a sample from a population with $\mu = 113$. H_1: $\overline{X}$ is computed on a sample from a population with $\mu \neq 113$. *Formula.*

$$t = \frac{\overline{X} - \mu}{s_{\bar{x}}} = \frac{\overline{X} - \mu}{s_x/\sqrt{N}}$$

Significance level. Assume $\alpha = .05$. *Critical values.* Given a nondirectional test, $df = N - 1 = 48$, $\alpha = .05$: $t_{crit} = \pm 2.01$ (approximately). *Decision rules.* If t_{obs} is between -2.01 and $+2.01$, do not reject H_0. If t_{obs} is less than -2.01 or greater than $+2.01$, reject H_0. *Computation.*

$$t_{obs} = \frac{\overline{X} - \mu}{s_x/\sqrt{N}} = \frac{117 - 113}{15/\sqrt{49}} = 1.87$$

Decision. Do not reject H_0. **12a.** $s_{\bar{x}} = 7/\sqrt{25} = 1.40$, 37.11 to 42.89, 36.08 to 43.92; **12b.** 24.69 to 67.31, 16.53 to 75.47.

CHAPTER 9   p. 233

1. See page 234.

2. $t = \dfrac{(a\ value)\ -\ (population\ mean\ of\ such\ values)}{(an\ estimate\ of\ the\ standard\ error\ of\ such\ values)}$

3. The assumption of normality is made so that the standard normal or the t distribution may be used to determine the required probability. This assumption can be made if the populations from which the two groups are drawn are normal or the sizes of the two samples are large. **4.** The correlated-groups design has the advantage of eliminating differences between subjects as a source of variability. Therefore, the correlated-groups design often is more sensitive to detecting experimental effects. However, practical considerations may prevent subjects from being tested twice. In general, if the test or treatment itself produces irreversible changes in the subject, an independent-groups design must be used. **5.** Verify that the following set of numbers satisfies the conditions of this problem.

A	B
2	0
4	3
6	4
8	7
10	8
12	9
14	13
16	15

6. Test for the difference between independent means. *Hypotheses. $H_0: \mu_1 \leq \mu_2$. $H_1: \mu_1 > \mu_2$. Assumptions.* The subjects are randomly and independently sampled, the groups are independent, variances are homogeneous, and the X's are normally distributed. *Decision rules.* Given the .05 level and $df = N_1 + N_2 - 2 = 17$, directional test. If $t_{obs} < 1.740$, do not reject H_0. If $t_{obs} \geq 1.740$, reject H_0. *Computation.*

$$t_{obs} = (\bar{X}_1 - \bar{X}_2) \bigg/ \sqrt{\left[\frac{(N_1 - 1)s_1^2 + (N_2 - 1)s_2^2}{N_1 + N_2 - 2}\right] \cdot \left[\frac{1}{N_1} + \frac{1}{N_2}\right]}$$

$$t_{obs} = (7.00 - 3.00) \bigg/ \sqrt{\left[\frac{(10 - 1)4.22 + (9 - 1)2.00}{10 + 9 - 2}\right] \cdot \left[\frac{1}{10} + \frac{1}{9}\right]} = 4.89$$

(t_{obs} may differ slightly depending on how many digits are used during computation.)

Decision. Reject H_0; the observed difference in means is too great to be a simple result of sampling error. The dissonance theory is supported. **7.** Test of the difference between two correlated groups. *Hypotheses. $H_0: \mu_1 = \mu_2$. $H_1: \mu_1 \neq \mu_2$. Assumptions.* The data are in the form of pairs of scores that were randomly and independently sampled and the population of the D_i is normally distributed. *Decision rules.* Given .05 level, $df = N - 1 = 9$, nondirectional test. If $-2.262 < t_{obs} < 2.262$, do not reject H_0. If $t_{obs} \leq -2.262$ or $t_{obs} \geq 2.262$, reject H_0. *Computation.* (Before $-$ After).

$$t_{obs} = \frac{\Sigma D}{\sqrt{[N\Sigma D^2 - (\Sigma D)^2]/(N - 1)}} = \frac{21}{\sqrt{10(217) - (21)^2]/(10 - 1)}} = 1.52$$

Decision. Do not reject H_0; the observed difference between means is too small and could be due to sampling error. In fact, performance is nonsignificantly *poorer* after the incentive plan than before. **8.** Test of difference between independent means. *Hypotheses. $H_0: \mu_1 = \mu_2$. $H_1: \mu_1 \neq \mu_2$. Assumptions.* The subjects are randomly and independently sampled, the groups are independent, variances are homogeneous, and X is normally distributed. *Decision rules.* Given .05 level, $df = N_1 + N_2 - 2 = 22$, nondirectional test. If $-2.074 < t_{obs} < 2.074$, do not reject H_0. If $t_{obs} \leq -2.074$ or if $t_{obs} \geq 2.074$, reject H_0. *Computation.*

$$t_{obs} = \frac{(4.25 - 1.33)}{\sqrt{\left[\frac{(12 - 1)8.93 + (12 - 1)5.88}{12 + 12 - 2}\right] \cdot \left[\frac{1}{12} + \frac{1}{12}\right]}}$$

$t_{obs} = 2.61$ to 2.63 (depending on rounding)

Decision. Reject H_0; the difference between observed means is too great to be simply a function of sampling error. There is a difference between the two therapeutic approaches. **9.** *Hypotheses. $H_0: \rho = 0$. $H_1: \rho \neq 0$. Assumptions.* The subjects are randomly and independently sampled and the population distributions of the two variables are normal. *Decision rules.* Given .05 level, $df = N - 2 = 8$, nondirectional test. If $-.6319 < r_{obs} < .6319$, do not reject H_0. If $r_{obs} \leq -.6319$ or $r_{obs} \geq .6319$, reject H_0. *Computation. $r_{obs} = .91$. Decision.* Reject H_0; the observed correlation is too large to be attributed to sampling error. Some people produce more than others regardless of which incentive system is used. **10.** *Hypotheses. $H_0: \rho_1 \leq \rho_2$. $H_1: \rho_1 > \rho_2$. Assumptions.* The subjects are randomly and independently sampled, the groups are independent, the population distributions for X and Y are normal, and N_1 and N_2 are both greater than 20. *Decision rules.* Given .05, directional test. If $z_{obs} < 1.645$, do not reject H_0. If $z_{obs} \geq 1.645$, reject H_0. *Computation.*

$$z_{obs} = \frac{z_{r_1} - z_{r_2}}{\sqrt{1/(N_1 - 3) + 1/(N_2 - 3)}} = \frac{1.256 - .662}{\sqrt{1/(35 - 3) + 1/(30 - 3)}}$$

$z_{obs} = 2.26$ or 2.27 (depending on rounding)

Decision. Reject H_0; the difference between the two correlations is too large to be simply due to sampling error. The correlation between the IQs of identical twins is larger than the correlation between the IQs of fraternal twins. **11a.** For the new curriculum, pretest $\overline{X} = 9.89$, $s = 2.52$; posttest $\overline{X} = 13.00$, $s = 2.06$; for the old curriculum, pretest $\overline{X} = 10.10$, $s = 2.60$; posttest $\overline{X} = 10.40$, $s = 2.41$. Test of difference between correlated means, for the new curriculum, $df = 8$, $t_{crit} = \pm 2.306$, $t_{obs} = -3.44$, reject H_0 (note that the t_{obs} is negative only because the pretest was considered group 1 in the statistical computations). For the old curriculum, $df = 9$, $t_{crit} = \pm 2.262$, $t_{obs} = -.76$, do not reject H_0. **11b.** For the new curriculum; *computation:*

$$t_{obs} = \frac{\overline{X} - \mu}{s_{\bar{x}}} = \frac{\overline{X} - \mu}{s_x/\sqrt{N}} = \frac{13.00 - 10.00}{2.06/\sqrt{9}} = 4.37$$

$df = 8$, $t_{crit} = \pm 2.306$, reject H_0. For the old curriculum; *computation:*

$$t_{obs} = \frac{\overline{X} - \mu}{s_{\bar{x}}} = \frac{\overline{X} - \mu}{s_x/\sqrt{N}} = \frac{10.40 - 10.00}{2.41/\sqrt{10}} = .52$$

$df = 9$, $t_{crit} = \pm 2.262$, do not reject H_0. **11c.** For the new curriculum: $df = 7$, $r_{crit} = .5822$, $r_{obs} = .31$, do not reject H_0. For the old curriculum, $df = 8$, $r_{crit} = .5494$, $r_{obs} = .88$, reject H_0. **11d.** $z_{crit} = \pm 1.96$, $z_{obs} = -1.89$ or -1.90 depending on rounding, do not reject H_0. **12.** $r = .22$, $z_r = .224$, $.224 \pm 1.96\sqrt{(1/59)} = -.031$ to .479 or $r = -.03$ to $r = .45$. Since the interval contains $r = .00$, the $r_{obs} = .22$ is not significant. **13.** 20%.

CHAPTER 10   p. 271

1. Ten separate t tests would be needed to make all possible comparisons. See pp. 272–273. **2a.** *Hypotheses.* H_0: $\mu_1 = \mu_2 = \mu_3 = \mu_4 = \mu$. H_1: Not H_0. *Assumptions.* Random and independent sampling, independent groups, homogeneity of group population variances, normality of population distribution of scores. *Decision rules.* Given .05 significance level and $df = 3, 19$, if $F_{obs} < 3.13$, do not reject H_0. If $F_{obs} \geq 3.13$, reject H_0. *Computation.*

$T_1 = 26$	$T_2 = 31$	$T_3 = 30$	$T_4 = 27$	$T_{total} = 114$
$n_1 = 5$	$n_2 = 6$	$n_3 = 6$	$n_4 = 6$	$N = 23$
$\overline{X}_1 = 5.20$	$\overline{X}_2 = 5.17$	$\overline{X}_3 = 5.00$	$\overline{X}_4 = 4.50$	
$\sum X_{1j}^2 = 158$	$\sum X_{2j}^2 = 189$	$\sum X_{3j}^2 = 186$	$\sum X_{4j}^2 = 175$	$\sum X_{ij}^2 = 708$
$\left(\dfrac{T_1^2}{n_1}\right) = 135.2000$	$\left(\dfrac{T_2^2}{n_2}\right) = 160.1667$	$\left(\dfrac{T_3^2}{n_3}\right) = 150.0000$	$\left(\dfrac{T_4^2}{n_4}\right) = 121.5000$	$\sum\left(\dfrac{T_j^2}{nj}\right) = 566.8667$

(I) = 565.0435	**(II)** = 708	**(III)** = 566.8667

$SS_{between} = 1.8232$	$df_{between} = 3$	$MS_{between} = .6077$
$SS_{within} = 141.1333$	$df_{within} = 19$	$MS_{within} = 7.4281$
$SS_{total} = 142.9565$	$df_{total} = 22$	

Source	df	SS	MS	F
Between groups	3	1.8232	.6077	.08
Within groups	19	141.1333	7.4281	
Total	22	142.9565		

Decision. Do not reject H_0; the observed differences between means could be attributable to sampling error. **2b.** Adding a constant to a group is analogous to introducing a treatment effect. Reanalyzing, MS_{within} remains the same ($MS_{within} = 7.4281$), whereas $MS_{between}$ reflects the introduction of mean differences ($MS_{between} = 38.4338$). $F_{obs} = 5.17$, which is greater than the critical value of 3.13 and leads to the rejection of H_0. **2c.** Adding 20 to the last score in each group greatly increases within–group variability ($MS_{within} = 79.2877$), but influences between–group variability much less ($MS_{between} = 36.6561$). The result is that the F_{obs} is much smaller: $F_{obs} = .46$; do not reject H_0. **3.** See pp. 276–278. **4.** See pp. 284–285. **5.** See pp. 296–297.

6. (I) $= 410.8889$, **(II)** $= 476.0000$, **(III)** $= 440.6667$.

Source	df	SS	MS	F
Between groups	2	29.7778	14.8889	6.32*
Within groups	15	35.3333	2.3556	
Total	17	65.1111		

$F_{crit} = 3.68$. Reject H_0.

7a. Year One: **(I)** $= 216,855.0476$, **(II)** $= 217,618$, **(III)** $= 217,264.9346$.

Source	df	SS	MS	F
Between groups	2	409.8870	204.9435	10.45**
Within groups	18	353.0654	19.6147	
Total	20	762.9524		

$F_{crit, .05} = 3.55$, $F_{crit, .01} = 6.01$. Reject H_0.

7b. Year Three: **(I)** $= 226,720.1905$, **(II)** $= 227,208$, **(III)** $= 226,857.4524$.

Source	df	SS	MS	F
Between groups	2	137.2619	68.6310	3.52
Within groups	18	350.5476	19.4749	
Total	20	487.8095		

$F_{crit, .05} = 3.55$. Do not reject H_0.

8. For Support vs. Inoculation:

$$F_{comp} = \frac{(5.33 - 3.00)^2}{\left(\frac{1}{6} + \frac{1}{6}\right)2.3556} = 6.91$$

$$q = \sqrt{2(6.91)} = 3.72$$

$q_{3.15(.05)} = 3.675$ (by interpolation), reject H_0.

For Support vs. Control:

$F_{comp} = .57, q = 1.07$, do not reject H_0.

For Inoculation vs. Control:

$F_{comp} = 11.46, q = 4.79$, reject H_0.

9a.

$F_{comp} = 11.38^{**}$ (or 11.31 or 11.33 depending on rounding).

$F_{crit(1,18;.05)} = 4.41, F_{crit(1,18;.01)} = 8.28$. Reject H_0.

9b.

$F_{comp} = 6.16^*$ (or 6.17^*, depending on rounding). $F_{crit(1,18;.05)} = 4.41$, reject H_0.

9c. Nondirective vs. Directive:

$F_{comp} = 6.16, q_{obs} = 3.51$. $q_{crit(3,18;.05)} = 3.61$. Do not reject H_0.

Nondirective vs. Control:

$F_{comp} = 4.20, q_{obs} = 2.90$. $q_{crit(3,18;.05)} = 3.61$. Do not reject H_0.

Directive vs. Control:

$F_{comp} = .3550, q_{obs} = .84$. $q_{crit(3,18;.05)} = 3.61$. Do not reject H_0.

The difference between the first test here and the test in **9b** is that the number of tests being calculated is taken into account here by using q to adjust the critical values.

10. $MS_{within} = 2.3556$, $df_{within} = 15$, $t_{.05,df=15} = 2.131$. Group 1: 3.99 to 6.67; Group 2: 1.66 to 4.34; Group 3: 4.66 to 7.34; est. $\omega^2 = .37$. See p. 262–263.

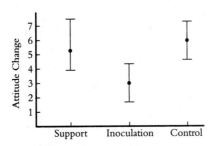

CHAPTER 11   p. 311

1. The steps in research design include formulating the question, operationalizing the variables, collecting data, analyzing data, and drawing conclusions. **2.** The value of the dependent variable depends on the independent variable; in experimental research the independent, but not the dependent, variable is under the control of the researcher. **3.** Observational research addresses the question of whether a relationship (either causal or noncausal) exists, whereas experimental research addresses the question of whether that relationship is causal. **4.** The literature review contributes to formulating a more pertinent research question, selecting particular variables for study, controlling other relevant parameters of the dependent variable, and selecting procedures that are likely to be appropriately effective. **5.** Situational parameters are characteristics of the situation or environment that influence the dependent variable; subject parameters are characteristics of the subjects that influence the dependent variable. **6.** Operational definitions are definitions of variables phrased in terms of operations. They make very specific what is meant by each variable and how it will be manipulated or measured. Operational definitions allow other researchers to use the variables in precisely the same way. **7.** Reliability refers to whether the measurement procedures assign the same value to a characteristic each time that it is measured under essentially the same circumstances. Validity refers to the extent to which the measurement procedures accurately reflect the variable being measured. **8.** Random assignment helps to distribute subject parameters equally across each group in an experiment. **9.** See p. 323. **10.** A double-blind experiment is one in which neither the subjects nor the observers know which group the subjects belong to. This type of experiment is used to minimize subject and observer biases. **11.** Conducting a pilot experiment helps to determine whether the procedures of a research design are likely to be effective. **12.** Research reports usually include introduction, methods, results, and discussion sections. **13.** See pp. 327–330.

CHAPTER 12 p. 332

1. An infinite number of tosses. **2a.** Mutually exclusive; **2b.** not mutually exclusive; **2c.** mutually exclusive; **2d.** not mutually exclusive. **3a, 3b,** and **3c.** independent; **3d.** independent (because the probability of drawing a red card is the same whether you draw an ace or not on the first draw; although it does depend on *which* one is drawn); **3e.** independent. **4.** See pp. 343–345 and 347–350. **5a.** Intersection; **5b.** union; **5c.** union; **5d.** intersection. **6a.** combinations; **6b.** combinations; **6c.** permutations. **7a.** 1/3; **7b.** 1/3; **7c.** 2/3. **8a.** 2/3; **8b.** 2/3. **9.** 1/36. **10a.** 2/3; **10b.** 16/49; **10c.** 2/7. **11a.** 720; **11b.** 12; **11c.** 20. **12.** 360. **13a.** 1; **13b.** 10; **13c.** 15. **14.** 70; 1680. **15.** 1/84; 1/504; 1/6. **16.** $1/[(35)(34)(33)] = 1/39,270 = .000025$. **17.** $21(1/3)^5(2/3)^2 = .038$; .044 or .045 depending on rounding.

CHAPTER 13   p. 360

1. All are possible. **2.** See pp. 381–382. **3.** An interaction occurs when the nature of the effect for one factor is not the same under all levels of the other factor. If an interaction is significant, that fact means that collapsing over levels of a factor is likely not meaningful. See p. 364. **4a.** *Hypotheses.* Factor *A.* $H_0: \alpha_1 = \alpha_2 = 0$. H_1: Not H_0. Factor *B.* $H_0: \beta_1 = \beta_2 = 0$. H_1: Not H_0. *AB* Interaction. $H_0: \alpha\beta_{11} = \alpha\beta_{12} = \alpha\beta_{21} = \alpha\beta_{22} = 0$. H_1: Not H_0. *Assumptions.* The groups are independent and randomly sampled with $n > 1$ and all n's equal from populations having normal distributions and homogeneous variances. The factors are fixed. *Decision rules.* For all tests $df = 1$, 12, and $\alpha = .05$. If $F_{obs} < 4.75$, do not reject H_0. If $F_{obs} \geq 4.75$, reject H_0. *Computation.* **(I)** = 451.5625, **(II)** = 511.0000, **(III)** = 453.1250, **(IV)** = 451.6250, **(V)** = 456.2500.

Source	df	SS	MS	F
A	1	1.5625	1.5625	.34
B	1	.0625	.0625	.01
AB	1	3.0625	3.0625	.67
Within	12	54.7500	4.5625	
Total	15	59.4375		

$F_{crit}(1, 12) = 4.75$.

Decision. Do not reject H_0 for any test. **4b.** *Computation.* **(I)** = 976.5625, **(II)** = 1161.0000,

(III) = 1103.1250, **(IV)** = 976.6250, **(V)** = 1106.2500.

Source	df	SS	MS	F
A	1	126.5625	126.5625	27.74**
B	1	.0625	.0625	.01
AB	1	3.0625	3.0625	.67
Within	12	54.7500	4.5625	
Total	15	184.4375		

$F_{crit}(1, 12) = 4.75$ (at .05) and 9.33 (at .01).

By adding the treatment effect to A, only the results for A and the total are changed. *Decision.* Reject H_0 for Factor A. **4c.** *Computation.* **(I)** = 976.5625, **(II)** = 1271.0000, **(III)** = 1053.1250, **(IV)** = 1071.6250, **(V)** = 1216.2500.

Source	df	SS	MS	F
A	1	76.5625	76.5625	16.78**
B	1	95.0625	95.0625	20.84**
AB	1	68.0625	68.0625	14.92**
Within	12	54.7500	4.5625	
Total	15	294.4375		

$F_{crit}(1, 12) = 4.75$ (at .05) and 9.33 (at .01).

Adding a treatment effect to one cell alters both main effects and the interaction. *Decision.* Reject H_0 for all tests. **5.** *Computation.* **(I)** = 661.2500, **(II)** = 745.0000, **(III)** = 662.5000, **(IV)** = 662.5000, **(V)** = 695.0000.

Source	df	SS	MS	F
Gender	1	1.2500	1.2500	.40
Task	1	1.2500	1.2500	.40
Gender × Task	1	31.2500	31.2500	10.00**
Within	16	50.0000	3.1250	
Total	19	83.7500		

$F_{crit}(1, 16) = 4.49$ (at .05) and 8.53 (at .01).

6. *Computation.* **(I)** = 174.0500, **(II)** = 199.0000, **(III)** = 178.1000, **(IV)** = 174.1000, **(V)** = 184.2000.

Source	df	SS	MS	F
Gender	1	4.0500	4.0500	4.38
Task	1	.0500	.0500	.05
Gender × Task	1	6.0500	6.0500	6.54*
Within	16	14.8000	.9250	
Total	19	24.9500		

$F_{crit}(1, 16) = 4.49$.

7. *Computation.* **(I)** = 370.0167, **(II)** = 525.0000, **(III)** = 408.0500, **(IV)** = 422.5833, **(V)** = 470.2500.

Source	df	SS	MS	F
Mental Ability	2	38.0333	19.0166	15.63**
Achievement	4	52.5666	13.1416	10.80**
MA × Ack	8	9.6334	1.2042	.99
Within	45	54.7500	1.2167	
Total	59	154.9833		

$F_{crit}(2, 45) = 3.205$ (at .05) and 5.11 (at .01); $F_{crit}(4, 45) = 2.575$ (at .05) and 3.77 (at .01).

Students attain years of education past high school as a positive function of their ability and an inverse function of their degree of underachievement.

8. est. $\omega^2_{Gender} = .12$ or 12%;

est. $\omega^2_{Task} = -.03$ or 0%;

est. $\omega^2_{SxT} = .20$ or 20%.

CHAPTER 14   p. 395

1a. Mann–Whitney U test; **1b.** Wilcoxon test; **1c.** Kruskal–Wallis test; **1d.** Spearman rank-order correlation. **2.** Power is the probability that the statistical test will correctly reject a false null hypothesis. Power-efficiency refers to the power of a test relative to the size of the sample, and allows one to compare the power of different tests. For a given N, parametric tests are more powerful than analogous nonparametric tests. **3.** For a Mann–Whitney test, the null hypothesis states that the two population distributions from which the samples were drawn are identical. The null hypothesis is false if the population distributions differ in means or in any other characteristic. **4.** Chi square 2 × 3 table, $\chi^2_{obs} = 6.75$, $df = 2$, $\chi^2_{crit} = 5.99$. Reject H_0. **5.** Chi square 3 × 2 table, $\chi^2_{obs} = .80$ (.77 with rounding), $df = 2$, $\chi^2_{crit} = 5.99$. Do not reject H_0. There is insufficient reason to reject the hypothesis that color preferences are similar for the two genders. **6.** Mann–Whitney U test, $U_{obs} = 78.5$, $U_{crit} = 20$ and 70 for N's of 9 and 10, .05 level, nondirectional test. Reject H_0. **7.** For group A, $\overline{X} = 39.57$, $M_d = 33$; for group B, $\overline{X} = 39.29$, $M_d = 39$. t test for the difference between independent groups, $t_{obs} = .04$. Do not reject H_0.

Mann–Whitney U test, $U_{obs} = 42$, $U_{crit} = 8, 41$. Reject H_0. The means of the groups are very similar but the medians are somewhat different. Because of the distributions (for example, $X = 82$), the medians are more appropriate. Although most scores in group A are relatively low, the extreme score of 82 elevates the mean of group A to close to that of group B. The t test of a difference between means is not significant while the U test of differences between distributions is significant. Both the median and the U test use predominantly the ordinal characteristics of the scale. **8.** Kruskal–Wallis test for independent samples. $H_{obs} = 7.63$. $H_{crit} = 5.99$, $df = 2$. Reject H_0. Reading makes a difference. **9.** Wilcoxon test for correlated groups. $W_{obs} = 4$, $W_{crit} = 5$ for $N = 9$ (one zero difference pair was eliminated), .05 level, nondirectional. Reject H_0 (W_{obs} is *less* than W_{crit}). **10.** Spearman rank-order correlations. $r_S = .53$. Critical value of $r_S = .649$, .05 level, nondirectional. Do not reject hypothesis that population relationship is zero. This finding indicates that either the self-esteem test has low reliability or that the fitness program, while generally improving self-esteem, did so in an uneven manner across individuals. **11a.** Wilcoxon test. $W_{obs} = 1$, $W_{crit} = 3$, $N = 8$, .05 level, nondirectional. Reject H_0. **11b.** Spearman rank-order correlation. $r_S = -.76$, $r_{S\,crit} = \pm.715$, $N = 8$, .05 level, nondirectional. Reject H_0. **11c.** For the third week: $H_{obs} = .06$, $H_{crit} = 5.99$, $df = 2$. Do not reject H_0. For the fifth week, $H_{obs} = 10.31$, $H_{crit} = 5.99$, $df = 2$. Reject H_0. **12.** Chi-square test of goodness of fit, $\chi^2_{obs} = .60$, $\chi^2_{crit} (df = 3) = 7.82$. Do not reject H_0. Chi-square test of a proportion, $\chi^2_{obs} = 1.54$, $\chi^2_{crit} (df = 1) = 3.84$. Do not reject H_0. The sample does not differ from the population more than expected by sampling error.

APPENDIX 1   p. 431

1a. $\frac{5}{7}$; **1b.** $1\frac{1}{3}$; **1c.** $\frac{2}{3}$; **1d.** $1\frac{8}{21}$; **1e.** $-\frac{2}{5}$; **1f.** $\frac{1}{2}$; **1g.** $\frac{1}{36}$; **1h.** $\frac{24}{77}$. **2a.** $\frac{1}{12}$; **2b.** $\frac{3}{20}$; **2c.** $\frac{45}{224}$; **2d.** 2;

2e. $1\frac{1}{2}$; **2f.** $4\frac{3}{8}$. **3a.** 5040; **3b.** 60; **3c.** 120. **4a.** 81; **4b.** 12; **4c.** 29; **4d.** 65; **4e.** -8; **4f.** 144;

4g. 2; **4h.** $\frac{1}{3}$; **4i.** 1; **4j.** $\frac{1}{9}$; **4k.** $1\frac{11}{16}$; **4l.** $\frac{2}{2187}$. **5a.** $x^2 + 2xy + y^2$; **5b.** $p^2 - 2pq + q^2$; **5c.**

$X^2 + 2X\overline{X} + \overline{X}^2$; **5d.** $(ab)^2 - 2ab^2c + (bc)^2$; **5e.** a; **5f.** $\left(\frac{a}{c}\right) - b$; **5g.** $(a + c)/(a - b)$; **5h.** $2X$;

5i. $X\sqrt{5}$; **5j.** $Y\sqrt{2Y}$. **6a.** sum = 13, square of the sum = 169, sum of squares = 65; **6b.** sum = 14, square of the sum = 196, sum of squares = 54; **6c.** sum = 8; square of the sum = 64, sum of squares = 78.

Index